Useful Physical Constants

Boltzmann's constant, k	8.62×10^{-5} ev/°K
Electron rest mass, m_0	9.11×10^{-28} gm
Energy associated with 1 eV	1.6×10^{-12} erg
Magnitude of the electronic charge, q	1.6×10^{-19} coulomb
Permittivity of free space, ϵ_0	8.86×10^{-14} farad/cm
Planck's constant, h	4.14×10^{-15} eV-sec
Room temperature value of kT	0.0259 eV
Velocity of light, c	3×10^{10} cm/sec
Angstrom unit, Å	10^{-8} cm
Micron, μm	10^{-4} cm
Thousandth of an inch, mil	25.4 μm

The Theory and Practice of Microelectronics

The Theory and Practice of Microelectronics

Sorab K. Ghandhi

Rensselaer Polytechnic Institute
Troy, New York

JOHN WILEY & SONS

New York • Chichester • Brisbane • Toronto • Singapore

20 19 18 17 16 15 14 13 12 11 10

Library of Congress Catalog Card Number: 68-28501
SBN 471 29718 6

Printed in the United States of America

PREFACE

In 1959 the original aim of microelectronics, or molecular engineering, as it was then called, was "the design of appropriate materials and appropriate electrodes to produce a desired effect." This proved to be far too ambitious a goal and over the next few years resulted in complex devices of great ingenuity but questionable performance, such as the "solar battery-tunnel diode oscillator."

During these early years significant developments were made in the technologies for batch processing of silicon devices. These developments culminated in the invention of the *p–n* junction isolation technique by Lehovec† and the planar process by Hoerni.‡

The Lehovec patent allowed the engineer for the first time to recast microelectronics as "the logical synthesis of complex electronic functions by the interconnection of separate components, in a single block of semiconductor material." This effectively transferred their implementation from the realm of new device invention to the application of well-developed active circuit synthesis techniques. Concurrently, the Hoerni patent allowed the practical realization of these functions by significantly increasing the fabrication yield. It is safe to say that these patents paved the way for the logical development of a large number of sophisticated, reliable microcircuits. The rapid growth of microelectronics from that time bears evidence to this fact.

The advent of microelectronics has not greatly altered the role of the electrical engineer as a circuit designer. A far more significant change is that it is now necessary for him to interact with the process engineer

†Pat. No. 3,029,366 assigned to the Sprague Electric Co., Massachussets.
‡Pat. No. 3,025,589 assigned to the Fairchild Camera and Instrument Corp., New York.

who will ultimately fabricate his circuit design. At the present time an increasingly large number of engineers is responsible for carrying a microcircuit design from its "idea" phase to its "product" phase. These engineers are involved in the actual design of the microcircuit as well as in the specification of its topology, physical dimensions, and doping profiles. In addition, they are involved in evaluating the end product, diagnosing faults at the process as well as at the device and circuit level, and providing the necessary feedback by which the errors may be corrected. Finally, they play an important role in generating the performance and test data that are required to specify the finished product.

Thus the emergence of this new field has accentuated the need for a broadly trained engineer, capable of understanding and solving problems at the circuit and systems level as well as at the devices and materials level. Such training is also of value to the engineer who confines his activities to circuit or system implementation. In increasing number he is required to work with microcircuits and to develop an understanding of their characteristics and limitations. Often he must plan for the time when his design will be fabricated as a custom microcircuit by an outside manufacturer. The involvement of the university in training this new breed of engineer goes hand in hand with this need. In fact, the rapidly changing state of technology imposes the additional responsibility of providing a depth of understanding on which the student can draw for many years after he has received his formal training.

It is hoped that this book will meet, in one volume, the apparently differing needs of the microcircuit-oriented engineer of today, and those of the university student who will become the engineer of tomorrow. For this reason emphasis is placed on laying the foundations for the many processes that are used in modern microelectronic technology. Next the physical limitations of these processes are explored and related to the characteristics of the ensuing devices. Finally, modern microelectronics practice is described in some detail. Often this serves to emphasize that to some extent the field is an art (hopefully not a black one!) and not a science.

My qualifications for attempting a book of this kind are based upon nine years of industrial experience in transistor circuit design and three years in the semiconductor device area, followed by five years of academic teaching and research in semiconductor devices and microelectronics. During this latter period I have been active as a consultant to the microelectronics industry. Thus I can claim an awareness of the needs of the graduate electrical engineering student on the one hand and of the microcircuit design engineer on the other.

The first nine chapters of this book describe the materials and processes required in the fabrication of microcircuits. Their contents are confined to monolithic circuits made of silicon, for these are by far the most important. Although many of the subjects treated in this section are relatively unfamiliar to the electrical engineer, their importance is emphasized by continual reference to their effect on the electrical properties of the resulting devices.

Chapter 10 describes the manner in which a sequence of processes may be combined to result in complete microcircuits made of interconnected ensembles of transistors, diodes, resistors, and capacitors. Although a variety of different fabrication schemes is considered, the emphasis is placed on the modern double-diffused epitaxial process. This process has been almost universally adopted by industry in the last few years. It does not appear that it will be superseded for some time to come.

Chapters 11 to 16 cover the basic physics of semiconductor devices as well as the electrical characterization of active and passive components fabricated by the monolithic process. Finally, Chapter 17 outlines the physics of semiconductor surfaces and describes the effects of surface phenomena on the electrical properties of microcircuit devices.

Whenever possible, problems have been provided, many of which were selected from practical situations and are intended to bring out points that have not been covered in detail in the text. In addition, references have been provided at the end of each chapter. No attempt has been made to use them as a means of giving credit to the persons who did the original research; rather their choice has been based entirely on the need to provide a useful means for further study.

I am indebted to the many people who have contributed, directly or indirectly, to the preparation of this book. In particular, I should like to thank my good friend and colleague Dr. K. E. Mortenson for giving me the freedom and encouragement to develop the graduate-level courses at Rensselaer on which this book is based. In addition, the heavy commitment of Rensselaer in the solid-state area has made available a number of colleagues with whom I have had many fruitful discussions. For these I must give special thanks to Drs. K. Rose, J. Borrego, J. Park, A. Armstrong, and P. Bakeman of the Electrophysics Division.

I should also like to thank my many friends at the Sprague Electric Co., North Adams, Mass., with whom I have been associated as a consultant. Much of the relevance of this book to modern microcircuit practice stems from my contacts in this capacity. In particular, I am grateful to Dr. R. S. Pepper, Dr. A. J. Saxena, J. Seacord, D. Rauscher, and E. Donovan for much help during the preparation of this material.

Many students have been subjected to different versions of this book over the last four years. They deserve my heartfelt thanks and sympathies as does Mrs. J. Grega, who typed the manuscript and its many revisions with great care and considerable patience. Finally, I am grateful to my wife and sons who provided much understanding and moral support during the preparation of the manuscript.

SORAB K. GHANDHI

Niskayuna, New York
April 1968

CONTENTS

The Theory and Practice of Microelectronics

Chapter 1

Material Properties

Although many elements and intermetallic compounds exhibit semiconducting properties, silicon is used almost exclusively in the fabrication of modern microcircuits. There are many reasons for this choice. Of these the most important are the following:

1. Silicon is an elemental semiconductor. Together with germanium, it can be subjected to a large variety of processing steps without the problems of decomposition that are ever present with compound semiconductors. For much the same reason its properties can be studied with considerably greater ease than those of compound semiconductors. As a consequence, perhaps more is known today about the preparation and properties of extremely pure, single-crystal germanium and silicon than for any other material in the periodic table.

2. Silicon has a wider energy gap than germanium. Consequently, it can be fabricated into microcircuits capable of operation at higher temperatures than their germanium counterparts. At the present time the upper operating ambient temperature for silicon microcircuits is between 125–175°C, and is entirely adequate for a large number of military applications.

3. Silicon readily lends itself to surface passivation treatments. This takes the form of a layer of thermally grown silicon dioxide which provides a high degree of protection to the underlying microcircuit. Although the fabrication of some of the newer devices (specifically, metal-oxide-semiconductor transistors) has shown that thermally grown silicon dioxide falls short of providing perfect control of surface phenomena, it is safe to say that the announcement of this technique by Atalla *et al.*[1] resulted in a decisive advantage for silicon over germanium as the starting material in microcircuits.

Silicon is not an optimum choice in every respect. For example, the mobility of holes and electrons in silicon is lower than in germanium, resulting in potentially poorer high-frequency performance. A more serious disadvantage lies in the fact that silicon (like germanium) is an indirect gap semiconductor. As a consequence, many important electro-optical applications are not possible with monolithic silicon microcircuits.

In this chapter some of the material properties of silicon are considered, since these have a bearing on both the fabrication processes that follow and on the electrical properties of the ensuing microcircuits.

1.1. ATOMIC STRUCTURE

Together with carbon, germanium, grey tin, and lead, silicon belongs in column IV of the periodic table. The silicon atom consists of a central nucleus with 14 electrons in various orbital configurations around it. The allowed configurations are determined on the basis of the well-known spectrographic selection rules, as follows:

$n =$ the principal quantum number for the orbit
$= 1, 2, 3, \ldots$
$l =$ quantum number specifying angular momentum
$= 0, 1, 2, \ldots n-1,$
$m_l =$ quantum number specifying orbit orientation
$= 0, \pm 1, \pm 2, \ldots, \pm l,$
$m_s =$ quantum number associated with electron spin,
$= \pm \frac{1}{2}.$

Each level may contain no more than two electrons, corresponding to spin up (↑) and spin down (↓). In addition to these rules, shell and subshell locations for the electrons are designated as follows:

1. Shell designations $K, L, M, N, \ldots$ are used for electrons having the principal quantum numbers $n = 1, 2, 3, 4, \ldots$ respectively.
2. Within each shell subshell designations $s, p, d, f, \ldots$ are used for electrons having the angular momentum number $l = 0, 1, 2, 3, \ldots$.

Using these rules and shell designations, the atomic structure of the silicon atom is built up as shown in Table 1.1. It is seen that the K and L shells are totally filled whereas the M shell is only partially filled. Electrons constituting this partially filled shell are the valence electrons, whose configuration is designated $(3s)^2(3p)^2$. In the silicon lattice the proximity of neighboring atoms causes a change in the valence-orbital configuration owing to highly directed forces from their nuclei. The result is a mixing or hybridization of these orbitals, referred to as the $(sp)^3$ or

Table 1.1
Atomic Structure of Silicon

n	l	m_l	m_s	Shell	Subshell	No. of Allowed States	States Actually Occupied
1	0		$\pm\frac{1}{2}$	K	1s	2	2
2	0	0	$\pm\frac{1}{2}$	L	2s	2	2
	1	$-1, 0, 1$	$\pm\frac{1}{2}$	L	2p	6	6
3	0	0	$\pm\frac{1}{2}$	M	3s	2	2
	1	$-1, 0, 1$	$\pm\frac{1}{2}$	M	3p	6	2
	2	$-2, -1, 0, 1, 2$	$\pm\frac{1}{2}$	M	3d	10	—

covalent-bonding orbitals. Because they represent a lower energy situation than that of the original configuration, the covalent bond is a stable one.

1.2. CRYSTAL STRUCTURE

Silicon belongs to the cubic class of crystals. Crystal types belonging to this class exhibit the following structures:

1. *Simple cubic (s. c.) crystals*. This is illustrated in Figure 1.1a. Very few crystals exhibit as simple a structure as this one; an example is polonium, which exhibits this structure within a specific range of temperatures.
2. *Body-centered cubic (b. c. c.) crystals*. This is illustrated in Figure 1.1b. Crystals such as those of tungsten and molybdenum exhibit this structure.
3. *Face-centered cubic (f. c. c.) crystals*. This is illustrated in Figure 1.1c. The structure is exhibited by a large number of elements, such as nickel, copper, gold, and platinum. The face-centered sites are shown different from the corner sites for illustrative purposes only.
4. *The diamond structure*, which consists of two interpenetrating f. c. c. sublattices, with one atom of the second sublattice located at one-fourth of the distance along a major diagonal of the first sublattice. This configuration is illustrated in Figures 1.2a and 1.2b. This lattice is exhibited by silicon as well by diamond, germanium, and grey tin.

The position of the various atoms in the diamond lattice can be readily given in multiples of the cube edge a. Thus for the f. c. c. structure the various coordinates for the corner lattice sites are $0, 0, 0$; $0, 1, 0$; $0, 0, 1$; $0, 1, 1$; $1, 0, 0$; $1, 1, 0$; $1, 0, 1$; and $1, 1, 1$. Coordinates for the face-centered sites are $\frac{1}{2}, \frac{1}{2}, 0$; $\frac{1}{2}, \frac{1}{2}, 1$; $0, \frac{1}{2}, \frac{1}{2}$; $1, \frac{1}{2}, \frac{1}{2}$; $\frac{1}{2}, 0, \frac{1}{2}$; and $\frac{1}{2}, 1, \frac{1}{2}$ respectively.

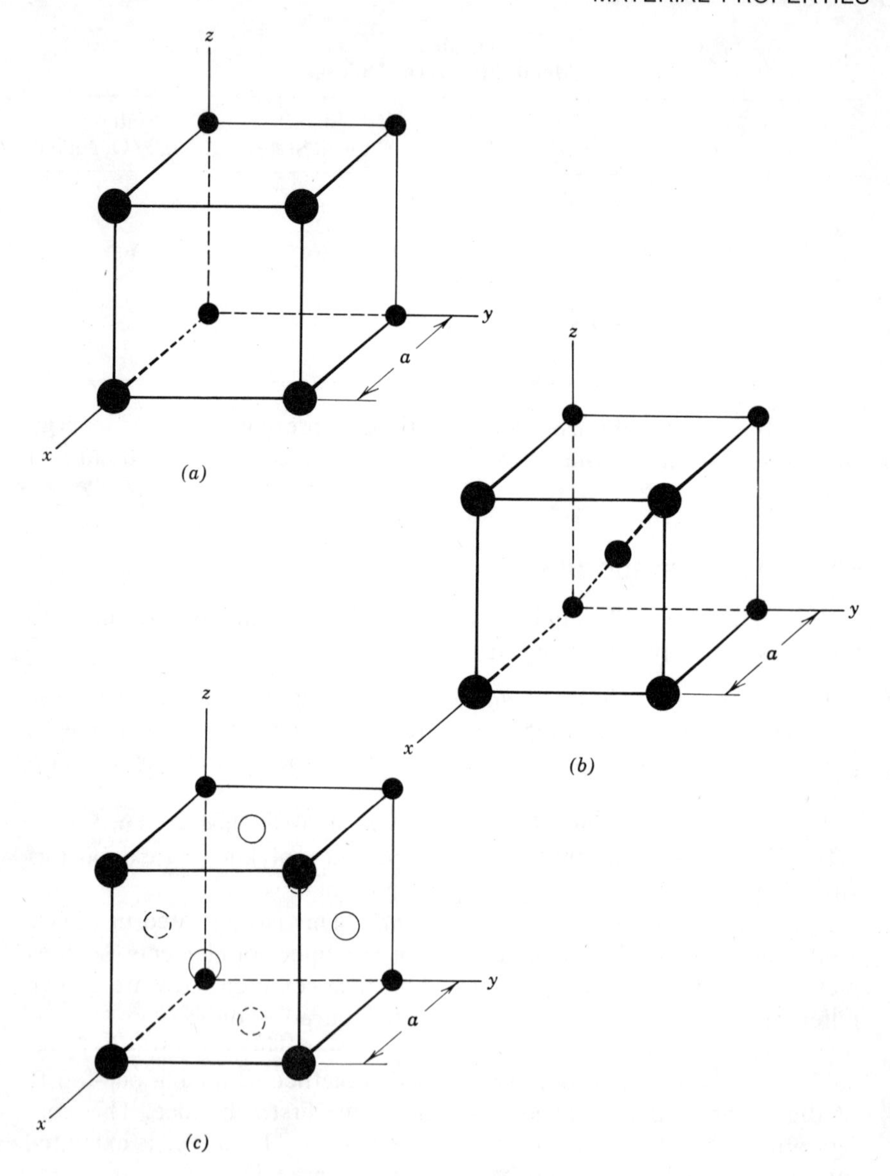

Figure 1.1 Cubic crystal lattices.

For the diamond lattice it is necessary to include the coordinates of the second sublattice, spaced at $\frac{1}{4}, \frac{1}{4}, \frac{1}{4}$ from those of the first. Within the unit cell these coordinates are $\frac{1}{4}, \frac{1}{4}, \frac{1}{4}$; $\frac{3}{4}, \frac{3}{4}, \frac{1}{4}$; $\frac{1}{4}, \frac{3}{4}, \frac{3}{4}$; and $\frac{3}{4}, \frac{1}{4}, \frac{3}{4}$ respectively. In Figure 1.2*b* these lattice sites are shown different from those of the

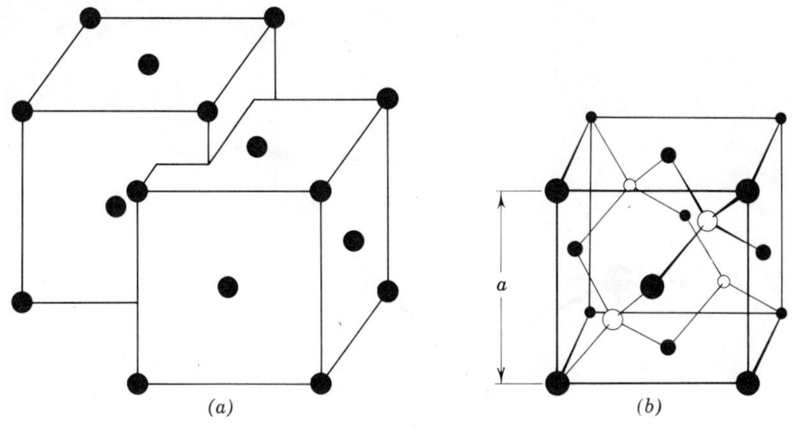

Figure 1.2 The diamond lattice.

original sublattice for illustrative purposes only. With reference to this figure, the following comments may be made pertaining to silicon:

1. Silicon has a coordination number of 4, i.e., each silicon atom has four nearest neighbors. In addition, each silicon atom has four valence electrons which provide covalent bonding with these nearest neighbors. Figure 1.3 shows an enlarged picture of a subcell with side $a/2$ in order to delineate a tetrahedral covalent bond.

2. The radius of the atom is given as one-half the distance between nearest neighbors. (This assumes a "hard sphere" picture for atoms.) For the diamond lattice this radius is $a\frac{\sqrt{3}}{8}$, where a is the cube edge. Since each atom is situated within a tetrahedron comprising its four nearest neighbors, this radius is often known as the *tetrahedral radius*. For silicon, $a = 5.428$ Å and the tetrahedral radius is thus 1.18 Å.

3. Using the hard-sphere picture for the atom, only $\pi\frac{\sqrt{3}}{16}$ (or about 34%) of the lattice is occupied by atoms. Thus the silicon lattice is a relatively loosely packed structure. (By way of comparison the packing density of a f. c. c. crystal is approximately 74%.)

Table 1.2 lists the various crystal properties of silicon. In addition, Table 1.3 lists the tetrahedral radii of various impurities that may be introduced into the lattice to control its electronic behavior. It should be noted that the radius of an impurity atom in a silicon lattice is not the same as the radius of the atom in its own lattice, since the internal field conditions are quite different for these two cases.

If r_0 is the tetrahedral radius of the silicon atom, the radius of the impurity atom may be written as $r_0(1 \pm \epsilon)$. The quantity ϵ is then defined as the misfit factor for the impurity and is indicative of the degree of strain present in the lattice as a result of its introduction.

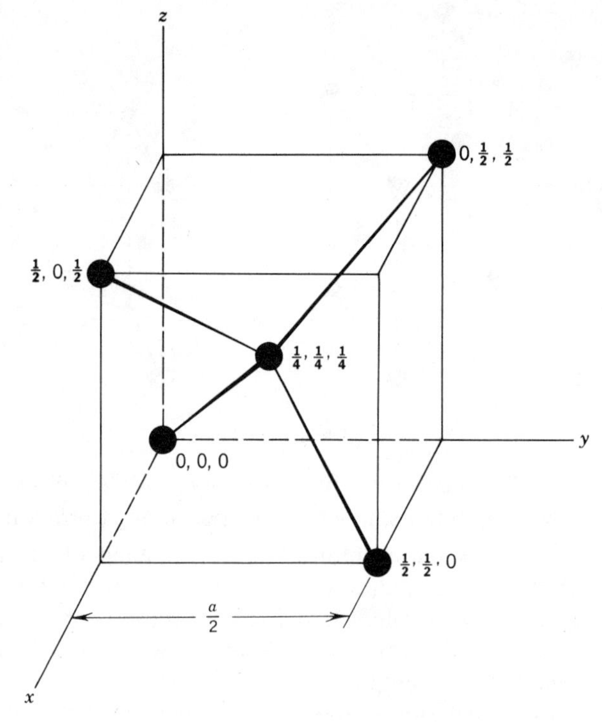

Figure 1.3 The diamond subcell.

Table 1.2
Crystal Properties of Silicon

Crystal type	Diamond
Coordination number	4
Tetrahedral radius	1.18 Å
Cube edge	5.428 Å
Atoms/cube of edge *a*	8
Packing density	34%

Table 1.3
Tetrahedral Radii and Misfit Factors for Various Dopants in Silicon

Dopant	P	As	Sb	B	Al	Ga	In	Au
Tetrahedral radius, Å	1.10	1.18	1.36	0.88	1.26	1.26	1.44	1.5
Misfit factor	0.068	0	0.153	0.254	0.068	0.068	0.22	0.272
Type of dopant		*n*-type			*p*-type			Deep-lying

1.3. CRYSTAL AXES AND PLANES

Directions in crystals of the cubic class are very conveniently described[2] in terms of Miller notation. Consider, for example, any plane in space, which may be described by

$$\frac{x}{a}+\frac{y}{b}+\frac{z}{c}=1. \tag{1.1}$$

Here a, b, and c are the intercepts made by the plane at the x, y, and z axes respectively. Writing h, k, and l as the reciprocals of these intercepts, so that $h=1/a$, $k=1/b$, and $l=1/c$, the plane may be described by

$$hx+ky+lz=1. \tag{1.2}$$

The Miller indices for such a plane† are (hkl). Integral values are usually chosen in multiples of the edge of the unit cell. Figure 1.4a shows a cubic crystal with some of its important planes indicated. Here the plane $ABCD$ is designated (110), while the plane EDC is designated (111). The (100) and (010) planes are also shown in this figure.

Figure 1.4b shows an example of how planes with negative indices may be described. Thus, the plane $PQRS$ is defined by $(0-10)$ and is commonly written as $(0\bar{1}0)$. The plane $RSTU$ is written in like manner as $(1\bar{1}0)$.

The atom configurations in many of the Miller planes in a cubic crystal are identical. Thus the planes (001), (010), (100), $(00\bar{1})$, $(\bar{1}00)$, and $(0\bar{1}0)$ are essentially similar in nature. For convenience they are written as the {001} planes.

Figure 1.4c shows examples of planes with higher indices. Thus the plane $GHKJ$ is denoted by $(1\tfrac{1}{2}0)$ or preferably by (210). In like manner the plane LHK is written in Miller notation as (212).

Planes with higher Miller indices may be sketched by extending these principles. They are not, however, often encountered in discussions of the material properties of semiconductors.

Indices of lattice plane direction (i.e., of the line normal to the lattice plane) are simply the vector components of the direction resolved along the coordinate axes. Thus the (111) plane has a direction written as [111], and so on. This is an extremely convenient feature of the Miller index system. For this notation the set of direction axes [001], [010], [100], $[00\bar{1}]$, $[0\bar{1}0]$, $[\bar{1}00]$ is written as $\langle 001 \rangle$.

†It should be noted that (hkl) refers to any one of a series of parallel planes in a cubic crystal. This may be seen by a simple shifting of the origin for the reference axes.

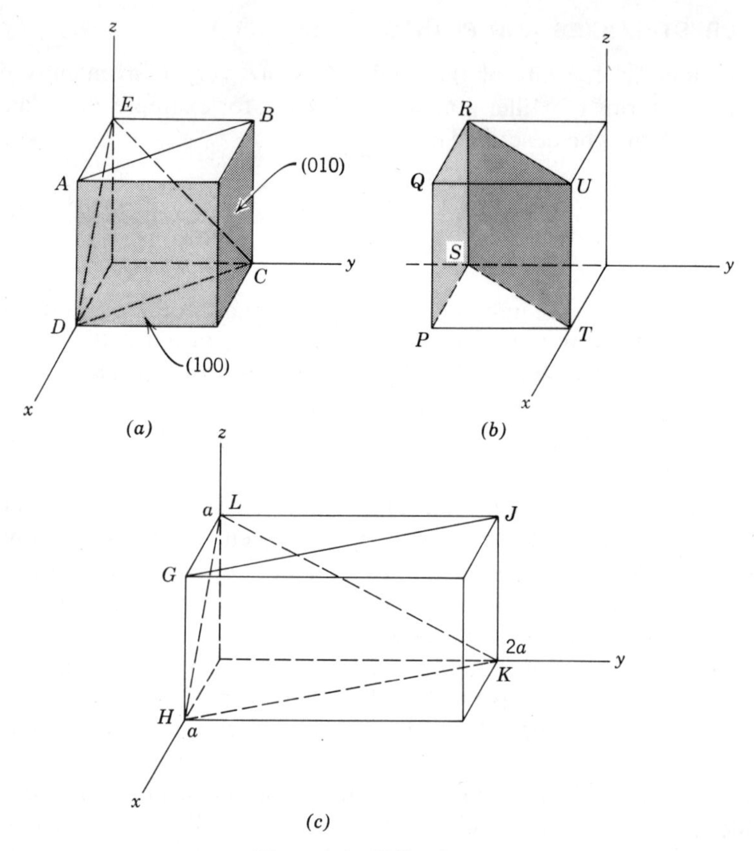

Figure 1.4 Miller planes.

The separation between two adjacent parallel planes (hkl) is given by

$$d = \frac{a}{\sqrt{h^2 + k^2 + l^2}}. \tag{1.3}$$

For silicon the separation between the various adjacent planes is as follows:

$$d(100) = 5.42 \text{ Å}. \tag{1.4}$$

$$d(110) = 3.83 \text{ Å} \tag{1.5}$$

$$d(111) = 3.13 \text{ Å}. \tag{1.6}$$

Thus it is seen that the {111} planes of silicon exhibit the smallest separation. Therefore growth of the crystal along a ⟨111⟩ direction is favored, since it results in the setting down of one atomic layer upon another in closest packed form.

The angle θ included between two crystallographic directions $[u_1v_1w_1]$ and $[u_2v_2w_2]$ is given by

$$\cos\theta = \frac{u_1u_2 + v_1v_2 + w_1w_2}{\sqrt{(u_1^2 + v_1^2 + w_1^2)(u_2^2 + v_2^2 + w_2^2)}}. \tag{1.7}$$

1.4. CRYSTAL DEFECTS

Any interruption in a perfectly periodic lattice may be called a *defect.* Defects may take various forms such as:

1. *Point defects.* These include vacancies, interstitials, and impurity atoms deliberately introduced for the purpose of controlling the electronic properties of the semiconductor.
2. *Dislocations or line defects.* These are one-dimensional defects, and are geometric faults or disturbances in the packing of atoms in the crystal lattice.
3. *Gross defects, such as slip and twinning.* Here the defect occurs along one or more planes in the crystal.

A study of defects is important since most mechanisms for diffusion and crystal growth are defect induced. In addition, all types of defects (chemical or otherwise) alter the electrical properties of the semiconductor in which they are present.

1.4.1. Point Defects

The most elementary point defect is the *vacancy.* This is present when, as a result of thermal fluctuations, an atom is removed from its lattice site and moved to the surface of the crystal. Defects of this type are known as *Schottky defects*, and are associated with an energy of formation of about 2.3 eV in silicon. Figure 1.5*a* shows the manner in which such a defect occurs in an otherwise regular silicon lattice.

A second elementary point defect that may be present in a crystal lattice in the *interstitial.* Such a defect occurs when an atom becomes located in one of the many interstitial voids within the crystal structure. Figure 1.5*b* shows schematically how this may occur in an otherwise regular crystal lattice. The energy of formation of an interstitial defect is relatively large in close-packed crystal structures. However, in the more

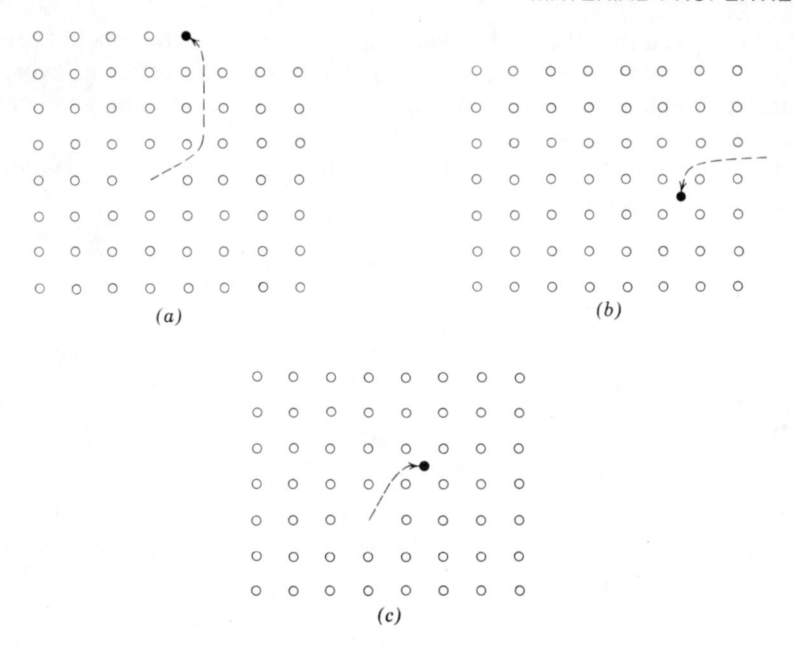

Figure 1.5 Point defects.

loosely packed silicon lattice this is not the case, and it is comparable to that for a vacancy.

The vacancy-interstitial pair, or *Frenkel defect*, occurs when an atom leaves its regular site in a crystal and takes up an interstitial position, as shown in Figure 1.5*c*. Since this interstitial is usually in the vicinity of the newly formed vacancy, the energy of formation of Frenkel defects is considerably lower than that for Schottky defects.

Figure 1.6 shows the unit cell for the diamond lattice. Within this unit cell are five intersitial voids[3] centered at $\frac{1}{2}, \frac{1}{2}, \frac{1}{2}$; $\frac{1}{4}, \frac{1}{4}, \frac{1}{4}$; $\frac{3}{4}, \frac{3}{4}, \frac{1}{4}$; $\frac{3}{4}, \frac{1}{4}, \frac{3}{4}$; and $\frac{1}{4}, \frac{3}{4}, \frac{3}{4}$. Each of these is large enough to contain a silicon atom (again assuming hard spheres of radius 1.18 Å) even though the constriction in passing from one void to another is about 1.05 Å in radius. Consequently, from a purely geometric viewpoint, the interstitial defect can be expected to be quite common in silicon.

Various combinations of these defects can also occur. Thus a single vacancy leads to the breaking of four covalent bonds, whereas two vacancies side by side require the breaking of only six bonds. Consequently, the energy of formation of a divacancy of this type is less than that required to form two separate vacancies. The divacancy is thus commonly encountered in the silicon lattice. On the other hand, the di-interstitial is considerably more difficult, if not impossible, to form.

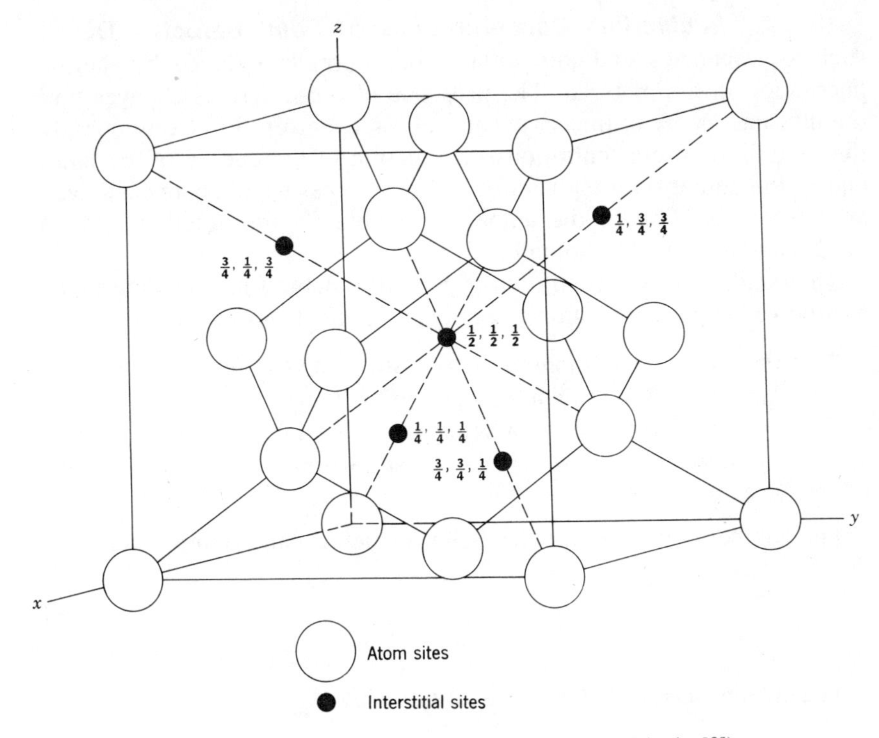

Figure 1.6 Interstitial sites in the diamond lattice (Rhodes [3]).

Chemical defects occur because of the presence of impurity atoms (intentionally or unintentionally introduced) in the silicon lattice. Impurity atoms that take up their locations at sites ordinarily occupied by silicon atoms (i.e. at lattice sites) are commonly referred to as *substitutional* impurities. Alternately, *interstitial* impurities are located in the many interstitial voids that are present in the silicon lattice.

At moderate doping levels, impurities which belong to columns III and V of the periodic table (the *p*-type and *n*-type dopants respectively) are substitutional in nature. The behavior of other impurities is more complex. Thus for gold in the silicon lattice, evidence indicates that a small fraction (about 10%) is in interstitial sites in the lattice while the rest is substitutional. With nickel, on the other hand, as much as 99.9% is in interstitial sites.

Impurities such as zinc, copper, cobalt, iron, and manganese are usually present unintentionally in the lattice. Their exact nature is unknown, but it is known that they occupy predominantly interstitial sites.

1.4.1.1. Equilibrium Concentrations of Point Defects. Defects such as vacancies and interstitials are internally induced by thermal fluctuation in the material. The presence of these defects changes both the internal energy of the crystal as well as its entropy [4]. Consequently, their equilibrium concentration is a function of the energy of formation and of the equilibrium temperature. On the other hand, chemical defects are also a function of the amount available for introduction into the crystal and of their solid solubility in silicon.

The equilibrium concentration of Schottky defects may be determined as follows, on the assumption that only these effects occur. Let

N = the total number of atoms in a crystal of unit volume,
n_s = number of Schottky defects per unit volume,
E_s = energy of formation of a Schottky defect, i.e., the energy required to move an atom from its lattice site within the crystal to a lattice site on its surface.

The number of ways in which a Schottky defect can occur is given by

$$C_{n_s}^{N} = \frac{N!}{(N-n_s)!\,n_s!}. \tag{1.8}$$

The entropy associated with this situation is

$$\begin{aligned} S &= k \ln (\text{number of ways}), \\ &= k \ln C_{n_s}^{N} \end{aligned} \tag{1.9}$$

where k is Boltzmann's constant (= 8.62×10^{-5} eV/°K). The internal energy E, is given by

$$E = n_s E_s \tag{1.10}$$

The change in free energy (neglecting volume changes) is given by

$$F = E - TS, \tag{1.11}$$

$$= n_s E_s - kT[\ln N! - \ln n_s! - \ln(N-n_s)!]. \tag{1.12}$$

The most probable equilibrium condition is the one where the free energy is a minimum with respect for changes in n_s, i.e., the case for

$$\left(\frac{\partial F}{\partial n_s}\right)_{T=\text{const}} = 0. \tag{1.13}$$

Differentiating (1.12) and setting to zero,

$$E_s = kT \frac{\partial}{\partial n_s}[\ln N! - \ln n_s! - \ln (N-n_s)!]. \tag{1.14}$$

The factorial terms may be simplified by using Stirling's formula for the factorial of a large number. Thus,

$$\ln x! \simeq x \ln x - x \tag{1.15}$$

and

$$\frac{\mathrm{d}}{\mathrm{d}x}(\ln x!) \simeq \ln x, \tag{1.16}$$

so that (1.14) reduces to

$$E_s = kT \ln \frac{N-n_s}{n_s} \tag{1.17}$$

or

$$n_s = \frac{N}{1+e^{E_s/kT}} \tag{1.18}$$

$$\simeq Ne^{-E_s/kT}. \tag{1.19}$$

The equilibrium concentration of Frenkel defects may be found by an analogous approach. Again it is assumed that only these defects are present. Let

N = number of atoms in a crystal of unit volume,
N' = number of available interstitial sites per unit volume,
n_f = number of Frenkel defects (i.e., vacancy-interstitial pairs) per unit volume,
E_f = energy of formation of a Frenkel defect.

A vacancy can occur in $C_{n_f}^{N'}$ ways, and an interstitial in $C_{n_f}^{N'}$ ways. Consequently, a Frenkel defect can occur in $C_{n_f}^{N} C_{n_f}^{N'}$ ways, if the events are assumed to be statistically independent.

The entropy associated with this situation is

$$S = k \ln C_{n_f}^{N} C_{n_f}^{N'}. \tag{1.20}$$

The internal energy is given by

$$E = n_f E_f. \tag{1.21}$$

The change in free energy is given by

$$F = n_f E_f - kT \ln C^{N}_{n_f} C^{N'}_{n_f}. \tag{1.22}$$

As before,

$$\left(\frac{\partial F}{\partial n_f}\right)_{T=\text{const}} = 0 \tag{1.23}$$

in thermal equilibrium. Differentiating (1.22) and using Stirling's formula gives

$$E_f = kT \ln \left\{ \frac{(N - n_f)(N' - n_f)}{n_f^2} \right\} \tag{1.24}$$

or

$$n_f \simeq \sqrt{NN'}\, e^{-E_f/2kT}. \tag{1.25}$$

Concentrations of point defects in excess of the equilibrium value may be obtained by subjecting the semiconductor to nonequilibrium processes. Thus excessively fast cooling (quenching) can result in a supersaturated concentration of these defects. Nuclear radiation damage also results in increasing the defect concentration over its equilibrium value.

1.4.2. Dislocations

A dislocation is a one-dimensional line defect in an otherwise perfect crystal, and results in a geometric fault in the lattice. It occurs when the crystal is subjected to stresses in excess of the elastic limit, e.g., during its growth from a melt. Although the nature of dislocations is quite complex, they are usually composed of combinations of two basic types—the screw dislocation and the edge dislocation. A simple cubic lattice is considered in the following sections. The diamond lattice is considerably more complex; the general properties of dislocation types, however, are very similar to those of the cubic lattice.

1.4.2.1. The Screw Dislocation. Figure 1.7 shows the manner in which a regular crystal lattice may be subjected to shear stresses in order to establish a screw dislocation. Imagine that the crystal is cut along the plane *ABCD*, which is one of its regular lattice planes, and let the two halves of the crystal on either side of this plane be subjected to shearing forces that are sufficiently large to cause them to be separated by one atomic spacing. The line of the screw dislocation so formed is *AD*, since this marks the boundary in the plane *ECBF* which divides the perfect crystal from the imperfect.

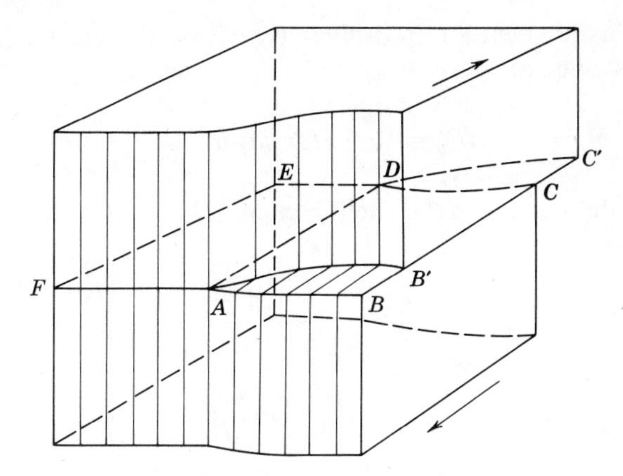

Figure 1.7 Screw dislocation.

The strain energy associated with a screw dislocation may be calculated [5] by considering a cylinder of material, with axis AD and inner and outer radii R_i and R_o respectively, as shown in Figure 1.8. It is assumed that the crystal behaves as an elastic solid within the cylinder defined by these radii. Let

$b =$ the amount of shear present in a shell of radius r and thickness dr,
$\mu =$ the shear modulus (7.9×10^{11} dynes/cm² for silicon).

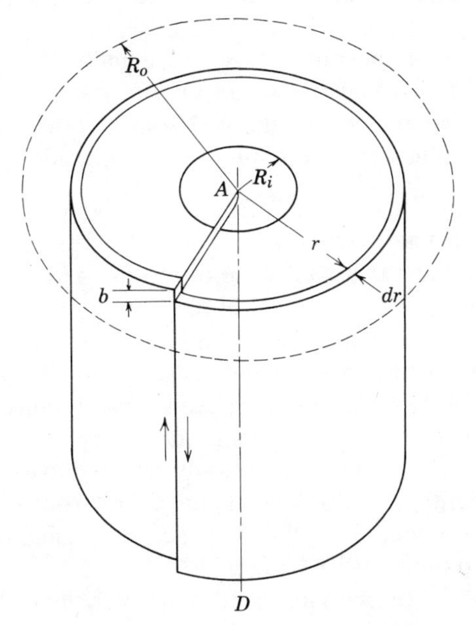

Figure 1.8 Details of a screw dislocation.

Then, the elastic shear strain is given by $b/2\pi r$. The elastic energy of the shell, dE_μ, is given by

$$dE_\mu = \tfrac{1}{2}\mu \,(\text{shear strain})^2 \, dV, \tag{1.26}$$

where dV is the volume of the shell. Hence,

$$dE_\mu = \left(\frac{\mu b^2}{4\pi}\right)\left(\frac{dr}{r}\right) \tag{1.27}$$

and

$$E_\mu = \left(\frac{\mu b^2}{4\pi}\right) \ln\left(\frac{R_o}{R_i}\right). \tag{1.28}$$

If this is the only dislocation in an infinite volume of material, $R_o = \infty$ and the energy associated with it must be infinite. Practical crystals, however, contain many dislocations randomly distributed. As a result, their strain fields are also randomly distributed and cancel each other at distances approximately equal to the mean distance between them. In typical crystals, R_o is about 10^4 atoms spacings. The inner-radius limit R_i is set by the fact that a region of atomic dimensions can no longer be considered as an elastic continuum and the theory of elasticity ceases to hold. As a consequence, it is reasonable to eliminate the inner one to two atoms from consideration.

Practical values of the ratio R_o/R_i are usually taken around 5×10^3. With the use of this value the strain energy for a screw dislocation in silicon may be calculated as about 10–19 eV/atom length. (By way of comparison, values for aluminum and diamond are 3.1 and 29 eV respectively.)

1.4.2.2. The Edge Dislocation. An edge dislocation is shown in Figure 1.9. Here, an extra half-plane of atoms, $ABCD$, is present in the otherwise regular lattice, with most of the distortion concentrated around the line AD. An edge dislocation of this type is created by applying a shearing force along the faces of the crystal, parallel to a major crystallographic plane. When this force exceeds that required for elastic deformation, the upper half of the crystal moves by a slip mechanism. The plane along which slip occurs is commonly referred to as a *slip plane*.

The strain energy associated with an edge dislocation is somewhat more difficult to compute than that for a screw dislocation. Its magnitude however, is approximately 50% larger.

In view of the large energies of formation for both basic dislocation types, it must be concluded that their equilibrium concentration is negli-

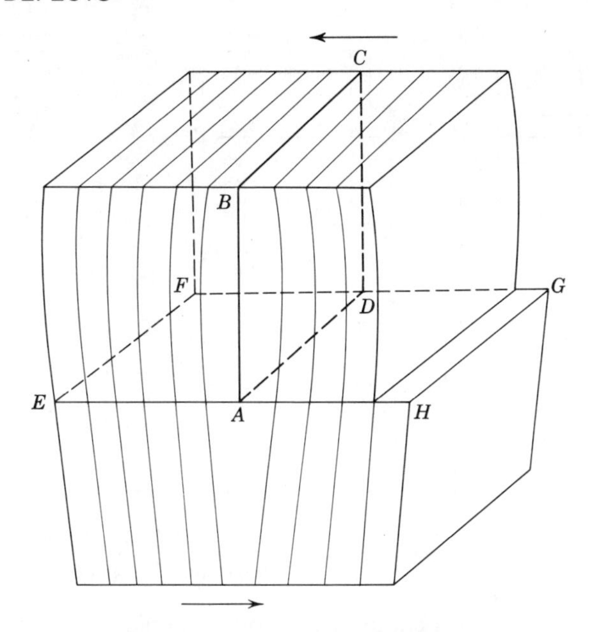

Figure 1.9 The edge dislocation.

gible, i.e., their presence is due to nonequilibrium processes such as those associated with the growth, freezing, or quenching of the crystal.

1.4.2.3. Movement of Dislocations. Figure 1.10 indicates the manner in which an edge dislocation may move completely through a crystal. The mechanism for such a movement is called *slip*. It is characteristic of the slip mechanism that it results in movement along planes of high atomic density where opposing forces are at a minimum.

The displacement of a screw dislocation also takes place along a slip plane. In Figure 1.7 this slip plane is given by *ABCD*. The end result of such a displacement is identical to the movement of an edge dislocation, even though the strain pattern is different.

In addition to slip, *climb* is an alternate method by which a dislocation can move in a crystal. For an edge dislocation, such as that shown in Figure 1.9, climb of the plane *ABCD* takes place at right angles to the slip plane *EFGH*. Figure 1.11 shows that this may occur as the result of the movement of atoms† out of the plane *ABCD*. Both substitutional or interstitial atoms may be involved in this process. Intuitively, it is seen that the energy of formation associated with such a process is on the same order of magnitude as that for a Schottky or Frenkel defect.

†Alternately, climb may also occur by atoms moving into the plane.

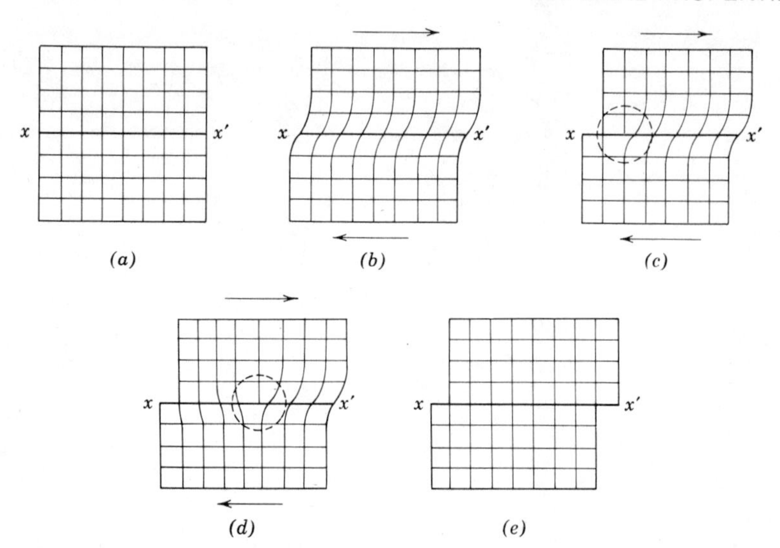

Figure 1.10 Crystal movement along a slip plane.

Climb in a screw dislocation occurs by a complex screw motion. Here the screw dislocation line twists itself into a helix, which can then climb. The actual movement of dislocations in a crystal is made up of combinations of these and other types of movements.

The energy of movement of a dislocation has been shown to be about 0.15 eV/atom spacing for silicon. This is the energy barrier that must be overcome in order for a dislocation to move in a crystal. A comparison with the energy of formation of a dislocation (about 10–19 eV) shows that it is extremely easy to induce dislocation motion in a crystal, even though it is difficult to create a dislocation. Thus, one of the more important problems of crystal growth is to avoid (or minimize) the formation of dislocations in the first place. Alternately, if such dislocations are formed, they can be induced to grow out of the crystal, leaving behind a relatively dislocation-free lattice.

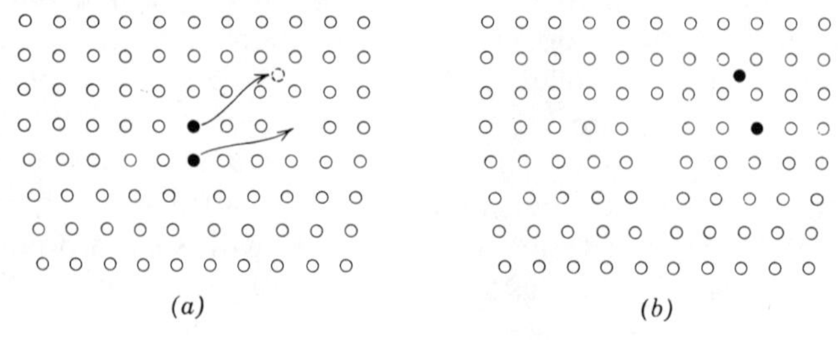

Figure 1.11 Climb of an edge dislocation.

1.4.2.4. The Frank-Read Source. There is considerable evidence indicating that there is not only dislocation movement in a crystal but also dislocation multiplication. Examination of deformed crystals has shown that this is indeed the case, and various mechanisms have been suggested for this multiplication. Figure 1.12 shows a model proposed by Frank and Read [5]. Consider a dislocation, as shown in Figure 1.12a, terminated at its end in xx'. Under the application of a stress F, the dislocation tends to bow out of its slip plane, as shown in Figure 1.12b. In so doing, it becomes longer and requires a greater stress to maintain its new radius. A critical condition is reached at which the dislocation line is semicircular. For a stress in excess of that required for this condition, the dislocation becomes unstable and progresses as shown in Figures 1.12c and 1.12d. Eventually it returns to its original form by the collapse of the cusp, as shown in Figure 1.12e, leaving an expanding loop in addition to it. The process now repeats itself, resulting in a number of dislocations from a single Frank-Read source of this type.

1.4.3. Dislocations in the Diamond Lattice

Figure 1.13 shows the spatial distribution of atoms within a unit cell in the diamond lattice. Characteristic of this distribution is the appearance of the hexagonal chair form, shown in heavy lines. The more complete

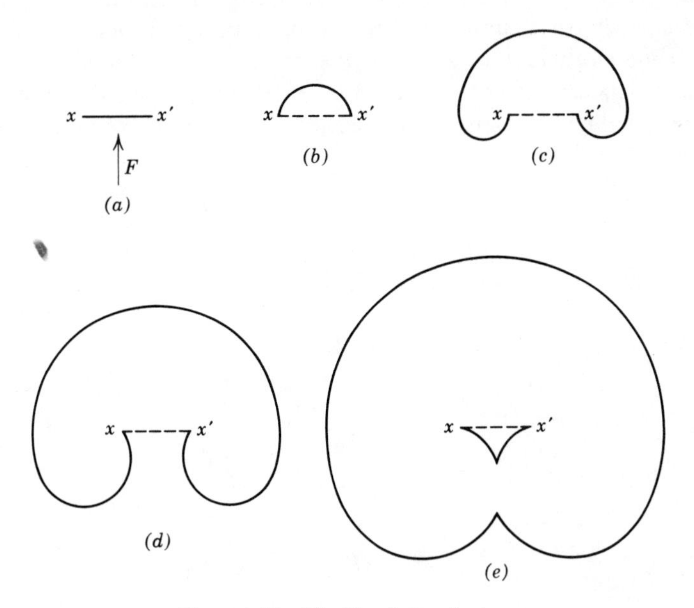

Figure 1.12 The Frank-Read source.

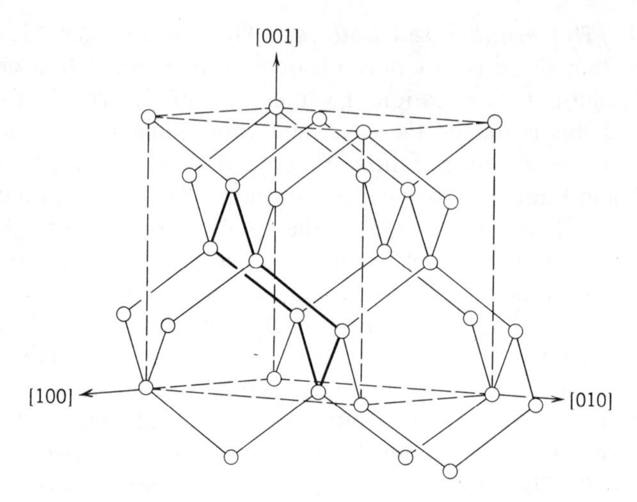

Figure 1.13 The diamond lattice (Hornstra[6]).

atomic arrangement is usually drawn as shown in Figure 1.14, with one of its ⟨111⟩ directions vertical. A distinguishing characteristic of this figure is the presence of the hexagonal ring type of structure, shown in heavy outline.

Figure 1.15 shows the simplest form of edge dislocation, the so-called 60° dislocation, that can be present in the diamond lattice. A dislocation of this type is characterized by a row of broken or dangling bonds and by the presence of an extra half-plane. Hornstra[6] has shown, however, that the dangling bonds in the lattice may be avoided by a diffusional re-

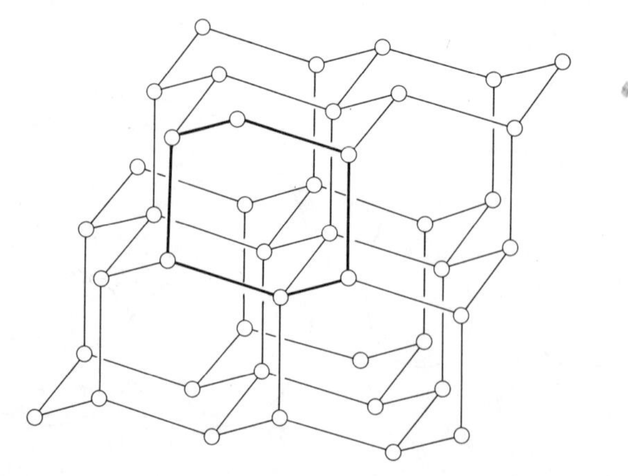

Figure 1.14 The diamond lattice (side view) (Hornstra[6]).

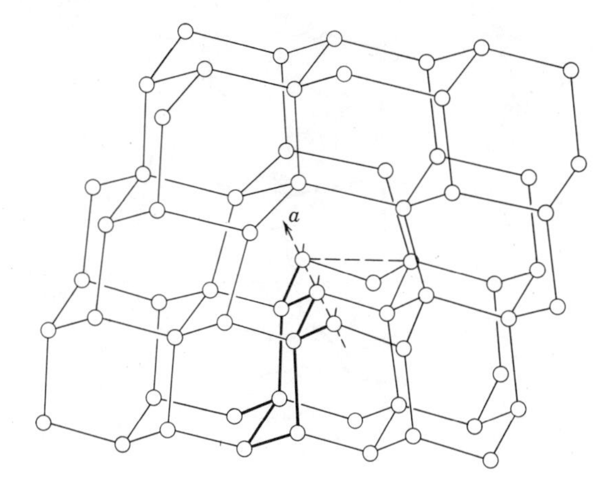

Figure 1.15 Edge dislocation in the diamond lattic; *a* is dislocation axis (Hornstra[6]).

arrangement such as that shown in Figure 1.16, disturbing the hexagonal atomic arrangement in the process. Although it is not clear which arrangement is actually present, there is certainly some reason to question the presence of dangling bonds at all.

Figure 1.17 shows a possible type of screw dislocation that may exist in the diamond lattice. The nature of this dislocation is seen in comparing a normal hexagonal chair sequence, labeled 8, 9, 10, 11, 12, 13, with the sequence 1, 2, 3, 4, 5, 6 involved in the screw structure.

More complex dislocation types are also possible in the diamond lattice. A discussion of these is beyond the scope of this chapter.

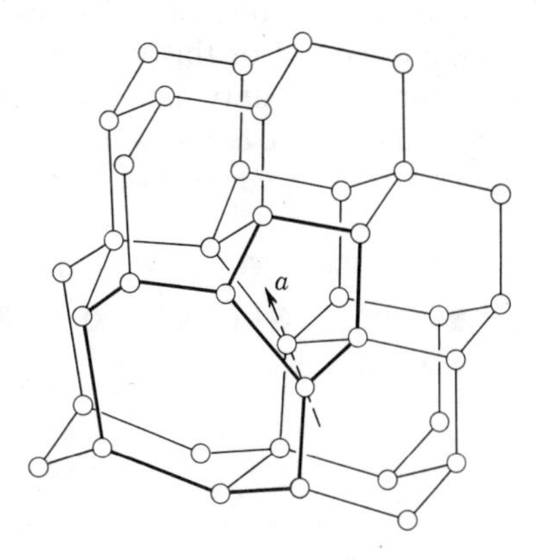

Figure 1.16 Diffusional rearrangement with no dangling bonds; *a* is dislocation axis (Hornstra[6]).

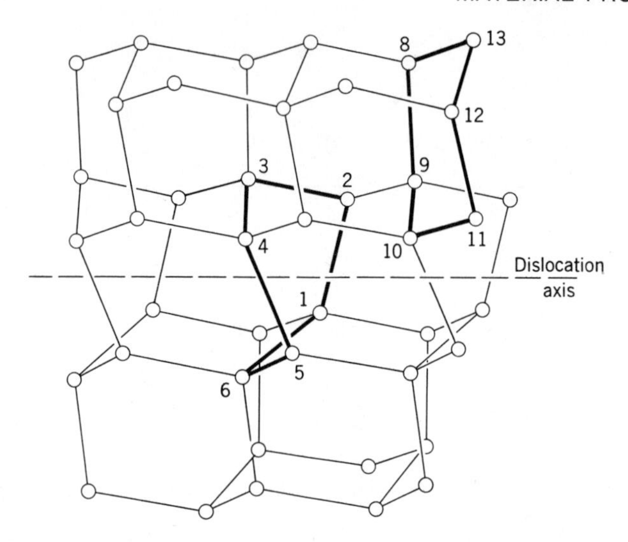

Figure 1.17 Screw dislocation in the diamond lattice (Hornstra [6]).

1.4.4. Twinning

Twinning is one form of gross defect that may occur in a crystal. Its presence is usually indicative of material that has a high dislocation content and is not suited for the fabrication of devices and microcircuits. Consequently, the subject will only be treated very briefly.

Twinning occurs when one portion of a crystal lattice takes up an orientation with respect to another, the two parts being in intimate contact over their bounding surfaces. This bounding surface is called the *twinning plane*. Figure 1.18 shows a two-dimensional representation of twinned and untwinned parts of a crystal. For this case, atoms along the line xx' are common to both twinned and untwinned sections, and the twinning plane is sometimes referred to as the *composition plane*.

Experimental evidence shows that excessive twinning is encountered

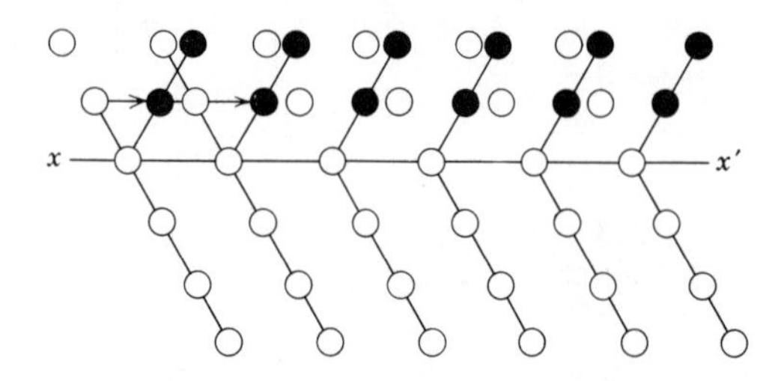

Figure 1.18 Twinned structure.

if the silicon is restricted during its growth from a melt. Thus crucible grown materials are highly prone to this defect.

1.5. ELECTRONIC PROPERTIES OF DEFECTS

The deliberate insertion of chemical defects into the silicon lattice is the basis for the fabrication and control of electrical properties of semiconductor devices and microcircuits. Although this subject is discussed in considerable detail in later chapters, attention must also be paid to the electronic behavior of defects that are unintentionally present in the crystal.

1.5.1. Effect on Resistivity

The presence of a vacancy in a crystal results in four unsatisfied bonds which would ordinarily be used to bind the atom to its tetrahedral neighbors. Thus a vacancy tends to be acceptor-like in behavior with a theoretical capability of having as many as four electrons added to it. The addition of each electron to this vacancy results in successively higher and higher values of energy level because of the large mutual electrostatic repulsion present between them. Consequently, it is reasonable to expect that a vacancy will exhibit one or two acceptor levels, but highly improbable that it will exhibit as many as four within the energy gap.

In like manner an interstitial has four valence electrons that are not involved in covalent binding with other lattice atoms and that may be lost to the conduction band. As a result, it exhibits donor-like behavior. Again it is probable that it exhibits only one or two donor levels within the energy gap.

There is some evidence to show that dislocations are acceptor-like in behavior. This observed behavior may be explained by the presence of dangling (or unfilled) bonds at the edge of the half-plane comprising the dislocation. As shown in Section 1.4.3, there is some question whether these dangling bonds are present in the diamond lattice. The current theory tends to favor the presence of a few such broken bonds, even though the majority have been altered by the diffusional rearrangement of Figure 1.16. This would explain the absence of strong acceptor-like properties for these dislocations.

In conclusion it is felt that dislocations do not, in themselves, affect the electronic properties of the crystal to any significant degree. However, they give rise to enhanced diffusion effects in their vicinity, and also result in the segregation of metallic impurities around them. These effects, in turn, lead to such problems as excessive leakage and premature breakdown in semiconductor junctions made on this material.

1.5.2. Effect on Mobility

It has been suggested that the presence of space charge cylinders around the dislocation leads to the scattering of carriers in their vicinity. However, such effects are ordinarily of second order; the most important property of a dislocation is that it interacts with chemical and other point defects in its neighborhood. This interaction exists between the localized disturbance, owing to an impurity atom, and the strain field in the vicinity of an edge dislocation. Its extent is directly proportional to the misfit factor. Thus the presence of a dislocation is usually associated with a concentration of impurities in its vicinity. These impurities, in turn, reduce the mobility of carriers in their neighborhood.

1.5.3. Effect on Lifetime

It has been postulated that both point defects and dislocations result in highly localized distortion of the energy-band structure in their vicinity, leading to the formation of trapping sites for free holes and electrons. Since there exists, in general, an inverse relationship† between the trap concentration and the minority carrier lifetime, it is reasonable to expect that this lifetime is also inversely related to the defect concentration. This result has been verified by numerous experiments on material with "as grown" as well as with induced defects (by nuclear radiation and plastic deformation techniques).

1.6. REFERENCES

[1] M. M. Atalla, *et al.*, "Stabilization of Silicon Surfaces by Thermally Grown Oxides," *Bell System Tech. J.*, **38**, 749–782 (1959).
[2] C. Kittel, *Introduction to Solid State Physics*, Wiley, New York, 1966.
[3] R. G. Rhodes, *Imperfections and Active Centres in Semiconductors*, Pergamon Press, Macmillan, New York, 1964.
[4] J. H. Brophy, R. M. Rose and J. Wulff, "The Structure and Properties of Materials," Vol. 11, *Thermodynamics of Structure*, Wiley, New York, 1964.
[5] A. H. Cottrell, *Theory of Crystal Dislocations*, Gordon and Breach, New York, 1964.
[6] J. Hornstra, "Dislocations in the Diamond Lattice," *J. Phys. Chem. Solids*, **5**, 129–141 (1958).

1.7. PROBLEMS

1. The gallium arsenide crystal consists of two interpenetrating f. c. c. sublattices, one of gallium and one of arsenic. One atom of the second sublattice is located one-fourth of the distance along a major diagonal of the first sublattice, resulting

†This relationship is explored in detail in the discussion of nonequilibrium processes.

in the zincblende structure. For this crystal, sketch the (100), (110), (111), and $(\overline{111})$ planes, identifying gallium and arsenic atoms in each case.

2. Compute the number of atoms/cm^2 for the different planes sketched in Problem 1.

3. A slice of silicon, cut from a crystal that is grown in the ⟨111⟩ direction, is found to fracture readily along certain well-defined cleavage planes. What are these planes?

4. Repeat Problem 3 for a slice cut from a ⟨100⟩ crystal.

5. A thin layer of single-crystal silicon is grown on a single-crystal substrate of ⟨111⟩ orientation. A tetrahedral stacking fault is initiated at a point on the substrate-layer interface and propagates along {111} planes as the layer is built up. It eventually terminates on the surface of the layer in the form of a triangle. Sketch the stacking fault and identify its various {111} planes. In addition, determine the height of the grown layer and the orientation of the sides of the triangle seen on the surface.

6. Sketch a {111} plane of silicon, identifying atoms in various layers. Thus, show that a crystal of this orientation is built up of layers of atoms in an $aa'bb'cc'$. . . sequence.

7. Calculate the vacancy concentration in a silicon crystal at 1200°C and at 27°C.

Chapter 2

Crystal Growth and Doping

Single-crystal silicon is obtained by controlled freezing from a melt. Since the melt is relatively close packed compared with the diamond lattice into which it freezes, this process is accompanied by a volume expansion, of about 10%. The growth of silicon within a crucible thus leads to considerable twinning and dislocations.

In this chapter two of the most common techniques for silicon crystal growth are described together with methods for doping and refining. Both of these techniques result in silicon of quality that is suitable for microcircuit fabrication.

2.1. GROWTH FROM THE MELT

The most commonly used technique for crystal growth is that proposed by Czochralski [1]. Here, as shown in Figure 2.1, the silicon is contained in a quartz or graphite crucible and kept in a molten condition by r.f. heating. A seed crystal is suspended over the crucible in a chuck. For growth the seed is inserted into the melt until its end is molten. It is now slowly withdrawn (at a rate of about 10 μm/sec), resulting in a single crystal which grows by progressive freezing at the liquid-solid interface. Provisions are often made to rotate both the crucible and the crystal during the pulling operation.

The entire assembly is enclosed within a fused silica envelope, the walls of which are sometimes water-cooled and flushed with an inert gas such as helium or argon. Since thermal conditions are continually changing during this operation, a feedback control system is used to maintain the temperature of the melt to within $\pm\frac{1}{2}$°C.

If H_i is the thermal input to the system and H_o is the heat loss, then

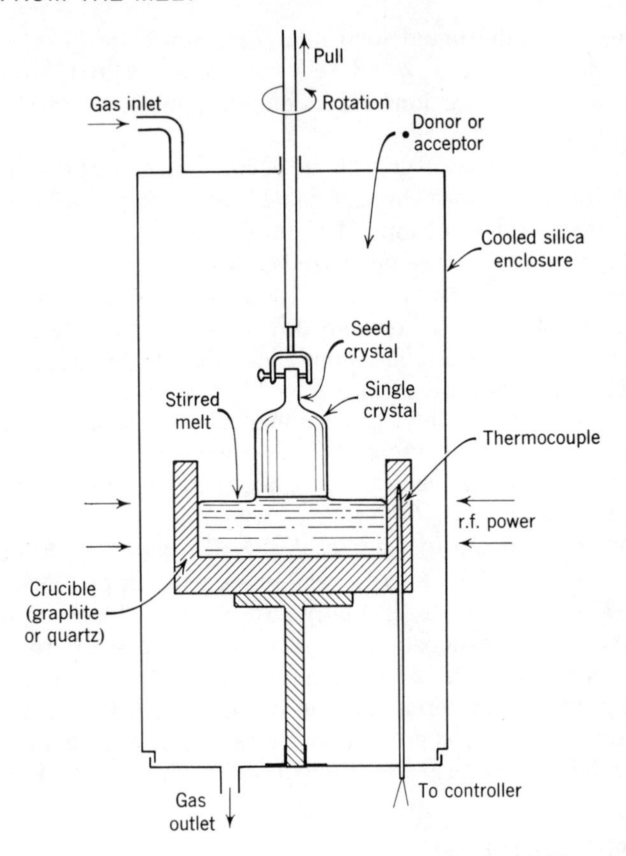

Figure 2.1 Czochralski crystal-growing apparatus (adapted from Bridges et al. [4]).

the heat difference $H_i - H_o$ is largely accounted for by the latent heat of crystallization L. Writing ρ_s as the density of silicon and A as the cross section of the grown crystal, the heat-balance equation is

$$H_o - H_i = AL\rho_s dx/dt, \tag{2.1}$$

where dx/dt is the pull rate. Thus it is seen that the crystal diameter, the thermal input, the heat loss, and the pull rate are interrelated.

2.1.1. Requirements for Proper Crystal Growth [2]

To grow suitable crystals relatively free from dislocations, it is important to pay attention to a number of factors during the growth period:

1. Crystal growth proceeds by the successive addition of layers of

atomic planes at the liquid-solid interface. Since the {111} planes are the most closely packed [$d = 3.13$ Å, as given in (1.6)], growth is preferred in the ⟨111⟩ directions. Consequently, this is the most commonly used orientation.

2. In order for crystal growth to occur along the {111} planes, the liquid-solid interface must be maintained as flat as possible, and at right angles to the ⟨111⟩ directions. The importance of accurate orientation of the seed crystal must be emphasized at this point.

3. In general, the interface takes the form of a meniscus which may either be concave up or concave down. Appropriate selection of the pull rate results in a nearly flat interface. This is highly desirable to minimize radial forces during freezing.

4. The flatness of the interface is enhanced by increasing the spin rate. Unfortunately, this also increases the corrosive effect of the molten silicon on the crucible walls. In practice, a relatively slow spin rate is used (about 1 to 3 rpm).

5. It has been noted in Chapter 1 that the energy of formation of a dislocation is in excess of 10 eV while its energy of propagation is only 0.15 eV. Consequently, every effort is made to begin with a small, accurately aligned, dislocation-free crystal as the seed. In addition, it has been found that, with a rapid growth rate, it is possible to make dislocations propagate preferentially out of the crystal. Hence it is common practice to start a crystal growth with a rapid pull rate and later to slow it down to the actual rate at which the majority of the crystal is grown.

2.2. DOPING IN THE MELT

This is usually done by adding a known mass of dopant to the melt in order to obtain the desired composition. Raw dopants are not used since the amounts to be added are unmanageably small. In addition, their physical characteristics are often quite different from those of the melt. (e.g., silicon melts at about 1412°C, while antimony melts at 630°C. Thus, elemental antimony cannot be added to molten silicon without disastrous results.) It is common practice to add the dopant in the form of a powder of highly doped silicon of about 0.01-Ωcm resistivity. In this manner both problems are avoided.

The addition of impurities to the melt, accompanied by stirring, results in a doped liquid from which the crystal is grown. In general, the concentration of the solute will be quite different in the solid and liquid phases of a crystal because of energy considerations.

Consider a crystal at any given point during its growth. Writing C_S and C_L as the concentration of solute (by weight) in the solid and the liquid phases in the immediate vicinity of the interface respectively, a distribution coefficient k may be defined, where

$$k = \frac{C_S}{C_L}. \tag{2.2}$$

Table 2.1 gives values[3] of k for the commonly used dopants. To an approximation, these values are independent of concentration. It is seen that this quantity is normally less than unity and ranges from as high as 0.72 for boron to as low as 2.25×10^{-5} for gold.

Table 2.1
Distribution Coefficients† for Various Dopants in Silicon

Dopant	P	As	Sb	B	Al	Ga	In	Au
Distribution coefficient, k	0.32	0.27	0.02	0.72	1.8×10^{-3}	7.2×10^{-3}	3.6×10^{-4}	2.25×10^{-5}
Type of dopant	n-type			p-type				Deep lying

†These are equilibrium values. Interface values are a function of the growth rate and are of comparable magnitude.

Since k is less than unity, excess solute is thrown off at the interface between the melt and the crystal. Consequently, the melt becomes increasingly solute-rich as crystal growth progresses, resulting in a crystal of varying composition. The precise nature of this composition may be determined for different growing conditions.

2.2.1. Rapid-stirring Conditions

Assume that rapid stirring is involved during very slow growth of the crystal. Then the solute in the immediate vicinity of the freezing interface is dispersed into the melt. In addition, it is assumed that there is no diffusion† of the solute within the crystal during the growth process. Let

W_M = initial weight of the melt,
C_M = initial concentration of the solute in the melt (by weight).

At a specified point in the growth process, when a crystal of weight W has been grown, let

C_L = concentration of solute in the liquid (by weight),
C_S = concentration of solute in the crystal (by weight),
S = weight of solute in the melt.

†This is reasonable, since the diffusion coefficient of impurities in the melt is about six orders of magnitude larger than in the grown crystal.

Consider an element of the crystal of weight dW. During its freezing, the weight of the solute lost from the melt is $C_S dW$. Thus

$$-dS = C_S dW. \tag{2.3}$$

At this point, the weight of the melt is $W_M - W$ and

$$C_L = \frac{S}{(W_M - W)}. \tag{2.4}$$

Combining these equations and substituting $C_S/C_L = k$

$$\frac{dS}{S} = -k\frac{dW}{W_M - W}. \tag{2.5}$$

The initial weight of the solute is $C_M W_M$. Consequently (2.5) may be integrated as

$$\int_{C_M W_M}^{S} \frac{dS}{S} = k\int_0^W -\frac{dW}{W_M - W}. \tag{2.6}$$

Solving,

$$S = C_M W_M\left(1 - \frac{W}{W_M}\right)^k. \tag{2.7}$$

Substituting in (2.4) gives

$$C_S = kC_M\left(1 - \frac{W}{W_M}\right)^{k-1}. \tag{2.8}$$

Figure 2.2 illustrates the crystal composition, described by this expression, for a range of values of k. It is seen that, as crystal growth progresses, the composition continually increases from an initial value of kC_M. In addition, high values of k are seen to result in considerably more uniform crystal composition during growth than are low values.

The validity of (2.8) breaks down as W/W_M approaches unity. In this region, the melt becomes excessively rich, and the crystal composition is determined by its ability to hold the solute in solution, i.e., by its solid solubility characteristic.

2.2.2. Partial-stirring Conditions

For realistic values of growth and stirring, the rejection rate of the solute atoms at the interface is higher than the rate at which they can be transported into the melt. Consequently the solute concentration at the inter-

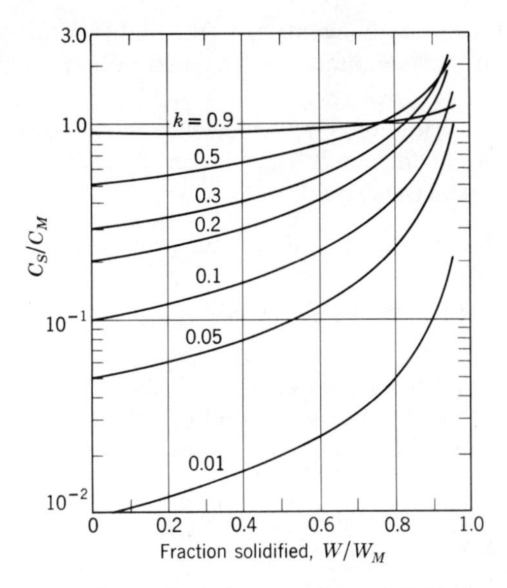

Figure 2.2 Doping profile during crystal growth (Bridges et al. [4]).

face builds up in excess of the concentration in the melt. This results in a crystal with a doping concentration in excess of that obtained for the case of full stirring.

The crystal composition may be determined by postulating that there is a thin stagnant layer of liquid immediately adjacent to the liquid-solid interface through which solute atoms flow by diffusion alone. Equilibrium conditions prevail beyond this layer. Figure 2.3 shows the liquid-solid

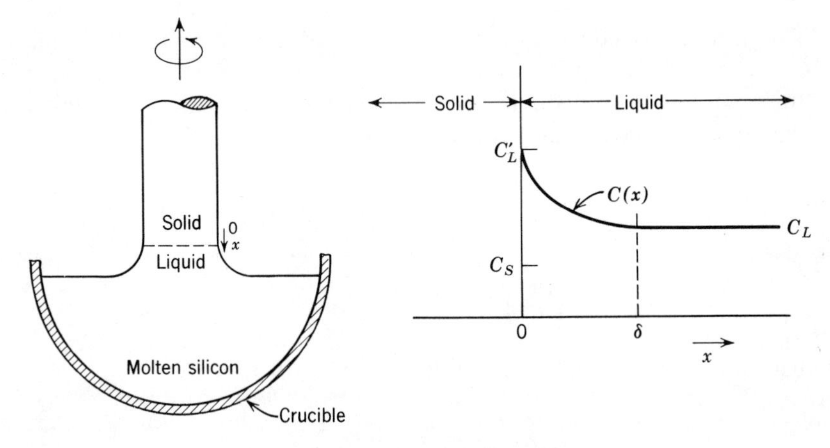

Figure 2.3 Partial-stirring conditions.

interface and the solute concentration beyond this interface. The diffusion-layer thickness is denoted as δ; the distribution coefficient k is given by C_S/C'_L.

An effective distribution coefficient may now be defined such that $k_e = C_S/C_L$. This coefficient is thus larger than k and depends on the growth parameters of the crystal. Let

R = growth rate for the crystal, i.e., the rate of movement of the liquid-solid interface,

D = diffusion constant for the solute atoms in the liquid, $\simeq 5 \times 10^{-5}$ cm²/sec.

The equation governing the diffusion of solute atoms in the layer may now be written. Noting that the amount of solute rejected from the solid is equal to that gained by the liquid, the stationary distribution is given by

$$D\frac{d^2C}{dx^2} + R\frac{dC}{dx} = 0, \tag{2.9}$$

hence

$$C = A\mathrm{e}^{-Rx/D} + B \tag{2.10}$$

and

$$\frac{dC}{dx} = -\frac{AR}{D}e^{-Rx/D}, \tag{2.11}$$

with A and B as constants of integration.

The first boundary condition is that

$$C = C'_L \quad \text{at} \quad x = 0. \tag{2.12}$$

The second boundary condition is obtained by noting that the sum of the impurity fluxes at a boundary must be zero. Again assuming that diffusion of solute atoms in the solid is neglected compared with diffusion in the liquid, this condition may be written as

$$D\left(\frac{dC}{dx}\right)_{x=0} + (C'_L - C_S)\,R = 0, \tag{2.13}$$

hence

$$\left(\frac{dC}{dx}\right)_{x=0} = -\frac{R}{D}(C'_L - C_S). \tag{2.14}$$

Substituting these boundary conditions into (2.10) gives

$$\frac{C - C_S}{C'_L - C_S} = e^{-Rx/D}. \tag{2.15}$$

But,

$$C = C_L \quad \text{at} \quad x = \delta, \tag{2.16}$$

hence

$$\frac{C_L - C_S}{C_L' - C_S} = e^{-R\delta/D}. \tag{2.17}$$

Substituting for k and k_e, and rearranging terms, gives

$$k_e = \frac{k}{k + (1-k)e^{-R\delta/D}}. \tag{2.18}$$

Finally, the crystal composition for partial stirring may be derived from the results for complete stirring by substituting k_e for k in (2.8), so that

$$C_S = k_e C_M \left(1 - \frac{W}{W_M}\right)^{k_e - 1}. \tag{2.19}$$

Values of k_e are higher than k and approach unity for large values of the normalized growth parameter $R\delta/D$. Uniform crystal composition is thus obtained for high rates of pull and spin (since δ is inversely related to the spin rate). In practice, these growth parameters are set by considerations outlined in Section 2.1.1. Thus, the pull rate is optimized to grow crystals of the desired diameter with low dislocation concentration. The spin rate is normally kept low to prevent corrosion of the crucible walls by the molten silicon.

It is common commercial practice to program the growth parameters so as to obtain uniform composition over a large fraction of the crystal growth. After an initial growth phase, during which time the crystal diameter is built up to its desired value, a programmed pull rate is used. If this is initiated when a crystal of weight W_1 has already been grown, then it may be shown[4] that the required program for the growth parameter is given by

$$R\delta/D = \ln\left(\frac{1 - W/W_M}{1 - W_1/W_M}\right). \tag{2.20}$$

This results in a crystal of constant solute concentration C_{S1}, where

$$C_{S1} = C_M k_e \left(1 - \frac{W_1}{W_M}\right)^{k_e - 1}. \tag{2.21}$$

2.2.3. Properties of Melt-grown Crystals

Single-crystal silicon, grown by the Czochralski method, is extensively used in the fabrication of both discrete devices and microcircuits. As shown, crystal growth is the result of a delicate balance of a number of interrelated (and sometimes conflicting) considerations. As a result, extreme variations in crystal quality are the rule rather than the exception.

Slices cut from single-crystal silicon are usually evaluated on the basis of the following properties:

Dislocation content. It is quite feasible to obtain Czochralski grown slices that are entirely free from dislocations. In practice the dislocation content ranges from zero to as high as 10^4 dislocations cm^2. In addition, crystals which freeze along a curved interface show corresponding patterns of dislocation concentration.

Resistivity. Values as high as 100 Ω-cm can be obtained. This upper limit presents no problems for microcircuit fabrication.

Radial Resistivity. Radial variations in resistivity may run as high as ± 30 to 50% on high-resistivity slices. For microcircuit applications a radial resitivity variation of ± 5 to $\pm 10\%$ is the maximum that can be tolerated.

Oxygen Content. The presence of oxygen in the crystal is caused by the corrosive action of the molten silicon on the silica walls. Oxygen concentrations of 10^{16} to 10^{18} atoms/cm^3 are obtained, the higher values corresponding to the higher spin rates. Most of the oxygen is present as SiO_2, which is electronically inactive, and segregates preferentially in the neighborhood of dislocations. Circuits that are fabricated in these regions are usually defective.

Oxygen tends to pin the extremities of dislocations, which can now readily multiply by the mechanism described in Section 1.4.2.4. Thus a high oxygen content is usually accompanied by a large dislocation concentration.

Some of the oxygen is present as SiO_4, which is donor-like. Heat treatment results in altering the SiO_4/SiO_2 proportion, thus resulting in reversible changes in the electronic properties of the slice. This problem is not serious in microcircuit applications where relatively low resistivity materials are used.

2.3. ZONE PROCESSES [5]

The zone process uses a rod of cast silicon as the starting material. The rod is usually maintained in a vertical position and is rotated during the operation. A small zone (typically 1.5 cm long) of the crystal is kept molten by means of r.f., and the r.f. coil moved so that this floating zone traverses the length of the bar. A seed crystal is provided at the starting point where the molten zone is initiated and arranged so that its end is just molten. As the zone traverses the bar, single-crystal silicon freezes at its retreating end and grows as an extension of the seed crystal.

Figure 2.4 shows a sketch of the necessary equipment. As with the Czochralski process, the bar is enclosed in a cooled silica envelope in which an inert atmosphere is maintained.

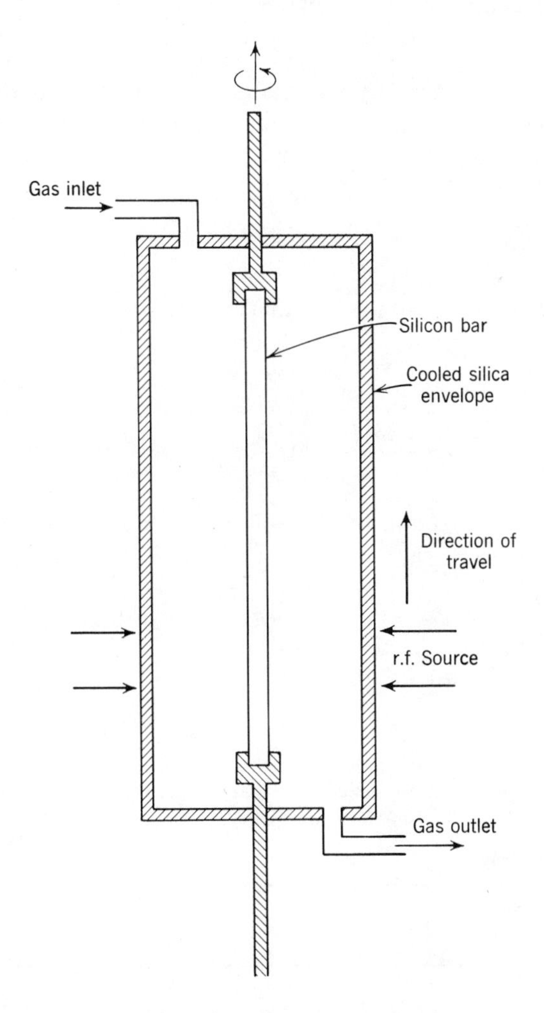

Figure 2.4 Zone apparatus.

2.3.1. Zone Refining

In this process the charge is usually in the form of a predoped, cast silicon rod, of doping concentration C_M, with a seed crystal juxtaposed at one end. The molten zone is initiated at this end, and passes along the bar. Referring to Figure 2.5, let

L = length of the molten zone,
x = distance along the bar,
s = amount of solute present in the molten zone at any given time,
A = cross section area of the bar,
ρ_s = specific gravity of silicon.

As the zone advances by dx, the amount of solute added to it at its advancing end is $C_M A \rho_s dx$. The amount of solute removed from it at the retreating end is $(k_e s/L)\, dx$, where k_e is the effective distribution coefficient, and

$$ds = C_M A \,\rho_s\, dx - \frac{k_e s}{L}\, dx. \tag{2.22}$$

This equation is subject to the boundary value

$$s = C_M A L \,\rho_s \quad \text{at} \quad x = 0. \tag{2.23}$$

Solving, gives

$$s = \frac{C_M A L \,\rho_s}{k_e}\,[1-(1-k_e)e^{-k_e x/L}]. \tag{2.24}$$

But C_S, the concentration of the solute in the crystal at the retreating end, is given by

$$C_S = \frac{k_e s}{A L \,\rho_s}. \tag{2.25}$$

Substituting (2.25) into (2.24),

$$C_S = C_M\,[1-(1-k_e)\,e^{-k_e x/L}]. \tag{2.26}$$

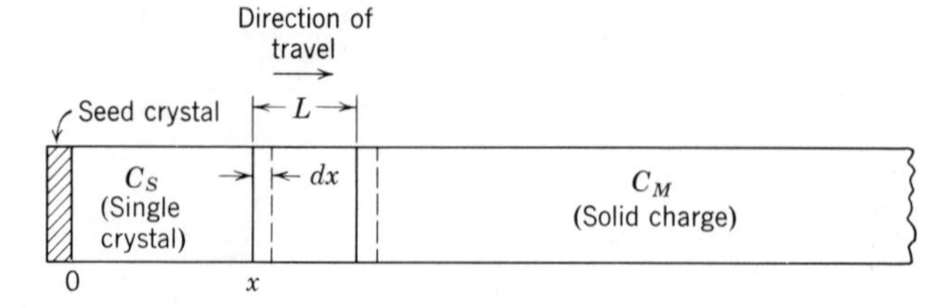

Figure 2.5 Model for zone refining.

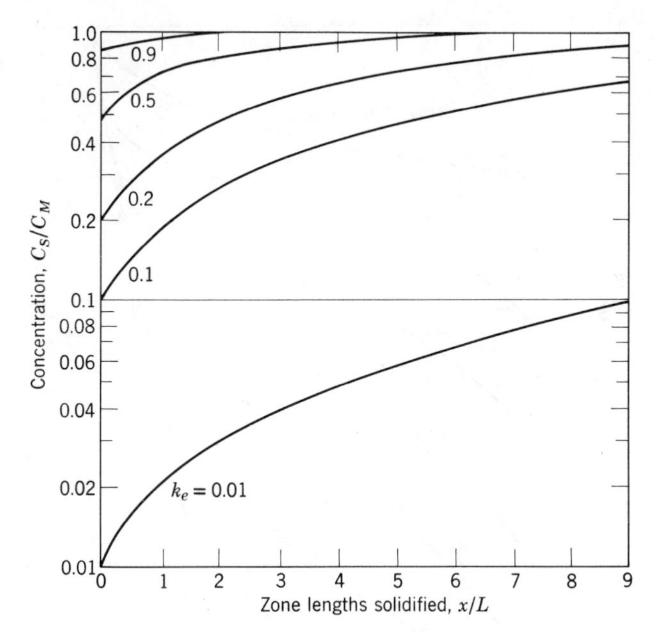

Figure 2.6 Doping profile during zone refining (Bridges et al. [4]).

Figure 2.6 shows the crystal concentration as a function of distance along the bar. As with crystal growth from the melt, it is seen that high values of distribution coefficient lead to more uniform compositions than low values.

The zone process is ideally suited to crystal refinement because of the ease with which successive passes can be made. Figure 2.7 illustrates the successive compositions of a crystal with a rather high value of k_e ($k_e = 0.1$) after a number of passes. It is seen that even for this case the degree of crystal refinement possible by this method is very high.

2.3.2. Zone Leveling

Zone leveling is a technique by which a zone-refined bar can be doped. Consider a bar of pure silicon, with a molten zone of length L. Initially, this zone contains a charge of doping concentration C_I. A seed crystal is also juxtaposed, as illustrated in Figure 2.8. Using the same notation as that of Section 2.3.1, (2.22) may be rewritten as

$$ds = -\frac{k_e s}{L}\,dx, \tag{2.27}$$

Since $C_M = 0$. At $x = 0$,

$$s = C_I A L \rho_s\ . \tag{2.28}$$

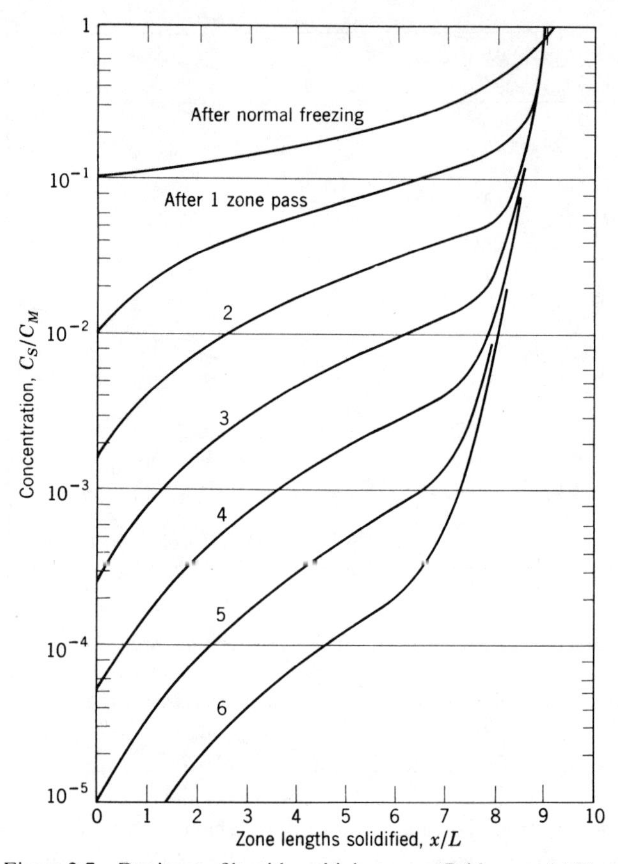

Figure 2.7 Doping profile with multiple passes (Bridges et al. [4]).

Solving (2.27) and inserting this boundary value gives

$$s = C_I A L \rho_s e^{-k_e x/L}. \tag{2.29}$$

But

$$C_S = k_e s / A L \rho_s, \tag{2.30}$$

hence

$$C_S = k_e C_I e^{-k_e x/L} \tag{2.31}$$

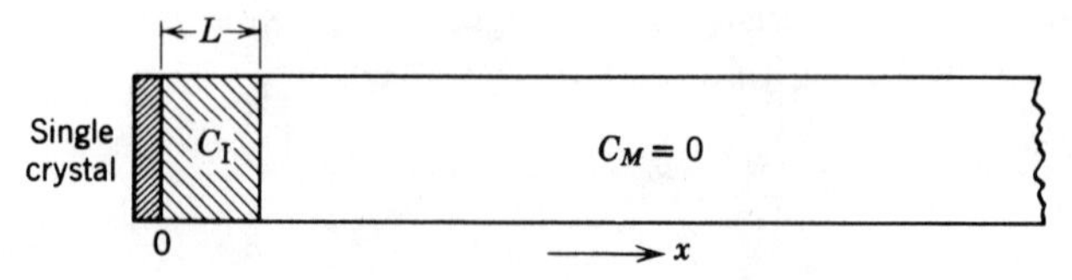

Figure 2.8 Model for zone doping.

Figure 2.9 shows the composition of the crystal as a function of zone lengths for various values of k_e. It is seen that for low values of k_e, very uniformly doped crystals may be obtained in a single pass by this method. In addition, if the zone length is programmed so as to be given by $L = k_1 x$, where k_1 is a constant, even better uniformity can be achieved.

The technique of zone reversal may also be used and provides a significant improvement in the uniformity of the composition. In this method the direction of the zone is reversed after it has traversed the bar. Thus deviations in composition resulting from the first pass are compensated

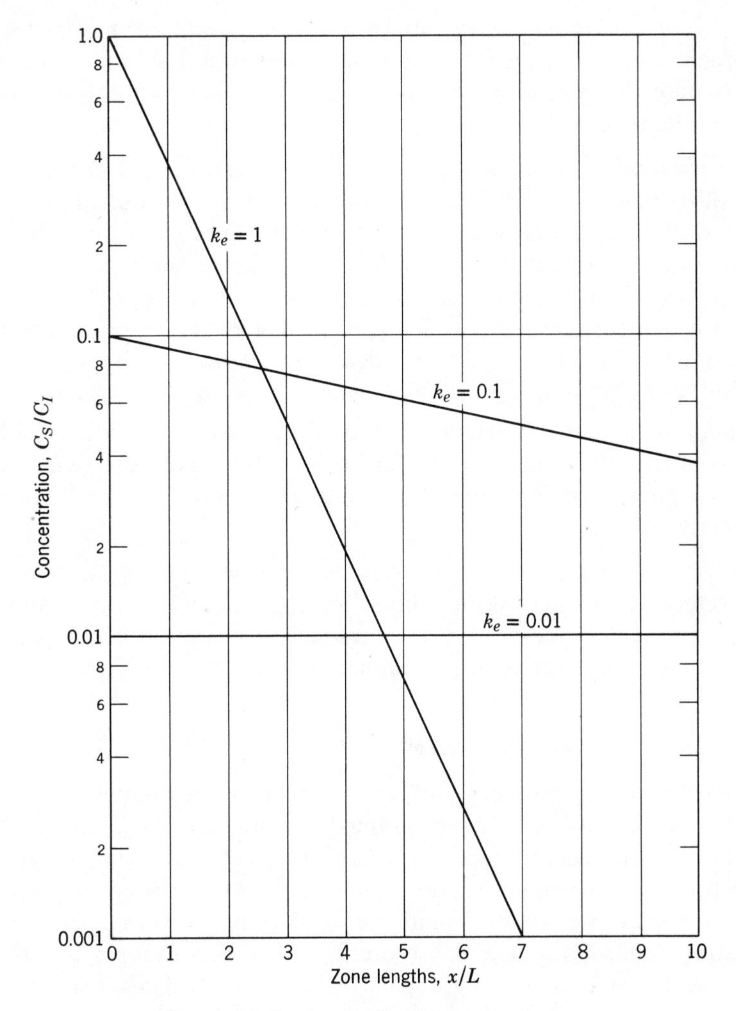

Figure 2.9 Doping profile during zone doping.

by the reverse pass. This results in very uniform compositions, except for the first and last zones where all the remaining impurities eventually freeze out.

2.3.3. Properties of Zone-processed Crystals

Because of its two step nature (i.e. from melt to cast rod to single crystal), the zone process is more expensive than the Czochralski process. Slices cut from the single crystal are evaluated on the basis of the following properties:

Dislocation Content. Typically, values range from 10^3 to 10^5 dislocations/cm^2. This is because the region near the retreating solid-liquid interface (where the crystal is grown) is highly stressed owing to the weight of the molten zone.

Resistivity. The composition of a zone-refined crystal can be better controlled than that of the Czochralski type, since no outside matter is added during crystal growth and the chamber can be made quite free from contamination. In fact, it is possible to conduct the process in a vacuum and purify the crystal of already present volatile impurities such as zinc. Finally, repeated processing is possible by successive passes of the molten zone. As a consequence, silicon slices with resistivities in excess of 2000 Ω-cm (for n-type material) can be obtained by this process.

Radial Resistivity. Variations of this parameter are comparable to those obtained with Czochralski crystals in the same resistivity range. In the higher resistivity ranges, variations as large as $\pm 50\%$ are not uncommon.

Oxygen Content. Vacuum-processed crystals can be made almost completely free from oxygen. The prime application for such material is in high-voltage devices where the presence of oxygen precipitates leads to degraded reverse-breakdown characteristics in devices.

2.4. SOME OVERALL COMMENTS

The underlying principles of two important techniques for crystal growth and doping have been outlined. In practice these are modified by many second-order effects, so that they can at best serve as a starting point for obtaining appropriately doped crystals. As a consequence, it must be exphasized that the final "zeroing-in" must be done under actual operating conditions, in which the effect of such variables as gas flow rates, heat input, and growth parameters can be determined for the particular crystal-growing apparatus.

2.5. REFERENCES

[1] G. K. Teal, and J. B. Little, "Growth of Germanium Single Crystals," *Phys. Rev.*, **78**(5), 647 (1950).
[2] R. G. Rhodes, *Imperfections and Active Centres in Semiconductors*, Pergamon Press, Macmillan, New York, 1964.
[3] F. A. Trumbore, "Solid Solubilities of Impurity Elements in Germanium and Silicon," *Bell System Tech. J.* **39**, 205–233 (1960).
[4] H. E. Bridges et al., *Transistor Technology*, Vol. 1, Van Nostrand, New York, 1958.
[5] W. G. Pfann, *Zone Melting*, Wiley, New York, 1958.

2.6. PROBLEMS

1. It is desired to grow gold-doped and boron-doped crystals by the Czochralski method, such that the concentration of dopant in each case is 10^{17} atoms/cm^3 when one-half of the crystal is grown. Assuming that growth is undertaken under rapid stirring conditions, compare the relative amounts of gold and boron that should be added to a pure silicon melt on an atom ratio basis.
2. An antimony-doped crystal is required to have a resistivity of 1.0 Ω-cm when one half of the crystal is grown. Assuming that a 100 gm pure silicon charge is used, what is the amount of 0.01-Ω-cm antimony-doped silicon that must be added. (Assume that the electron mobility is 1500 cm^2/volt-sec).
3. A small silicon bar is doped with both boron and antimony. One end of the bar is inserted into a furnace and melted. The bar is now withdrawn. Sketch the doping profile that results from the regrowth process and calculate the base width of the resulting transistor. Assume equilibrium values for k and that the original antimony/boron atom ratio was 33.
4. Neglecting the effects of impurity distribution in the final zone, show that the limiting value of the solute concentration in a zone-refined crystal is given by

$$C_S = A\, e^{Bx},$$

where

$$A = \frac{C_M Bl}{e^{Bl} - 1},$$

$$k = \frac{BL}{e^{Bl} - 1}.$$

Here L is the length of the molten zone and l is the total length of the bar.

Chapter 3

Phase Diagrams and Solid Solubility

A number of different materials are used in the fabrication of microcircuits. As a consequence, compositions of two or more of these are often encountered at various points within the circuit structure. By way of example the electronic properties of the silicon are controlled by the introduction of small amounts of donor and acceptor elements into the lattice, whereas ohmic contacts are made by the alloying of aluminum to the silicon. Occasionally, combinations of materials are inadvertently formed during heat treatment or during the storage of devices at elevated temperature. The most notorious of these results in the formation of the so-called *purple plague*, which sets an upper limit to the temperature at which many microcircuits can be stored.

In this chapter, the behavior of these combinations is described by means of phase diagrams [1, 2]. This behavior is important since it often determines the nature and choice of the fabrication process. In addition, it provides a clue as to the problems that may arise when certain combinations of materials are used together.

A *phase* is defined as a state in which a material may exist, which is characterized by a set of uniform properties. If these phases are presented for equilibrium conditions, the resulting diagram is called an *equilibrium diagram*. Since equilibrium conditions are attained at rates that are much slower than the freezing rate, most diagrams involving one or more solid phases are usually called *phase diagrams* and represent quasi-equilibrium conditions.

3.1. UNARY DIAGRAMS

These are diagrams showing the phase change in a single element as a function of temperature and pressure. They also apply to compounds† that undergo no chemical change over the range for which the diagram is constructed.

In its simplest form, the unary diagram consists of three lines that intersect at a common point, thus delineating three areas on a two-dimensional plot. Figure 3.1 shows such a diagram for water. The common point, referred to as a *triple point*, is invariant for the system and defines the temperature and pressure at which solid, liquid, and gaseous phases are all in equilibrium with one another. At all other temperature and pressure combinations, either one or two phases are present.

3.2. BINARY DIAGRAMS

These are phase diagrams showing the relationship between two components as a function of temperature. The second variable pressure is usually set at one atmosphere. In this way a relatively complex three-dimensional representation is avoided.

Figure 3.2 shows one of many different types of binary diagrams that are encountered in practice. Common to all such diagrams are the following features:

1. The abcissa represents various compositions of two components

†The term *component* is often used to denote elements and compounds of this type interchangeably.

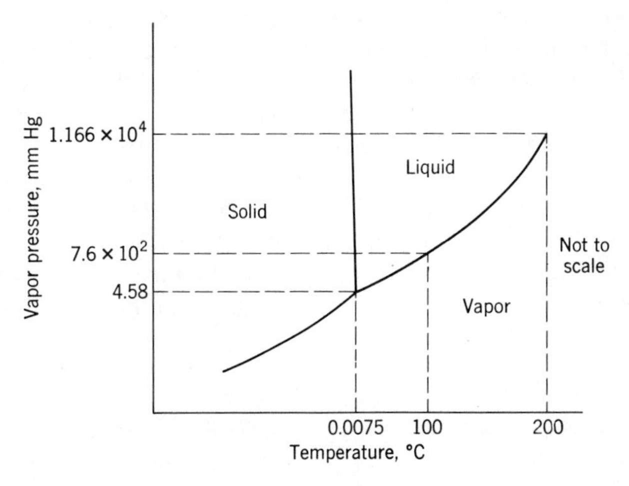

Figure 3.1 A unary phase diagram.

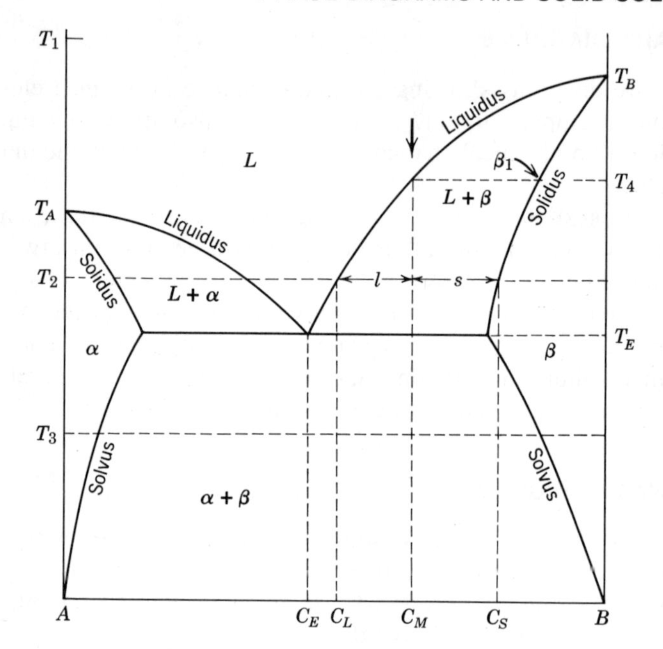

Figure 3.2 A binary phase diagram.

A and B, usually specified in weight percent or atom percent. Each end represents a pure component. This component may be an element or a compound.

2. In traversing a phase diagram from one side to another at any constant temperature, all single-phase regions are separated by two-phase regions as the composition is varied. In addition, no two two-phase regions can be adjacent. By way of example, the phases that exist at T_2 are α, $L+\alpha$, L, $L+\beta$, and β, in succession. On the other hand, the only phases that are present at T_3 are α, $\alpha+\beta$, and β.

At any temperature, the equilibrium composition of the two single phases that make up a two-phase region may be determined as follows (see Figure 3.2): Consider a melt of initial composition C_M (the percentage weight of B in the melt). Let this melt be cooled from some temperature T_1 to a temperature T_2, corresponding to a point in the two-phase region. Let

W_L = weight of liquid at this temperature,
W_S = weight of solid (in the β phase, for this example),
C_L, C_S = composition of the liquid and solid respectively (percentage amount of B by weight).

Then $W_L C_L$, $W_S C_S$ = weight of B in the liquid and solid respectively. But

$$(W_L + W_S)C_M = \text{total weight of } B.$$

Hence by the conservation of matter

$$W_S/W_L = \frac{C_M - C_L}{C_S - C_M} = l/s$$

where l and s are the length of the two lines measured from the C_M ordinate to the boundaries of the two-phase region. This is known as the *lever rule* and is directly applicable to the analysis of compositional changes during the freezing of a crystal from the melt.

Depending on the components involved, various types of binary phase diagrams are encountered in practice. Some of these are now described.

3.2.1. Isomorphous Diagrams

This type of diagram is characteristic of components that are completely soluble in each other. In such systems, as shown in the germanium-silicon phase diagram of Figure 3.3, the final state of any composition of the two elements is a single phase. It has been empirically found that isomorphous phase diagrams are only exhibited by binary systems in which the components are within 15% of each other in atomic radius, have the same valence and crystal structure, and have no appreciable difference in electronegativity. As a consequence, not many binary systems belong to this class; some examples are Cu-Ni, Ag-Pd, and Au-Pt, in addition to the Ge-Si system. Limited solubility of components is by far the more common occurrence in binary systems.

3.2.2. Eutectic Diagrams

An eutectic diagram results when the addition of either component to a melt lowers its over-all freezing point, as shown in Figure 3.2. Here the freezing point of the molten mixture has a minimum value T_E, below T_A and T_B. This minimum value is known as the *eutectic point*, and the corresponding mixture C_E is called the *eutectic composition*. Eutectic systems usually occur when two components are completely miscible in the liquid phase but are only partly soluble in the solid state. Most semiconductor systems fall into this class.

Referring to Figure 3.2, consider the cooling of a melt of initial composition C_M. As the temperature is reduced below T_4, a solid of composition β_1 first freezes out. With falling temperature, the liquid composition

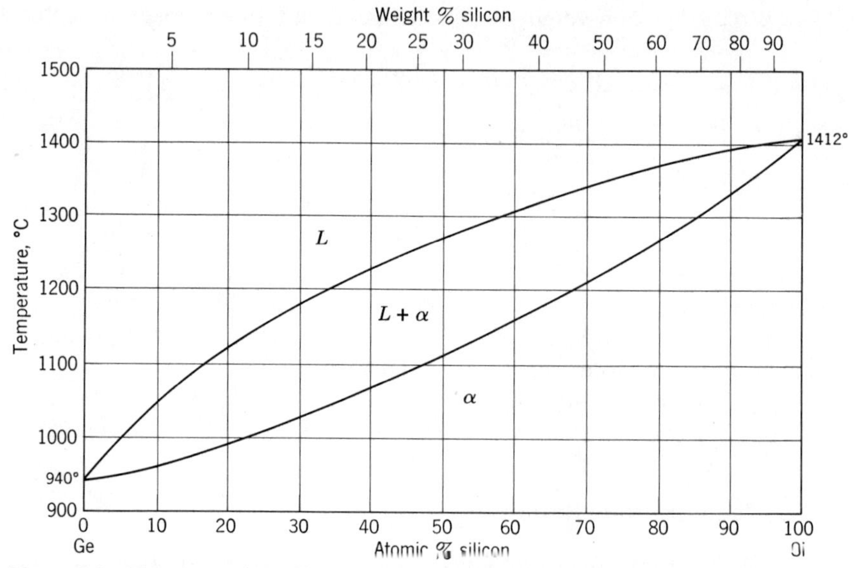

Figure 3.3 The germanium-silicon system. (From M. Hansen, and A. Anderko, *Constitution of Binary Alloys*, McGraw-Hill, New York, 1958.)

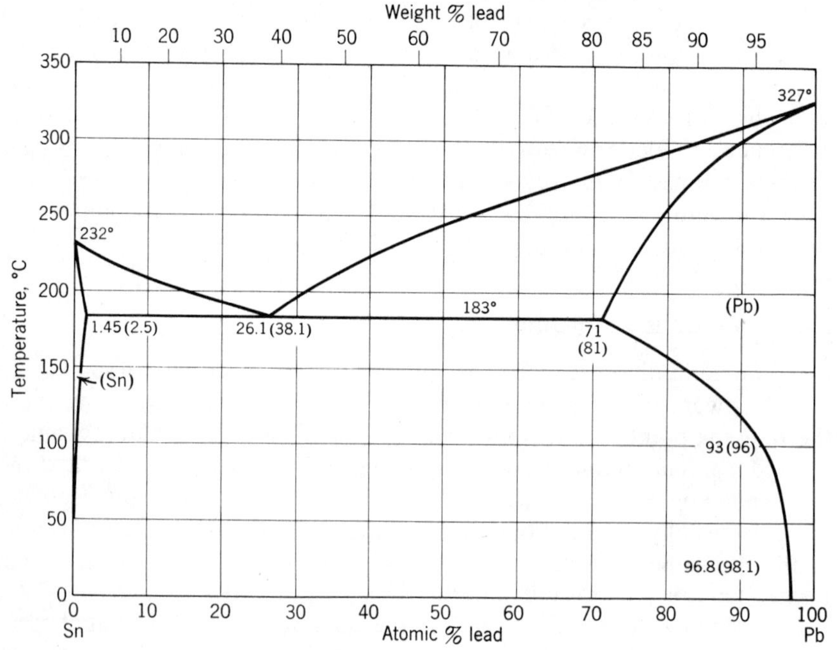

Figure 3.4 The lead-tin system. (From M. Hansen, and A. Anderko, *Constitution of Binary Alloys*, McGraw-Hill, New York, 1958.)

moves along the liquidus line, becoming richer in A, until the eutectic temperature T_E is reached. At this point the melt is of eutectic composition C_E, and freezes isothermally to form the $\alpha+\beta$ phase. Thus the final solid consists of β-phase aggregates in an $(\alpha+\beta)$-phase mixture of eutectic composition.

Figure 3.4 shows the lead-tin system which exhibits this type of

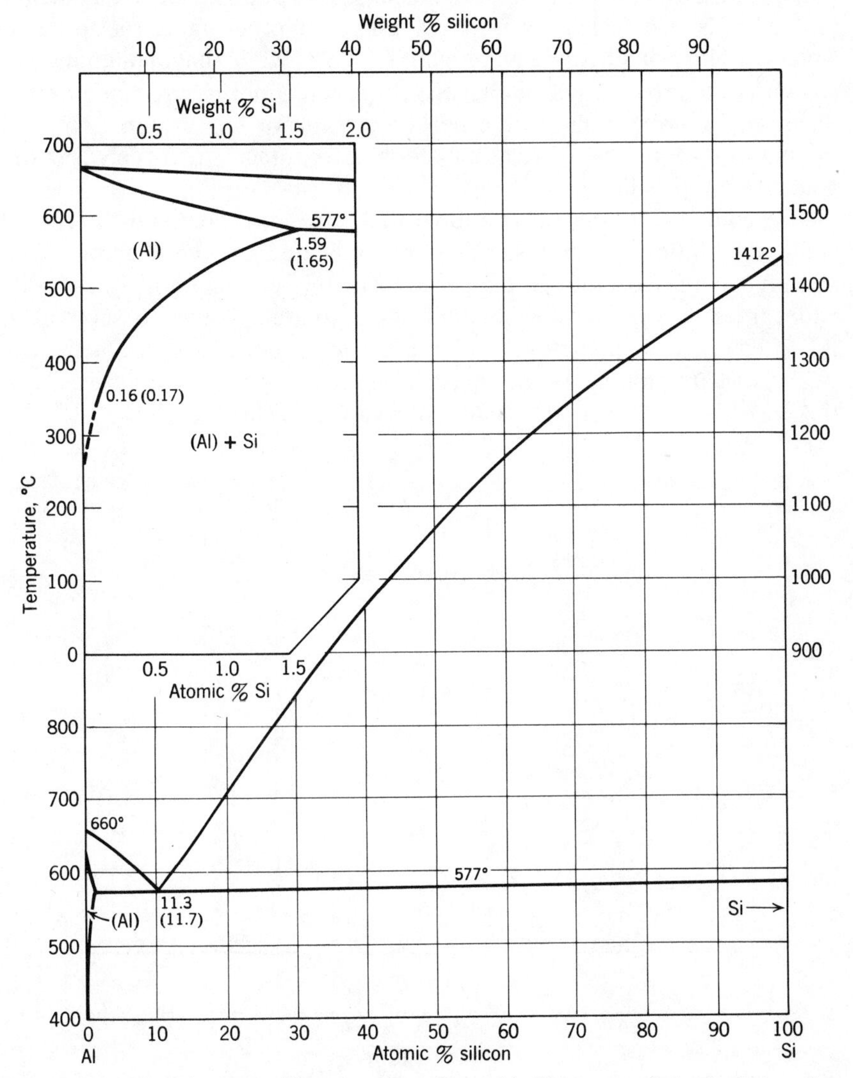

Figure 3.5 The aluminum-silicon system. (From M. Hansen, and A. Anderko, *Constitution of Binary Alloys*, McGraw-Hill, New York, 1958.)

characteristic. Here the eutectic point has a 38% lead-62% tin composition by weight and a eutectic temperature of 183°C. On the other hand, the freezing points of pure lead and pure tin are 327°C and 232°C respectively.

Variations of the eutectic diagram of Figure 3.2 are often seen in practice, depending on the nature of the terminal solid solutions. Thus in the Pb-Sn system the terminal solid solubility of tin in lead is significant (19%), as is that of lead in tin (2.5%). On the other hand, the terminal solid solubility of silicon in aluminum is 1.65%, while that of aluminum in silicon is so small ($< 0.1\%$) that the β phase cannot be indicated in this diagram. Figure 3.5 shows the phase diagram for this system, which is of importance in the fabrication of ohmic contacts to *p*-type and degenerate *n*-type silicon.

The gold-silicon system is shown in Figure 3.6. Here both terminal solid solubilities are too small to be indicated on the diagram. The sharply depressed eutectic temperature of this combination leads to its widespread use in the die bonding of microcircuit chips to substrates. An alternative combination, which provides a better wetting action, is gold and germanium, whose phase diagram is shown in Figure 3.7. In this case the resulting bond is a ternary Au-Ge-Si alloy.

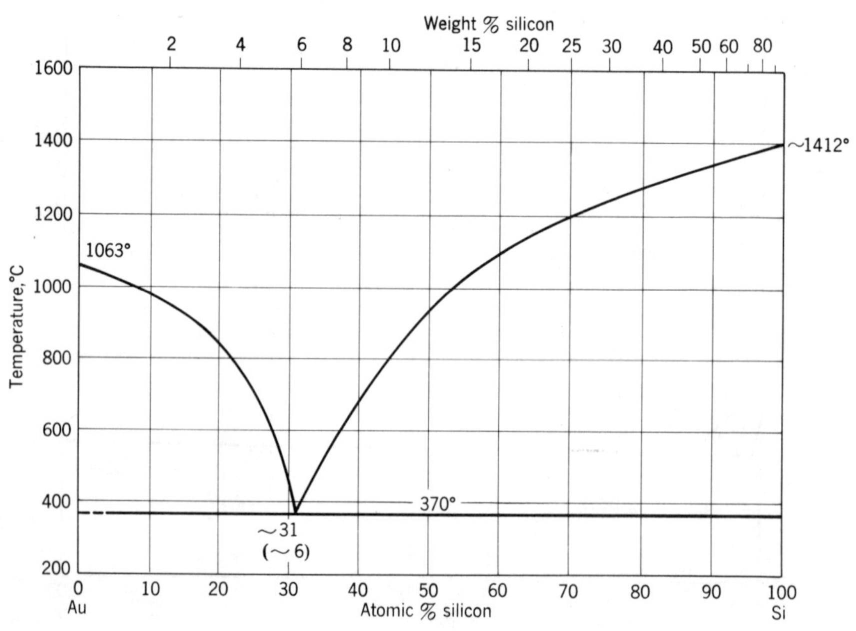

Figure 3.6 The gold-silicon system. (From M. Hansen, and A. Anderko, *Constitution of Binary Alloys*, McGraw-Hill, New York, 1958.)

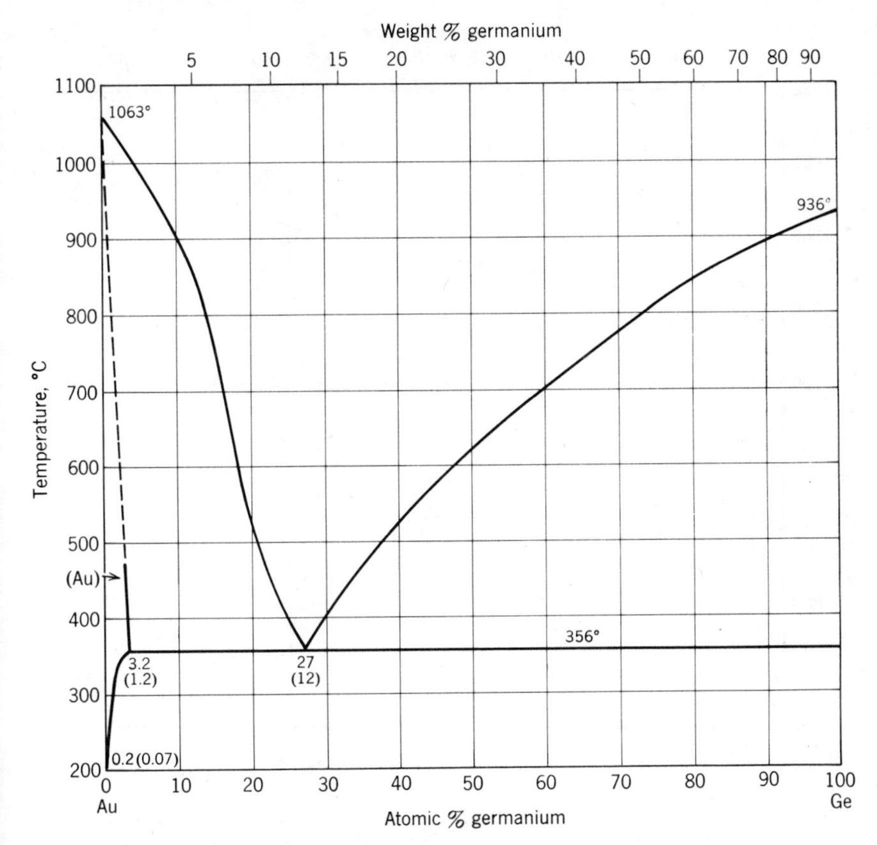

Figure 3.7 The gold-germanium system. (From M. Hansen, and A. Anderko, *Constitution of Binary Alloys*, McGraw-Hill, New York, 1958.)

Occasionally, binary systems may exhibit a eutectic composition that is very close to one of the components. Figure 3.8 shows the antimony-silicon system which belongs to this class.

3.2.3. Congruent Transformations

In many of the more complex systems, one or more intermediate compounds may be formed at specific temperatures and compositions. In contrast to mixtures, which exist at the microscopic level, these intermediate compounds exist on an atomic scale. They usually occur within an extremely narrow compositional range, corresponding to very small departures from stoichiometry. Consequently, they are indicated on the phase diagram in the form of discrete vertical lines, as shown for the indium-antimony system of Figure 3.9. Since these lines represent a

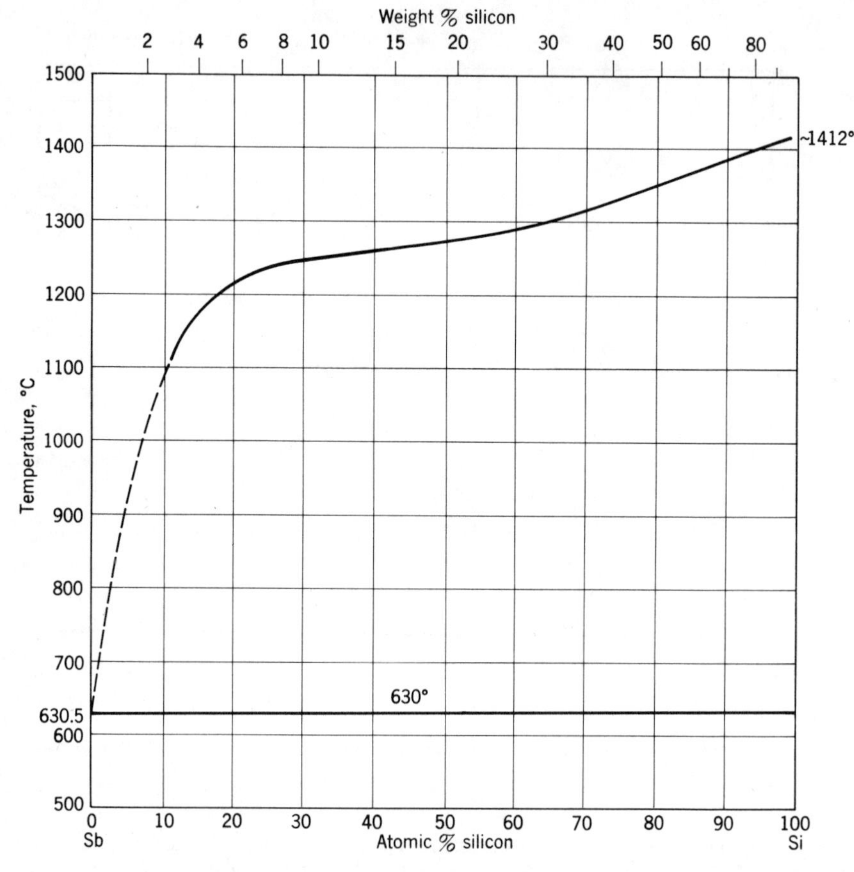

Figure 3.8 The antimony-silicon system. (From M. Hansen, and A. Anderko, *Constitution of Binary Alloys*, McGraw-Hill, New York, 1958.)

change from one phase directly into another, without any apparent alteration in composition, they are considered to represent congruent transformations.

A congruent transformation effectively isolates the system on each side of it. Thus Figure 3.9 can be considered as a In-InSb system and a separate InSb-Sb system, both of which are seen to be eutectic in nature. In this manner it is possible to break up relatively complex phase diagrams into a number of simpler ones.

3.2.4. Peritectic Reactions

A peritectic reaction is yet another way in which an intermediate compound can occur in a binary system. By way of example Figure 3.10

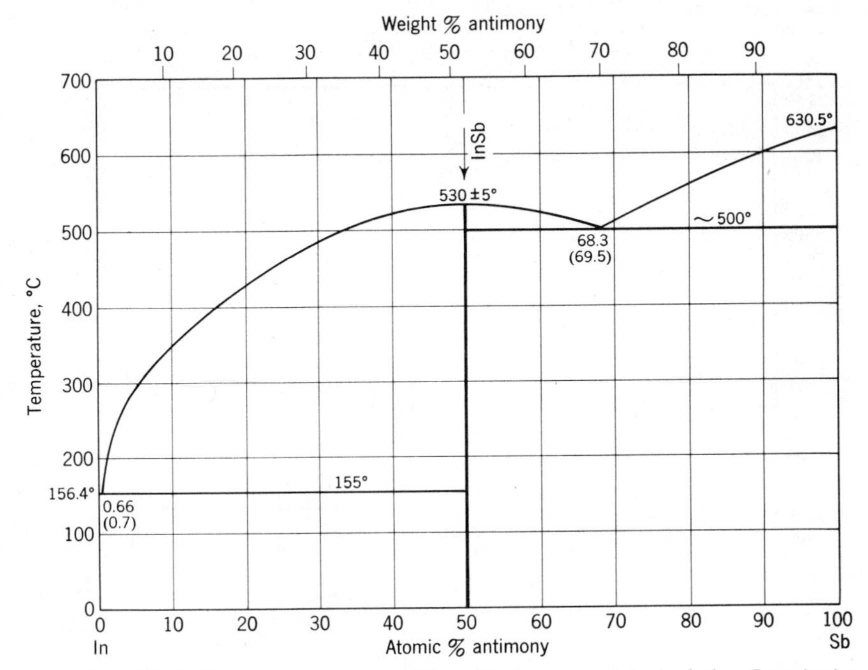

Figure 3.9 The indium-antimony system. (From M. Hansen, and A. Anderko, *Constitution of Binary Alloys*, McGraw-Hill, New York, 1958.)

show the arsenic-silicon system, in which the compound $SiAs_2$ is formed by a peritectic reaction, whereas SiAs represents a congruent transformation. An eutectic point also exists for the SiAs-As system, corresponding to a 96% As-4% Si composition by weight.

Consider what happens if a mixture of 86% As-14% Si by weight is cooled from some high temperature. At and below 1020°C solid SiAs is precipitated from the melt, which becomes arsenic-rich until the temperature reaches 944°C. At this point, the liquid composition is 90% As-10% Si by weight. With a further reduction in temperature, the solid SiAs combines with some of this excess liquid to form a liquid + $SiAs_2$ phase. At 786°C and lower, both the $SiAs_2$ and the β phase precipitate from the liquid to form a solid $\beta + SiAs_2$ phase as shown.

Cooling through a peritectic temperature causes the formation of a nonequilibrium structure by the process of "surrounding". Here the $SiAs_2$ is formed by the surrounding of a solid SiAs core by the liquid. Under normal cooling conditions, this $SiAs_2$ layer creates a barrier to the diffusion of liquid to the SiAs, resulting in a reaction that proceeds at an ever decreasing rate. Thus a peritectic reaction is usually accompanied by the formation of relatively large (micron-size) precipitates.

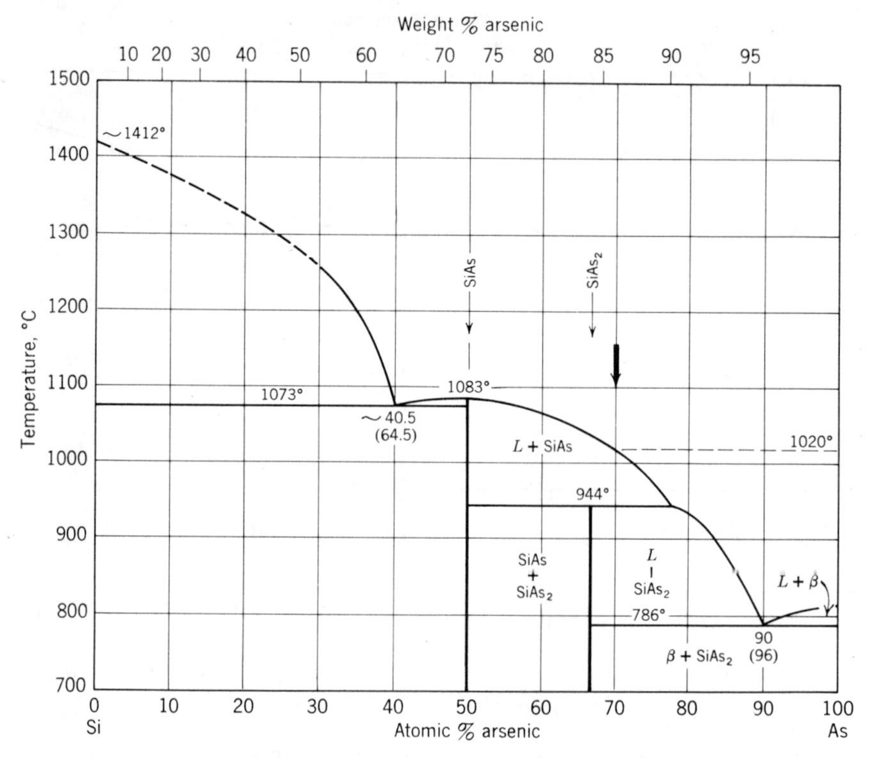

Figure 3.10 The arsenic-silicon system. (From M. Hansen, and A. Anderko, *Constitution of Binary Alloys*, McGraw-Hill, New York, 1958.)

Peritectic phases usually occur as part of more complex phase diagrams. One such, for the nickel-silicon system, is shown in Figure 3.11. In this system Ni_3Si is formed by a peritectic reaction at 1165°C, and exists in three structural modifications, β_1, β_2, and β_3. In addition, $NiSi_2$ is formed by a peritectic reaction at 993°C, and exists in low- and high-temperature versions, labeled *L* and *H* respectively. The compounds Ni_5Si_2 and Ni_2Si melt congruently at 1282 and 1318°C respectively. The compound Ni_3Si_2 is formed at 845°C by a peritectoid reaction, to be described in Section 3.2.5. Finally, nickel-silicon eutectics are present at 1152, 1265, 964 and 966°C respectively. As may be expected, the electronic properties of silicon doped with nickel are highly sensitive to heat treatment.

Figure 3.12 shows the phase diagram of the aluminum-gold system, which is of importance in the bonding of gold leads to aluminum contact pads on microcircuits. Of the many possible intermetallic compounds that are indicated here, the most significant are [3] $AuAl_2$—a dark purple, strongly bonding, highly conductive compound, and Au_2Al—a tan-

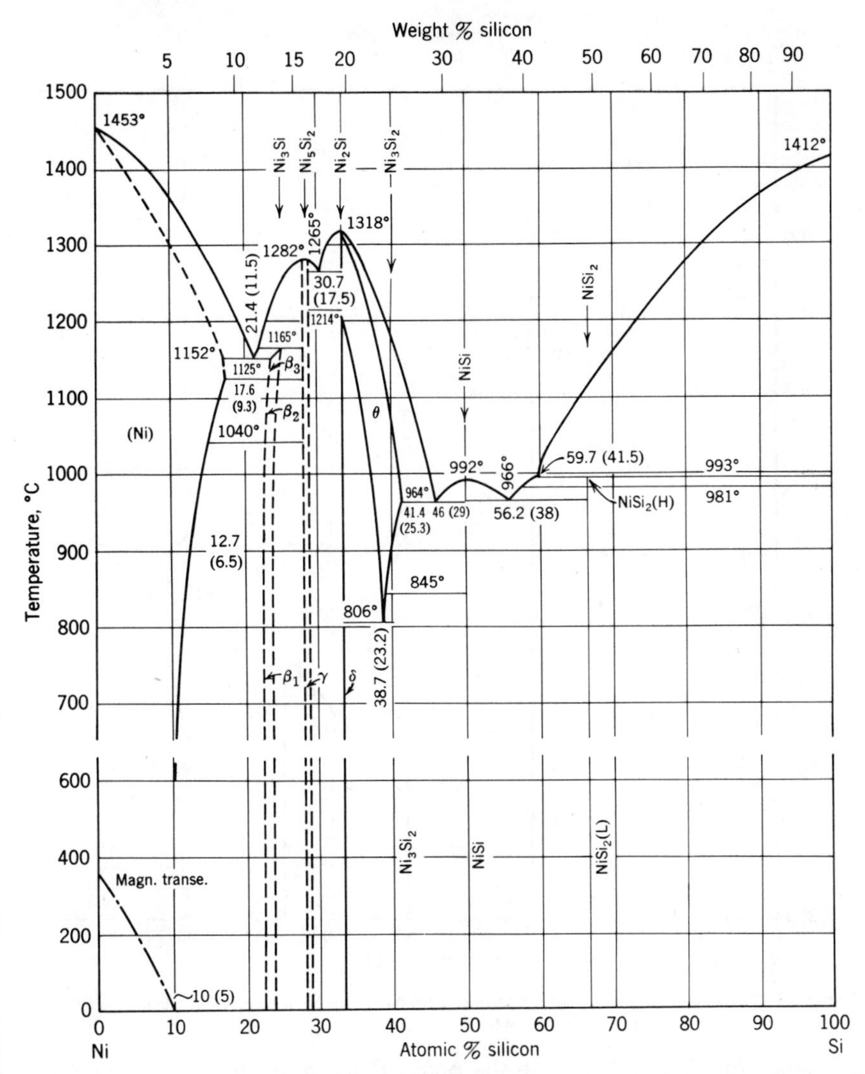

Figure 3.11 The nickel-silicon system. (From M. Hansen, and A. Anderko, *Constitution of Binary Alloys*, McGraw-Hill, New York, 1958.)

colored, brittle, poorly conducting compound. The formation of this latter compound† is a serious cause for lead-attachment failure in microcircuits which are stored for periods of time at 300°C or higher.

Experiments have shown that the formation of Au_2Al is enhanced in the

†For many years, the tan Au_2Al was not noticed in the presence of the purple $AuAl_2$. Thus the term *purple plague* was wrongly given to this phenomenon, and still persists in the literature.

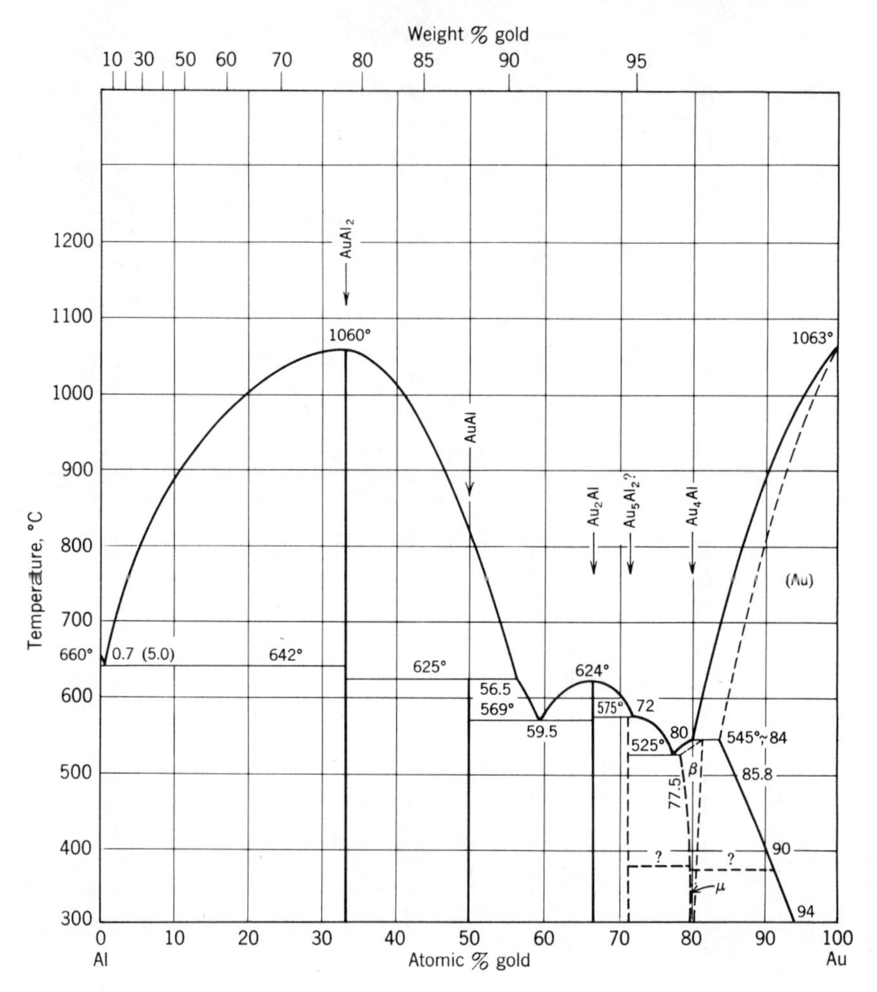

Figure 3.12 The gold-aluminum system. (From M. Hansen, and A. Anderko, *Constitution of Binary Alloys*, McGraw-Hill, New York, 1958.)

presence of silicon. Although the phase diagram of the ternary Al-Au-Si system has not been established, it has been proposed[4] that the effect of silicon addition to the Al-Au system produces simple eutectic lowering of invariant points. Thus the formation of these compounds may be expected to be accelerated in the presence of the silicon. This has been observed experimentally.

3.2.5. Other Reactions

In addition to eutectic and peritectic behavior, a number of other less common reactions may occur in binary systems. Thus a single phase can cool to form two phases in one of three possible ways, as follows:

(a) Monotectic: $L_1 \xrightarrow{\text{cooling}} \alpha + L_2$

(b) Eutectic: $L \xrightarrow{\text{cooling}} \alpha + \beta$

(c) Eutectoid: $\gamma \xrightarrow{\text{cooling}} \alpha + \beta$

Here L, L_1, and L_2 represent liquid phases and α, β, γ represent solid phases, including compounds.

In addition, three possibilities occur when two phases react to form a third, different phase, as follows:

(a) Syntectic: $L_1 + L_2 \xrightarrow{\text{cooling}} \beta$

(b) Peritectic: $L + \alpha \xrightarrow{\text{cooling}} \beta$

(c) Peritectoid: $\alpha + \gamma \xrightarrow{\text{cooling}} \beta$

Examples of some of these reactions are shown in the phase diagrams for the As-Si system (Figure 3.10—eutectic and peritectic), and the Ni-Si system (Figure 3.11—peritectoid and other reactions).

3.3. SOLID SOLUBILITY

It has been noted that many binary systems exist in which the terminal solid solubility of one component in the other is extremely small. This is usually the case for donor and acceptor impurities in silicon. As a consequence, it is necessary to expand greatly the scale of the phase diagram in order to show this important region. Figure 3.13 shows this part of the phase diagram, and is typical of most impurities in silicon. The solid solubility of these impurities is seen to increase with temperature, reach a peak value, and fall off rapidly as the melting point of the silicon is approached. This is commonly referred to as a *retrograde solid-solubility characteristic*.

An even more highly expanded version of this diagram in the vicinity of the melting point of silicon is shown in Figure 3.14. The process of freezing from a melt may be described with its aid. Starting with a melt of composition C_M (percentage weight of the solute in the melt), the crystal begins to freeze out initially at a composition kC_M, where k is the ratio of the slopes of the liquidus and solidus curves (assumed straight lines). With further cooling, the crystal composition moves along the solidus line until the concentration of the solid solution is C_M. During this freezing, the melt concentration is initially C_M and moves along the liquidus curve until its concentration becomes C_M/k. It should be noted that k

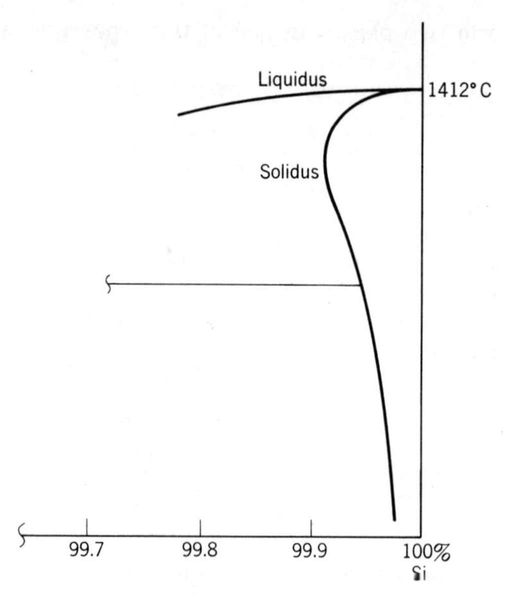

Figure 3.13 Retrograde solubility characteristic.

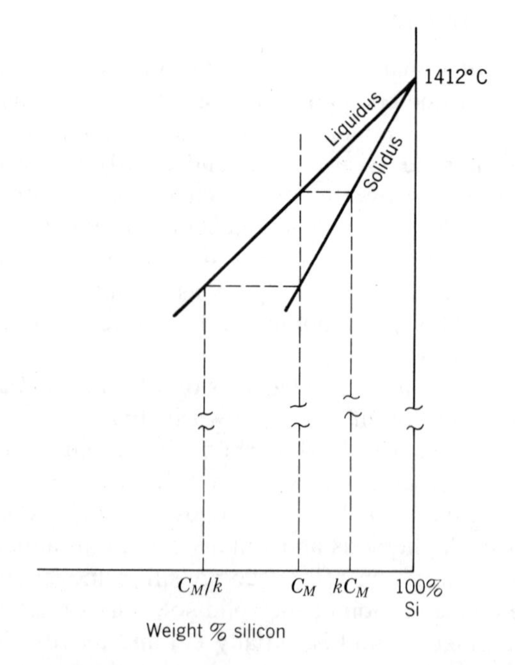

Figure 3.14 Phase diagram for crystal doping.

is the distribution coefficient described in Section 2.2 and is thus given by the ratio of the slopes of the liquidus and solidus curves in the vicinity of the melt composition.

Information concerning the solid solubility of various elements in silicon[5] is derived from cooling and freezing curves of this type. Figure 3.15 shows values for various commonly used dopants in silicon. As

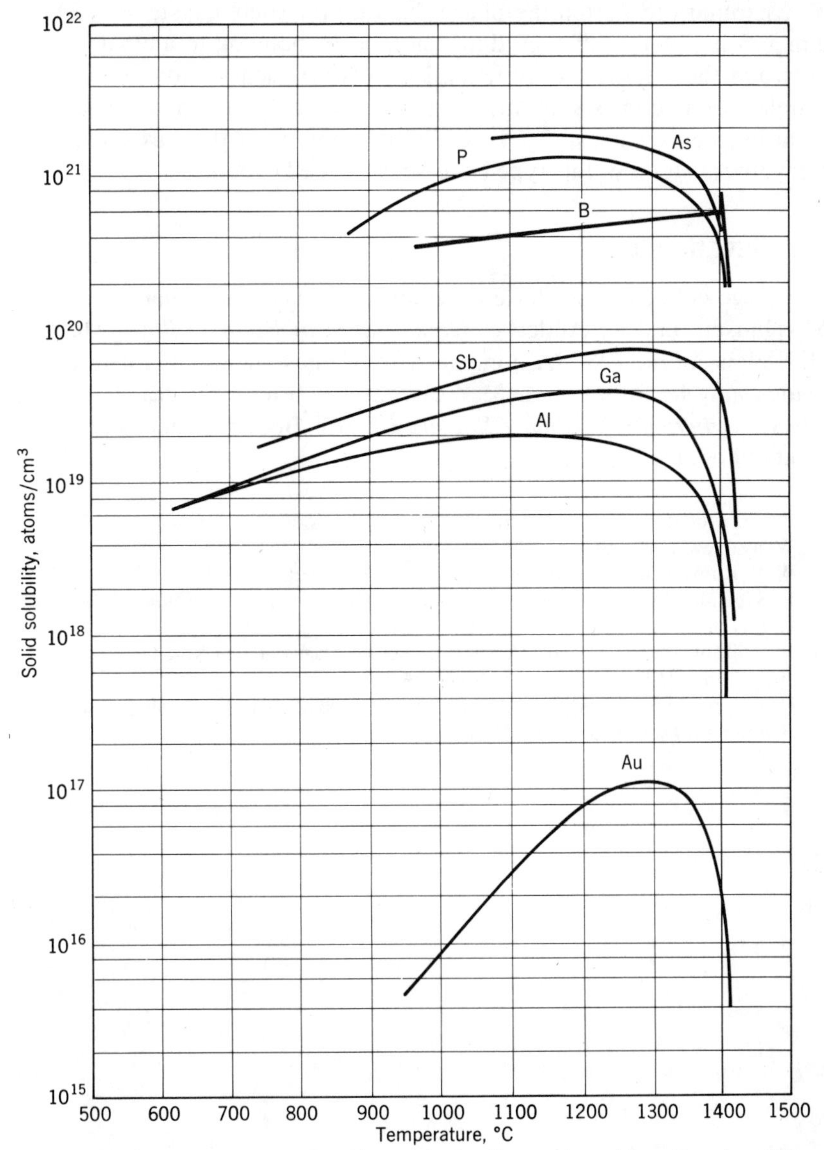

Figure 3.15 The solid solubility of impurities in silicon (adapted from Trumbore[5]).

seen from this figure, nearly all of these elements indicate a retrograde behavior in their solubility characteristics.

3.4. TERNARY DIAGRAMS

These diagrams are needed when three components are involved. It is customary to sketch the phase diagram for such a system as a three-dimensional plot, with each dimension corresponding to a binary system of two of these three components. In many practical situations a common simplification consists of drawing the ternary diagram as a series of binary systems, each with a fixed, different amount of the third component. Very few systems of this type have been studied in detail.

3.5. REFERENCES

A large collection of phase diagrams for binary alloys may be found in M. Hansen, and K. Anderko, *Constitution of Binary Alloys*, McGraw-Hill, New York, 1958. A compendium of phase diagrams for refractory oxides may be found in E. M. Levin et al., "Phase Diagrams for Ceramists," *American Ceramics Society* (1956). Both references are somewhat out of date.

[1] W. G. Moffat et al., "The Structure and Properties of Materials," Vol. 1, *Structures*, Wiley, New York, 1964.

[2] W. R. Runyan, *Silicon Semiconductor Technology*, McGraw-Hill, New York, 1965.

[3] R. Schmidt, "Mechanism of Lead Failures in Thermocompression Bonds," *IRE Trans. on Electron Devices*, **ED-9**, 506 (1962).

[4] B. Selikson, and T. Longo, "A Study of Purple Plague and its Role in Integrated Circuits," *Proc. IEEE*, **52**(12), 1638–1641 (1964).

[5] F. A. Trumbore, "Solid Solubiltiy of Impurity Elements in Germanium and Silicon," *Bell System Tech. J.*, **39**, 205–233 (1960).

Chapter 4

Diffusion

Diffusion provides the single most powerful means of introducing controlled amounts of chemical impurities into the silicon lattice[1]. In addition, diffusion techniques are ideally suited to batch processing, where many slices are handled in a single operation.

The laws governing the diffusion of impurities[2] were established many years before the development of semiconductors. It is only since reasonably defect-free materials of high purity have become available, however, that the method has become generally accepted for semiconductor-device fabrication. At the present time, diffusion is a basic process step in the fabrication of all silicon microcircuits.

A number of simplifying assumptions can be made in the development of diffusion theory as it applies to semiconductor-device fabrication. Thus the single-crystal nature of the solid in which diffusion takes place allows the effects of grain boundaries to be ignored. Because a very small number of impurities are involved, dimensional changes during the diffusion process are not considered. Finally, the almost exclusive use of parallel-plane device and circuit structures results in considerable simplification in the mathematics. As a consequence, diffusion theory is a reasonably good approximation to diffusion practice.

4.1. THE NATURE OF DIFFUSION

Diffusion describes the process by which atoms move in a crystal lattice. Although this includes self-diffusion phenomena, our interest is in the diffusion of impurity atoms that are introduced into the silicon lattice for the purpose of altering its electronic properties. In addition to concentration gradient and temperature, geometrical features of the crystal lattice (such as crystal structure and defect concentration) play an important part in this process.

The wandering of an impurity in a lattice takes place in a series of random jumps[4]. These jumps occur in all three dimensions, their net flow being the statistical average over a period of time. The mechanisms by which jumps can take place are described as follows:

Interstitial Diffusion. This is illustrated in Figure 4.1*a*. Here impurity atoms move through the crystal lattice by jumping from one interstitial site to the next. They may start at either lattice or interstitial sites and may finally end up in either type of site. However, interstitial diffusion requires that their jump motion occur from one interstitial site to another adjacent interstitial site.

Substitutional Diffusion. Here (see Figure 4.1*b*), impurity atoms wander through the crystal by jumping from one lattice site to the next, thus substituting for the original host atom. However, it is necessary that this adjacent site be vacant; that is, vacancies must be present to allow substitutional diffusion to occur. These vacancies are provided by thermal fluctuations in the lattice, as discussed in Section 1.4.1.1. Since the equilibrium concentration of vacancies is quite low, it is reasonable to expect substitutional diffusion to occur at a much slower rate than interstitial diffusion. This is indeed the case.

Interchange Diffusion. This occurs when two or more atoms diffuse by an interchange process. Such a process is known as a *direct interchange* process when it involves two atoms and as a *cooperative interchange* when a larger number is involved. Figure 4.1*c* illustrates both types of mechanisms. The probability of occurrence of interchange-diffusion effects is relatively low.

Combination Effects. Combinations of the aforementioned mechanisms may occur within a crystal. Thus, for example, a certain fraction of impurity atoms may diffuse substitutionally and the rest interstitially. In addition, some of the diffusing atoms may finally end in substitutional sites, with others in interstitial sites.

4.1.1. Interstitial Diffusion

There are five interstitial voids in a unit cell of the diamond lattice (see Figure 1.6). Although some of these are occupied by point defects, as discussed in Section 1.4.1.1, their equilibrium concentration is low, even at normal diffusion temperatures (1000 to 1200°C). Consequently nearly all interstitial sites are available for receiving impurity atoms as they wander through the lattice.

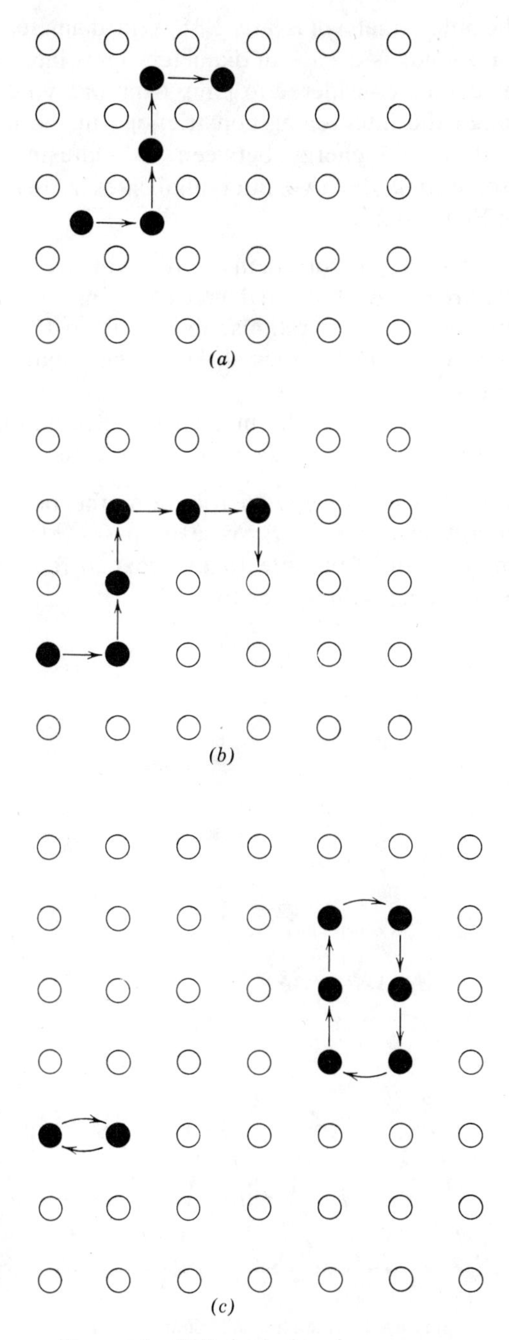

Figure 4.1 Diffusion by jumping processes.

In silicon the interstitial voids are 2.36 Å in diameter, and the constriction between voids is 2.10 Å in diameter. Thus interstitially located impurity atoms can be considered to jump from one void to the next by squeezing through the intervening constrictions in the lattice. For this situation the interaction energy between the diffusing atom and the saddle point separating the two interstitial sites is periodic in nature and is shown in Figure 4.2. Let

E_m = interaction energy involved, in electron volts,
T = temperature of the lattice, in degrees Kelvin,
ν_0 = frequency of lattice vibrations, about 10^{13} to 10^{14}/sec; this is the frequency with which atoms strike the potential barrier depicted in Figure 4.2,
ν = frequency with which thermal energy fluctuations occur with sufficiently large magnitude to overcome the potential barrier.

Assuming a Boltzmann energy distribution, the probability that an atom has an energy in excess of E_m is given by $e^{-E_m/kT}$. Since the atom can jump from one interstitial site to the next in four different ways, the frequency of jumping is thus given by

$$\nu = 4\nu_0 \, e^{-E_m/kT}. \tag{4.1}$$

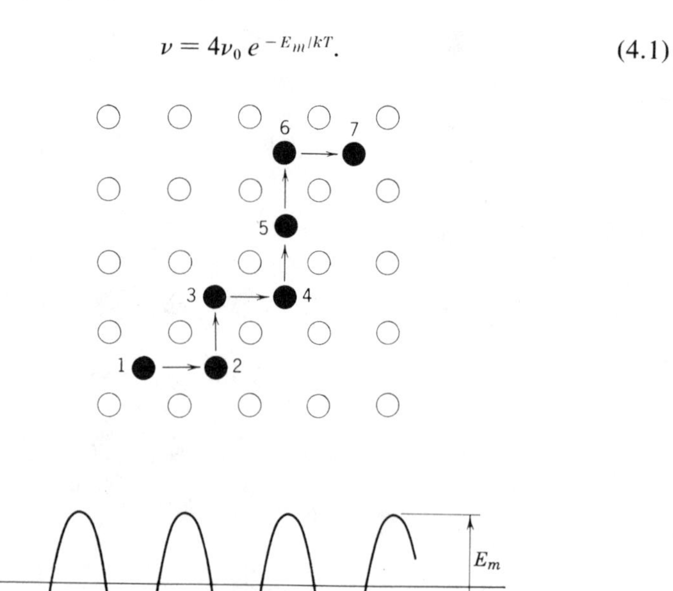

Figure 4.2 Interstitial diffusion by jumping.

Typical values for E_m are about 1.0 eV. Thus an interstitial impurity atom will jump from one void to another at a rate of about once every minute at room temperature. The rate is considerably higher at typical diffusion temperatures (900 to 1200°C).

4.1.2. Substitutional Diffusion

Here a similar type of reasoning applies. In jumping from one lattice site to another, the impurity atom experiences a periodic potential owing to the breaking of certain covalent bonds and the making of yet other adjacent ones, as shown in Figure 4.3. If the height of the potential barrier is E_n, the probability that one of these atoms will have a thermal energy in excess of this value is $e^{-E_n/kT}$.

The number of available lattice sites to which the impurity atom can jump is given by the number of vacancies in the lattice, that is, by the equilibrium concentration of Schottky defects. It has been shown in Section 1.4.1.1 that the fraction of such defects in a crystal is $e^{-E_s/kT}$, where E_s is the energy of formation of a Schottky defect. Thus the probability that a neighboring site will be vacant is given by $e^{-E_s/kT}$. Finally, since each lattice site has four tetrahedrally situated nearest neighbors,

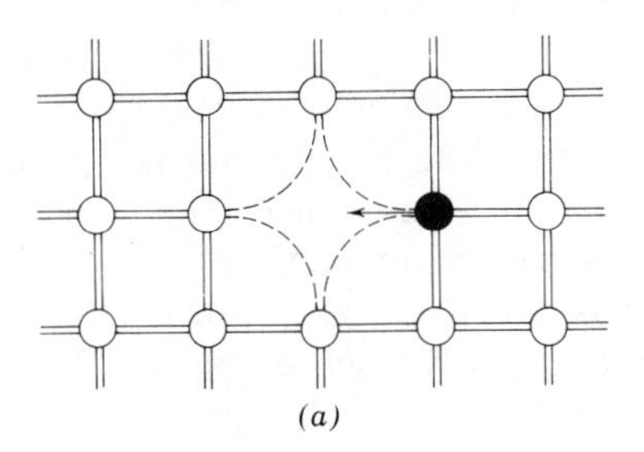

(a)

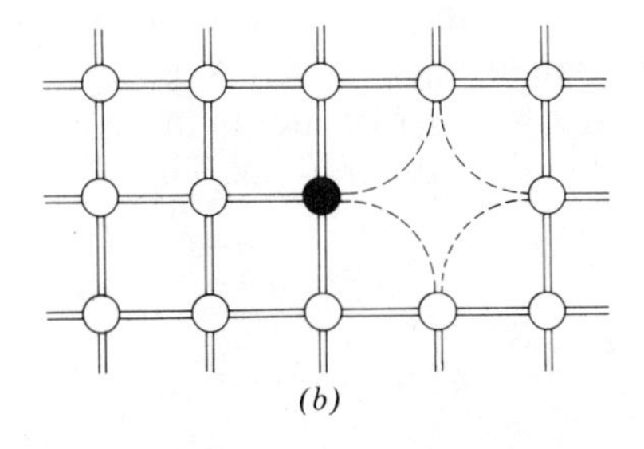

(b)

Figure 4.3 Substitutional jumps.

each jump can be made in one of four different ways. Consequently,

$$\nu = 4\,\nu_0 e^{-E_s/kT}\, e^{-E_n/kT}, \tag{4.2}$$

$$= 4\nu_0 e^{-(E_n+E_s)/kT}. \tag{4.3}$$

Values of E_n+E_s predicted from this simple theory are found to be slightly larger than those actually observed. This is because the binding energy between an impurity atom and its neighboring silicon atom is less than that between two adjacent silicon atoms in the lattice. Consequently the energy of formation of a vacancy next to an impurity atom is less than that of forming a vacancy at any other site. This is borne out by the experimental fact that values of E_n+E_s range from 3 to 4 eV for substitutional impurities in silicon while the activation energy of self-diffusion in silicon is 5.5 eV. Note that the jump rate of a substitutional impurity at room temperature is about once every 10^{45} years (as compared to the jump rate of about once every minute for interstitial diffusers)!

4.2. DIFFUSION IN A CONCENTRATION GRADIENT

So far we have only considered the statistics of random jump motion in the crystal lattice. In the presence of a concentration gradient, this motion results in a net transport of impurities. The law governing this motion is now considered for impurities which diffuse substitutionally in the silicon lattice. All of the impurities belonging to columns III and V of the periodic table (i.e., the *p*-type and *n*-type impurities†) fall into this class and constitute the greater majority of cases that arise in practice.

In the silicon lattice substitional diffusion takes place by jump motion between tetrahedral sites of spacing d. For the diamond lattice it is seen that a single jump has projections of length $d/\sqrt{3}$ along each of the crystal axes.

Consider a crystal of cross section A, divided up by a series of parallel planes at right angles to a major crystal axis. Let the spacing between planes be $d/\sqrt{3}$. Figure 4.4 depicts this arrangement, with the crystal divided into layers 1, 2, etc. Then each layer contains atoms whose tetrahedral neighbors are in adjacent layers on either side.

Let n_1 and n_2 be the number of atoms in layer 1 and 2 respectively, and N_1 and N_2 the volume concentrations. Then

$$N_1 = \frac{\sqrt{3}n_1}{Ad}, \tag{4.4a}$$

$$N_2 = \frac{\sqrt{3}n_2}{Ad}. \tag{4.4b}$$

†Deep-lying impurities do not fall into this class. They will be discussed later.

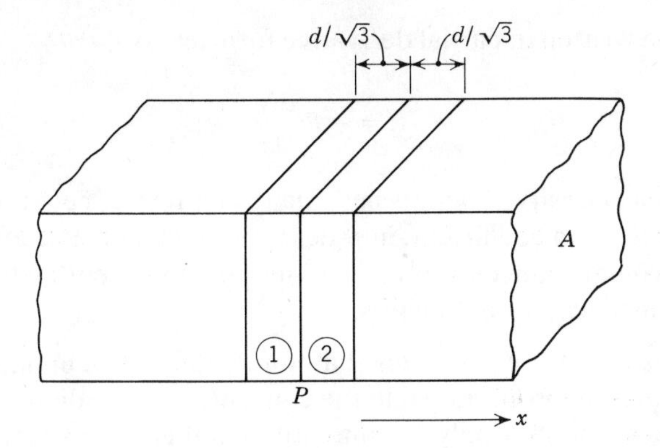

Figure 4.4 Diffusion due to a concentration gradient.

For diffusion in the silicon lattice atoms in any one layer must jump into neighboring layers. Each atom has four such neighbors, two in the layer to the left and two in the layer to the right. Thus in a single jump period of time $1/\nu$, half of the moving atoms jump right while the other half jump left, on the average.

Then the net flow of atoms across the plane P in a direction x is given by

$$\frac{\Delta n}{\Delta t} = \frac{(n_1 - n_2)/2}{1/\nu},$$

$$= \frac{\nu A d}{2\sqrt{3}}(N_1 - N_2). \tag{4.5}$$

But

$$\frac{\Delta N}{\Delta x} = \frac{N_2 - N_1}{d/\sqrt{3}}, \tag{4.6}$$

hence

$$\frac{\Delta n}{\Delta t} = -A \frac{\nu d^2}{6} \frac{\Delta N}{\Delta x}. \tag{4.7}$$

Writing the flux density j as the time rate of change of number of impurities per unit area, (4.7) reduces to

$$j = -\frac{\nu d^2}{6} \frac{\Delta N}{\Delta x}. \tag{4.8}$$

Defining a diffusion constant D, such that

$$D \equiv \frac{\nu d^2}{6}, \tag{4.9}$$

(4.8) may be written in partial derivative form as

$$j = -D\frac{\partial N}{\partial x}, \tag{4.10}$$

where j = flux density, in atoms per square centimeter per second,
D = diffusion coefficient, in square centimeters per second,
N = volume concentration, in atoms per cubic centimeter,
x = distance, in centimeters.

This is Fick's first law. It states that, under diffusion conditions, the flux density is directly proportional to the concentration gradient.

Similar arguments apply to interstitial diffusers, except that jumps between interstitial voids must be considered. Since each void also has four nearest neighbors, the same line of reasoning applies.

4.2.1. Field-aided Motion

In addition to diffusion, ionized impurities can move in the presence of a drift field. This motion results in a drift component of velocity in the direction of the electric field. If the random scattering of ionized impurities in the lattice is represented by a restraining force that is directly proportional to the drift velocity, the equation of motion may be written as

$$F = Zq\mathscr{E} = m^*\frac{dv}{dt} + \alpha v, \tag{4.11}$$

where F = the force on the impurity ion,
Zq = its charge,
$\mathscr{E}$ = the electric field,
v = the average velocity,
m^* = the effective mass,
α = a factor of proportionality.

Note that the sign of the charge on the impurity ion determines the direction of this force. Thus a positive force is exerted on a positively charged impurity in a positive $\mathscr{E}$ field.

In steady state a drift velocity v_d is attained, such that

$$v_d = \frac{Zq}{\alpha}\mathscr{E},$$
$$\equiv \mu\mathscr{E} \tag{4.12}$$

where μ is defined as the mobility of the impurity ion in square centimeters per volt per second.

The movement of impurities due to drift and diffusion may be assumed to be independent events. Consequently, in the presence of a field, the net flow of atoms across the plane P in the direction x (see Figure 4.4) is given by modifying (4.5) so that

$$\frac{\Delta n}{\Delta t} = \frac{\nu Ad}{2\sqrt{3}}(N_1 - N_2) + \mu N\mathscr{E}A. \tag{4.13}$$

Thus, the flux density j is given by

$$\begin{aligned} j &= -\frac{\nu d^2}{6}\frac{\Delta N}{\Delta x} + \mu N\mathscr{E}, \\ &= -D\frac{\partial N}{\partial x} + \mu N\mathscr{E}, \end{aligned} \tag{4.14}$$

in partial-derivative form.

4.3. THE DIFFUSION COEFFICIENT

The behavior of the diffusion coefficient may now be determined. Substituting (4.9) into (4.1), the diffusion coefficient for an interstitial impurity is given by

$$\begin{aligned} D &= \frac{\nu d^2}{6} = \frac{4\nu_0 d^2}{6} e^{-E_m/kT} \\ &= D_0 e^{-E_m/kT}. \end{aligned} \tag{4.15}$$

In like manner the diffusion coefficient for a substitutional impurity is obtained by combining (4.9) and (4.3). Thus

$$\begin{aligned} D &= \frac{4\nu_0 d^2}{6} e^{-(E_n + E_s)/kT} \\ &= D_0 e^{-(E_n + E_s)/kT}. \end{aligned} \tag{4.16}$$

Both D_0 and the exponential term are functions of temperature. However, the variation of D with temperature is largely controlled by changes in the exponential factor. As a result, D_0 is usually considered to be constant over a range of a few hundred degrees of diffusion temperature.

At normal doping levels, the diffusion coefficient is independent of concentration. This coefficient, however, is found to increase at concentrations in excess of 10^{20} atoms/cm^3. The reason lies in the fact that extrinsic conduction can still occur at diffusion temperatures (900 to 1200°C) for silicon doped to these concentration levels. For this case the mobile carriers (electrons or holes) have a much higher diffusion rate than the impurity atoms and tend to outrun them, creating a space charge which causes a field to build up. This is shown in Figure 4.5*a*, where an impurity gradient is established from A to B by introducing column V impurities at A. The direction of the $\mathscr{E}$ field due to the space charge associated with the positively charged impurity ions and the mobile electrons is seen to aid the drift of these ions in the direction from A to B. Figure 4.5*b* shows the situation for column III impurities. Once again the $\mathscr{E}$ field is seen to aid the drift of impurities from A to B. For both situations the resulting enhanced motion may be interpreted in terms of an increased effective diffusion coefficient. Figure 4.6 shows this effective diffusion coefficient for phosphorus as a function of doping concentration, and serves to illustrate this point.

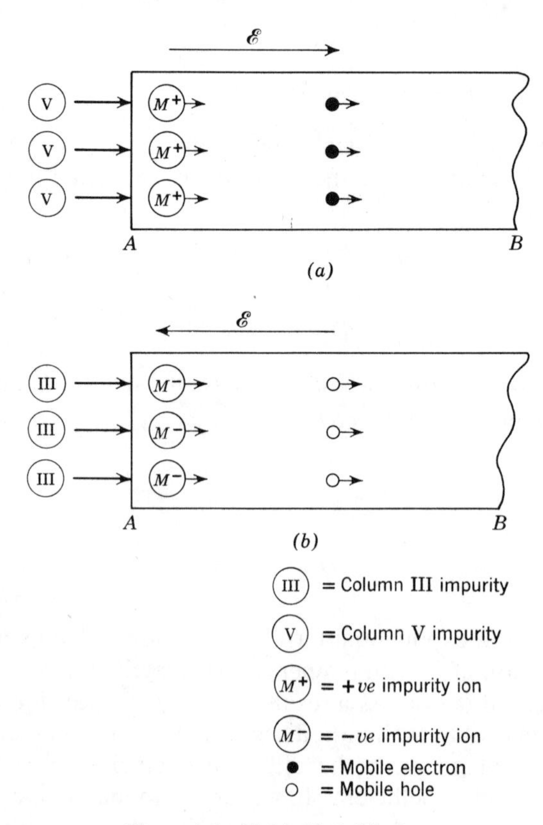

Figure 4.5 Field-aided diffusion.

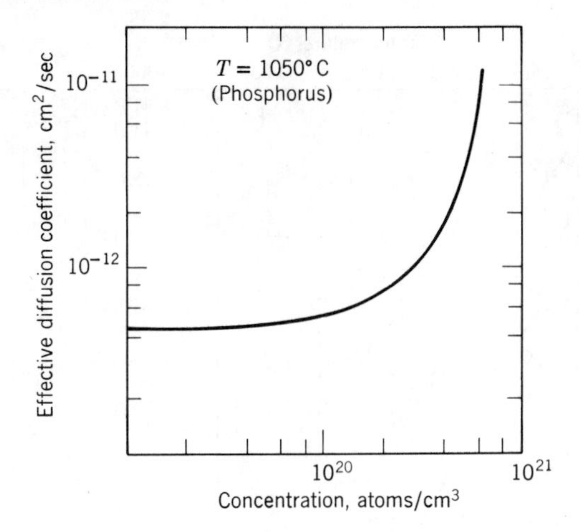

Figure 4.6 Effective diffusion coefficient at high concentrations.

4.3.1. Substitutional Impurities

Values of D_0 and the activation energy of diffusion are determined by experimental methods. The most definitive results to date have been those obtained for substitutional impurities by Fuller and Ditzenberger[4]. Figure 4.7 and Table 4.1 display their results, together with the temperature range over which they were obtained. The data were obtained by diffusing the impurities into silicon wafers of known (but opposite type) background concentration and by making accurate measurements of the *p-n* structures thus formed. Their experimental technique essentially duplicates the fabrication process for *p-n* junctions; consequently, it is reasonable to expect that their results are most reliable when applied to this end.

4.3.2. Interstitial Impurities

Reliable values for diffusion constant and activation energy of diffusion of interstitial impurities are quite difficult to obtain for the following reasons:

1. These impurities usually diffuse about five to six orders of magnitude faster than the substitutional diffusers. Hence considerable error can occur as a result of out-diffusion effects while the specimens are cooled to room temperatures. Attempts at rapidly quenching the wafers usually result in the generation of a large number of crystal defects and tend to obscure the interpretation of data.

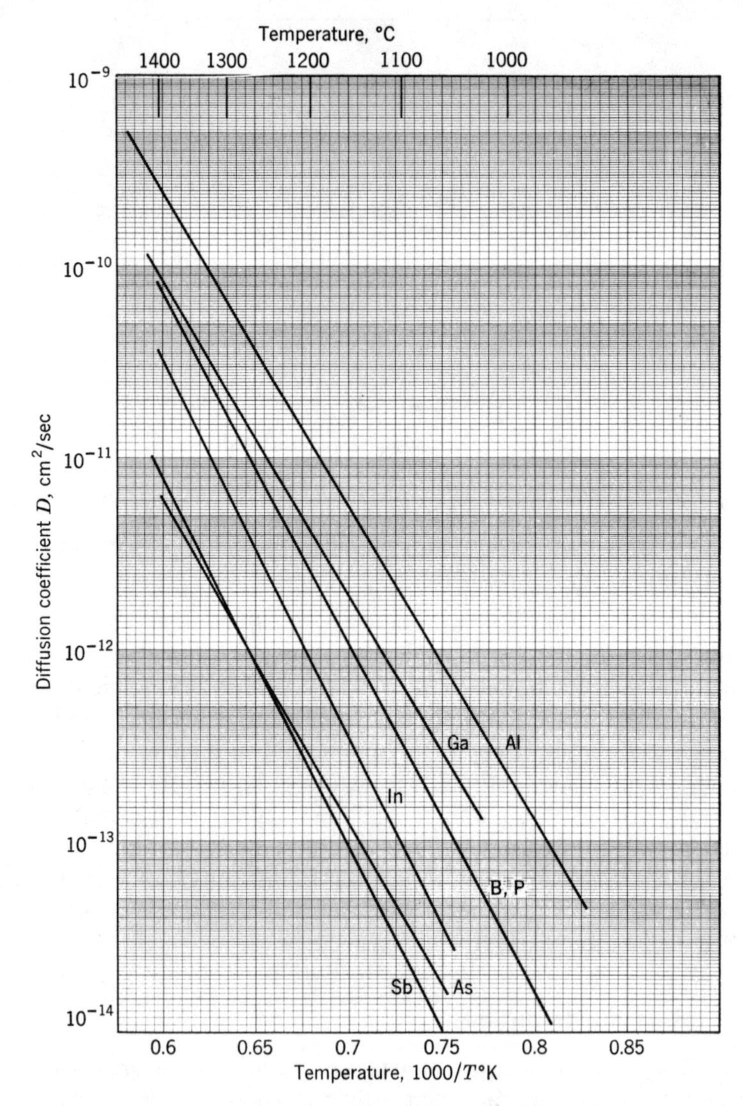

Figure 4.7 Diffusion coefficients of substitutional diffusers in silicon.

2. Many interstitial diffusers are partially substitutional in nature. Consequently, their movement is governed by an effective diffusion coefficient, the behavior of which is a function of the concentration of substitutional diffusers, as well as of temperature.

3. On freezing, most interstitial diffusers end up in both electronically active and inactive sites. The fractions of each type differ widely from

Table 4.1
Diffusion Constant [4] and Activation Energy of Diffusion for Substitutional Diffusers in Silicon

Impurity	P	As	Sb	B	Al	Ga	In
D_0, cm²/sec	10.5	0.32	5.6	10.5	8	3.6	16.5
Activation energy, eV	3.69	3.56	3.95	3.69	3.47	3.51	3.9
Temperature, range of measurement, °C	950–1235	1095–1380	1095–1380	950–1275	1080–1375	1105–1360	1105–1360
Type of impurity	*n*-type			*p*-type			

element to element. Thus about 90% of gold terminates in active sites, while the corresponding figure for nickel is only 0.1%. The analysis of the results is thus quite complex, since some analytical techniques (such as resistivity measurements) only provide information on the part† that is electronically active. On the other hand, radioactive tracer techniques result in information on the entire impurity content.

4. The interaction energy between an impurity and the strain field associated with a dislocation tends to favor the clustering of interstitial diffusers in the neighborhood of this dislocation. Thus the diffusion processes is dominated by the defect nature of the crystal. As a consequence, these impurities generally do not obey Fick's law, and the experimental data are difficult to interpret meaningfully.

5. Many interstitial diffusers form compounds with silicon over certain ranges of diffusion temperature. These tend to segregate in clusters in the silicon lattice and are often electronically inactive. Thus material doped with these impurities is usually sensitive to heat treatments. Of particular importance is the behavior of oxygen in silicon, as described in Section 2.2.3.

Table 4.2 and Figure 4.8 show the diffusion constant and the activation energy of diffusion for a number of interstitially diffusing impurities that are often found in silicon. It must be emphasized that the data here are by no means definitive, much of it being in the nature of average values and estimates. Because of its importance in fabrication technology, the behavior of gold in silicon is treated in detail in the next section.

†It is often assumed that the active impurities are those that end up in substitutional sites, since they can enter into covalent bonding with the silicon atoms. This is not always true, however, the notable exception being lithium.

Table 4.2
Diffusion Constant and Activation Energy of Diffusion for Interstitial Diffusers in Silicon

Impurity	[5][1] Li	[6] S	[7] Fe	[8] Cu	[9] Ag	[7] Au	[10] O	[11] Ni	[12] Zn
D_0, cm²/sec	2.5×10^{-3}	0.92	6.2×10^{-3}	4×10^{-2}	2×10^{-3}	1.1×10^{-3}	0.21	$D = 10^{-5}$ cm²/sec	$D = 10^{-6}$–10^{-7} cm²/sec
Activation energy of diffusion, eV	0.655	2.2	1.60	1.0	1.6	1.12	2.44		
Temperature range of measurement, °C	25–1350	1050–1370	1100–1350	800–1100	1100–1350	800–1200	1300	1100–1360°C	900–1360°C
Type of impurity	Deep-lying, usually multiple-charge states								

[1]Numbers in brackets pertain to references at end of chapter.

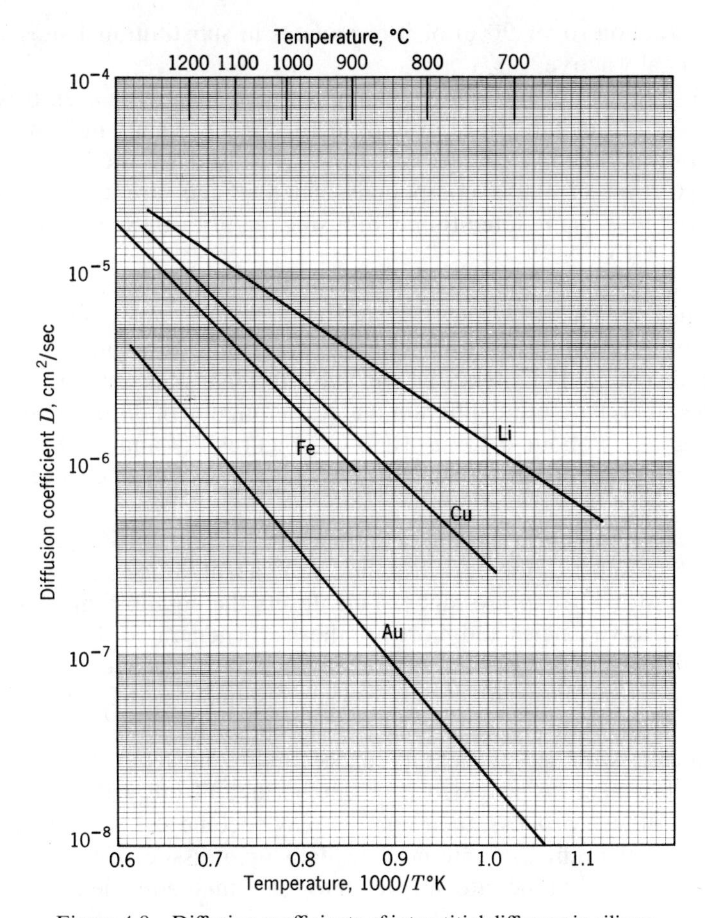

Figure 4.8 Diffusion coefficients of interstitial diffusers in silicon.

4.3.2.1. Gold in Silicon. Most interstitially diffusing impurities are known to exhibit deep-lying levels in silicon. In high-speed digital circuits their intentional introduction results in the reduction of minority carrier lifetime and constitutes an important fabrication step. Of the many available materials gold is today the most commonly used impurity for this purpose. The reasons for its choice over other impurities are as follows:

1. Gold has a relatively high solid solubility (10^{17} atoms/cm^3 at 1280°C) compared with other deep-lying impurities. Thus the amount of gold introduced into a microcircuit can be controlled over a wide range of values.
2. Although gold diffuses predominantly by an interstitial mechanism,

a large fraction (over 90%) of it terminates in substitutional sites and is electronically active.

3. Gold does not form any compounds with silicon (as seen from the phase diagram of the Au-Si system in Figure 3.6). Thus its behavior is free from anomalous effects that may be caused by the formation or decomposition of these compounds. Such effects are commonly encountered with many interstitially diffusing impurities. (See, for example, Section 3.2.4.)

Gold diffuses into the silicon lattice by two mechanisms—a fast interstitial mechanism with $D_i \simeq 10^{-5} cm^2/sec$ at 1050°C and a slow substitutional process with $D_s \simeq 10^{-10} cm^2/sec$ at 1050°C. In addition, the concentrations of gold in interstitial and substitutional sites (defined as c_i and c_s respectively) are quite different and are a function of diffusion temperature (e.g., c_i is about $0.1 c_s$ at 1050°C). Figures 4.9 and 4.10 show the diffusion coefficients and concentrations of both types of gold as a function of temperature[13].

On introduction into the silicon lattice, interstitially located gold atoms can become substitutional by dropping into vacant lattice sites when they come within their capture range. This upsets the equilibrium concentration of these vacancies and more are thermally generated. In silicon with a high defect concentration, this readily occurs in the vicinity of a dislocation and results in its climb.† As a consequence, the concentration of vacant lattice sites tends to preserve its equilibrium value.

In silicon with a low defect concentration, however, these vacant lattice sites must be provided from the surface of the wafer. Since this is a much slower process, their number may become less than the equilibrium value; the diffusion rate now depends on a dissociative mechanism whereby gold in lattice sites moves into interstitial sites, leaving behind vacancies.

As a result of these processes, the diffusion of gold in silicon is a strong function of the defect concentration for that particular crystal, in addition to the temperature.

Consider a piece of silicon with a vacancy concentration of n_v and interstitial and substitutional gold concentrations of c_i and c_s respectively. Then the equilibrium reaction can be written as

$$c_s \rightleftharpoons c_i + n_v. \tag{4.17}$$

In a crystal with many defects, the concentration of vacancies readily maintains its equilibrium value, and the gold spends $c_i/(c_i + c_s)$ of its time

†The presence of oxygen in the silicon tends to pin the dislocations and, hence, is an additional complicating factor in the diffusion of gold.

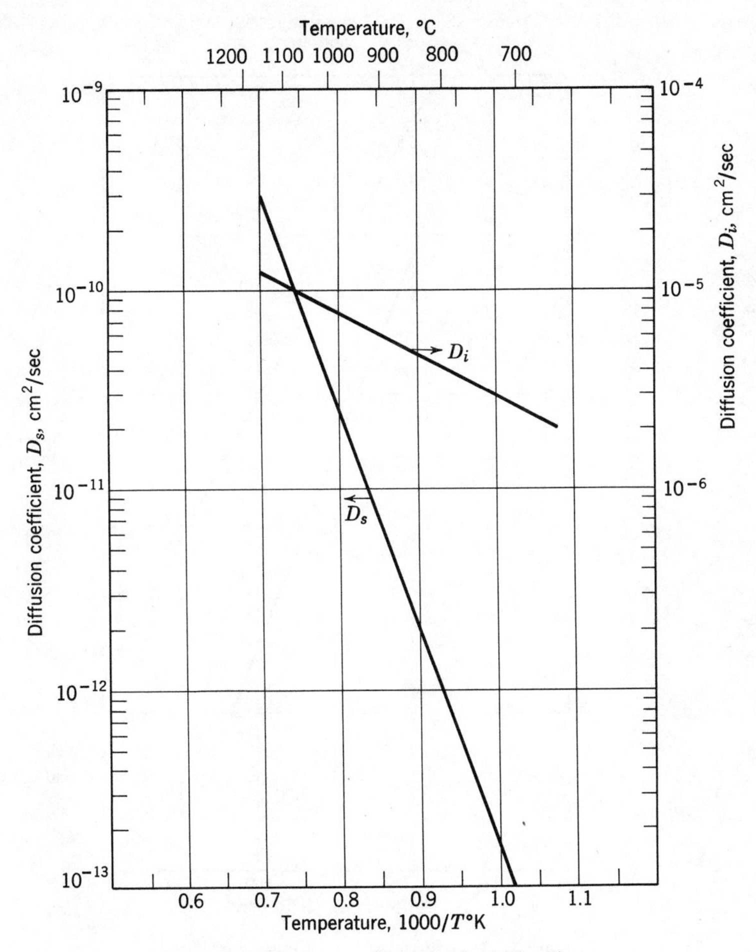

Figure 4.9 Diffusion coefficients of gold in silicon.

in a high state of diffusivity. This results in an effective diffusion constant

$$D' = \frac{D_i c_i}{c_i + c_s}. \tag{4.18}$$

In a crystal with few defects, however, the fraction of the vacancies that are not filled with gold atoms is $n_v/(n_v + c_s)$ and the effective diffusion constant is thus

$$D'' = \frac{D_i n_v}{n_v + c_s}. \tag{4.19}$$

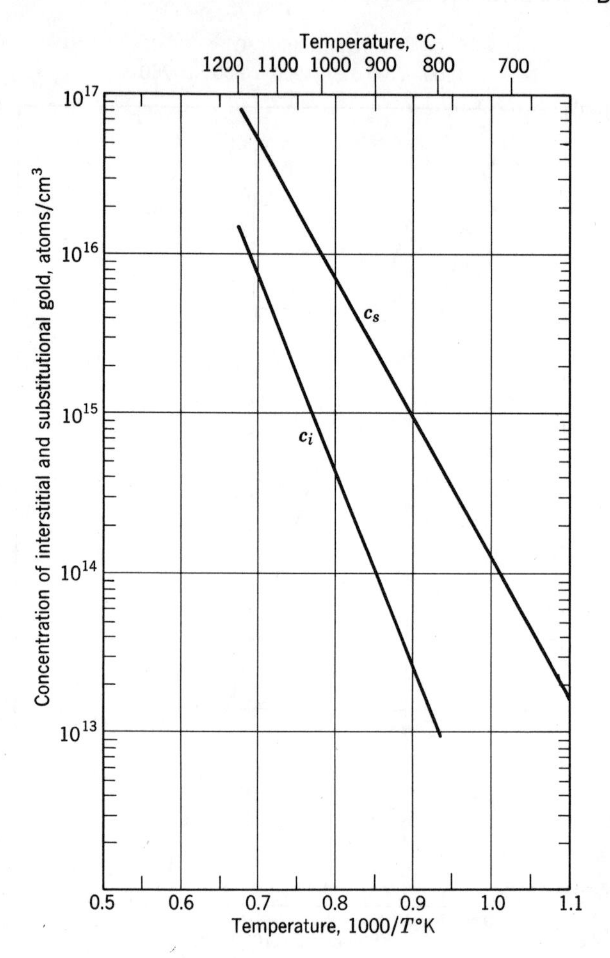

Figure 4.10 Solid solubilities of gold in silicon.

Gold diffusion is generally the last step in the fabrication of high-speed digital microcircuits. Thus diffusion takes place into a wafer in which are already present various doped regions, each of differing defect concentration depending on the type and amount of impurity in it. Although the diffusion constant for gold thus varies from region to region, the value given by (4.18) is realistic for typical situations, since interstitial diffusion is the dominant process.

In actual practice no attempt is made to establish a specific doping profile, and the wafer is doped by maintaining it at an elevated temperature for what is, for all practical purposes, an infinitely long time. Thus varying amounts of gold are present in the entire microcircuit. The diffusion temperature is altered to adjust the gold concentration in the silicon; however, the actual concentration is also a function of such variables as the manner in which the wafer is cooled to room temperature. The choice of the precise diffusion time and temperature is determined by trial and error, with a desired end effect in mind.

4.4. FICK'S SECOND LAW

This law may be derived by applying considerations of continuity to (4.10). Consider the flow of particles in a crystal of cross section A, between planes P_1 and P_2, separated by dx, as shown in Figure 4.11. The rate of accumulation of particles in the region between planes is $A(\partial N/\partial t)\,dx$. This can also be written as the difference between the fluxes flowing into and out of the region. The flux entering the region at P_1 is Aj and the flux leaving the region at P_2 is $A(j+dj)$. The net flux entering the region is thus $-A\,dj$. Hence

$$A\frac{\partial N}{\partial t}dx = -A\,dj. \tag{4.20}$$

But $dj = (\partial j/\partial x)\,dx$. Hence, applying Fick's first law [see (4.10)],

$$\frac{\partial N}{\partial t} = \frac{\partial}{\partial x}\left(D\frac{\partial N}{\partial x}\right). \tag{4.21}$$

If D is assumed to be independent of concentration, this reduces to

$$\frac{\partial N}{\partial t} = D\frac{\partial^2 N}{\partial x^2}, \tag{4.22}$$

where N = volume concentration, in atoms per cubic centimeter,
D = diffusion coefficient, in square centimeters per second,
x = distance, in centimeters,
t = diffusion time, in seconds.

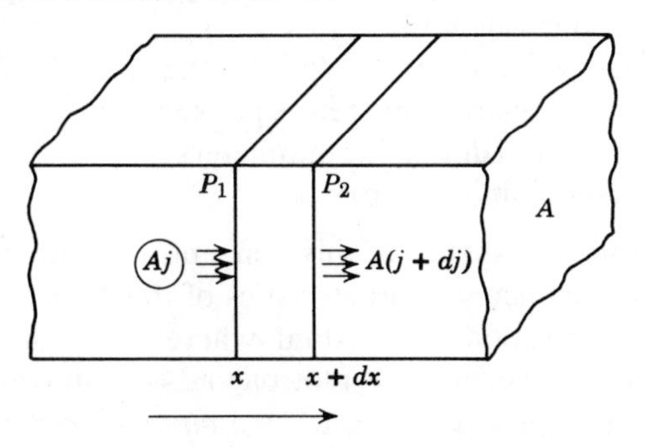

Figure 4.11 Diffusing flux.

This is Fick's second law. It states that the time rate of change of volume concentration of impurities is directly proportional to the second derivative of the volume concentration with respect to distance. Since only volume concentrations are involved, the expression may be used to obtain impurity distributions for a variety of diffusion situations.

A similar reasoning applies to diffusion in the presence of an electric field. For this situation it is seen from (4.14) that the diffusion equation is modified to

$$\frac{\partial N}{\partial t} = D\frac{\partial^2 N}{\partial x^2} - \mu\mathscr{E}\frac{\partial N}{\partial x}. \tag{4.23}$$

4.5. APPLICATION OF FICK'S LAW

Fick's law may be solved for a number of situations which arise during the fabrication of microcircuits. The mathematics of diffusion is considered in detail in Appendix A. In this section the results are taken directly from this appendix.

4.5.1. Diffusion from a Constant Source

This situation occurs when a silicon slice, suitably prepared, is exposed to an impurity source of constant concentration during the diffusion period. The concentration of the impurity is such as to result in a surface concentration of N_0 in the silicon.

For this situation, the impurity concentration at any given distance and time is $N(x, t)$ where [see (A.39)]

$$N(x, t) = N_0 \operatorname{erfc}\frac{x}{2\sqrt{Dt}}, \tag{4.24}$$

where N_0 = impurity concentration at the silicon surface, in atoms per cubic centimeter,
D = value of diffusion constant for the specific diffusion temperature, in square centimeters per second,
x = penetration depth, in centimeters,
t = diffusion time, in seconds.

Figure 4.12 shows a sketch of this concentration for various diffusion times. The most significant characteristics of this type of diffusion is that the surface concentration is constant whereas the diffusion depth increases with time. A normalized plot of (4.24) is shown in Figure 4.13. A diffusion of this type is referred to as a *complementary error function* (erfc) diffusion.

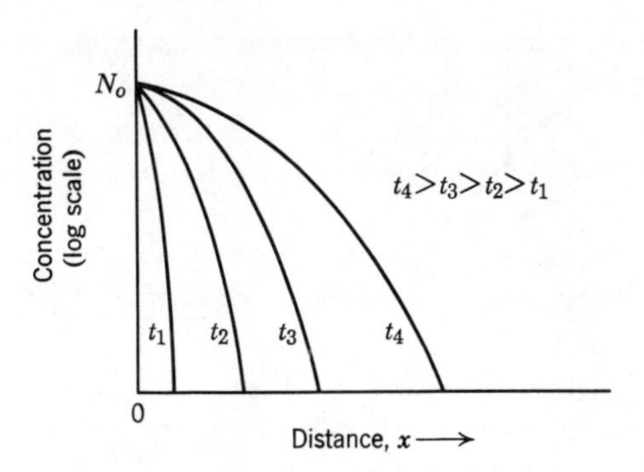

Figure 4.12 Constant-source diffusion profiles.

The erfc diffusion is performed by exposing the silicon to a constant concentration of the impurity during the entire process. At low levels, the surface concentration is given by that value which is in equilibrium with the surrounding gas. With increasing source concentration, the surface concentration rises until it is ultimately set by the solid solubility of the impurity in silicon for that specific diffusion temperature. The emitter diffusion of a transistor is usually of this type. Here, a high surface concentration is desired, necessitating the use of high diffusion temperatures (1100 to 1200°C).

4.5.2. Diffusion from an Instantaneous Source

Here, a finite quantity of the diffusing matter is first placed on the surface of the silicon wafer. Diffusion proceeds from this fixed source, and it is assumed that all of this matter is consumed instantaneously by the silicon. The impurity concentration resulting from this type of diffusion is given by [see (A.47)]

$$N(x, t) = \frac{Q}{\sqrt{\pi D t}} e^{-(x/2\sqrt{Dt})^2}, \tag{4.25}$$

where Q = amount of matter placed on the surface prior to diffusion, in atoms per square centimeter,
D = diffusion coefficient, in square centimeters per second,
x = diffusion distance, in centimeters,
t = diffusion time, in seconds.

Figure 4.13 shows a normalized plot of (4.25). Figure 4.14 shows a sketch of the diffusion profile for various diffusion times. A diffusion of this type is referred to as a *gaussian diffusion*.

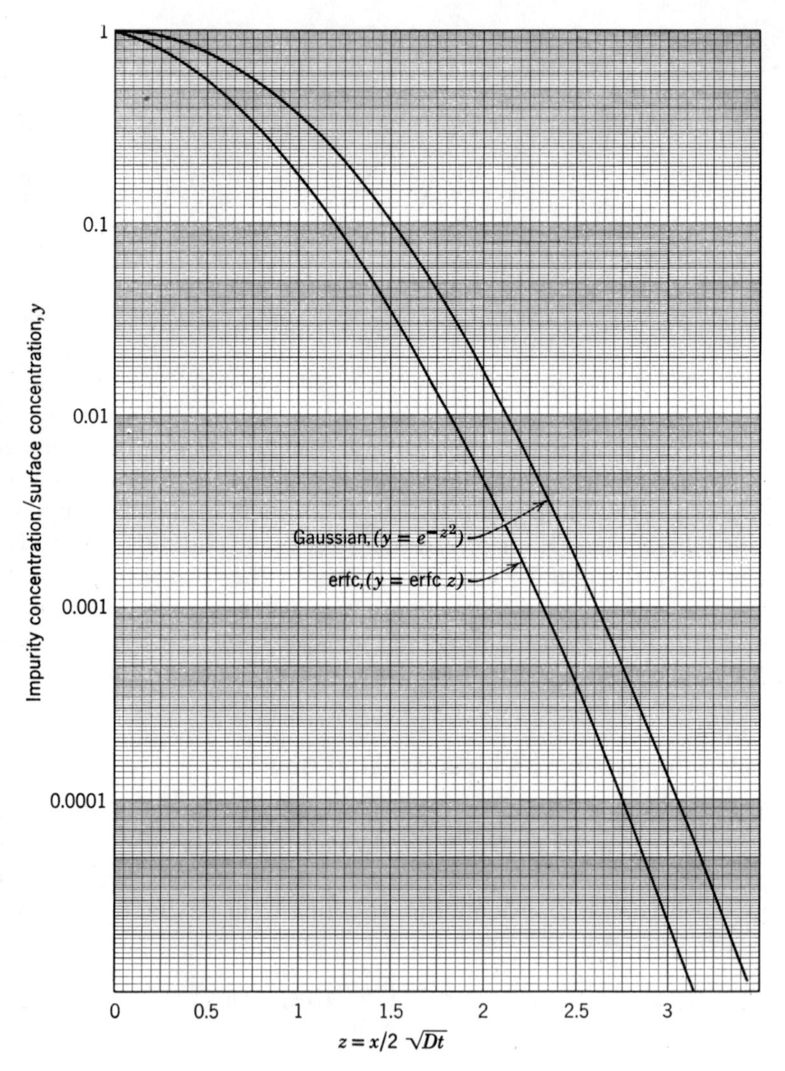

Figure 4.13 The erfc and gaussian profiles.

As seen from these figures, the significant difference between these two types of diffusion lies in the fact that the surface concentration of the gaussian diffusion changes for varying diffusion times whereas that of the erfc diffusion does not. Otherwise they are essentially similar and can be approximated by exponential profiles at concentration levels that are two or more orders of magnitude below the surface concentration.

Instantaneous source diffusion is ideally suited for those cases where

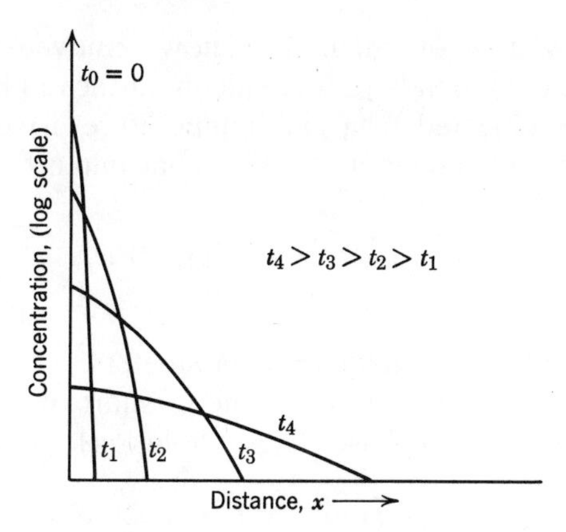

Figure 4.14 Instantaneous-source diffusion profiles.

a relatively low value of surface concentration is required in conjunction with a high diffusion depth.† The base diffusion of a transistor normally presents this type of situation. Surface concentrations encountered here are below those needed for the emitter diffusion, whereas the diffusion depth is higher.

It may be shown that the amount of matter required for a gaussian diffusion is considerably less than a monolayer. Thus a special technique must be used to place it on the silicon surface. This is done by first conducting a constant-source diffusion at a low temperature ($\simeq$ 900°C) for a short time. By this process, known as *predeposition*, a small quantity of the impurity is transported into an extremely thin layer *in* the silicon. Since the penetration depth is negligibly small, this provides a suitable approximation to placing the matter *on* the surface. The actual amount of matter transported during the predeposition phase is given [see (A.45)] by

$$Q = 2N_{01}\left(\frac{D_1 t_1}{\pi}\right)^{1/2}, \tag{4.26}$$

where Q = amount of matter entering the silicon during predeposition, in atoms per square centimeter,

N_{01} = surface concentration at the predeposition temperature, in atoms per cubic centimeter,

D_1 = diffusion coefficient at the predeposition temperature, in square centimeters per second,

t_1 = predeposition time, in seconds.

†Although the low surface concentration can also be obtained by diffusion from a constant source at low diffusion temperature, the amount of time required for the desired penetration depth would be prohibitive.

The external source of impurity is now removed by dissolving all impurity-bearing compounds on the silicon surface in hydrofluoric acid, and the slice is subjected to a high-temperature drive-in phase for an appropriate time and temperature. The final impurity concentration is thus given by

$$N(x, t_1, t_2) = \frac{2N_{01}}{\pi}\left(\frac{D_1 t_1}{D_2 t_2}\right)^{1/2} e^{-(x/2\sqrt{D_2 t_2})^2} \tag{4.27}$$

where the subscript 2 refers to drive-in parameters.

Equation 4.27 assumes an extremely short predeposition phase compared with the drive-in phase. Thus it holds as long as

$$\sqrt{D_1 t_1} \ll \sqrt{D_2 t_2}. \tag{4.28}$$

4.5.3. The Two-step Diffusion

This diffusion technique is an outgrowth of the last process. It presents so many advantages over the previous methods, however, that it is the most common diffusion process today.

The two-step diffusion is initiated by first conducting a constant-source diffusion at a low temperature for a short time. The impurity supply is shut off and the drive-in phase initiated in an oxidizing ambient. This results in the formation of a surface oxide on the silicon, preventing the further in-diffusion of impurities which are present *on* the slice and inhibiting the out-diffusion of impurities which were already predeposited *in* the slice. The final impurity concentration is a function of the predeposition and drive-in parameters. If the subscripts 1 and 2 are used to denote predeposition and drive-in respectively, an erfc diffusion results when $D_1 t_1 \gg D_2 t_2$. Conversely, a gaussian profile results if $D_1 t_1 \ll D_2 t_2$.

In practical situations neither of these inequalities hold. For this case the resulting distribution has been obtained by Smith [14] for the analogous heat-flow problem. His results may be modified to give the final impurity distribution as

$$N(x, t_1, t_2) = \frac{2N_{01}}{\pi} \int_0^U \frac{e^{-\beta(1+u^2)}}{1+u^2}\, du, \tag{4.29}$$

where

$$U = \left(\frac{D_1 t_1}{D_2 t_2}\right)^{1/2} \tag{4.30}$$

and

$$\beta = \left(\frac{x}{2\sqrt{D_1 t_1 + D_2 t_2}}\right)^2. \tag{4.31}$$

In addition, it can be shown that the final surface concentration is given by

$$N_{02} = \frac{2N_{01}}{\pi} \tan^{-1} U, \tag{4.32}$$

hence

$$\frac{N(x, t_1, t_2)}{N_{02}} = \frac{1}{\tan^{-1} U} \int_0^U \frac{e^{-\beta(1+u^2)}}{1+u^2} du. \tag{4.33}$$

Table 4.3 gives values of the integral of (4.33). In addition, Figure 4.15 shows normalized concentration profiles for two-step diffusions with various values of U as well as the limiting cases of erfc and gaussian diffusions.

The two-step process may be used to approximate a erfc diffusion profile without the accumulation of large amounts of impurity on the surface of the silicon. Thus it avoids the possibility of surface damage. In addition, this process can be used to approximate a gaussian diffusion while avoiding the necessity of exposing the surface of the slice at any time. Junctions are always formed beneath this protective layer and show excellent breakdown characteristics. Finally, the glass may be used as the basis for the next diffusion step.

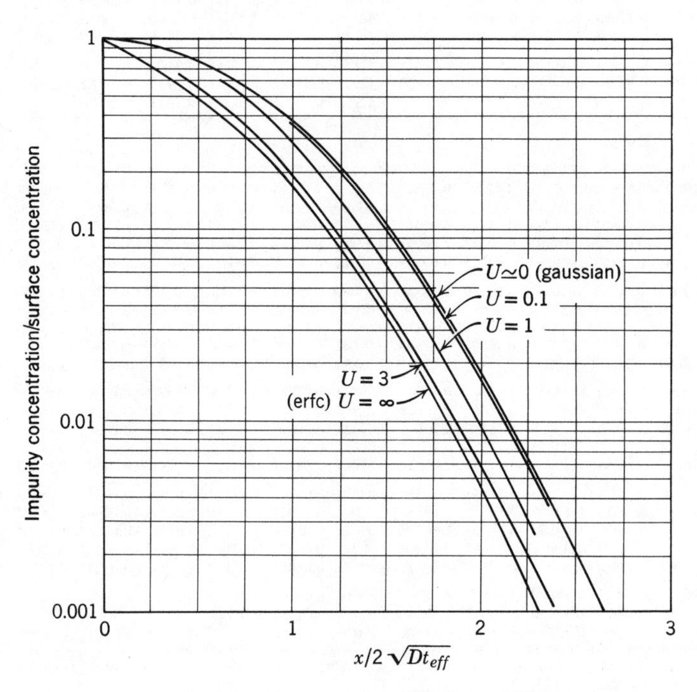

Figure 4.15 The two-step diffusion profiles.

Table 4.3
Values of the Integral

$$\int_0^U \frac{e^{-\beta(1+u^2)}}{1+u^2}\,du$$

U \ β	0·1	0·2	0·3	0·4	0·5	0·6	0·7	0·8	0·9	1·0	1·1	1·2
0.1	0.09015	0.08155	0.07376	0.06672	0.06035	0.05459	0.04938	0.04467	0.04040	0.03655	0.03306	0.02990
0.2	0.17838	0.16119	0.14566	0.13162	0.11894	0.10748	0.09713	0.08777	0.07931	0.07167	0.06477	0.05853
0.3	0.26295	0.23723	0.21403	0.19310	0.17422	0.15719	0.14182	0.12795	0.11545	0.10416	0.09398	0.08479
0.4	0.34254	0.30837	0.27761	0.24993	0.22501	0.20259	0.18240	0.16422	0.14786	0.13314	0.11988	0·10794
0.5	0.41626	0.37374	0.33557	0.30132	0.27058	0.24299	0.21822	0.19599	0.17603	0.15812	0.14203	0.12759
0.6	0.48366	0.43290	0.38751	0.34692	0.31062	0.27814	0.24908	0.22308	0.19982	0.17900	0.16036	0.14368
0.7	0.54464	0.48580	0.43340	0.38673	0.34515	0.30809	0.27505	0.24562	0.21937	0.19596	0.17508	0.15645
0.8	0.59940	0.53264	0.47347	0.42100	0.37447	0.33317	0.29652	0.26398	0.23508	0.20940	0.18657	0.16628
0.9	0.64829	0.57380	0.50812	0.45017	0.39903	0.35385	0.31393	0.27864	0.24742	0.21979	0.19532	0.17365
1.0	0.69176	0.60975	0.53784	0.47475	0.41935	0.37066	0.32783	0.29013	0.25693	0.22765	0.20183	0.17903
1.1	0.73033	0.64100	0.56318	0.49529	0.43600	0.38415	0.33877	0.29900	0.26411	0.23348	0.20655	0.18286
1.2	0.76448	0.66808	0.58465	0.51232	0.44950	0.39486	0.34726	0.30574	0.26946	0.23772	0.20991	0.18553
1.3	0.79470	0.69148	0.60276	0.52634	0.46035	0.40327	0.35377	0.31078	0.27336	0.24074	0.21225	0.18734
1.4	0.82144	0.71164	0.61797	0.53781	0.46901	0.40979	0.35870	0.31449	0.27616	0.24286	0.21385	0.18855
1.5	0.84509	0.72899	0.63069	0.54714	0.47586	0.41482	0.36238	0.31720	0.27815	0.24431	0.21492	0.18933
1.6	0.86601	0·74388	0.64130	0.55469	0.48123	0.41865	0.36511	0.31914	0.27953	0.24530	0.21562	0.18983
1.7	0.88454	0.75666	0.65010	0.56076	0.48542	0.42153	0.36710	0.32051	0.28048	0.24595	0.21607	0.19014
1.8	0.90095	0.76759	0.65739	0.56562	0.48865	0.42369	0.36854	0.32147	0.28112	0.24638	0.21636	0.19033
1.9	0.91549	0.77693	0.66340	0.56948	0.49114	0.42529	0.36956	0.32213	0.28154	0.24665	0.21653	0.19045
2.0	0.92838	0.78491	0.66833	0.57254	0.49303	0.42646	0.37029	0.32258	0.28182	0.24682	0.21664	0.19051
2.5	0.97404	0.81009	0.68228	0.58029	0.49735	0.42887	0.37165	0.32335	0.28225	0.24707	0.21678	0.19059
3.0	0.99920	0.82094	0.68698	0.58234	0.49825	0.42928	0.37183	0.32343	0.28229	0.24708	0.21679	0.19059
∞	1.02834	0.82795	0.68892	0.58291	0.49843	0.42933	0.37184	0.32343	0.28229	0.24709	0.21679	0.19059

U \ β	1.3	1.4	1.5	1.6	1.7	1.8	1.9	2.0	2.5	3.0	4.0	5.0
0.1	0.02705	0.02446	0.02213	0.02002	0.01811	0.01638	0.01481	0.01340	0.00811	0.00491	0.00180	0.00066
0.2	0.05289	0.04779	0.04319	0.03903	0.03527	0.03187	0.02880	0.02603	0.01568	0.00945	0.00343	0.00125
0.3	0.07651	0.06903	0.06228	0.05620	0.05071	0.04575	0.04128	0.03725	0.02228	0.01333	0.00477	0.00171
0.4	0.09720	0.08752	0.07881	0.07097	0.06391	0.05756	0.05183	0.04668	0.02766	0.01640	0.00577	0.00204
0.5	0.11462	0.10297	0.09251	0.08312	0.07468	0.06711	0.06030	0.05419	0.03178	0.01866	0.00645	0.00224
0.6	0.12875	0.11538	0.10340	0.09268	0.08308	0.07448	0.06678	0.05988	0.03475	0.02021	0.00688	0.00236
0.7	0.13982	0.12499	0.11174	0.09992	0.08936	0.07993	0.07150	0.06398	0.03677	0.02120	0.00712	0.00242
0.8	0.14824	0.13219	0.11790	0.10519	0.09387	0.08379	0.07481	0.06680	0.03806	0.02180	0.00724	0.00244
0.9	0.15444	0.13741	0.12230	0.10889	0.09699	0.08642	0.07702	0.06867	0.03885	0.02213	0.00730	0.00245
1.0	0.15889	0.14109	0.12535	0.11141	0.09907	0.08814	0.07844	0.06985	0.03931	0.02231	0.00733	0.00246
1.1	0.16200	0.14361	0.12739	0.11307	0.10041	0.08923	0.07933	0.07056	0.03956	0.02240	0.00734	0.00246
1.2	0.16411	0.14529	0.12872	0.11412	0.10125	0.08989	0.07985	0.07098	0.03969	0.02244	0.00735	0.00246
1.3	0.16552	0.14638	0.12956	0.11478	0.10176	0.09028	0.08016	0.07122	0.03976	0.02246	0.00735	0.00246
1.4	0.16643	0.14706	0.13008	0.11517	0.10205	0.09051	0.08033	0.07134	0.03979	0.02247	0.00735	0.00246
1.5	0.16700	0.14749	0.13039	0.11540	0.10222	0.09063	0.08042	0.07141	0.03980	0.02247	0.00735	0.00246
1.6	0.16736	0.14774	0.13057	0.11552	0.10231	0.09070	0.08046	0.07144	0.03981	0.02247	0.00735	0.00246
1.7	0.16757	0.14789	0.13067	0.11559	0.10236	0.09073	0.08049	0.07146	0.03981	0.02247	0.00735	0.00246
1.8	0.16770	0.14797	0.13073	0.11563	0.10239	0.09075	0.08050	0.07147	0.03982	0.02247	0.00735	0.00246
1.9	0.16777	0.14802	0.13076	0.11565	0.10240	0.09075	0.08050	0.07147	0.03982	0.02247	0.00735	0.00246
2.0	0.16781	0.14804	0.13078	0.11566	0.10240	0.09076	0.08051	0.07147	0.03982	0.02247	0.00735	0.00246
2.5	0.16786	0.14807	0.13079	0.11567	0.10241	0.09076	0.08051	0.07147	—	—	—	—
3.0	0.16786	0.14807	0.13079	0.11567	0.10241	0.09076	0.08051	0.07147	—	—	—	—
∞	0.16786	0.14807	0.13079	0.11567	0.10241	0.09076	0.08051	0.07147	0.03982	0.02247	0.00735	0.00246

From R. C. T. Smith, "Conduction of Heat in the Semi-infinite Solid with a Short Table of an Important Integral," *Australian J. Phys.*, vol. 6, pp. 127–130 (1953).

4.5.4. The Effect of Successive Diffusions

It is often required to calculate the total effect of diffusion during a series of temperature cycles. By way of example the emitter-diffusion step takes place after the base drive-in. Thus the impurities in the base are subjected to one set of time and temperature values during the base drive-in phase and to a second set during the emitter-diffusion phase. If a gold-diffusion step follows, the base is subjected to yet another time and temperature cycle. To compute the total effect of these cycles, it is necessary to obtain an effective Dt product for this region. The effective Dt product is given by

$$(Dt)_{\text{eff}} = \sum D_1 t_1 + D_2 t_2 + D_3 t_3 + \ldots, \tag{4.34}$$

where $t_1, t_2, t_3 \ldots$ are the different diffusion times, and $D_1, D_2, D_3 \ldots$ are the appropriate diffusion constants in effect during these times.

4.5.5. Other Diffusion Conditions

Computations have been made by several authors for a variety of diffusion conditions. Some have been listed in the references at the end of this chapter [15–19].

4.6. JUNCTION FORMATION

All of the equations of Section 4.5 have been established on the basis that diffusion takes place into regions of zero initial concentration. Since the diffusion equation is linear, superposition may be used to calculate the effect of finite background concentrations which are normally encountered in practical situations; for example, a *p-n* junction is fabricated by diffusing a *p*-type impurity into an *n*-type wafer of background concentration N_C, as shown in Figure 4.16*a*. For this situation, the effective doping concentration at any point is given by the difference of acceptor and donor concentrations, whichever is larger. If the *p*-type impurity profile is given by

$$N(x, t) = N_0 \operatorname{erfc} \frac{x}{2\sqrt{Dt}}, \tag{4.35}$$

the wafer is *p*-type as long as $N(x, t) > N_C$, and *n* type for $N_C < N(x, t)$. The junction is thus located at x_j, such that

$$N_0 \operatorname{erfc} \frac{x_j}{2\sqrt{Dt}} = N_C. \tag{4.36}$$

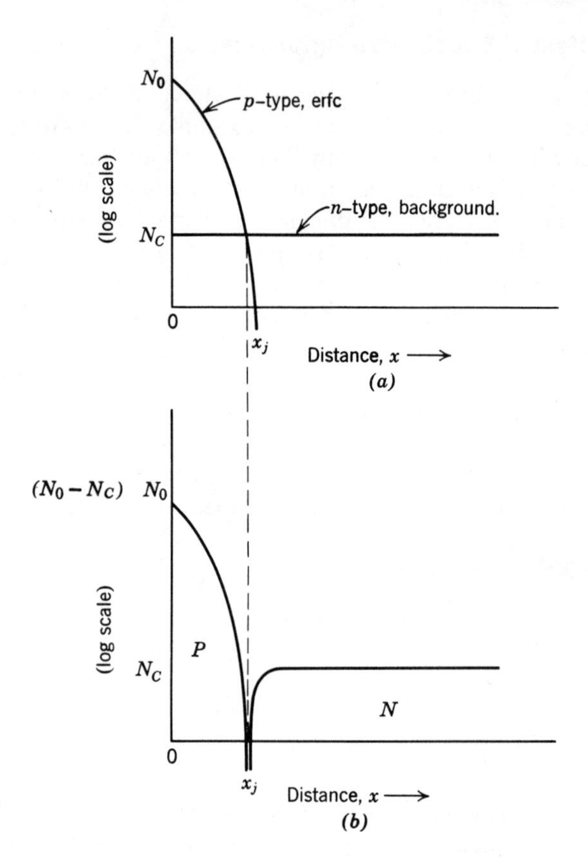

Figure 4.16 Junction formation by diffusion.

Figure 4.16*b* illustrates the net impurity concentration for the *p-n* junction formed in this manner.

The successive diffusion of two impurities into a doped wafer can result in the formation of an *n-p-n* transistor structure, as shown in Figure 4.17*a*. Here the background concentration is *n*-type and the two diffusions are made with *p*- and *n*-type impurities respectively. As seen from the figure, it is necessary that $N_{0E} \gg N_{0B} \gg N_C$ in order to form this *n-p-n* structure. Thus the surface concentration for the *p*-type base diffusion must be less than that for the *n*-type emitter diffusion, even though its diffusion depth must be greater. As explained in Section 4.5.2, this is done by making the base diffusion of the gaussian type while the emitter diffusion is of the erfc type.

Figure 4.17*b* shows the net impurity concentrations for the three regions of an *n-p-n* transistor formed by this process. Although the

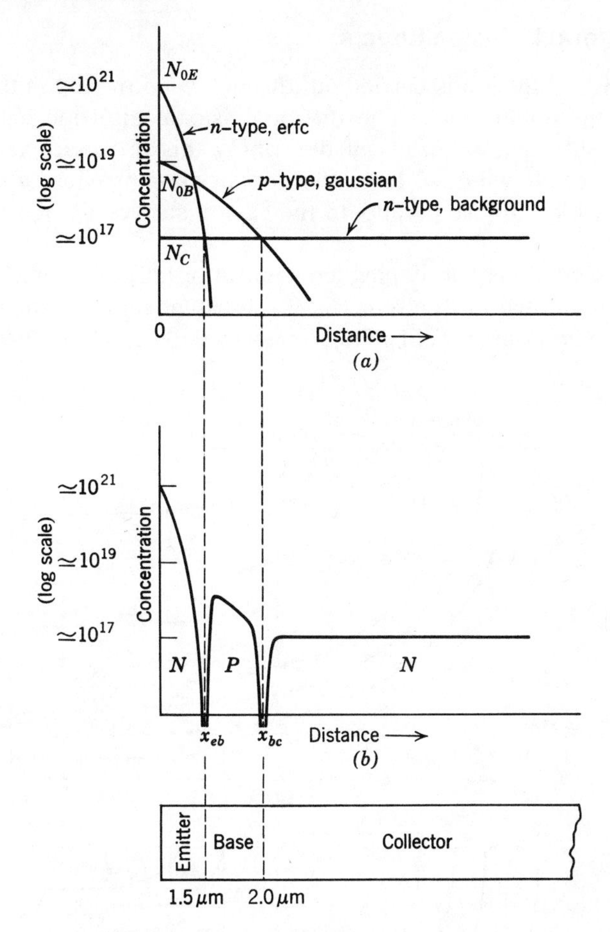

Figure 4.17 Transistor formation by diffusion.

choice of concentration and diffusion depth is a function of the device design, typical values are presented for a high-speed switching transistor to give an idea of the magnitudes involved.

In theory, successive diffusions can be used to fabricate structures with as many regions as desired. Each diffusion, however, results in the addition of extra impurities, so that the total number of impurities in the diffused regions becomes excessive (even though the electronic properties of the region are controlled by the net difference between acceptors and donors). Consequently, the defect concentration in successive regions increases, resulting ultimately in unusable material. This places an upper limit of about three to four on the number of successive diffusions that can be performed in practice.

4.6.1. Lateral Diffusion Effects

In practice, diffusion is carried out through windows cut in the surface oxide on the silicon slice. The one-dimensional diffusion equation represents a satisfactory means of describing this process, except at the edge of the oxide window. Here the dopant glass provides a source for impurities which diffuse parallel to the silicon surface as well as at right angles to it.

Contours of constant doping concentration[20] resulting from this situation are shown in Figure 4.18*a* for constant-source diffusing conditions. These contours are, in effect, a map of the location of the junctions

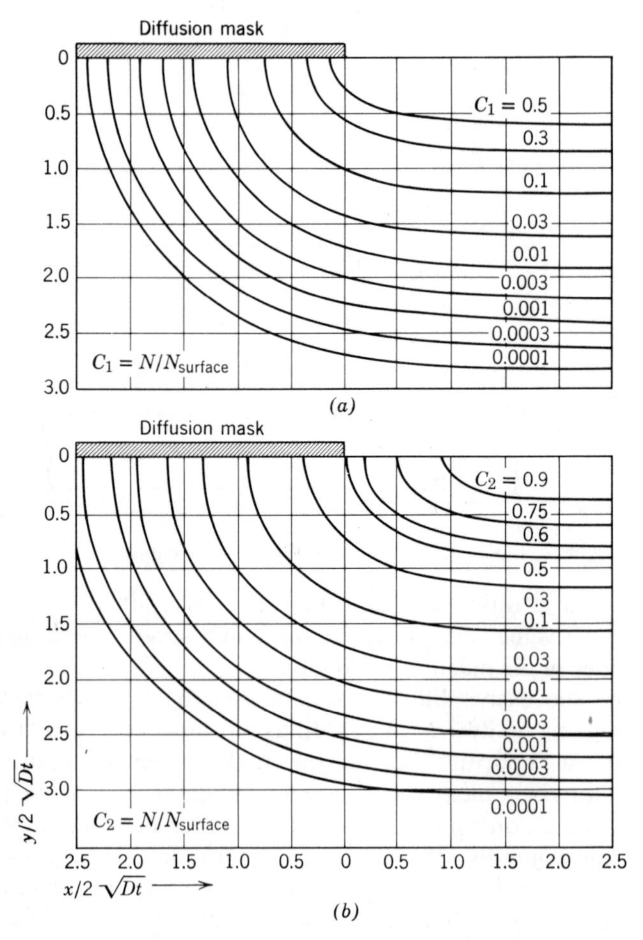

Figure 4.18 Lateral diffusion profiles (Kennedy and O'Brien[20]).

created by diffusing into various background concentrations. It is seen from this figure that the penetration parallel to the silicon surface is about 75 to 85% of the penetration at right angles, for concentrations that are two or more orders of magnitude below the surface concentration.

Figure 4.18*b* shows the contours of constant doping concentration for instantaneous-source diffusions. Here too, the penetration ratio (parallel to right angles) is about 75 to 85% for concentrations that are two or more orders of magnitude below the surface concentration. Note that the depletion of the instantaneous source results (for some cases) in actually terminating the contours within the window. This is of no consequence in practice, since a junction is usually formed by doping into a background that is considerably below the surface concentration.

4.7. DIFFUSION SYSTEMS

The basic requirements of any diffusion system are that a means be provided for bringing the diffusing impurity in contact with a suitably prepared silicon wafer, and that this process be maintained for a specific time and at a specific temperature. Within the broad framework of these requirements, the following additional features are also desired:

1. The surface concentration should be capable of being controlled over a wide range of values, up to the solid solubility limit of the impurity in silicon.
2. The diffusion process should not result in damage to the surface of the silicon wafer. This is an extremely important requirement, because the entire microcircuit is fabricated within the first few microns below this surface.
3. After diffusion, material residing on the surface of the wafer should be capable of easy removal, if so desired.
4. The system should give reproducible results from one diffusion run to the next and from slice to slice within a single run.
5. The system should be capable of processing a number of slices at a time.

Diffusion systems may be of either sealed- or open-tube type. In the sealed system the slices and dopant are enclosed in a clean, evacuated quartz tube prior to heat treatment. After diffusion the slices are removed by breaking the tube. Systems of this type can be easily maintained free from contamination. Their use, however, is commonly restricted to the laboratory because of the inconvenience of sealing and unsealing these tubes.

In the open-tube method slices are placed in a clean diffusion tube made of high-purity fused quartz. A separate diffusion furnace, diffusion tube, and slice carrier are reserved specifically for each impurity because they are contaminated with it during the process. Insertion is done from one end of the tube, whereas the other end is used for the flow of gases or impurities in vapor form. This method is capable of handling a number of slices at one time and is considerably more convenient than the sealed tube system. Consequently it is universally used in commercial microcircuit fabrication.

Diffusion furnaces are usually operated at temperatures ranging from 800 to 1200°C and are equipped with electronic controls to maintain a central flat zone with a $\pm\frac{1}{2}$°C tolerance. The length of this zone varies from $\pm$2 in. for laboratory systems to $\pm$12 in. for industrial units.

A number of techniques may be used to place the impurity atoms in contact with the slices. Perhaps the simplest method consists of coating the slice with the element to be diffused. This is done by electroplating, vacuum evaporation, or sputtering. Diffusion of the impurity takes place from the impurity-silicon interface. This interface may remain solid during diffusion; alternatively, if the impurity-silicon system exhibits a eutectic point below the diffusion temperature, an alloy interface will result. In either case the diffusion takes place from a constant source of essentially infinite concentration. Thus the surface concentration is set at the solid solubility limit of the impurity in silicon at the diffusion temperature.

A second technique consists of transporting an impurity-bearing compound in an inert gas to the surface of the silicon. At operating temperature a reversible reaction occurs between the compound and the silicon, resulting in an equilibrium concentration of impurity atoms from which diffusion takes place. This results initially in a surface concentration directly proportional to the partial pressure of the impurity in the transport gas. With increasing vapor pressure a point is eventually reached when the source concentration becomes excessive. At this point the surface concentration is set by the solid-solubility limit at the diffusion temperature. Diffusion processes of this type proceed out of a liquid layer that provides a considerable degree of protection to the silicon surface.

Impurities may also be introduced as a result of a direct vapor-phase reaction with the silicon. In practice such processes are difficult to control. Furthermore, most diffusions from gaseous sources are, in reality, due to reaction processes at the silicon surface, the use of a vapor serving only to transport the impurity compound to the silicon slices.

A number of diffusion techniques are now described insofar as they apply to the fabrication of modern microcircuits. The list, however, is by no means exhaustive.

4.7.1. Choice of p-Type Impurity

The *p*-type impurities are boron, aluminum, gallium, and indium. Table 4.4 gives the maximum solid solubilities that are attainable using these impurities. Of the different choices, the number can be reduced to one in view of the following considerations:

1. Indium is actually a moderately deep-lying impurity in silicon with an acceptor level at 0.16 eV above the valence band.
2. Aluminum is a highly reactive metal and combines with oxygen in the silicon lattice. This results in a number of anomalous effects with heat treatment.
3. Gallium has an extremely large diffusion constant† in silicon dioxide. Thus an oxide mask cannot be used to provide the selective area diffusions that are central to present-day microcircuit processes. Although other forms of masking (such as the use of silicon nitride) can perform this function, they have not received widespread commercial use up to now.

As a result of these considerations, boron is at present the exclusive choice for *p*-type diffusions in modern microcircuits.

4.7.1.1. Boron Diffusion Systems. Boron has a diffusion coefficient of 10^{-12} cm^2/sec at 1150°C, which is typical for substitutional impurities in silicon. It has a high solid solubility and can be diffused with a surface concentration as large as 2.5×10^{20} atoms/cm^3. Thus it is suited for a wide range of diffusion requirements.

The tetrahedral radius of boron in silicon is 0.88 Å (corresponding to a misfit factor $\epsilon = 0.254$). Thus the presence of large amounts of boron in the silicon lattice is accompanied by strain-induced defects which lead to considerable crystal damage. This sets an upper limit of about 5×10^{19} atoms/cm^3 to the surface concentration that can actually be achieved in practical structures. Any boron in excess of this amount is found to be electronically inactive and leads to even further damage of the crystal structure.

Elemental boron is quite inert at temperatures in excess of the melting point of silicon (1412°C). Consequently, diffusion is usually performed out of a surface reaction between a boron compound and the silicon. The most suitable compound for this purposes is boron trioxide, B_2O_3, the surface reaction at diffusion temperatures being given by

$$2B_2O_3 + 3\,Si \rightleftharpoons 4B + 3SiO_2. \tag{4.37}$$

†About 100 times larger than for diffusion in silicon!

Table 4.4
Maximum Solid Solubilities for Diffusing Impurities in Silicon [21]

Impurity	P	As	Sb	B	Al	Ga	In	Au
Max. solid solubility	1.3×10^{21}	2×10^{21}	6×10^{19}	2.5×10^{20}	$10^{19} - 10^{20}$	4×10^{19}	10^{19}	10^{17}
Temp. for max. solid solubility, °C	1150	1150	1300	1200	1150	1250	1300	1300
Type of dopant		*n*-type			*p*-type			Deep-lying

Boron trioxide is a liquid at normal diffusion temperatures. In addition, it mixes readily with the small amount of silicon dioxide always present on the exposed surface of the silicon wafer to form a boro-silicate glass, as seen from the phase diagram of Figure 4.19.

The simplest application technique consists of directly painting a slurry of B_2O_3 in alcohol on the silicon surface. This method is too crude for microcircuit fabrication and results in extremely poor control of the silicon surface concentration. All of the techniques that follow are essentially refinements in the manner in which B_2O_3 is delivered to the wafer. The systems have the following points in common:

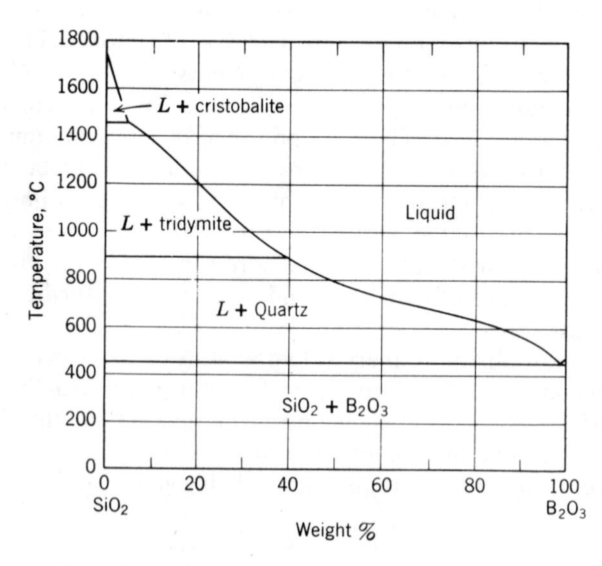

Figure 4.19 The B_2O_3-SiO_2 system. [From T. J. Rockett, and W. R. Foster, "Phase Relations in the System Boron Oxide-Silica," *J. Am. Ceram. Soc.*, **48**, 75 (Feb. 1965).]

1. They are of the open-tube type. A number of suitably masked silicon wafers are processed at a time by insertion into one end of the diffusion tube.

2. In many cases a preliminary reaction occurs in the diffusion tube at some intermediate temperature. The end product of this reaction is B_2O_3, which is then delivered to the silicon slices.

3. During the diffusion process, slices are exposed to the B_2O_3 source for a short period of time to form a glassy layer on the silicon surface. Once this is formed the source is removed and the main diffusion (drive-in) is carried out in the presence of an oxidizing ambient (dry or wet oxygen). This technique prevents the buildup of excessive amounts of impurity on the wafer, which may lead to the formation of silicides and other hard-to-remove compounds. In addition, the SiO_2 so formed prevents pitting of the silicon surface and leaves the slice with a protective layer at the conclusion of the diffusion. Diffusion is thus a two-step process, the actual diffusion profile being given by Figure 4.15.

With these features in mind, the following systems are now described:

Solid-source systems. Figure 4.20 shows a sketch of the diffusion system used for this method. Here a platinum boat is used to hold a source of boron trioxide upstream from the carrier with the silicon wafers. In operation the carrier gas transports vapors from the B_2O_3 source and deposits them on the silicon slices.

Figure 4.21 shows a typical experimentally derived[22] curve of surface concentration as a function of source temperature for this system. As seen from this curve, the source boat and the silicon slices may be conveniently maintained at the same temperature, avoiding the need for a two-zone furnace.

Predeposition is carried out in an inert gas such as nitrogen. The addition of 2–3% oxygen to the carrier gas during this phase is sometimes done because the resulting SiO_2 aids in the formation of the borosilicate

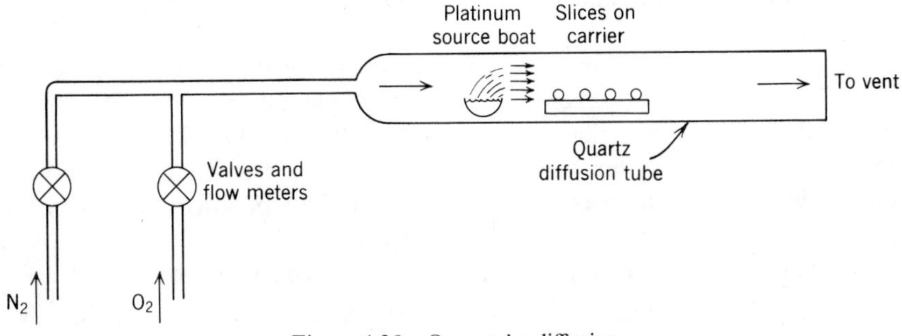

Figure 4.20 Open-tube diffusion.

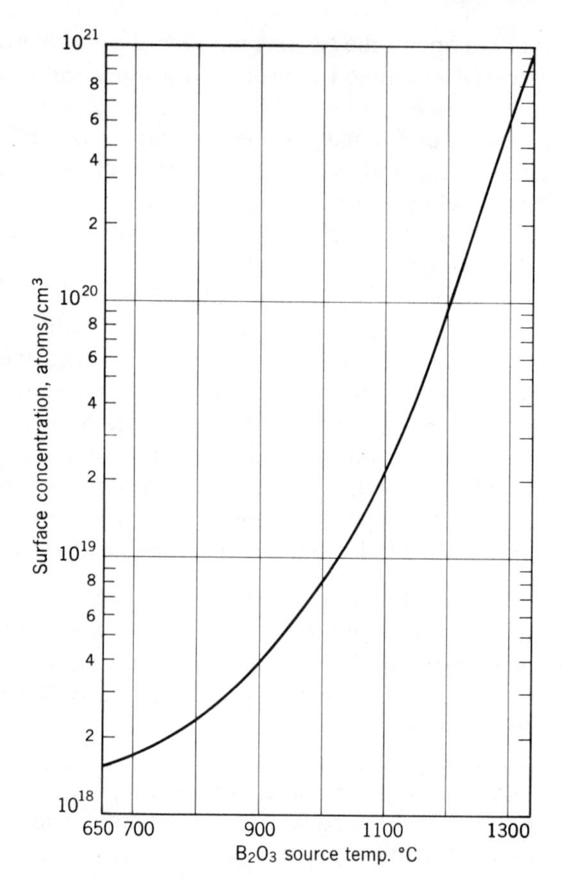

Figure 4.21 Surface concentration versus temperature for a B_2O_3 system (Kurtz and Yee[22]).

glass which protects the slices during diffusion. After predeposition, the source boat is removed, and drive-in proceeds in a strongly oxidizing atmosphere.

A refinement to this method consists of performing the predeposition in a closed platinum box[23]. The source is placed on the bottom of this box and the slices are placed on a quartz† shelf so that the B_2O_3 vapors must circulate in order to be deposited on their surface. The box is closed with a lid which is not gastight, and predeposition occurs in the presence of the carrier gas. After predeposition the wafers are removed from the box for the drive-in phase, which proceeds as before.

†A platinum shelf cannot be used in direct contact with the silicon, since the Pt-Si system has a eutectic temperature of 830°C.

The box method provides considerably greater uniformity than the open-source scheme, which is subject to gradient problems (i.e., wafers that are downstream receive less dopant than those closer to the source, and have a lower surface concentration). Highly reproducible results are obtained by this method and it is used by some microcircuit manufacturers. However, other techniques are preferred because of their greater convenience.

Liquid-source Systems. Here a volatile liquid is used as the dopant source. A carrier gas is bubbled through the liquid which is transported in vapor form to the surface of the silicon. It is common practice to saturate the carrier gas with the vapor so that the concentration is relatively independent of gas flow. Surface concentration is thus entirely set by the temperature of the bubbler and of the diffusion system.

Various liquid sources may be used in the system shown in Figure 4.22*a*. In each case a preliminary reaction results in the deposition of B_2O_3 on the silicon slices. Some of the liquid sources in use are the following:

1. *Trimethylborate(TMB)*—The preliminary oxidizing reaction is

$$2(CH_3O)_3B + 9O_2 \xrightarrow{900°C} B_2O_3 + 6CO_2 + 9H_2O. \tag{4.38}$$

The vapor pressure of TMB is usually controlled by refrigeration since it is extremely volatile at room temperatures.

2. *Boron tribromide.* Here the reaction is

$$4BBr_3 + 3O_2 \longrightarrow 2B_2O_3 + 6Br_2. \tag{4.39}$$

Since bromine is a reaction product, provision must be made for venting the system. In addition, there is a tendency for pitting to occur if excessive vapor concentrations are used (the so-called *halogen pitting*) or if the carrier gas is not sufficiently oxidizing in nature.

Gaseous Systems. These are even more convenient than the liquid. Although it is possible to control the surface concentration by adjusting the gas flow, it is common practice to conduct the predeposition in an excess gas concentration. Thus these systems are relatively insensitive to gas-flow rate. As with the liquid system, an initial reaction results in the formation of B_2O_3 which is transported on to the silicon slice where deposition and subsequent drive-in take place. Some of the gaseous sources in use are as follows:

1. *Diborane.* This is a highly poisonous, explosive gas. It is used in a 99.9% argon dilution by volume, which is reasonably safe to handle. The preliminary oxidizing reaction for the system is

$$B_2H_6 + 3O_2 \xrightarrow{300°C} B_2O_3 + 3H_2O. \tag{4.40}$$

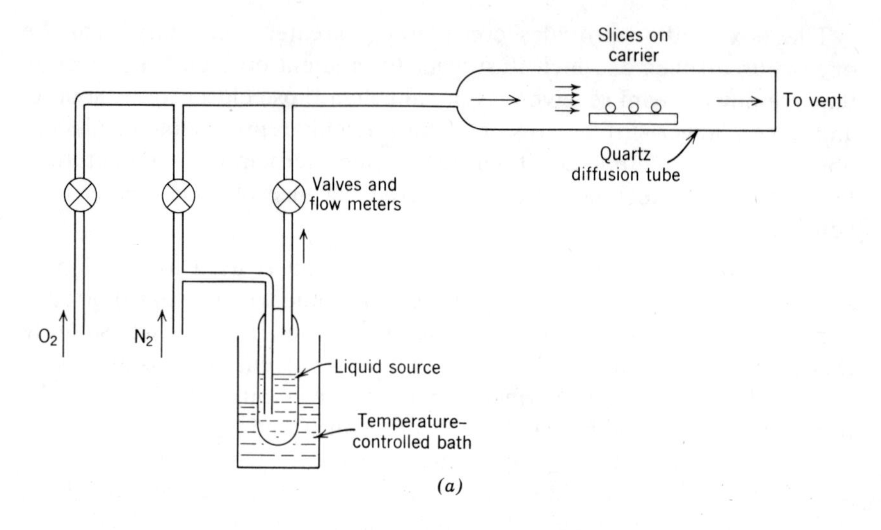

(a)

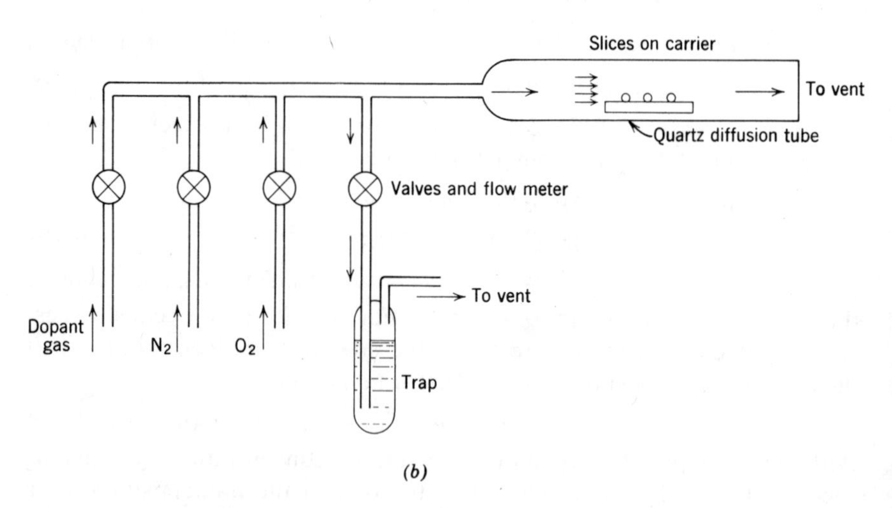

(b)

Figure 4.22 Liquid- and gaseous-source systems.

Since only water is released from this reaction, the method is not prone to the pitting effects experienced with boron tribromide. In addition, venting is not required. It is necessary however, to install an input trap with a weak HCl solution to capture any unused diborane in the gas lines. The system is shown in Figure 4.22*b*.

2. *Boron trichloride*. Here the preliminary reaction is

$$4BCl_3 + 3O_2 \longrightarrow 2B_2O_3 + 6Cl_2. \tag{4.41}$$

Again, halogen pitting is a problem at high surface concentrations. An alternate reaction that has been tried with BCl_3 is a reducing one in the presence of hydrogen. Thus,

$$2BCl_3 + 3H_2 \longrightarrow 2B + 6HCl. \tag{4.42}$$

The reproducibility of this system has been found to be poor in practice, which has been attributed to the absence of an intermediate boro-silicate glass. In addition, the resulting HCl and the silicon react in the absence of an oxidizing atmosphere to form volatile silicon tetrachloride, as follows:

$$4HCl + Si \rightleftharpoons SiCl_4 + 2H_2. \tag{4.43}$$

Halogen pitting of the silicon surface is caused in this manner.

Over-all results obtained by these various methods have been summarized in Table 4.5. In addition, experimentally noted features of the different systems are also listed.

4.7.1.2. Gallium Diffusion Systems. Successful gallium diffusion systems have been built with the use of gallium trioxide as the impurity compound. It has been found, however, that the diffusion must be carried out in the presence of a reducing atmosphere. A solid-source system is usually used, with the Ga_2O_3 held at the same temperature as the silicon slices. In operation it is presumed that Ga_2O_3 is reduced to the more volatile Ga_2O, which is then transported to the silicon surface. The reducing reaction for this situation is

$$Ga_2O_3 + 2H_2 \longrightarrow Ga_2O + 2H_2O. \tag{4.44}$$

As mentioned earlier, oxide masking cannot be used with gallium. Consequently, the status of gallium systems has not advanced as far as that for boron systems.

4.7.2. Choice of *n*-Type Impurity

The *n*-type impurities are phosphorus, antimony, and arsenic. None of these exhibits any undesirable characteristic that precludes their use in microcircuit fabrication. As seen from Table 4.4, they are all highly soluble in silicon and can be used to meet a wide range of diffusion requirements. Consequently, the choice of impurity depends on the specific application for which each is uniquely suited.

1. The diffusion constant of phosphorus is identical to that of boron and is about ten times larger than that of antimony or arsenic. It is

Table 4.5
A Comparison[24] of Boron Diffusion Systems

Original Impurity Source	Room Temperature State	Temperature range of Source During Impurity, °C	Impurity Concentration Range	Advantages	Disadvantages
Boric acid	Solid (converts to B_2O_3 upon heating)	600–1200	High and low	Readily available; proven source	Source contaminates tube; control is difficult
Boron tribromide	Liquid	10–30	High and low	Clean system; good control over wide range of impurity concentration; can use in nonoxidizing diffusion	Geometry of system is important
Methyl borate	Liquid	10–30	High	Simple to prepare and operate	Restricted to high surface concentration
Boron trichloride	Gas	Room temperature	High and low	Same as boron tribromide; accurate control by gas-metering equipment; easy installation and operation	
Diborane	Gas	Room temperature	High and low	Same as boron trichloride	Highly toxic

From A. M. Smith and R. P. Donovan, "Impurity Diffusion in Silicon," paper presented at the third Annual Microelectronics Symposium, St. Louis (April 1964).

used in the majority of diffusion applications, because it is both uneconomical and undesirable to operate diffusion systems for longer periods of time than necessary.

2. The low diffusion constant of antimony and arsenic make these impurities ideal for use in early phases of device fabrication. This is because they tend, once they are introduced into the lattice, to be relatively less affected by subsequent fabrication steps than are the faster substitutional diffusers such as phosphorus.

3. In applications where slow diffusers are required, antimony is preferred over arsenic because of its ease of handling. Most arsenic systems utilize highly poisonous, volatile sources and require elaborate precautions for safe handling. In addition, depletion of the arsenic source is often a serious problem during diffusion.

4. Arsenic systems are capable of considerably higher useful surface concentration than their antimony counterparts. This is because the arsenic atom provides an excellent fit to the silicon lattice.

4.7.2.1. Phosphorus Diffusion Systems. Phosphorus has a diffusion coefficient similar to that of boron. Its misfit factor is considerably lower than that of boron (0.068 as compared to 0.254) and sets an upper limit of 5×10^{20} atoms/cm^3 to the surface concentration that can actually be achieved in practical structures.

The most commonly used compound for phosphorus diffusion is phosphorus pentoxide. At diffusion temperatures the surface reaction from which diffusion proceeds is given by

$$2P_2O_5 + 5Si \rightleftharpoons 4P + 5SiO_2. \tag{4.45}$$

Phosphorus pentoxide readily combines with SiO_2 to form a phosphosilicate glass which is liquid at diffusion temperatures. The phase diagram for the SiO_2-$P_2O_5 \cdot SiO_2$ system is presented in Figure 4.23 and shows those regions that are relevant to the diffusion process.

In addition to the paint-on method, phosphorus diffusion can be conducted in a number of systems similar to those used for boron. A brief description of these follows.

Solid-source Systems. Because of its high vapor pressure it is necessary to maintain the phosphorus pentoxide source at temperatures between 200 to 300°C. This is usually done with a two-zone furnace, where the low-temperature zone is upstream from the diffusion zone. Considerations for the choice of carrier gases are similar to those for boron. Here, also, a two-step diffusion is the preferred method.

Phosphorus pentoxide is a hygroscopic solid, the vapor pressure of which is a strong function of its water content. Thus it is quite difficult

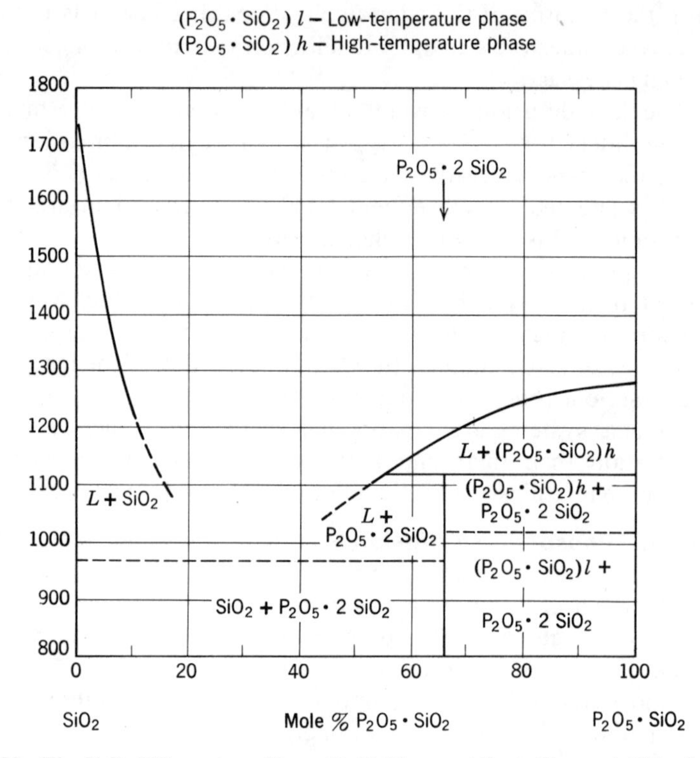

Figure 4.23 The P_2O_5-SiO_2 system. [From T. Y. Tien, and R. A. Hummel, "The System SiO_2-P_2O_5," *J. Amer. Ceramic Soc.*, **45**, 422 (1962).]

to obtain close control of surface concentration with the use of this compound. Other sources, such as ammonium monophosphate ($NH_4H_2PO_4$) and ammonium diphosphate $(NH_4)_2H_2PO_4$, do not suffer from this problem and are used at source temperatures from 450 to 900°C. Again the preliminary reaction appears to result in the delivery of P_2O_5 to the silicon slices.

A closed-box method has been used very successfully[25] with phosphorus diffusion. Here calcium oxide is added to the phosphorus pentoxide to allow operating of the source at diffusion temperatures. In addition, the mixture is relatively insensitive to moisture and to changes in composition. Since calcium is electronically inactive in silicon, its presence has no effect on the resulting device characteristics.

Liquid-source Systems. The most popular of these systems uses phosphorus oxychloride, $POCl_3$, in the temperature range of 0 to 40°C. An oxidizing gas mixture is used during the predeposition phase, resulting in the formation of P_2O_5 at some point in the system before the diffusion zone. The preliminary reaction is

$$4POCl_3 + 3O_2 \longrightarrow 2P_2O_5 + 6Cl_2. \tag{4.46}$$

The presence of oxygen during the predeposition phase aids in preventing halogen pitting effects, which are noticeable only at surface concentrations in excess of 10^{21} atoms/cm^3. The system is free from moisture problems and allows a high degree of control of surface concentration by adjusting the bubbler temperature.

Alternate liquid sources are phosphorus trichloride and phosphorus tribromide. These may be used with a nonoxidizing atmosphere if desired.

Gaseous Systems. The most commonly used gas is phosphine, which is both highly toxic and explosive. In dilute form (with 99.9% N_2) it is relatively safe to handle.

A slightly oxidizing carrier gas is used, the preliminary reaction being

$$2PH_3 + 4O_2 \longrightarrow P_2O_5 + 3H_2O. \tag{4.47}$$

As with the other reactions, P_2O_5 is delivered to the silicon slices. By-products of the reaction are water and the excess carrier gas (usually a nitrogen-oxygen gas mixture). Although it is not necessary to provide system venting to the output, an acidic CuCl trap must be used in the inlet gas lines to remove any unused gas. Characteristics of the system are quite similar to those of the diborane system.

A comparison of the properties of various phosphorus diffusion systems is made in Table 4.6.

4.7.2.2. Antimony Diffusion Systems. The diffusion coefficient of antimony in silicon is about one-tenth that of phosphorus. Consequently, its use is dictated for those cases in which it is important that the dopant be relatively immobile with further processing.

The misfit factor for antimony is 0.153. This is intermediate to that of boron and phosphorus; the useful surface concentration that can be attained is about 5×10^{19} atoms/cm^3.

The diffusion of antimony in silicon is usually performed from the trioxide (Sb_2O_3) or from the tetraoxide (Sb_2O_4). The tetraoxide has been used in solid-source systems at a temperature of about 900°C. Liquid-source systems have also been constructed, using antimony pentachloride (Sb_3Cl_5) in a bubbler arrangement. Stibine (SbH_3) is used as a source in gaseous systems. In all cases the antimony is transported to the silicon slices in the form of an oxide, and diffusion occurs from a glassy layer as a result of the surface reaction with the silicon.

4.7.2.3. Arsenic Diffusion Systems. Arsenic is similar to antimony in its diffusion characteristics and is thus used as an alternative to it. Its chief advantage over antimony lies in the fact that its tetrahedral radius is a perfect fit to the silicon lattice (i.e., the misfit factor is 0). Thus large amounts of arsenic can be introduced into the lattice without the generation of strain damage. This is expecially advantageous in the fabrication of intermediate low-resistivity layers on which it is required to grow additional single-crystal material. This process, known as epitaxy, is discussed in Chapter 5.

Arsenic diffusion is usually done from the reaction of arsenic trioxide (As_2O_3) and silicon. In solid-source systems a source temperature of 150 to 250°C is used. Gaseous systems using arsine (AsH_3) have also been operated. Their features are similar to those of the other gaseous systems previously described.

4.7.3. Gold Diffusion Systems[26]

The reasons for the choice of gold over other impurities exhibiting deep-lying levels have already been outlined in Section 4.3.2.1. Since gold is an extremely rapid diffuser (with a diffusion coefficient about five orders of magnitude larger than that of boron or phosphorus), it is usually the last impurity to be introduced into the microcircuit. Sometimes the gold diffusion and emitter drive-in are carried out simultaneously to obtain a saving in process time.

Gold diffusion is usually carried out from the element, which is vacuum-evaporated on the silicon in a layer about 500 Å thick. Diffusion proceeds from a liquid gold-silicon alloy, as seen by the phase diagram of Figure 3.6, and results in silicon damage to a depth of many microns. For this reason, gold diffusion is always performed from the back of the wafer, remote from the actual microcircuit.

Diffusion is normally carried out in the temperature range from 800 to 1050°C. The concentration is controlled by means of this temperature while the time (about 10 to 15 min) is more than enough to cause the gold to diffuse through the entire microcircuit. The precise diffusion temperature is selected on a cut-and-try basis to obtain the desired end result.

Gold-doped silicon slices must be removed rapidly from the diffusion furnace and brought to room temperature in a manner that is repeatable from run to run. The repeatibility of this quenching cycle is important in ensuring consistent results because out-diffusion effects are significant with fast diffusers.

Table 4.6
A Comparison[24] of Phosphorus Diffusion Systems

Original Impurity Source	Room Temperature State	Temperature Range of Source During Diffusion, °C	Impurity Concentration Range	Advantages	Disadvantages
Red phosphorus	Solid	200–300	Low ($<10^{20}$ cm^{-3})	Low surface concentration	Variable composition and vapor pressure
Phosphorus pentoxide	Solid	200–300	High($>10^{20}$ cm^{-3})	Proven source for high surface concentration	Sensitive to water vapor; requires frequent tube cleaning
Ammonium phosphate	Solid	450–1200	High and low	Avoids water vapor dependence	Purification is marginal
Phosphorus oxychloride	Liquid	2–40	High and low	Clean system; good control over wide range of impurity concentrations	Geometry of system is important
Phosphorus trichloride	Liquid	170	High and low	Same as phosphorus oxychloride; can be used in nonoxidizing diffusion	
Phosphine	Gas	Room temperature	High and low	Same as phosphorus trichloride; accurate control by gas metering; easy installation and operation	Highly toxic

From A. M. Smith and R.P. Donovan, "Impurity Diffusion in Silicon," paper presented at the Third Annual Microelectronics Symposium, St. Louis (April 1964).

4.8. SPECIAL PROBLEMS IN DIFFUSION

In this section are discussed a number of problems that arise in diffusion processes. Many of these have not be solved analytically; solutions for them are experimentally developed for each particular situation as it arises in practice.

4.8.1. Redistribution During Oxide Growth [27]

In microcircuit fabrication diffusions are made through windows cut in a mask on the surface of the silicon. The simplest, most convenient mask is one of silicon dioxide, grown out of the slice itself by subjecting it to an oxidizing gas at elevated temperatures. This mask may be grown in a separate step after a diffusion. Alternately, the two-step diffusion process may be used to grow it simultaneously with the drive-in phase. As mentioned in Section 4.5.3, this latter technique is superior to the first, because it not only provides immediate protection for the silicon surface but also saves an additional masking step. In addition, it minimizes the time the slice is maintained at elevated temperatures and, hence, reduces the undesired movement of already diffused impurities.

Common to both of these processes is the fact that some of the impurity-doped silicon is consumed to form an oxide. This effect has its parallel in the freezing of a doped crystal from a melt, as described in Chapter 2, and results in a redistribution of impurities. The extent of this redistribution is a function of the rate at which the silicon is consumed and of the relative diffusion coefficients and solid solubilities of the impurities in silicon and silicon dioxide.

The ratio of the equilibrium concentration of the impurity in silicon to its equilibrium concentration in the oxide is the distribution coefficient for the impurity in the Si-SiO_2 system. Experimentally derived values[28] for this distribution coefficient k are about 0.3 for boron, 20 for gallium, and about 10 for phosphorus, antimony and arsenic.

For the case in which $k < 1$ the growing oxide takes up the impurity from the silicon. Thus for boron diffusion, impurity depletion occurs on the silicon side of the interface. The extent of this depletion also depends on the rate at which the boron diffuses through the SiO_2 layer to its surface and escapes into the gaseous ambient. The diffusion rate in SiO_2 is orders of magnitude below that in silicon; hence the impurity depletion is dominated by distribution effects.

For the case in which $k > 1$ the growing oxide rejects the impurity. If the diffusion through the oxide is slow, as is the case for phosphorus, this results in a pileup of the impurity at the silicon surface . If, however, the impurity diffuses rapidly through the oxide (as is the case with gallium), the net result can still be an impurity depletion.

Finally, even if $k = 1$, the process of oxidation results in impurity depletion in the silicon. This is because the silicon roughly doubles in volume on forming silicon dioxide. Thus diffusion will occur from the highly doped silicon into the relatively lightly doped oxide.

Figure 4.24 shows the equilibrium concentrations of impurity that may occur in uniformly doped silicon in the presence of a growing oxide surface. For practical diffusions the problem is further complicated by the fact that the concentration is not uniform. In addition, the impurity is both diffusing into the silicon and into the growing oxide at the same time. An analytical solution is available for the uniformly doped case but not for the practical case of simultaneous in-diffusion and oxide growth.

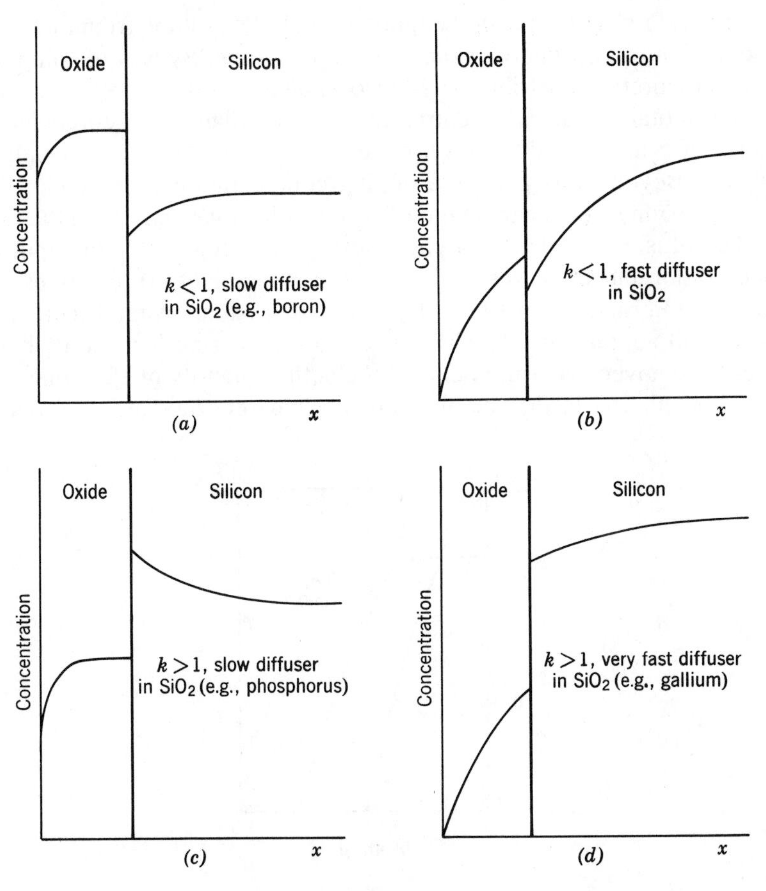

Figure 4.24 Redistribution during oxidation.

In practice the pileup effect in phosphorus diffusion is relatively small and may be ignored. This is caused by the fortuitous cancellation of the distribution and diffusion effects under typical processing conditions.

The effect of redistribution cannot be ignored for boron, however. In a typical practical situation involving a 45-min drive-in at 1200°C (15 min in dry oxygen followed by 30 min in wet oxygen) the surface concentration is approximately 50% of its value if redistribution effects were absent. In practice such departures from theoretical predictions are usually handled with the aid of correction curves developed for the specific conditions under which the diffusion run is made. These methods are discussed in Section 4.9.4.

4.8.2. Shallow Diffusion

Figure 4.25 shows an actual doping profile[29] resulting from the shallow diffusion of phosphorus from a constant source. By way of comparison the theoretical distribution is also shown for this case. It is seen that the actual surface concentration is lower than the theoretical by a factor of 5 and that diffusion for the first 0.5 μm is quite independent of depth. Beyond this point the doping concentration drops off in the expected manner. One explanation for this behavior is based on the fact that the diffusion coefficient of phosphorus is not constant with impurity concentration but increases by as much as a factor of 10 to 1 over the concentration range from 10^{20} to 10^{21} atoms/cm^3, as shown in Figure 4.6.

A second anomalous effect is noted if drive-in is performed after the source is removed. Measurements of the actual quantity of electronically active phosphorus in the sample show an apparent increase as a result

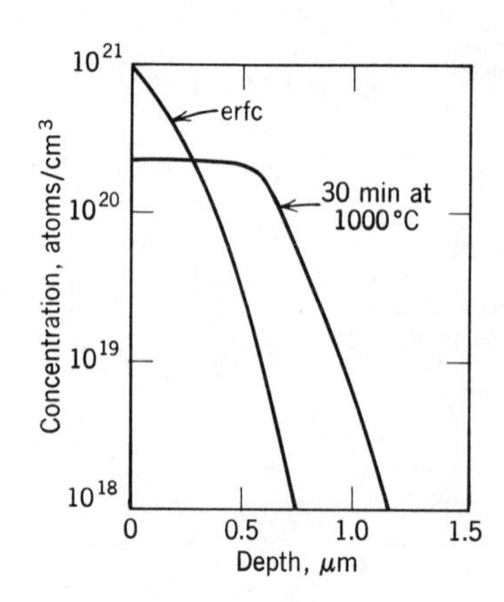

Figure 4.25 Shallow diffusion profile (Tannenbaum[20]).

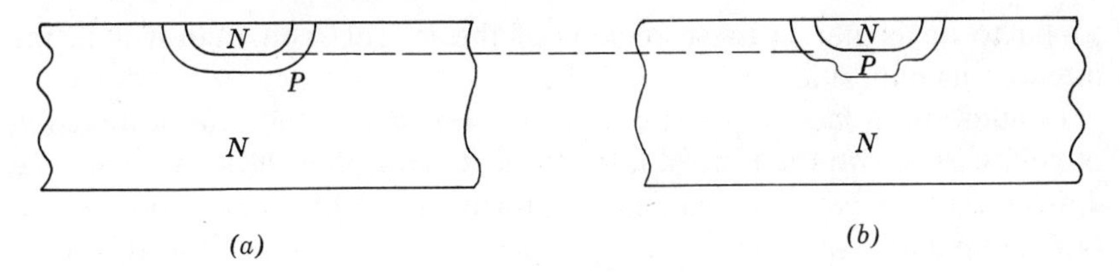

Figure 4.26 The EDE effect.

of this process! Thus a second explanation can be based on considerations of misfit factor. At very high concentrations some of the phosphorus is inactive. On drive-in this phosphorus moves into active sites as the concentration falls.

It is generally felt that both of these explanations contribute to the anomalous behavior of highly doped, shallow diffusions of the type used for the emitter regions of some high-speed microcircuits.

4.8.3. Enhanced Diffusion under the Emitter[30]

This phenomenon, sometimes referred to as EDE or the emitter-push effect, results in a transistor the cross section of which is shown in Figure 4.26. It has recently occupied a considerable amount of attention, because it is particularly apparent in shallow structures of the type used in high-speed digital circuits.

The EDE effect has only been observed in structures in which the emitter surface concentration is in excess of 10^{20} atoms/cm^3. Thus it would appear that field-induced diffusion is necessary for this phenomenon to occur.

No completely satisfactory theory has been offered for this effect. It is generally felt, however, that the phenomenon is caused by the combined effects of the emitter and base concentrations on the diffusion coefficient of the base impurity (boron).

It has been experimentally found that the EDE effect is enhanced by slow cooling after the emitter diffusion, and reduced by quenching. It is customary to take advantage of this fact during emitter diffusion.

4.8.4. Preferential Doping with Gold[31]

As described in Section 4.3.2.1, the diffusion coefficient of gold is a strong function of the defect content of the silicon into which it is introduced. Consequently, gold diffuses preferentially in those regions of the slice which have a high defect concentration, that is, in the region in which the microcircuit is actually fabricated. Even though gold is introduced from the opposite side of the wafer, its concentration is often found to be higher in these regions of the microcircuit than it is in the intervening bulk silicon.

In addition, it has been experimentally noted that the rate of diffusion of gold varies with the type of layer through which it is diffusing. Thus the diffusion of gold is inhibited in moving through a highly doped boron layer. No completely satisfactory reason has been advanced for this phenomenon.

4.9. THE EVALUATION OF DIFFUSED LAYERS

A number of techniques are available for evaluating the properties of a diffused layer. Perhaps the most straightforward method is to conduct the diffusion with a radioactive isotope of the impurity along with the impurity itself. The evaluation of the final diffused sample consists of measuring its integrated radiation intensity as successive layers are removed, usually by chemical etching. In this manner the detailed nature of the diffusion profile can be obtained for the entire impurity content, under the assumptions that the isotope atoms disperse uniformly among the normal atoms of the diffusing impurity and that the diffusion coefficient of the isotope is the same as that of the impurity.

The primary disadvantage of this technique is that it is not suited for the rapid on-line monitoring of diffusion processes during microcircuit fabrication. In addition, the method gives information on the entire impurity content and not on the electronically active part with which the engineer is concerned.

The properties of the diffused layer are readily determined by the *p-n* junction method. Here the dopant is diffused into a slice of known background concentration but opposite impurity type, and an evaluation is undertaken of the *p-n* junction so formed. This method gives results which bear directly on the end goal of the diffusion process, and is thus in common use. Details of the evaluation technique now follow.

4.9.1. Sheet Resistance

It is difficult to specify the specific resistivity of a diffused layer because it is inhomogeneous. For a layer of this type a sheet resistance is more appropriate.

Consider a rectangular layer of diffused material of length l, width w, and thickness t. If its resistance is measured across the faces of width w and thickness t, then

$$R = \frac{\rho(t)l}{wt} \tag{4.48}$$

where $\rho(t)$ is the specific resistivity of the material in ohm-centimeters and varies with depth. This equation may be more conveniently rewritten in the form

$$R = \rho_{\square} l/w \tag{4.49}$$

where $\rho_{\square}$ is defined as the sheet resistance of the layer, in ohms. If the layer takes the form of a square sheet, its resistance is given by $\rho_{\square}$, regardless of its actual dimensions. Hence the sheet resistance is usually specified in "ohms per square."

In microcircuit fabrication, the sheet resistance of a diffused layer is most conveniently obtained by the direct measurement of a resistor which is fabricated as part of a test pattern. About four or five such test patterns are strategically located on each silicon slice; these provide for a number of different experiments which are used to monitor the progress of the circuit during various fabrication steps.

Figure 4.27 shows a test pattern by which a measurement can be made of the sheet resistance of the base-diffusion (p-type) layer on an n-type substrate. After the p-type diffusion is conducted, a constant current is applied across the points AB, and the voltage developed across CD is read with the aid of a high-impedance voltmeter. With reference to Figure 4.27,

$$\rho_{\square} = \frac{V}{I}\frac{w}{l}. \tag{4.50}$$

Since w/l is known for a specific test pattern, the sheet resistance is directly found.

Figure 4.28 shows the manner in which the sheet resistance of the emitter diffusion (n-type) is determined. Here an initial p-type region is

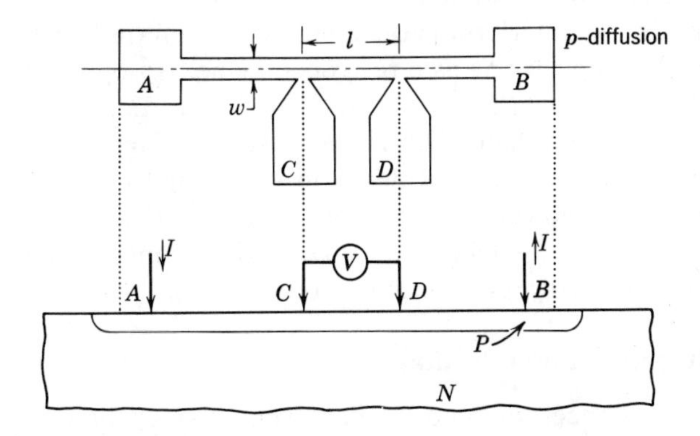

Figure 4.27 Base sheet-resistance test pattern.

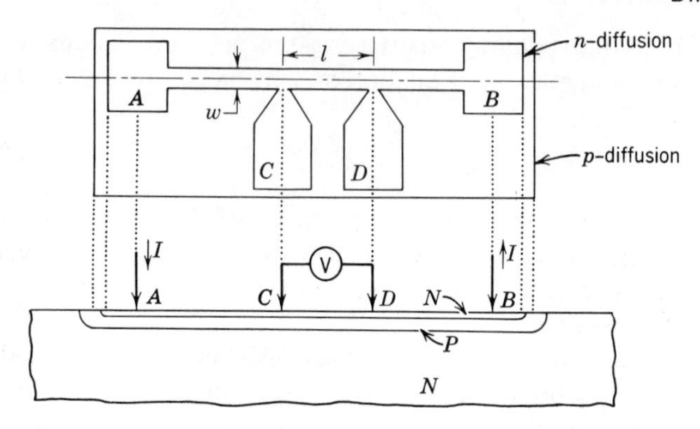

Figure 4.28 Emitter sheet-resistance test pattern.

formed during the base-diffusion step, and the n-type emitter diffusion is made within it. The sheet resistance of this layer is measured as described for the base diffusion.

4.9.2. Junction Depth

To determine the junction depth a small chip of the diffused slice is lapped on an angle so as to expose the actual junction. This angle is on the order of $\frac{1}{2}$ to 1°, so that the junction region is visually magnified. Next the junction is delineated by means of a selective etch. A large number of such etches are in use, a common one consisting of a solution of copper sulfate in dilute hydrofluoric acid. In operation the acid serves to dissolve surface oxides; the copper sulfate selectively plates the region so as to delineate the junction.

Junction-depth measurements based on geometric considerations can lead to considerable error if the lapping angle is not precisely known. Consequently, an interferometric method is preferred, with the upper surface of the chip serving as a reference plane. An optical flat is placed on this chip, as shown in Figure 4.29, and is vertically illuminated by collimated monochromatic light. The resulting fringe pattern gives a direct measure of the vertical depth in wavelengths of the illuminating source. For a sodium vapor lamp whose spectral radiation is concentrated at 5895.93 Å and 5889.96 Å (the D_1 and D_2 lines respectively) the distance between fringes is approximately 0.29 μm.

4.9.3. Surface Concentration

The surface concentration can be determined from the sheet resistance and the junction depth. With reference to the diffusion profile of Figure

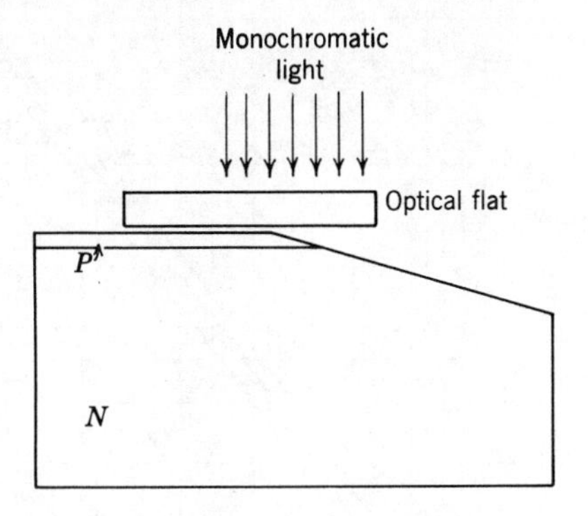

Figure 4.29 Junction-depth measurement.

4.30, the average conductivity of the layer is given by

$$\bar{\sigma} = \frac{1}{\rho_{\square} x_j} = \frac{1}{x_j} \int_0^{x_j} q\mu(x)\,[N(x) - N_C]\,dx, \tag{4.51}$$

where $\mu(x)$ is the mobility for the majority carrier type in the diffused layer and is a function of the concentration at any given depth.

With the use of appropriate values for the mobility, computer solutions [32] may be derived for both p- and n-type layers with erfc and gaussian profiles in silicon for different values of background concentration. Figures 4.31 and 4.32 show curves obtained in this manner for p-type

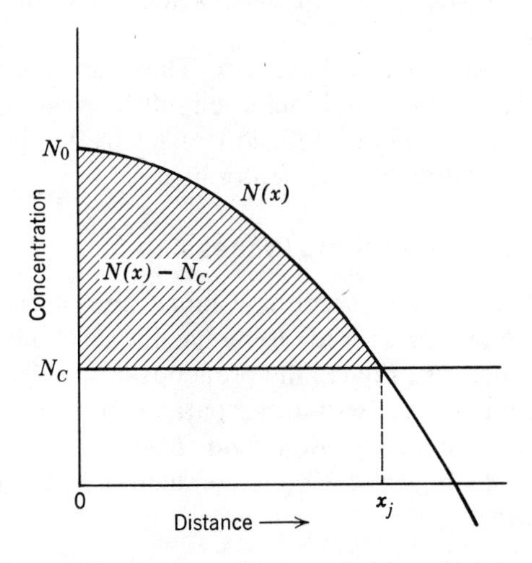

Figure 4.30 Doping profile for resistivity calculations.

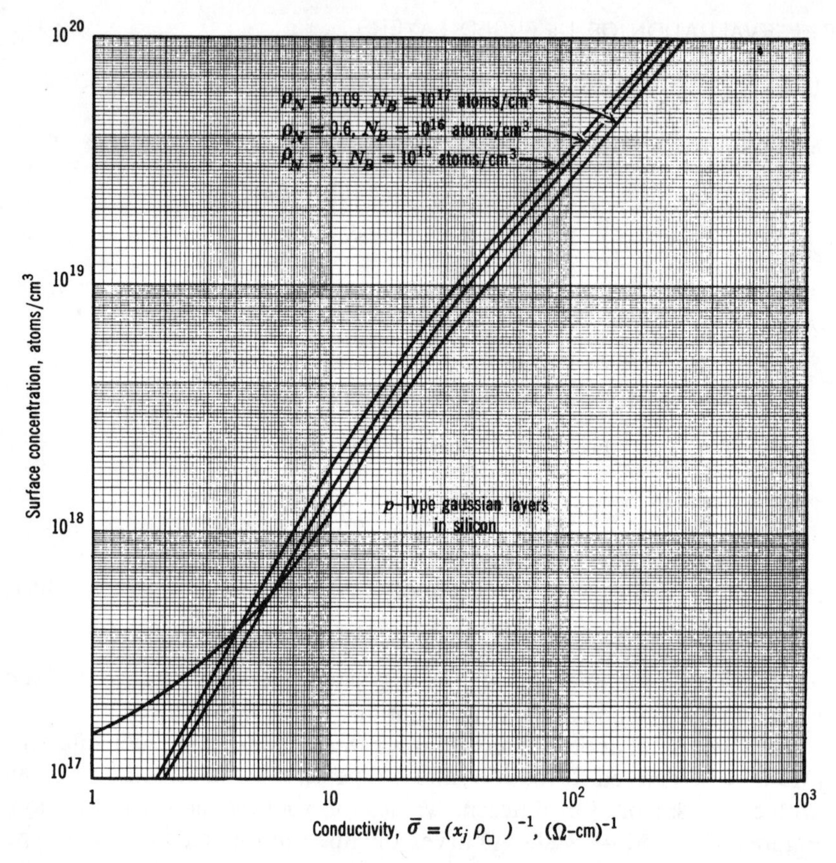

Figure 4.31 Resistance of p-type gaussian layers (adapted from Irvin[32]).

gaussian layers and n-type erfc layers. These are the most commonly encountered diffusion profiles in microcircuit fabrication and are representative of base and emitter diffusion respectively. With these curves the surface concentration may be determined.

4.9.4. Slice Evaluation During Processing

In practice, the diffusion process is monitored on test slices with the aid of a series of graphs prepared for the specific set of diffusion conditions. The first of these graphs relates the predeposition sheet resistance to the predeposition temperature, with other parameters (such as predeposition time and background concentration) held constant.

With reference to Figure 4.30, the conductivity of the layer after predeposition is given by

$$\bar{\sigma}_1 = \frac{q}{x_{j1}} \int_0^{x_{j1}} \mu(x)\,[N_1(x) - N_C]\,dx \tag{4.52}$$

where the subscript 1 refers to the predeposition phase. Making the approximation that mobility is independent of the impurity concentration and that $N_1(x) \gg N_C$ over most of the range of integration, (4.52) reduces to

$$\bar{\sigma}_1 \, x_{j1} \simeq q\mu 2N_{01}\sqrt{D_1 t_1/\pi}. \tag{4.53}$$

But

$$D_1 = D_{01} e^{-E/kT_1}. \tag{4.54}$$

Substituting into (4.53) gives

$$\bar{\sigma}_1 \, x_{j1} = 2q\mu N_{01}\left(\frac{D_{01}t_1}{\pi}\right)^{1/2} e^{-E/2kT_1}. \tag{4.55}$$

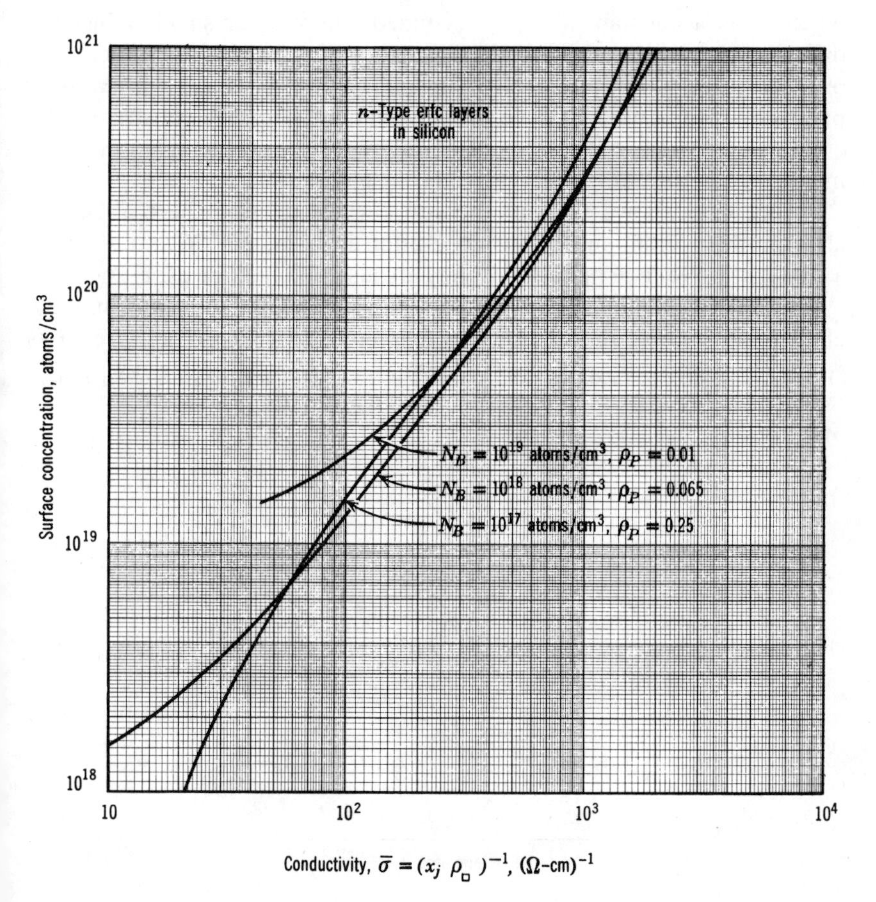

Figure 4.32 Resistance of n-type erfc layers (adapted from Irvin[32]).

The junction depth after predeposition is given by [see (4.36)]

$$x_{j1} = 2\sqrt{D_1 t_1}\,\mathrm{erfc}^{-1}\frac{N_C}{N_{01}}, \tag{4.56}$$

$$= 2\sqrt{D_{01} t_1}\,\mathrm{erfc}^{-1}\frac{N_C}{N_{01}}\quad e^{-E/2kT_1}. \tag{4.57}$$

The sheet resistance is given by

$$\rho_\square = \frac{1}{x_{j1}\bar{\sigma}_1} = Ae^{E/2kT_1}, \tag{4.58}$$

where A is a constant if N_{01} is assumed constant for small changes in diffusion temperature.† Thus the logarithm of the sheet resistance after predeposition is approximately a linear function of the reciprocal temperature. In practice, a curve of this relation is readily constructed by conducting two or three test runs at different diffusion temperatures and measuring the resulting sheet resistance in each case.

Figure 4.33 shows a set of drive-in curves, relating sheet resistance after predeposition, sheet resistance after drive-in, and drive-in time for a constant drive-in temperature. They are also obtained by experiment.

It is customary to check the diffusion process after predeposition and after about 90% of the drive-in is completed. At each point corrections

†This is a reasonable approximation, as is seen from the solid solubility curves of Chapter 3.

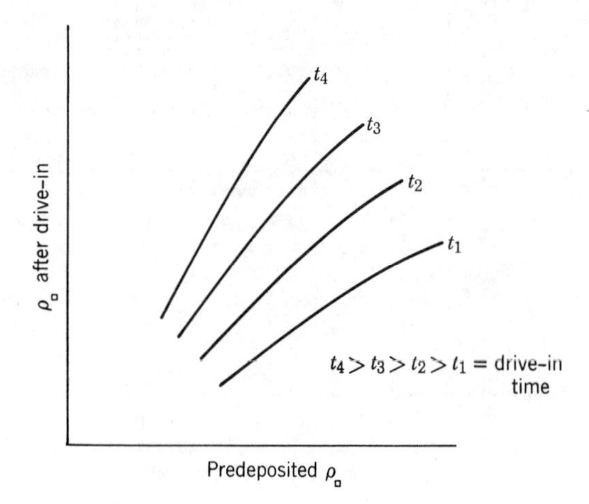

Figure 4.33 The effect of drive-in.

can be made to the process with the aid of the curves described here, in conjunction with the curves of Figures 4.31 and 4.32, and a knowledge of the junction depth.

4.10. REFERENCES

[1] B. I. Boltaks, *Diffusion in Semiconductors,* Academic, New York, 1963.
[2] W. Jost, *Diffusion in Solid, Liquid and Gases*, Academic, New York, 1962.
[3] N. B. Hannay (Ed.), *Semiconductors,* Reinhold, New York, 1960.
[4] C. S. Fuller, and J. A. Ditzenberger, "Diffusion of Donor and Acceptor Elements in Silicon," *J. Appl. Phys.*, **27**, 544–553 (1956).
[5] E. M. Pell, "Diffusion Rate of Lithium in Silicon at Low Temperatures," *Phys. Rev.*, **119**, 1222–1225 (1960).
[6] R. O. Carlson et al., "Sulphur in Silicon," *J. Phys. Chem. Solids,* **8**, 81–83 (1959).
[7] J. D. Struthers, "Solubility and Diffusivity of Gold, Iron and Copper in Silicon," *J. Appl. Phys.*, **27**, 1409–1414 (Dec. 1957); also **28**, 516 (Apr. 1957).
[8] B. I. Boltaks, and I. I. Sosinov, "The Diffusion of Copper in Silicon," *Zh. Tech. Fiz.*, **28**, 679 (1958).
[9] B. I. Boltaks and H. Shih-yin, "Diffusion, Solubility and the Effect of Silver Impurities on Electrical Properties of Silicon," *Soviet Phys. Solid State,* **2**, 2303 (1961).
[10] C. Haas, "The Diffusion of Oxygen into Silicon," *J. Phys. Chem. Solids*, **15**, 108–111 (1960).
[11] Y. Tokomaru, "Properties of Silicon Doped with Nickel," *Jap. J. Appl. Phys.*, **2**, 542–547 (1963).
[12] C. S. Fuller, and F. J. Morin, "Diffusion and Electrical Behavior of Zinc in Silicon," *Phys. Rev.*, **105**, 379–384 (1957)
[13] W. R. Wilcox, and T. J. LaChapelle, "The Mechanism of Gold Diffusion in Silicon," *J. Appl. Phys.*, **35**, 240 (1964).
[14] R. C. T. Smith, "Conduction of Heat in the Simi-infinite Solid with a Short Table of an Important Integral," *Australian J. Phys.*, **6**, 127–130 (1953).
[15] F. M. Smits, and R. C. Miller, "Rate Limitation at the Surface for Impurity Diffusion in Semiconductors," *Phys. Rev.*, **104**, 1242–1245 (1956).
[16] W. R. Rice, "Diffusion of Impurities during Epitaxy," *Proc. IEEE*, **52**, 284–295 (1964).
[17] T. I. Kucher, "The Problem of Diffusion in an Evaporating Solid Medium," *Soviet Phys. Solid State,* **3**, 401–404 (1961).
[18] R. L. Batdorf and F. M. Smits, "Diffusion of Impurities into Evaporating Silicon," *J. Appl. Phys.*, **30** (2), 259–264 (1959).
[19] R. B. Allen, H. Bernstein and A. D. Kurtz, "Effect of Oxide Layers on the Diffusion of Phosphorus into Silicon," *J. Appl. Phys.*, 31, 334–337 (1960).
[20] D. P. Kennedy and R. R. O'Brien, "Analysis of the Impurity Atom Distribution Near the Diffusion Mask for a Planar *p-n* Junction," *IBM J. Res. Dev.*, **9**(3), 179–186 (May 1965).
[21] F. A. Trumbore, "Solid Solubility of Impurity Elements in Germanium and Silicon," *Bell System Tech. J.*, **39**(1), 205 (1960).
[22] A. D. Kurtz and R. Yee, "Diffusion of Boron in Silicon," *J. Appl. Phys.* **31**, 303–305 (Feb. 1960).
[23] L. A. D'Asaro, "Diffusion and Oxide Masking in Silicon by the Box Method," *Solid State Electron.* **1** 3–12 (1960).

[24] R. M. Burger and R. P. Donovan, Fundamentals of Silicon Integrated Device Technology—Volume 1, Prentice-Hall, New Jersey, 1967.

[25] I. M. Mackintosh, "The Diffusion of Phosphorus in Silicon," *J. Electrochem. Soc.*, **109**, 1065–1067 (Nov. 1962).

[26] C. J. Uhl, "A Gold Diffusion Process for Reducing Switching Times in Diffused Silicon Transistors," *The Western Elec. Eng.*, **7**, 18–24 (Jan. 1963).

[27] M. M. Atalla and E. Tannenbaum, "Impurity Redistribution and Junction Formation in Silicon by Thermal Oxidation," *Bell System Tech. J.*, **39**, 933–946 (July 1960).

[28] A. S. Grove et al., "Redistribution of Acceptor and Donor Impurities during Thermal Oxidation of Silicon," *J. Appl. Phys.*, **35**,(9), 2695–2701 (1964).

[29] E. Tannenbaum, "Detailed Analysis of Thin Phosphorus-diffused Layers in *p*-Type Silicon," *Solid-State Electron.*, **2**, 123–132 (Mar. 1961).

[30] R. Gereth et al., "Localized Enhanced Diffusion in NPN Silicon Structures," *J. Electrochem. Soc.*, **112**(3), 323–329 (1965).

[31] W. R. Wilcox et al., "Gold Diffusion in Silicon: Effect on Resistivity and Diffusion in Heavily Doped Layers," *J. Electrochem. Soc.*, **111**, 1137 (1964).

[32] J. C. Irvin, "Resistivity of Bulk Silicon and Diffused Layers in Silicon," *Bell System Tech. J.*, **41**(2), 387 (1962).

4.11. PROBLEMS

1. A diffusion furnace operating at 1000°C has a ±1°C tolerance. What is the corresponding tolerance on the diffusion depth, assuming a gaussian diffusion?

2. A p diffusion is made into an n region such that $N_0 \geqslant 1000N_C$. Show that the junction depth is approximately proportional to $\sqrt{Dt}$ and determine the factor of proportionality. A constant-source diffusion may be assumed.

3. The following diffusions are performed into n-type silicon with a background concentration of 3×10^{16} atoms/cm^3: (*a*) 15-min constant-source boron predeposition at 900°C, followed by a 30-min drive-in at 1100°C. (*b*) a 15-min constant-source phosphorus diffusion at 1100°C. Determine the surface concentrations and junction depths that result from this process.

4. Sketch the doping profiles of Problem 3 on semilog paper. Hence determine the net impurity gradients at the emitter and the collector in atoms/cm^4.

5. A base diffusion is made into n-type silicon ($N_C = 3 \times 10^{16}$ atoms/cm^3) using the same schedule as in Problem 3. Ignoring redistribution effects, determine the surface concentration and junction depth, and develop a schedule for a constant-source diffusion which will result in the same surface concentration and junction depth. Hence verify the statement that the base diffusion should be carried out from an instantaneous source.

6. Determine the effective diffusion coefficient of gold as a function of temperature. Highly intrinsic material may be assumed.

7. Gold doping is performed at 1000°C into what may be considered an infinitely thick slice of highly extrinsic silicon. Determine the diffusion time needed for the gold concentration to reach 50% of its surface concentration at a point 100 μm from the diffusing source. Thus verify the basic assumption used in setting the time for a practical gold diffusion.

Chapter 5

Epitaxy

Epitaxy is the process of setting down an amount of material on top of a crystalline substrate while still preserving the over-all single-crystal structure. As applied to microcircuit fabrication, it describes the process of growing a thin (5–15 μm) single-crystal layer of suitably doped silicon upon a single-crystal silicon substrate. The silicon layer is usually of different doping concentration and/or resistivity type to the substrate upon which it is grown. Thus epitaxial growth provides an alternative to diffusion as a process for fabricating appropriately doped semiconductor regions. Unlike diffusions, each new layer is quite independent of the last and not the result of a counterdoping process. Consequently, the total impurity concentration of successive layers can be maintained at reasonable levels and does not limit the number of layers that can be grown.

In microcircuits the substrate serves the function of a mechanical support member, with electronic processes confined to the grown layers. Consequently, a high degree of crystalline perfection is required of these layers. This places a limitation on their number, since crystal quality deteriorates with the formation of each successive layer. At the present time, microcircuit-fabrication processes are based on no more than one epitaxial growth step.

5.1. THE GROWTH PROCESS

Fundamental to this process is a means for transporting atoms of the material to be grown to the substrate on which epitaxy is desired. On arriving on this substrate, these atoms move around until they find a region to which they can attach. This movement is enhanced by the fact that the substrate is maintained at an elevated temperature.

Attachment of the arriving atoms occurs preferentially at a nucleation center. Crystal growth is initiated at a number of such centers and spreads laterally until a layer is completed. In principle fresh nucleation centers are formed on this layer and the process repeats. In this manner the epitaxial film is built up as a series of atomic planes. The growing layer follows the substrate plane sequence, since this results in an energy minimum for the system as a whole.

In practice it is impossible to confine the epitaxial growth to an atomic layer at a time. In fact, better crystal quality is obtained if the growth is performed on a slightly misoriented crystallographic plane. This encourages the nucleation process to occur at the step corners formed by the (111) plane surfaces and allows a higher deposition rate without the development of surface defects.

It is common practice to dope the layer during the growth process. As a result, the silicon atoms are accompanied by impurity atoms which are usually of different ionic radii. This may lead to disorder in the crystal structure of the epitaxial layer.

The atoms may be transported directly to the substrate by vacuum evaporation or sputtering. The most common technique, however, is an indirect process involving the hydrogen reduction of a silicon compound at the substrate surface. Silicon tetrachloride ($SiCl_4$) is generally used for this purpose, even though trichlorosilane ($SiHCl_3$) and silicon tetrabromide ($SiBr_4$) have also been used. An alternate method consists of the pyrolytic decomposition of silane (SiH_4) into silicon and hydrogen.

5.1.1. Chemistry of Growth

The hydrogen reduction of $SiCl_4$ is thought to proceed along the following lines:

1. A gas-phase mixture of H_2 and $SiCl_4$ is delivered into a quartz chamber in which the heated silicon slices are located. At some point upstream from these slices, partial reduction of the $SiCl_4$ takes place by means of the reversible reaction

$$SiCl_4 + H_2 \xrightleftharpoons{800\text{–}1000°C} SiCl_2 + 2HCl. \tag{5.1}$$

2. On arrival at the silicon slice, which is maintained at about 1250 to 1275°C, the $SiCl_2$ is adsorbed on its surface. The reaction taking place on the surface is given by

$$2SiCl_2 \longrightarrow Si + SiCl_4. \tag{5.2}$$

For this second reaction to take place, it is necessary that two molecules of $SiCl_2$ be involved in the presence of a host body (the substrate). Thus the reaction occurs on the surface of the silicon, and not in the gas phase surrounding it. In practice reactions occurring in the gas phase result in premature nucleation of the silicon atoms and give rise to polycrystalline growth.

3. The resulting $SiCl_4$ is desorbed into the gas stream, where it again undergoes reduction. The over-all reaction may be written as

$$SiCl_4 + 2H_2 \rightleftharpoons Si + 4HCl. \tag{5.3}$$

It is important to note that this is a reversible reaction, the end product of which is either the growth of a silicon layer or the etching of the substrate. The exact nature of this reaction is a function of the mole fraction of $SiCl_4$ in H_2. Experimental data on a typical system [1] are shown in Figure 5.1 and illustrates this effect. As seen from these data, etching occurs at concentrations in excess of 0.28 mole fraction of $SiCl_4$.

Impurity doping is done at the same time as the epitaxial growth. Although solid and liquid dopants can be used, gaseous diborane and phosphine are far more convenient. These are mixed with the other gases and dissociate to elemental phosphorus and boron in the vicinity of the slices.

An alternate process for epitaxial growth consists of the pyrolytic decomposition of silane, SiH_4. Here the reaction is given by

$$2SiH_4 \xrightarrow{1000°C} 2Si + 2H_2. \tag{5.4}$$

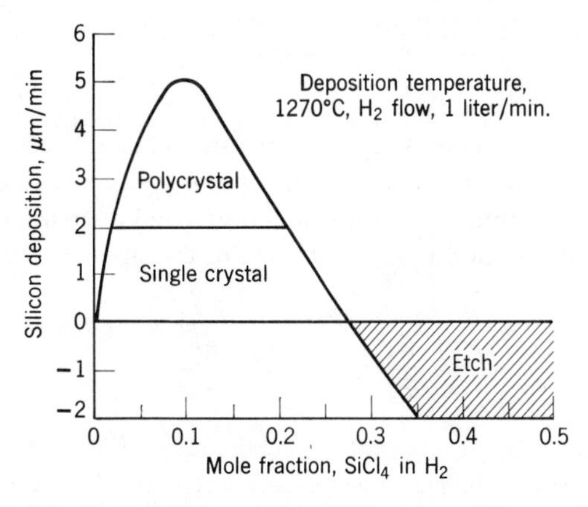

Figure 5.1 Deposition rate for the $SiCl_4$ process (Theuerer[1]).

Unlike the previous situation, this is not a reversible reaction. Furthermore the complete absence of hydrogen chloride gas as an end product avoids the necessity of elaborate venting. An additional advantage is that the process can be performed at relatively low temperatures, thus avoiding many of the out-diffusion problems that arise with the $SiCl_4$ process.

Although pyrophoric and extremely dangerous in high concentrations, silane is reasonably safe to handle in dilutions of 5% or less (with hydrogen gas as the diluent). Its chief disadvantage lies in the fact that it is extremely difficult to avoid the gas-phase nucleation of silicon particles when this process is used. Thus it usually results in polycrystalline growth. Although some success has been obtained in growing single-crystal layers by this process, it is generally not used at the present time.

5.1.2. Kinetics of Growth

The kinetics of growth may be determined by assuming that N_g is the concentration of the reacting species in the gas-phase, resulting in a surface concentration N_0 for this species. The flux arriving at the silicon surface is given by j, the number of molecules of reactant per unit area per unit time, where

$$j = h(N_g - N_0). \tag{5.5}$$

Here h is the gas-phase mass-transfer coefficient, and it is assumed that Henry's law† holds for this system.

If the reaction at the silicon surface is assumed to be linear, steady-state growth conditions demand that the flux at the surface of this growing film be given by

$$j = kN_0, \tag{5.6}$$

where k is the surface reaction rate constant. This is a reasonable assumption since the growth rate (about 1 μm/min) is relatively low.

Writing n as the number of atoms in a unit volume of silicon ($n \simeq 5 \times 10^{22}$ atoms/cm^3) and combining (5.5) and (5.6), the growth rate of the layer is given by

$$\frac{dx}{dt} = \frac{j}{n} = \frac{N_g}{n} \frac{hk}{h+k}. \tag{5.7}$$

†Henry's law states that the mole fraction of a species that is dissolved in a gas at any temperature is proportional to the partial pressure exerted by that species at that temperature.

The reaction rate constant k varies exponentially with temperature. Consequently, growth is limited by this term at low temperatures. The limiting growth rate for this condition is

$$\frac{dx}{dt} = \frac{kN_g}{n}. \tag{5.8}$$

At high temperatures, epitaxial growth is limited by the gas-phase mass-transfer coefficient, which is temperature independent. The limiting growth rate of the layer is given for this condition by

$$\frac{dx}{dt} = \frac{hN_g}{n}. \tag{5.9}$$

It should be recognized that a very simple model has been presented here which involves many linearizing assumptions. Nevertheless, it serves to explain the experimentally observed characteristics [2] of epitaxial growth, as shown in Figure 5.2.

5.2. EPITAXIAL SYSTEMS AND PROCESSES

Epitaxial growth is carried out in a vessel called a *reactor*. Two types of reactors are in common use, the simplest taking the form of a hori-

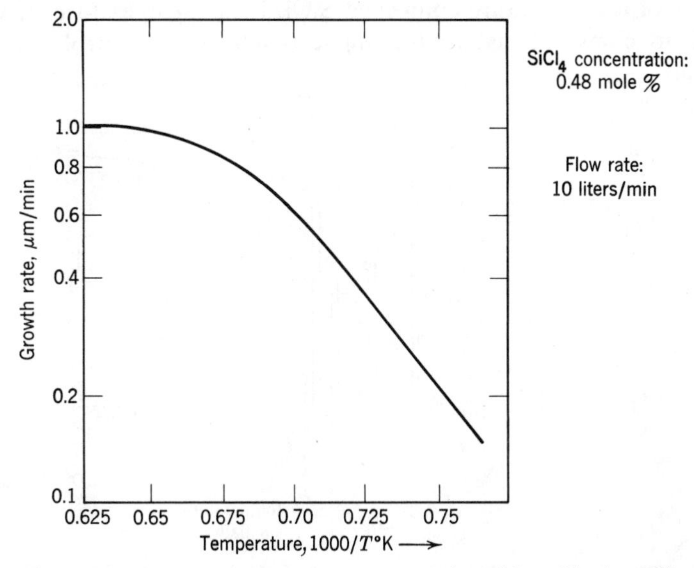

Figure 5.2 Growth rate versus temperature (adapted from Shepherd [2]).

zontal quartz tube. In this reactor the flow of gases is at right angles to the axis of the wafers. Thus some wafers are downstream from the gas source, yet others are upstream, resulting in a variation of growth rate from slice to slice. The problem is reduced, however, by tilting the slice carrier, as shown in Figure 5.3.

The vertical reactor avoids this problem by making the gas flow parallel to the axes of the slices. In addition, the slice holder is capable of rotation during the growth. Although greater uniformity has been claimed for this system, the simplicity of the horizontal reactor makes it the more commonly used type at the present time.

Figure 5.3 shows a schematic diagram for both horizontal and vertical reactor systems. Common to both systems are the following features:

1. In order for the reaction to take place on the silicon slices, it is necessary to maintain the walls of the reactor relatively cool. Thus r. f. or induction heating is used. This has the added advantage of establishing a thermal gradient from the slices toward the reactor wall. The transport of foreign matter from the walls to the slices is avoided in this manner.
2. Slices are placed on an electrically conductive *susceptor* which is heated by the r. f. The commonest form of susceptor is a graphite block, which is coated with one or more layers of pyrolytic graphite, silicon carbide, boron nitride, or quartz to prevent contamination of the reactor vessel.
3. The $SiCl_4$ is introduced into the system by passing H_2 gas over the surface of a temperature-controlled $SiCl_4$ bath. Temperatures from 0 to −30°C are commonly used in this application to obtain control of the vapor pressure.

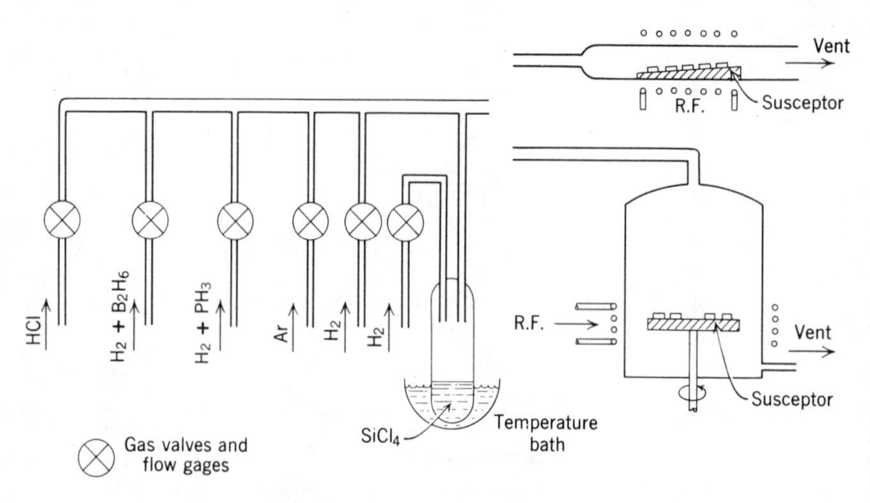

Figure 5.3 Epitaxial reactor systems.

4. Dopant gases are also provided. These are usually diborane (B_2H_6) and phosphine (PH_3). In contrast with diffusion techniques, reactor doping is carried out to the total exclusion of oxygen or nitrogen. This is necessary to avoid the formation of any silicon nitrides or oxides during the growth of the layer.

5. Anhydrous HCl gas is also provided. This is used in the cleaning steps before growth.

6. Adequate venting of the system is provided in the various lines as well as in the exit port.

5.2.1. Pregrowth Cleaning and Etching

Perhaps the single most important requirement for good epitaxial growth is that the layer be grown on a completely damage-free, oxide-free surface. Although nothing can be done to reduce the impurity-induced defects that are always present in a doped semiconductor, every effort is made to eliminate other sources of defects. Thus the silicon slices are mechanically lapped and chemically polished to remove, as much as possible, the regions of surface damage. After chemical polishing, slices are inserted into the reactor and flushed in dry hydrogen gas at a temperature of about 1200°C. This serves to reduce all traces of oxide that may have been formed while transferring the slices into the reactor.

Once the hydrogen cleanup is over, anhydrous HCl is introduced into the system. Etching proceeds by the conversion of surface silicon to volatile $SiCl_4$ by a reaction of (5.3). However, the excess HCl concentration leads to etching of the silicon slice.

Etching is performed in an HCl-H_2 gas mixture. Correct adjustment of this mixture results in extremely high-quality, optically flat finishes. A 1 to 5% mole concentration of HCl is typical, and results in an etch rate of about 0.5 to 2 μm/min.

The success of the etching process is a critical function of the quality of the HCl gas. Thus traces of water, nitrogen, or methane in the gas can lead to the formation of oxides, nitrides, and carbides of silicon respectively, which, in turn, serve as sites for the initiation of defects in the epitaxial layer.

Alternate etching techniques have also been proposed. The commercial availability of semiconductor-grade HCl gas, however, has resulted in the almost universal adoption of this technique.

5.2.2. Growth Parameters

The growth of the epitaxial layer is determined by a number of parameters which are adjusted individually for each system. These parameters are now considered briefly.

5.2.2.1. *Bubbler Temperature.* Figure 5.4 shows the vapor pressure of $SiCl_4$ as a function of bubbler temperature. In practice a bubbler temperature of 0°C is preferred because it can be accurately maintained with ease.

5.2.2.2. *Silicon Tetrachloride Concentration.* Hydrogen gas is fed to the system, partly through the bubbler and partly bypassing it. The combination of the flow rates is adjusted to give the desired concentration of $SiCl_4$ in the gas phase. The rate of layer growth with $SiCl_4$ concentration has been shown for a typical system in Figure 5.1. Note that etching can occur at high concentrations. In addition, polycrystalline growth results over a range of values of $SiCl_4$ concentration, corresponding to excessively high growth rates. In practice a growth rate of about 1.0 μm/min has been found to result in reasonably defect-free single-crystal layers.

5.2.2.3. *Substrate Temperature.* The growth rate of the epitaxial layer as a function of temperature has been shown in Figure 5.2 for a specific set of experimental conditions. Here, both the mass-transfer limited and surface-reaction-rate limited regions are seen to occur. In typical operation, substrate temperatures are maintained around 1250 to 1300°C.

5.2.2.4. *Doping Concentrations.* Doping can be carried out by adding liquid dopant sources to the $SiCl_4$. A more flexible system results, however, from the use of gaseous sources independently fed to the reactor after being diluted with H_2. In epitaxial systems the resistivity of the grown layer is varied by controlling the dopant concentration in the gas phase. Thus accurate control of the gas flow rate is critical to successful doping.†

†In contrast the impurity concentration in diffused layers is usually set by the solid solubility of the dopant at the diffusion temperature. Thus the temperature control is critical in diffusion, which gas flow is not.

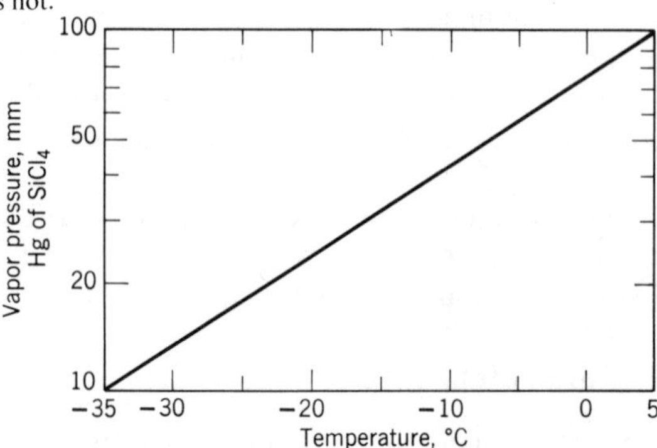

Figure 5.4 Vapor pressure of $SiCl_4$ versus temperature.

Figure 5.5 shows typical results for doping concentration in the epitaxial layer as a function of the ratio of gas phase impurity of silicon that is used. Curves of this type are experimentally determined for each system when it is put into operation.

5.3. REDISTRIBUTION DURING GROWTH

Redistribution of impurities occurs during epitaxial growth, resulting in departures from the ideal, abruptly discontinuous doping profile. Two different redistribution mechanisms will be considered independently in the following sections. It is understood, however, that they are present simultaneously during epitaxial growth.

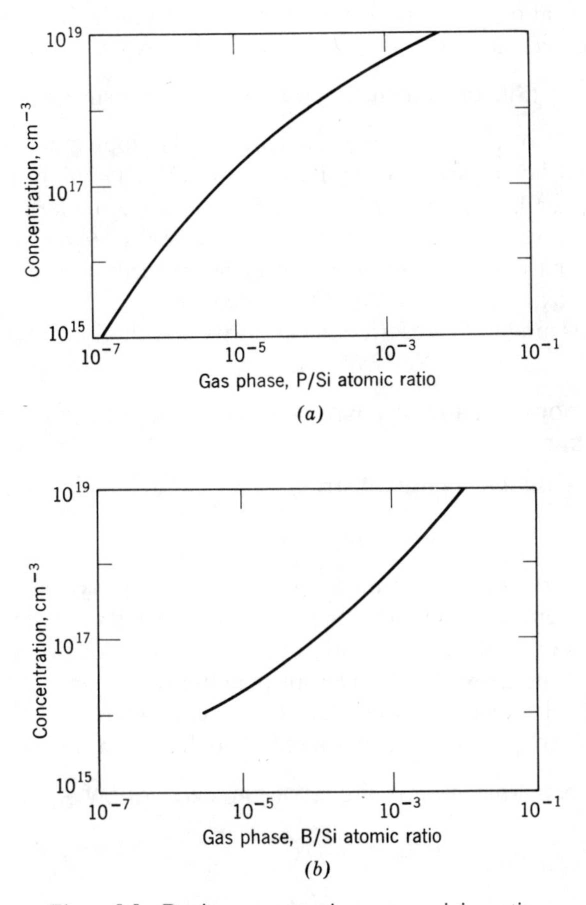

Figure 5.5 Doping concentration versus mixing ratio.

5.3.1. Etch-back Effects

The hydrogen reduction of $SiCl_4$ has been seen to be a reversible process. This leads to a redistribution of impurities during the growth of the layer by a process known as *auto-doping* or *etch-back* [3, 4]. The model for this process may be considered in terms of a sequence of steps (which actually occur together) as follows:

1. During growth some silicon and dopant are removed from the layer as a result of the etching reaction. Their rate of removal is roughly proportional to their concentration in the solid phase.
2. The resulting silicon and dopant in the gas phase mix with the incoming gas ambient and modify its impurity concentration.
3. This modified gas mixture results in the deposition of silicon and dopant in a ratio that is proportional to the new concentrations in the gas phase. The over-all process may be represented by

$$\text{(Silicon, dopant)}_{\text{solid}} \rightleftharpoons \text{(Silicon, dopant)}_{\text{gas}}. \tag{5.10}$$

As a result of this redistribution process, the doping concentration of the epitaxial layer varies during its growth until an equilibrium situation is reached for layers of sufficient thickness. Since modern microcircuit technology favors the use of thin epitaxial layers ($\simeq 6\ \mu m$), it is possible for the layer to have a continually varying impurity concentration over its entire thickness as a result of this mechanism.

Using this model, we describe the impurity distribution for the following cases:

CASE 1. DOPED SUBSTRATE WITH EPITAXIAL GROWTH FROM AN INTRINSIC GAS

For this case the impurity distribution in the grown layer is given by [3]

$$N_x = N_S e^{-\phi x}, \tag{5.11}$$

where N_x = impurity concentration in the grown layer, in atoms per cubic centimeter at x centimeters from the interface.

N_S = impurity concentration at the interface of the substrate and the grown layer, in atoms per cubic centimeter,

x = distance from the interface, in centimeters,

ϕ = growth factor, experimentally determined, in cm^{-1}.

CASE 2. INTRINSIC SUBSTRATE WITH EPITAXIAL GROWTH FROM A DOPED GAS

For this situation,

$$N_x = N_\infty (1 - e^{-\phi x}), \tag{5.12}$$

where N_∞ is the doping concentration if equilibrium conditions are reached during the epitaxial growth (i.e., for an infinitely thick layer).

In actual situations the final impurity distribution is given by the superposition of these two cases.

Typical values of ϕ, experimentally determined for a specific system, are given in Figure 5.6 as a function of growth temperature. It is seen that etch-back effects are reduced† by operating the system at high temperatures (i.e., at high growth rates).

Figure 5.7*a* shows the application of this theory to the growth of a lightly doped layer on a heavily doped substrate of the same impurity type, and is representative of a situation that is encountered during epitaxial growth over a buried layer.‡ It is assumed that the substrate concentration is N_S and the concentration in the epitaxial layer, if made infinitely thick, is N_∞. As seen from this figure, the ideal step of impurity concentration is not realized. In addition, it is possible that the final steady-state value of N_∞ may not be attained for thin epitaxial layers.

Figure 5.7*b* shows the situation when a heavily doped layer is epitaxially grown on a lightly doped substrate of opposite impurity type. This too, is a situation that is commonly encountered in microcircuit-fabrication processes. In this case departure from the ideal results in a shift in the position of the junction formed by the two layers. This shift, known as *junction lag* [4], may be shown to be inversely proportional to the growth parameter ϕ. Thus the junction lag may be reduced by growing the epitaxial layer at elevated temperatures.

5.3.2. Diffusion Effects

Diffusion processes are also present during epitaxial growth. Here too, the problem can be broken up into two case and the final distribution

†Note that high values of ϕ correspond to relatively abrupt doping profiles.
‡The use of these buried layers is treated in Section 10.4.

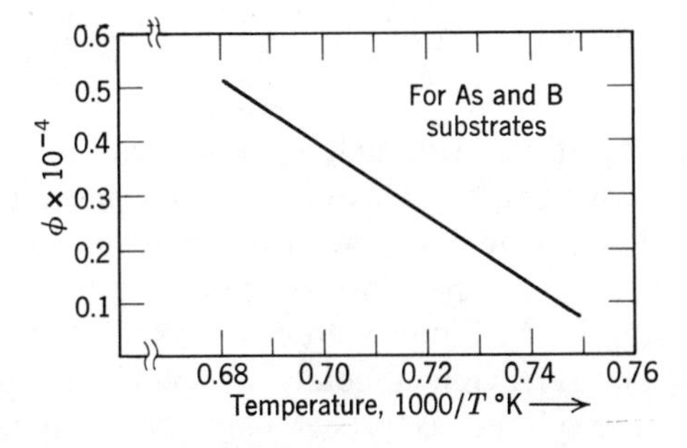

Figure 5.6 Etch-back effects in epitaxial growth (Kahng[4]).

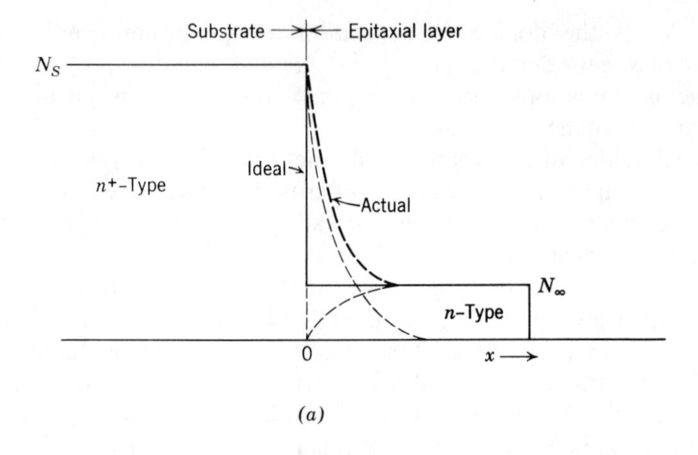

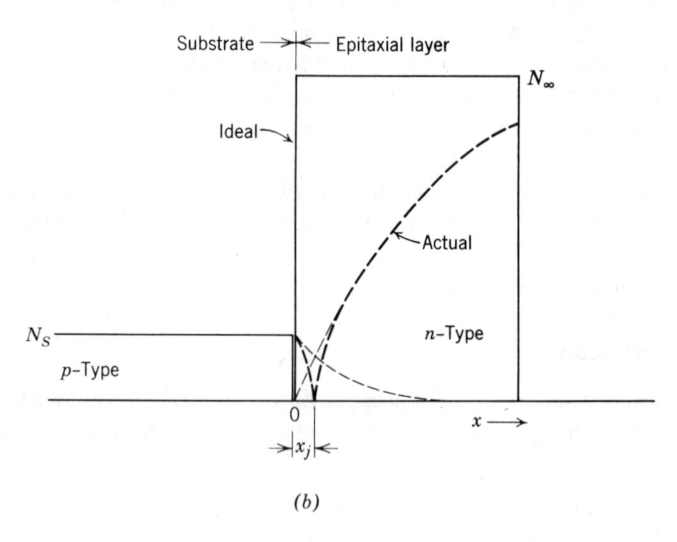

Figure 5.7 The effect of etch-back on doping profile.

obtained by superposition. Although a general solution of this situation, applicable to all growth rates, is available in the literature [5], only the practical situation is considered here. For this situation the growth rate V is so much higher than the diffusion rate (i.e., $Vt \gg \sqrt{Dt}$) that the epitaxial film can always be considered to be infinitely thick compared with the extent of the region affected by solid-state diffusion. Hence diffusion can be considered as proceeding into an infinite solid. With this in mind, the following cases are considered.

CASE 1. DIFFUSION DURING THE GROWTH OF AN INTRINSIC LAYER ON A DOPED SUBSTRATE

Let

$N_1(x)$ = impurity concentration, in atoms per cubic centimeter,
N_S = initial impurity concentration in the substrate, in atoms per cubic centimeter,
x = distance along the epitaxial layer, in centimeters, with the origin taken at the substrate-layer interface.

Then

$$N_1(x) = \frac{N_S}{2}\left(1 - \operatorname{erf}\frac{x}{2\sqrt{D_S t}}\right), \tag{5.13}$$

where D_S is the diffusion constant for the substrate impurity.

CASE 2. DIFFUSION DURING THE GROWTH OF A DOPED LAYER ON AN INTRINSIC SUBSTRATE

Let

$N_2(x)$ = impurity concentration,
N_E = impurity concentration at the surface of the epitaxial layer.

Then,

$$N_2(x) = \frac{N_E}{2}\left(1 + \operatorname{erf}\frac{x}{2\sqrt{D_E t}}\right), \tag{5.14}$$

where D_E is the diffusion coefficient for the epitaxial layer. At distances removed from the substrate-film interface, such that $x \gg 2\sqrt{D_E t}$, this equation can be further simplified to

$$N_2(x) \simeq N_E. \tag{5.15}$$

The final impurity distribution is given by the superposition of (5.13) and (5.14), with due attention paid to the impurity type.

5.4. DIFFUSION DURING SUBSEQUENT PROCESSING

In addition to diffusion effects during growth, the doping profile of epitaxial layers is also subject to change during subsequent processing. Thus, if an abrupt junction between two impurity concentrations is considered, with N_S and N_E as the initial concentrations of the substrate and the epitaxial layer respectively, the final impurity concentration after various heat treatments is given by

$$N(x,t) = \frac{N_S}{2}\left[1 - \operatorname{erf}\frac{x}{2\sqrt{(Dt)_S}}\right] + \frac{N_E}{2}\left[1 + \operatorname{erf}\frac{x}{2\sqrt{(Dt)_E}}\right], \tag{5.16}$$

where $(Dt)_S$ and $(Dt)_E$ are the effective Dt products for the substrate and epitaxial layers (see Section 4.5.3), and distance is measured from the substrate-layer interface towards the silicon surface.

Two cases are of particular importance in microcircuit fabrication:

CASE 1.

N_S and N_E are of the same impurity type and $N_S \gg N_E$. This corresponds to the growth of an n-type film on an n^+ buried layer, and is shown in Figure 5.8a. It is seen that successive processing results in a con-

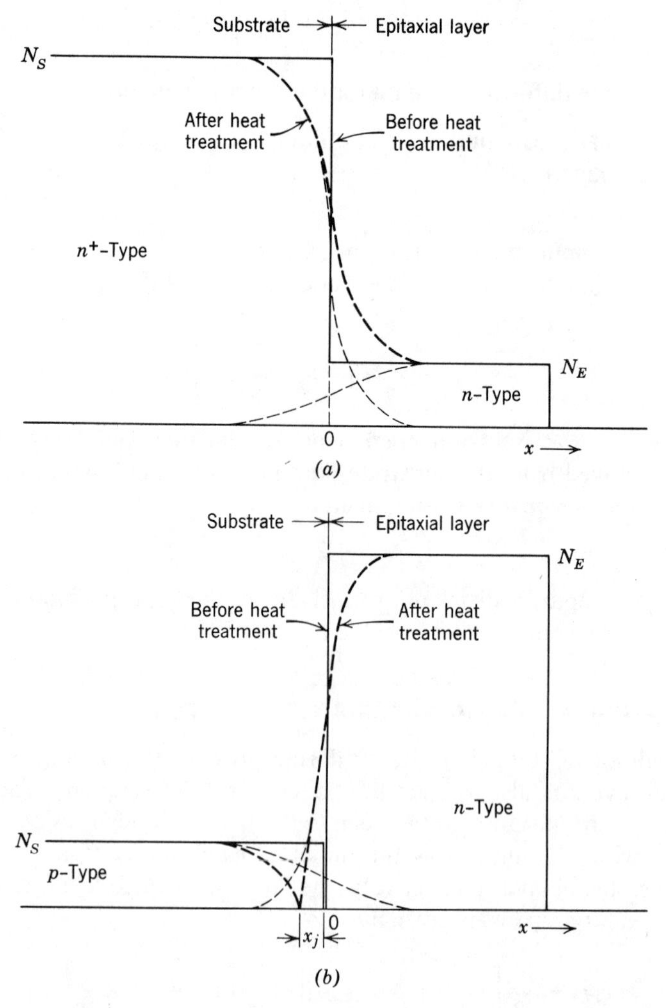

Figure 5.8 The effect of heat treatment on doping profile.

siderable amount of out-diffusion from the buried layer. In microcircuit fabrication, it is common practice to use a slower diffusing impurity (antimony or arsenic) for this buried layer in order to reduce the out-diffusion effect.

CASE 2.

N_S and N_E are of opposite conductivity type so as to form a *p-n* junction. If the diffusion coefficients of the two types are reasonably close (as is true for boron and phosphorous), the position of the junction is found to move in the direction of the lighter doped region. This situation is encountered in many microcircuits where an *n*-type epitaxial layer is grown on a comparatively lightly doped *p*-type substrate. As shown in Figure 5.8*b*, subsequent processing results in the receeding of the *p-n* junction into the substrate.

A final case is worth mention. If the two concentrations are initially the same, but of opposite conductivity types, the position of the junction remains fixed on subsequent processing, regardless of the values of the diffusion constants.

5.5. THE EVALUATION OF EPITAXIAL LAYERS

The epitaxial growth of a suitably doped, single-crystal layer forms the starting point for the most commonly used microcircuit fabrication technique. Since the entire microcircuit is eventually fabricated within this layer, it is important that its suitability be assessed for this purpose. With this in mind, it is necessary to evaluate such layer parameters as sheet resistance, thickness, and doping profile.

5.5.1. Sheet Resistance

In microcircuits, a layer of *n*-type silicon is epitaxially grown on a *p*-type substrate. The resistance of this layer may be characterized in terms of its sheet resistance, as described in Section 4.9.1. In this case, however, this measurement cannot be made directly, since it is not possible to mask a suitable pattern on the slice.

The sheet resistance of such a layer may be measured by the four-point probe configuration shown in Figure 5.9*a*. Here, four equally spaced collinear probes are placed on the layer; a current I is passed through the outer probes, and the voltage developed across the inner probes, V, is measured. The value of current is so chosen that V and I are linearly interrelated.

Since the epitaxial layer is of opposite impurity type to its substrate, it may be assumed that the current flow is restricted within it. Consider a

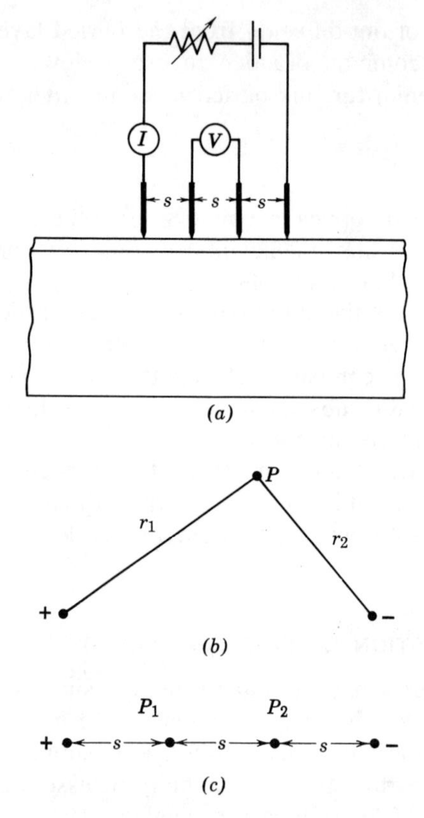

Figure 5.9 Four point probe.

layer of infinite dimensions compared to the probe spacing. For the configuration of Figure 5.9*b*, comprising a + source and a − source of current, the potential at any point P is given by

$$\Psi_P = \frac{I\rho_\square}{2\pi} \ln \frac{r_2}{r_1} + A, \tag{5.17}$$

where r_1 and r_2 are the distances of the point P from the + and − source respectively, and A is a constant of integration. For the four-point probe configuration of Figure 5.9*c*,

$$\Psi_1 = \frac{I\rho_\square}{2\pi} \ln 2 + A \tag{5.18}$$

and

$$\Psi_2 = \frac{-I\rho_\square}{2\pi} \ln 2 + A, \tag{5.19}$$

where Ψ_1 and Ψ_2 are the potentials at points 1 and 2 respectively. Thus,

$$\Psi_1 - \Psi_2 = V = \frac{I\rho_\square}{\pi} \ln 2. \tag{5.20}$$

Rearranging (5.20) gives

$$\rho_\square = \left(\frac{\pi}{\ln 2}\right)\frac{V}{I}, \tag{5.21}$$

$$= 4.5324\frac{V}{I}. \tag{5.22}$$

Thus the sheet resistance may be directly calculated from the V/I ratio.

Formulas for other, more practical, probe configurations are available [6] in the literature. For probes located centrally on a diameter of a circular sample with insulating edges, the sheet resistance can be shown to be given by

$$\rho_\square = C\frac{V}{I}, \tag{5.23}$$

where C is a function of the probe spacing and the slice diameter, as shown in Figure 5.10.

Four-point probe measurements are usually made on monitor slices that are placed in the reactor along with those on which the circuits are to be fabricated. This avoids the necessity for interpreting data on slices which already have diffused regions (buried layers) prior to epitaxial growth.

5.5.2. Layer Thickness

This is measured by angle lap and interferometric techniques, as described in Section 4.9.2. Here, too, staining techniques are used to delineate the p–n junction so formed.

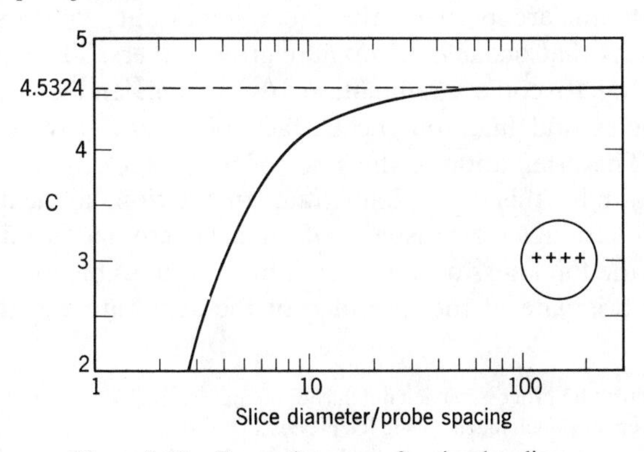

Figure 5.10 Correction curve for circular slices.

5.5.3. Doping Profile

The commonly used method for determining the doping profile of an epitaxially grown film consists of successively removing layers of the film and measuring the sheet resistance of the film with each successive removal. This removal is usually done by the anodic oxidation† of the slice and by the removal of the oxide so formed in hydrofluoric acid. Let

ρ_1 = sheet resistance of the epitaxial film before the removal of a thin layer,
ρ_2 = sheet resistance of the film after the removal of this layer, and
δ = thickness of the layer that is removed.

Then the average specific resistivity of the removed layer $\bar{\rho}$ is given by

$$\bar{\rho} = \frac{\rho_1 \rho_2}{\rho_2 - \rho_1} \delta, \tag{5.24}$$

where $\bar{\rho}$ is in ohm–centimeters. By successive repetition of this process the resistivity profile of the wafer can be obtained. This, in turn, may be converted to an impurity-concentration profile.

5.6. IMPERFECTIONS IN EPITAXIAL LAYERS

A number of different types of imperfections may be present in epitaxially grown layers. Some of these are crystallographic in nature and result in a loss of periodicity in the crystal lattice. Yet others are gross defects, usually arising from improper handling or cleaning procedure.

5.6.1. Stacking Faults

Stacking faults are perhaps the most important types of crystallographic defects that occur in epitaxially grown layers. They can be made suitable for microscopic examination by etching, and appear as equilateral triangles and lines on the surface of layers grown on a {111} substrate.‡ The orientation of the sides of these stacking faults are in the ⟨110⟩ direction for this type of substrate. In addition, all the lines and the sides of these triangles are usually of equal length and are directly proportional to the thickness of the layer. Thus it is possible to conclude that these faults originate at the interface of the substrate and the epitaxial layer.

†This is an electrolytic process carried out at room temperature. Consequently, the impurity concentration remains unchanged during the process.
‡Silicon crystals are most commonly available in this orientation (see Chapter 2).

The mechanism of growth of a stacking fault is now considered for {111} layers in an f. c. c. lattice. For this case the stacking sequence of atomic planes is *abcabc*..., whereas that of the diamond lattice is *aa'bb'cc'aa'bb'*.... Thus the growth mechanisms are essentially similar, although somewhat easier to visualize for the f. c. c. lattice.

Epitaxial growth has been described as the ordered deposition of atomic planes, one at a time, with nucleation centers for a fresh plane being formed only after the last plane is completed. If, however, such a center is formed in an incorrect sequence with respect to its neighbors, the regularity of the layers is disrupted. On disruption the layers continue to grow in this new sequence within the fault. Thus, as seen in Figure 5.11, the defect propagates from layer to layer, retaining its shape and increasing in size by one atom spacing at a time. The various faces of the defect all fall upon inclined {111} planes, resulting in the equilateral tetrahedron of Figure 5.12.

From geometrical considerations,

$$d = \sqrt{\tfrac{2}{3}}\, s \tag{5.25}$$

where d is the thickness of the grown layer and s is the length of one side of the triangle on the surface. This relationship is often used to determine the depth of the epitaxial layer.

It has been proposed [8] that patches of SiO_2 on the surface of the crystal are the prime cause of stacking faults. Since the Si—Si spacing in the lattice is 2.35 Å, and the Si—O—Si spacing for silicon polyhedra is 3.05 Å, the presence of this oxide results in steps which are not equal to

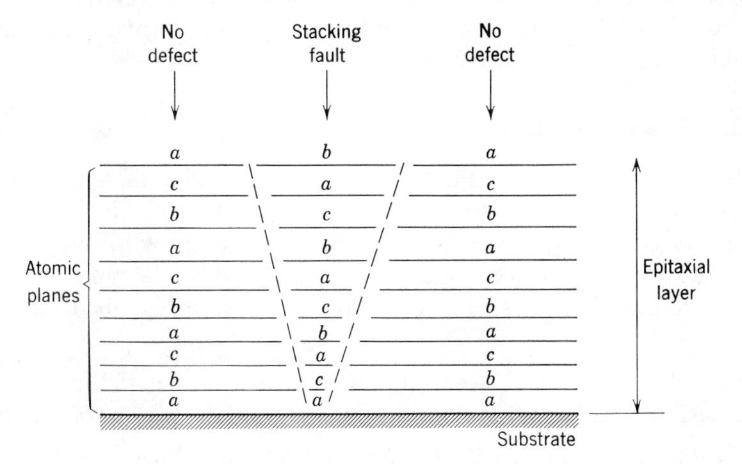

Figure 5.11 Formation of a stacking fault.

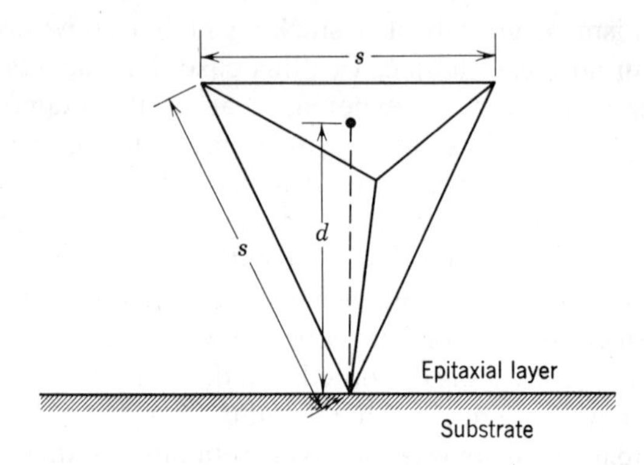

Figure 5.12 Detail of a stacking fault.

an integral number of atomic plane spacings. Periodicity is maintained in the lattice during epitaxial growth by the introduction of stacking faults that start at the edge of the oxide and grow into the depositing layer. Experimental verification of this theory has been noted in the fact that the concentration of stacking faults is critically dependent on the effectiveness of the hydrogen reduction step (note that the subsequent HCl etch does not remove SiO_2).

Stacking faults, taken by themselves, do not alter the electronic properties of the epitaxial layer. They give rise, however, to inhomogeneties in the diffusion of impurities and act as nucleation centers for metal precipitates. The *p–n* junctions built from this material are often short-circuited, or exhibit soft reverse-breakdown characteristics[9]. In addition, breakdown is found to be initiated by the formation of microplasmas at the edge of these faults.

5.6.2. Gross Defects

There are many gross defects that can occur during the growth of an epitaxial layer. The majority of these are caused by leaks in the system, the presence of foreign matter during epitaxial growth, and surface scratches. Their presence can usually be detected by the naked eye and leads to rejection of the slice. Defects of this type are described here very briefly.

Substrate Scratches. These scratches act as sites for the initiation of stacking faults and result in a line of such faults on the surface of the layer.

Pits, Voids, and Spikes. These are caused by small particles of silicon or silicon dioxide in the reactor during the epitaxial growth.

Pyramids. These sometimes occur as a result of a high growth rate or of an improperly oriented substrate.

Haze. This takes the form of a cloudy appearance and results from a leaky system or from improper cleaning and solvent removal procedures prior to epitaxial growth.

All of these gross defects are indicative of poor fabrication technique and can be corrected by taking suitable precautions.

5.7. REFERENCES

[1] H. C. Theuerer, "Epitaxial Silicon Films by the Hydrogen Reduction of $SiCl_4$," *J. Electrochem. Soc.*, **108**(7), 649–653 (1961).
[2] W. H. Shepherd, "Vapor Phase Deposition and Etching of Silicon," *J. Electrochem.* Soc., **112**(10), 988–994 (1965).
[3] C. O. Thomas *et al.*, "Impurity Distribution in Epitaxial Silicon Films," *J. Electrochem. Soc.*, **109**(11), 1055–1061 (1962).
[4] D. Kahng *et al.*, "Epitaxial Silicon Junctions," *J. Electrochem. Soc.*, **110**(5), 394–400 (1963).
[5] A. S. Grove *et al.*, "Impurity Distribution in Epitaxial Growth," *J. Appl. Phys.*, pt. 1, **36**(3), 802–810 (1963).
[6] F. M. Smits, "Measurement of Sheet Resisitivities with the Four-Point Probe," *Bell System Tech. J.*, **37**, 711–718 (1958).
[7] G. R. Booker and R. Stickler, "Crystallographic Imperfections in Epitaxially Grown Silicon," *J. Appl. Phys.*, **33**(11), 3281–3290 (1962).
[8] R. H. Finch *et al.*, "Structure and Origin of Stacking Faults in Epitaxial Silicon," *J. Appl. Phys.*, **34**(2), 406–415 (1963).
[9] H. Kressel, "A Review of the Effect of Imperfections on the Electrical Breakdown of P–N Junctions," *RCA Review*, **28**(2), 175–207 (1967).

5.8. PROBLEMS

1. A phosphorus-doped layer, having a doping concentration of 10^{16} atoms/cm^3, is grown on a boron-doped substrate having a background concentration of 10^{15} atoms/cm^3. Determine the junction lag resulting from this growth, assuming that $\phi = 0.5 \times 10^{-4}$.

2. A phosphorus-doped epitaxial layer of concentration N_E is grown on an arsenic-doped buried layer of concentration N_S. An abrupt doping profile results from this process. The slice is subjected to heat treatment at 1200°C for a period of 10 min. Sketch the doping profile resulting from this heat treatment for $N_S = 1000N_E$.

3. Repeat Problem 2, assuming a heat treatment period of 100 min. Compare the results for the two cases.

4. In Reference 8 the energy of formation of a stacking fault on a {111} surface has been quoted as 50 ergs/cm^2. Compare this number with the energy of movement of an edge dislocation. Do you think it is possible to remove stacking faults by annealing?

5. Describe the growth of a stacking fault in an epitaxial layer on a {100} substrate. Along what planes does this fault propagate and what is its appearance on the surface?

Chapter 6

Oxides and Oxide Masking

The surface of a silicon slice is covered at all times with a layer of one or more forms of silica glass. This is even true for freshly cleaved pieces of silicon, which become covered with a few monolayers ($\simeq$ 10 to 15 Å) of oxide on exposure to air for a few milliseconds. In microcircuit fabrication, oxide layers are intentionally placed[1] on the slice to protect it during the various processing steps. Diffusions are made in selected regions with the aid of windows cut in this layer, which otherwise acts as a barrier to the impurities. The edges of *p-n* junctions grown in this manner are formed under the oxide layer and thus enjoy a considerable amount of protection. In addition, the oxide layer is an insulator and allows the placement of evaporated metal interconnection patterns on its surface.

In this chapter the nature of the oxide layer is described as well as the manner in which its properties are modified by the introduction of impurities. This is followed by a discussion of means for obtaining this layer and for evaluating its thickness. The masking properties of this layer are also described.

6.1. INTRINSIC SILICA GLASS

Intrinsic silica glass consists of fused silicon dioxide. It is thermodynamically unstable below 1710°C and tends to return to its crystalline form at temperatures below this value. However, at temperatures below 1000°C the rate of this devitrification is usually considered negligible.

The model for pure silica in the vitreous state[2] consists of a random three-dimensional network of silicon dioxide, constructed from polyhedra (tetrahedra or triangles) of oxygen ions. The centers of these polyhedra are occupied with Si^{+4} ions. The tetrahedral distance between the silicon and oxygen ions is 1.62 Å while the distance between oxygen ions is 2.65 Å.

Silica polyhedra are joined to one another by *bridging* oxygen ions, each of which is common to two such polyhedra. In crystalline silica all such oxygen ions play this role and all vertices of the polyhedra are tied to their nearest neighbors by these ions. In fused silica or silica glass however, some of the vertices have *nonbridging* oxygen ions which belong to only one polyhedron. The degree of cohesion between the polyhedra and, hence, of the network as a whole is thus a function of the percentage of bridging to nonbridging oxygen ions.

In pure silica glass movement of the silicon atom is accomplished by the rupture of four Si—O bonds, whereas the movement of a bridging oxygen atom requires the rupture of only two Si—O bonds. As a consequence, oxygen is freer to move in this glass than is silicon. The movement of this oxygen from its polyhedral site gives rise to the formation of an oxygen ion *vacancy*; both bridging and nonbridging vacancies may be formed, although the latter is more probable from binding energy considerations. Vacancies of this type represent positively charged defects in the structure.

Diffusion processes in intrinsic silica glass are similar to those described in Chapter 4, even though a pseudoperiodic structure is involved. As may be expected, the process can be described by a diffusion coefficient D such that

$$D = D_0 e^{-E_d/kT}, \tag{6.1}$$

where D_0 is the diffusion constant and E_d is the activation energy of the diffusing species.

It is also necessary to consider the permeation [3] rate of gases through the oxide layer. This describes the flow of gas from one side of a layer to the other, as opposed to diffusion, where the movement of impurities within the layer are considered.

Let V be the volume of gas flowing through unit area of the layer, in cubic centimeters per second. Then V varies linearly with gas-pressure differential across the layer, and inversely with layer thickness. Thus

$$V = K\frac{\Delta P}{d}, \tag{6.2}$$

where ΔP = the differential gas pressure, in centimeters,
d = the layer thickness, in millimeters,
K = the permeability, in cubic centimeters (STP) per second per square centimeter per millimeter thickness per centimeter differential pressure.

In general, it is also possible to write the permeability in the form

$$K = K_0 e^{-E_p/kT} \tag{6.3}$$

where E_p is the activation energy of permeation. Values for permeability and diffusion constant are given in Table 6.1 for hydrogen, oxygen, and water vapor in silica glass. In addition, Figure 6.1 shows the diffusion constant of these three species over a range of temperature. Note that these data represent order of magnitude values only, since the diffusion constant is critically dependent on the nature of the glass network, that is, on the ratio of bridging to nonbridging oxygen ions.

It is seen from these data that hydrogen permeates rapidly through the silica glass. In addition, the activation energy of diffusion of hydrogen and water vapor is considerably smaller than that of oxygen.

Diffusion experiments in the presence of an electric field have established that a charged species of the diffusant is involved. Thus movement in a silicon dioxide layer is accomplished by drift as well as by diffusion.

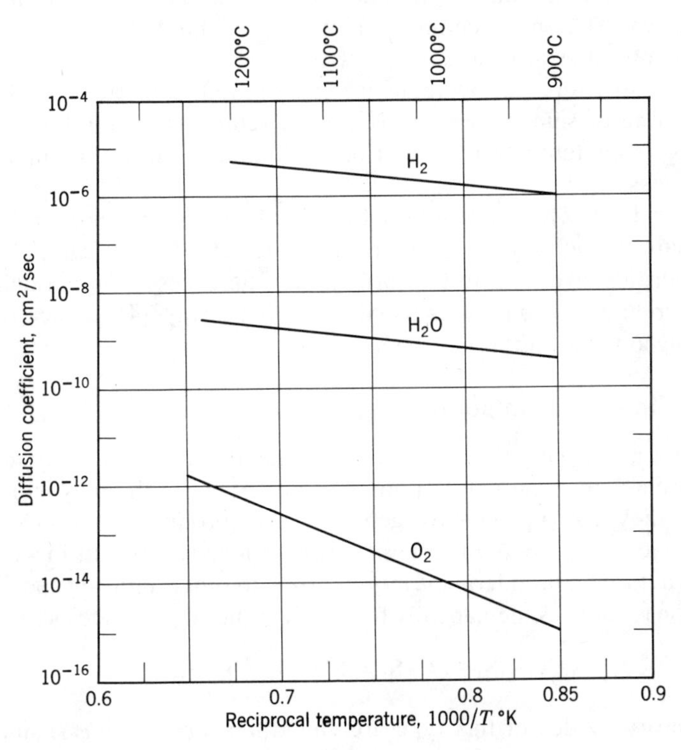

Figure 6.1 Diffusion constants in SiO_2.

Table 6.1
Diffusion and Permeation Rates in Silica Glass

Diffusing Species	K, 700°C	D_0, cm²/s	E_d, eV
H_2	2.1×10^{-9}	9.5×10^{-4}	0.685
O_2	10^{-15}	1.5×10^{-2}	3.09
H_2O	10^{-15}	1×10^{-6}	0.794

6.2. EXTRINSIC SILICA GLASS

The properties of silica glass are greatly modified [4] by the introduction of impurities. These impurities are of two types – substitutional and interstitial.

6.2.1. Substitutional Impurities

A substitutional impurity is one which replaces the silicon in a silica polyhedron. The most common impurities of this type are B^{3+} and P^{5+}. These impurities are called *network formers*, since it is possible to build vitreous structures using them instead of SiO_2 (i.e. glasses which are totally free of silicon dioxide). In microcircuit fabrication, however, our interest is centered on the effect of small amounts of these impurities in silica glass.

The valence of a substitutional cation is usually either 3 or 5. In the silica lattice, such cations result in charge defects. The presence of column V impurities gives rise to the formation of an excess of nonbridging ions while column III impurities usually (but not always [4]) reduce the nonbridging ion concentration.

6.2.2. Interstitial Impurities

These are usually the oxides of large metal ions of low positive charge which enter into the network interstitially between the polyhedra. In so doing, they give up their oxygen to it, thus producing two nonbridging oxygen ions in place of the original bridging ion. This results in weakening the structure and rendering it more porous to other diffusing species. The reaction is shown schematically for Na_2O in the silica lattice as follows:

$$Na_2O + Si{-}O{-}Si = Si{-}O + O{-}Si + 2Na^+$$

Impurity oxides of this type are called *network modifiers*, since they are not capable of forming glasses by themselves. Ions such as Na^+, K^+,

Pb^{++}, and Ba^{++} fall into this class. Aluminum sometimes plays a dual role of network modifier as well as network former.

6.2.3. Water Vapor

Water vapor can also be present in silica glass. On entering it combines with bridging oxygen ions to form pairs of stable nonbridging hydroxyl groups. The reaction is described schematically by

$$H_2O + Si—O—Si = Si—OH + OH—Si$$

The presence of these hydroxyl groups also tends to weaken the silicon network and render it more porous to diffusing species. Thus their behavior is similar to that of interstitial impurities.

Figure 6.2 shows a schematic representation of fused silica glass together with the various types of defect structures that may be present.

6.3. OXIDE FORMATION

Various methods may be used to obtain a layer of oxide on the surface of the silicon slices. These include such techniques as reactive sputtering,

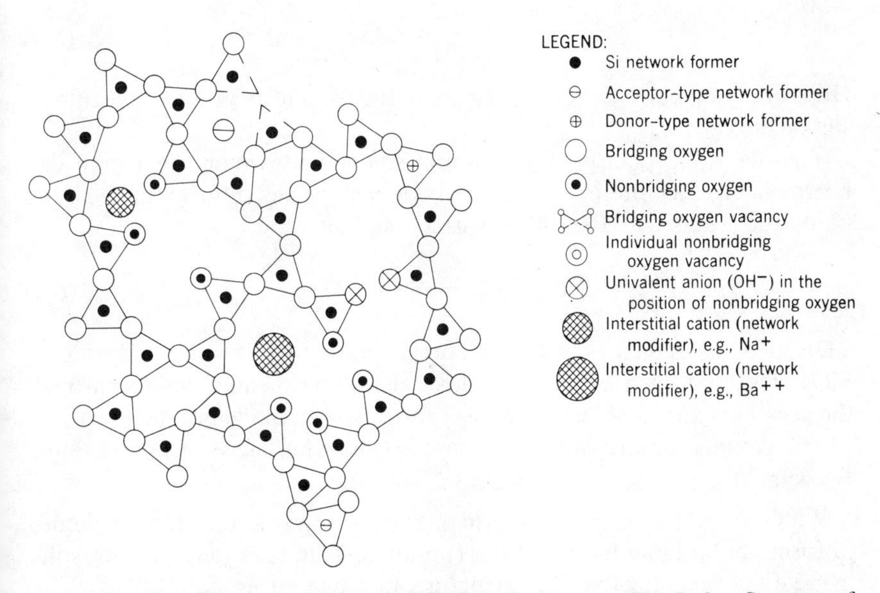

Figure 6.2 The silica network. [Adapted from A. G. Revesz, "The Defect Structure of Grown Silicon Dioxide Films," *IEEE Trans. on Electron Devices*, **ED-12** (3), 97–102 (1965)].

anodic deposition, and the pyrolytic decomposition of oxysilane compounds. To date the most convenient method consists of direct oxidation of the silicon surface. From considerations of molecular weight and density it can be shown that the thickness of the oxide layer thus formed is 2.27 times the thickness of the consumed silicon.

Oxidation of the silicon slice is conveniently carried out by subjecting it to dry oxygen, wet oxygen, or live steam while the wafer is maintained at an elevated temperature.

6.3.1. Chemistry of Oxidation

The mechanism of oxide formation is based on the fact that the oxidizing species moves through the growing oxide layer in order to reach the silicon surface. Thus, it is reasonable to expect the process to proceed at an ever decreasing rate as the thickness of the intervening oxide layer increases.

The chemistry of oxidation with dry oxygen is relatively straightforward. It is assumed that the species diffusing through the growing layer are oxygen ions. Since the permeability of oxygen in silicon dioxide is low ($< 10^{-15}$), the role of permeation is not considered significant.

The chemical reaction at the silicon surface is

$$Si + O_2 \rightarrow SiO_2. \tag{6.4}$$

Here one molecule of oxygen results in the formation of one molecule of silicon dioxide.

The over-all process of oxidation of silicon with water vapor may also be considered as one in which the oxidizing species diffuses through the oxide and reacts with the silicon surface, so that

$$Si + 2H_2O \rightarrow SiO_2 + 2H_2. \tag{6.5}$$

Thus two molecules of water vapor are used to form one molecule of SiO_2. The hydrogen evolved by this reaction permeates rapidly through the growing oxide and leaves the system at the gas-oxide interface.

The detailed nature of this reaction is somewhat more complex [5] and is assumed to proceed in the following manner:

Water vapor reacts with the bridging oxygen ions in the silica structure to form nonbridging hydroxyl (OH) groups. This reaction, which results in greatly weakening the silica structure, may be written as

$$H_2O + Si{-}O{-}Si \rightarrow Si{-}OH + OH{-}Si.$$

2. At the oxide-silicon interface, the hydroxyl groups react with the silicon lattice to form silica polyhedra and hydrogen. The reaction is

$$\begin{matrix} \text{Si—OH} \\ \text{Si—OH} \end{matrix} + \text{Si—Si} \rightarrow \begin{matrix} \text{Si—O—Si} \\ \text{Si—O—Si} \end{matrix} + H_2.$$

3. Hydrogen leaves the oxide layer, reacting further with bridging oxygen ions in the silica structure to form hydroxyl groups, as shown by

$$\tfrac{1}{2}H_2 + \text{O—Si} \rightleftharpoons \text{OH—Si}$$

Thus the hydrogen further aids in weakening the structure. As the oxide layer thickens, these reactions proceed at an ever decreasing rate.

6.3.2. Kinetics of Oxide Growth

The kinetics of oxide growth on silicon may be determined by means of the relatively simple model of Figure 6.3. Assume that a silicon slice is brought in contact with the oxidant, resulting in a surface concentration of N_0 molecules/cm^3 for this species. The magnitude of N_0 depends on the solid solubility of the species in the oxide as well as on such parameters as the gas-flow rate and the temperature. At 1000°C the solid solubility of these species is 5.2×10^{16} molecules/cm^3 for dry oxygen and 3×10^{19} molecules/cm^3 for water vapor at a pressure of 1 atm.

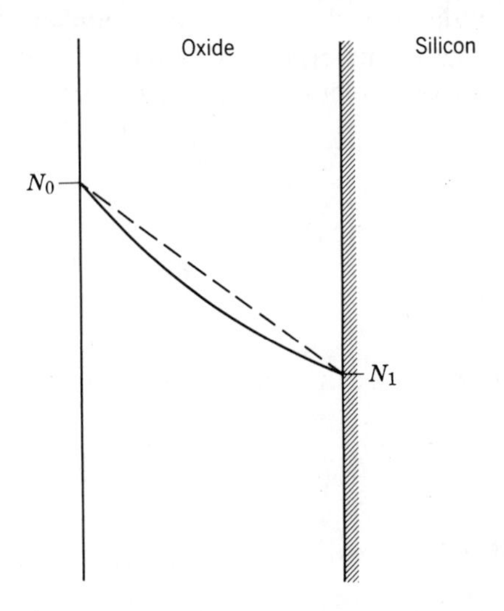

Figure 6.3 Diffusion of reactant during oxidation.

The oxidizing species diffuses through the silicon dioxide layer, resulting in a concentration N_1 at the surface of the silicon. Transport of the species occurs by both drift and diffusion. Writing D as the effective diffusion constant for this situation, the flux density of oxidizing species arriving at the gas-oxide interface is given by j, where

$$j = D\frac{\partial N}{\partial x}, \tag{6.6}$$

$$\simeq \frac{D(N_0 - N_1)}{x}. \tag{6.7}$$

On arrival at the silicon surface the species enters into chemical reaction with it. If it is assumed that this reaction proceeds at a rate proportional to the concentration of the oxidizing species, then

$$j = kN_1, \tag{6.8}$$

where k is the reaction-rate constant. These fluxes must be equal under steady-state diffusion conditions. Hence combining (6.7) and (6.8) gives

$$j \simeq \frac{DN_0}{x + D/k}. \tag{6.9}$$

The reaction of the oxidizing species with the silicon results in the formation of silicon dioxide. Writing n as the number of molecules of the oxidizing impurity that are incorporated into unit volume of the oxide, the rate of change of the oxide layer thickness is given by

$$\frac{dx}{dt} = \frac{j}{n} \tag{6.10}$$

$$= \frac{DN_0/n}{x + D/k}. \tag{6.11}$$

Solving this equation, subject to the boundary value that $x = 0$ at $t = 0$, gives

$$x = \frac{D}{k}\left[\left(1 + \frac{2N_0k^2t}{Dn}\right)^{1/2} - 1\right] \tag{6.12}$$

Equation 6.12 reduces to

$$x = \frac{N_0k}{n}t \tag{6.13}$$

for small values of t, and to

$$x = \left(\frac{2N_0D}{n}t\right)^{1/2} \tag{6.14}$$

for large values. Thus in the early stages of growth, in which the reaction is rate-limited, the oxide thickness varies linearly with time. In later stages the reaction is diffusion-limited and the oxide thickness is directly proportional to the square root of time.

6.3.3. Initial Growth Phase

The theory for the kinetics of oxide formation has been found to apply very well to growth in wet oxygen and steam. There is consistent evidence, however, to indicate the presence of an extremely rapid initial growth phase with dry oxygen.

This process has been explained by Grove [6]. In his model he has postulated that the molecular oxygen, on entering the oxide, dissociates to form a negative-charged "superoxide ion." O_2^-, and a hole. The hole, having considerably higher mobility than the oxygen ion, runs ahead of it; the result is the formation of a space-charge region. The resulting field enhances the diffusion† of the oxygen in the layer.

Cabrera and Mott [7] have shown that the region of high space-charge density is near the gas-oxide interface, the rest of the oxide layer being almost space-charge neutral. The thickness of this region is on the order of the extrinsic Debye length and depends inversely on the square root of the concentration of the oxidizing species in the layer. This Debye length is about 150 to 200 Å in dry oxygen, but only 5 Å for water vapor.‡ Consequently, an accelerated oxidation rate is seen for a depth of about 200 Å in dry-oxygen processing, but unnoticed for wet-oxygen or steam processing.

6.4. OXIDATION SYSTEMS

Oxide growth is carried out in a quartz diffusion tube, in which the silicon slice is maintained at a temperature between 900 to 1200°C. Experience has shown that the use of live steam leads to extremely poor grades of oxide because of the etching action of the excess water [8].

†The process is in effect quite similar to that of enhanced impurity diffusion at high concentration levels, described in Section 4.3.

‡This is because the solid solubility of water in silicon is three orders of magnitude larger than that of oxygen.

Consequently, provision is normally made for only dry- and wet-oxygen processes, the latter being accomplished by flowing the oxygen through a water bubbler. The water temperature is usually maintained below the boiling point (about 95°C) to prevent its undue depletion.

The density of the resulting oxide layer is a function of the oxidizing species employed, the use of dry oxygen resulting in the denser structure. Table 6.2 compares the densities of the oxide layer formed by the two processes as well as their growth rates and dielectric strengths. In commerical practice a combination of these processes is often used, the oxidation procedure being both initiated and concluded in dry oxygen with an intervening wet-oxygen step.

Figures 6.4 and 6.5 show the growth rates for oxide layers formed by these two processes. It is observed that the reaction is diffusion limited over most of the oxidation range. It should be noted that oxide growth with wet oxygen is considerably more rapid than with dry oxygen. This is caused by the considerably higher solid solubility of water vapor in silicon dioxide as well as by its higher diffusion constant (see Figure 6.1).

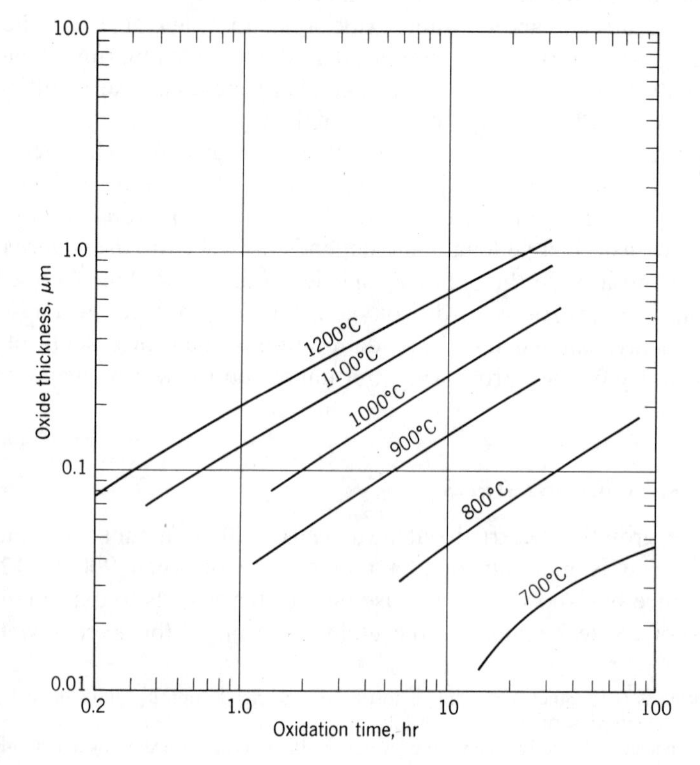

Figure 6.4 Oxide growth rate for dry oxygen (adapted from Deal and Grove [6]).

Table 6.2
Properties and Growth Rates of Silica Oxides

Oxidation Type	Temperature, °C	Density gm/cm^3	Rate, $\mu m^2/min$	Dielectric Strength, V/mil
Dry oxygen	1000	2.27	1.48×10^{-4}	550
	1200	2.15	7.64×10^{-4}	517
Wet oxygen ($95°C\ H_2O$)	1000	2.18	38.5×10^{-4}	525
	1200	2.21	117.5×10^{-4}	534
Steam	1000	2.08	54.5×10^{-4}	498
	1200	2.05	159×10^{-4}	490

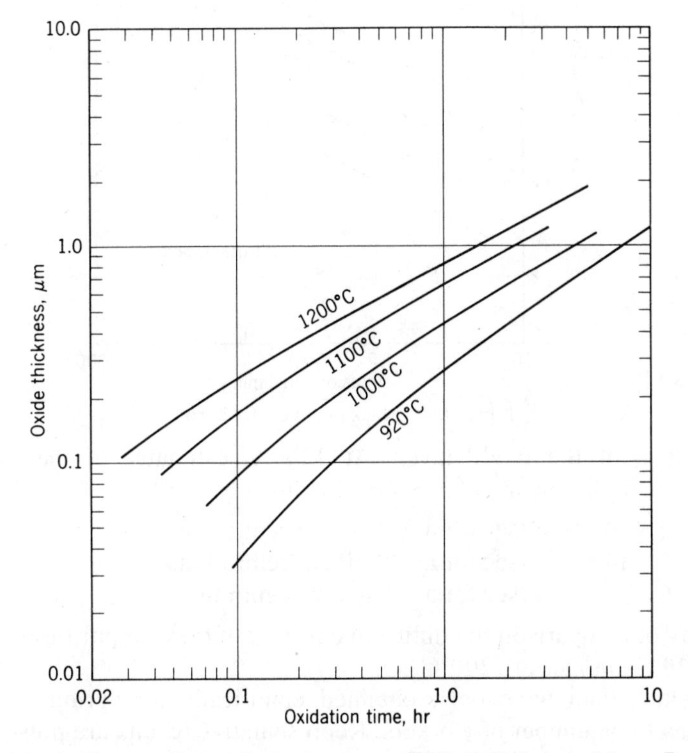

Figure 6.5 Oxide growth rate for wet oxygen ($95°C\ H_2O$) (adapted from Deal and Grove[6]).

6.5. MASKING PROPERTIES OF OXIDE LAYERS

An important property of a silicon dioxide layer is its ability to mask against those impurities commonly used in microcircuit fabrication. This masking property may be explained in terms of the ability of the impurity to behave as a network former in the silica structure. Thus both B_2O_3 and P_2O_5 form mixed (borosilicate and phosphosilicate) glasses with the silica layer with which they come into contact. The boundary between the mixed glass phase and the silica phase is quite sharp; the masking properties of the layer are excellent until this boundary extends down to the silicon-oxide interface.

The diffusion process for network formers is concentration dependent, since their presence leads to the creation of charge defects and alters the

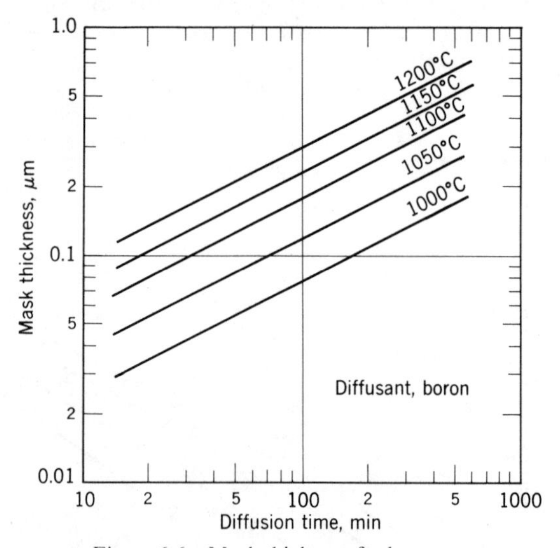

Figure 6.6 Mask thickness for boron.

properties of the oxide layer. At 1200°C estimates of the diffusion constants for boron and phosphorus in SiO_2 are as follows:

$D = 2 \times 10^{-16}$ cm²/sec for a 0.1% P concentration.
$D = 3 \times 10^{-13}$ cm²/sec for a 5.0% P concentration.
$D = 6 \times 10^{-15}$ cm²/sec for a 0.1% B concentration.

By way of comparison the diffusion constant of these impurities in silicon is 2×10^{-12} cm²/sec at 1200°C.

Masking data have been obtained empirically for various diffusion systems by a number of workers. Representative results are presented in curves of the type shown in Figures 6.6 and 6.7 and are typical of commercial practice in microcircuit fabrication.

The diffusion coefficient of gallium in SiO_2 has been found to be about 10^{-10} cm^2/sec at 1200°C, whereas that for aluminum is even larger. Thus these impurities cannot be masked by the use of silica layers. Silicon nitride has been found suitable in this situation. The material, however, is quite difficult to deposit as well as to etch. Its application is thus restricted to special situations.

6.6. ELECTRONIC PROPERTIES OF OXIDE LAYERS

The silicon dioxide layer is in contact with exposed *p-n* junction edges in a microcircuit and is thus subjected to conditions of high electric field intensity. In addition, it has been noted that there are regions of

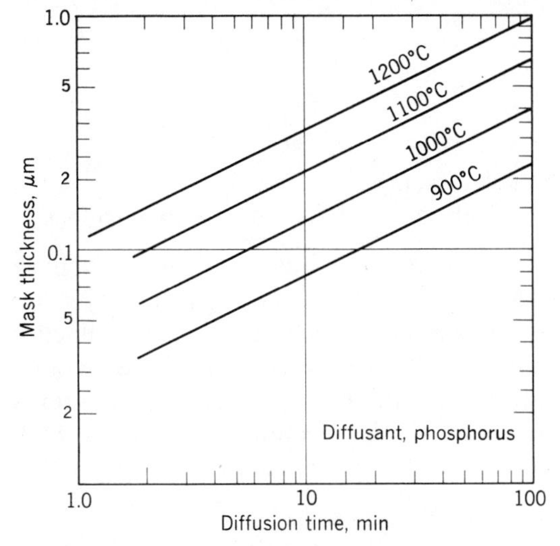

Figure 6.7 Mask thickness for phosphorus.

high space-charge density within the layer itself. As a consequence, the properties of the microcircuit are affected by the presence of this layer.

The detailed nature of the electronic properties of the SiO_2 layer is considered in the later chapter. For the present it is sufficient to state that this layer is a significant improvement over alternate surface treatments and is universally used in microcircuit fabrication.

6.7. EVALUATION OF OXIDE LAYERS

The quality of oxide layers may be evaluated in a number of ways. Thus dielectric breakdown strength is one measure of this quality. Tests of this type usually entail the evaporation of a metal layer (aluminum or gold) on the surface to form a capacitor, whose breakdown voltage is experimentally determined. Oxide layers thickness is then measured by the angle-lap and interferometric technique described in Section 4.9.2. An estimate of the oxide density may thus be obtained, since it is linearly related to breakdown strength (see Figure 6.8).

The most commonly used methods of evaluation are visual in nature. Thus a check is made to ensure that the oxide is smooth and free from crystallites and other surface blemishes. Oxide thickness may also be determined to a fair degree of accuracy by visual means. If a thin film with a reflecting back surface is viewed by monochromatic light in an almost perpendicular direction, it can be shown that the interference maxima will be located at wavelengths λ_k, such that

$$\lambda_k = \frac{4nd}{2k+1}, \tag{6.15}$$

where $k = 0, 1, 2, \ldots,$
$d =$ film thickness, in Å,
$n =$ index of refraction of the film (1.459 for SiO_2).

Alternatively if the film is viewed in white light, it will exhibit a brightly colored appearance at one of the wavelengths given by (6.15). Thus once the film color is known, its thickness can be determined from Table 6.3, which has been derived for SiO_2 layers[9]. (A more complete color table with more detailed descriptions is available in Reference 10). The uncertainty on the value of k is generally resolved on the basis of known growth conditions, which allow a rough estimate of film thickness.

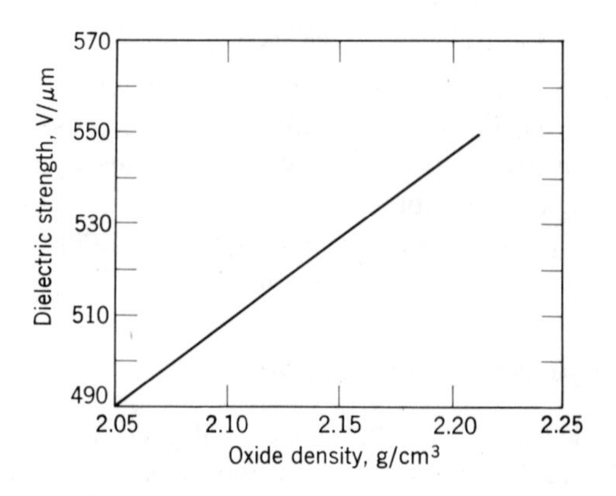

Figure 6.8 Oxide density versus dielectric strength (Deal[8]).

Table 6.3
Silicon Dioxide Interference Color-Thickness Relationship [9]

Color	Thickness, μm			
	$k=0$	$k=1$	$k=2$	$k=3$
Gray	0.01			
Tan	0.03			
Brown	0.05			
Blue	0.08			
Violet	0.10	0.28	0.46	0.65
Blue	0.15	0.30	0.49	0.68
Green	0.18	0.33	0.52	0.72
Yellow	0.21	0.37	0.56	0.75
Orange	0.22	0.40	0.60	
Red	0.25	0.44	0.62	

6.8. REFERENCES

[1] C. J. Frosh and L. Derick, "Surface Protection and Selective Masking During Diffusion in Silicon," *J. Electrochem. Soc.,* **104** (9), 547–552 (1957).

[2] J. M. Stevels and A. Kats, "The Systematics of Imperfections in Silicon-Oxygen Networks," *Philips Res. Rept.*, **11**, 103–114 (1956).

[3] R. M. Barrer, *Diffusion in and Through Solids*," Macmillian, New York, 1941.

[4] J. M. Stevels, "New Light on the Structure of Glass," *Phillips Tech. Review*, **22** (9/10), 300–311 (1960/61).

[5] J. R. Ligenza, "Oxidation of Silicon by High Pressure Steam," *J. Electrochem. Soc.*, **109**, 73–76 (1962).

[6] B. E. Deal and A. S. Grove, "General Relationship for the Thermal Oxidation of Silicon," *J. Appl. Phys.*, **36** (12), 3370–78 (1965).

[7] N. Cabrera and N. F. Mott, *Rept. Progr. Physics*, **12**, 143 (1948).

[8] B. E. Deal, "The Oxidation of Silicon in Dry Oxygen, Wet Oxygen, and Steam," *J. Electrochem. Soc.*, **110** (6), 527–533 (1963).

[9] R. M. Burger and R. P. Donovan, *Fundamentals of Silicon Integrated Device Technology, Volume 1,* Prentice-Hall, New Jersey, 1967.

[10] W. A. Pliskin and E. E. Conrad, "Nondestructive Determination of Thickness and Refractive Index of Transparent Films," *I.B.M. J. Res. Dev.*, **8**, 43–51 (1964).

Chapter 7

Chemical Processes

Chemical processes are involved at various steps in microcircuit fabrication. Thus the initial handling of a wafer involves lapping and chemical etching. During successive diffusions, chemical techniques are used to open windows over appropriate regions of the protective oxide. Finally, contacts are made to the wafer by metallization through holes that are etched in the oxide.

Chemical processes were generally considered to be the single most important reason for the lack of reproducibility and stability in semiconductor devices. This is still the case, although the use of planar techniques has gone a long way towards alleviating these problems.

7.1. THE ETCHING PROCESS

The chemical etching[1] of a semiconductor usually proceeds in two steps. The first of these is an oxidation-reduction (so-called *redox*) reaction in which the semiconductor is promoted from its zero oxidation state to some higher oxidation state, as shown by

$$M^0 \rightleftharpoons M^{x+} + xe^-, \tag{7.1}$$

where M represents the semiconductor. This reaction may also be written in the form

$$M^0 + xh^+ \rightleftharpoons M^{x+}. \tag{7.2}$$

In this manner it is emphasized that the oxidation reaction requires holes for its execution.

The reduction reaction, which occurs simultaneously, is accompanied by the liberation of holes. Writing N as the oxidizing species,

$$N \rightleftharpoons N^{x-} + xh^{+}. \tag{7.3}$$

The entire reaction, which is charge neutral, is given by

$$M^{0} + N \rightleftharpoons M^{x+} + N^{x-}. \tag{7.4}$$

The second reaction involves the dissolution of oxidation products in the form of complex soluble ions. This is done by means of a complexing agent. For silicon, nitric acid is the most commonly used oxidizing agent and hydrofluoric acid is the complexing agent.

Both oxidation and reduction reactions described by (7.2) and (7.3) take place at localized areas on the silicon surface. The oxidation reaction is an *anodic reaction* and takes place at localized anodes. Conversely, reduction of the oxidizing species is a *cathodic reaction* and occurs at localized cathodes. The over-all reaction proceeds from the action of these localized electrolytic cells and gives rise to the flow of relatively large (in excess of 100 A/cm^2) corrosion currents.

Over a period of time each localized area (which is large compared to atomic dimensions) adopts the role of both anode and cathode. If the proportion of time allocated to each role is roughly equal, uniform etching occurs. Conversely, selective etching occurs if these times are very different. Such factors as crystallographic orientation, defect structure, impurities in the etchant, temperature, and hydrodynamics of the semiconductor-etchant interface play an important role in determining the degree of selectivity of the etchant as well as its etch rate.

7.1.1. The Cathodic Reaction

The cathodic reduction of HNO_3 is an autocatalytic reaction[2], in which the reaction products promote the reaction itself. This is verified by the experimentally observed fact that the enhanced removal of reaction products by rapid stirring leads to a lower reaction rate. Autocatalysis proceeds in the presence of trace impurities of HNO_2, as shown:

$$HNO_2 + HNO_3 \rightarrow N_2O_4 + H_2O \tag{7.5a}$$
$$N_2O_4 \rightleftharpoons 2NO_2, \tag{7.5b}$$
$$2NO_2 \rightleftharpoons 2NO_2^{-} + 2h^{+}, \tag{7.5c}$$
$$2NO_2^{-} + 2H^{+} \rightleftharpoons 2HNO_2. \tag{7.5d}$$

The HNO_2 generated in (7.5*d*) reenters into reaction with HNO_3 in (7.5*a*), and the process is thus autocatalyzed. The first of these reactions is the rate-limiting one; in some cases NO_2^- ions are deliberately added (in the form of ammonium nitrite) to induce the reaction. Since the HNO_2 is regenerated in this reaction, the oxidizing power is a function of the amount of undissociated HNO_3.

One end product of this reaction is the generation of holes, as shown by (7.5*c*). These holes are a necessary component of the anodic reaction.

7.1.2. The Anodic Reaction

It has been postulated that the anodic reaction of silicon is a two-step process, as follows:

1. A surface silicon atom loses x electrons which go into the conduction band. This is accompanied by the decomposition of water molecules, as shown by

$$Si + xH_2O \rightarrow Si(OH)_x + xH^+ + xe^-. \tag{7.6}$$

2. Holes provided by the cathodic reaction cooperate with water molecules in the reaction

$$Si(OH)_x + (4-x)h^+ + (2-x)H_2O \rightarrow SiO_2 + (4-x)H^+. \tag{7.7}$$

The over-all anodic reaction is given by

$$Si + 2H_2O + (4-x)h^+ \rightarrow SiO_2 + 4H^+ + xe^-. \tag{7.8}$$

In these reactions values of x range from 1 to 2.

7.1.3. The Complexing Reaction

Fluorine is normally used as the complexing anion for silicon. This is introduced in the form of hydrofluoric acid, the reaction being given by

$$SiO_2 + 6HF \rightarrow H_2SiF_6 + 2H_2O. \tag{7.9}$$

Stirring serves to remove the soluble H_2SiF_6 from the vicinity of the silicon slice.

7.1.4. The Over-all Reaction

The over-all etching process is preceeded by an induction period during which the autocatalysis of HNO_3 is initiated. This induction period is one of uncertain duration and is usually avoided by the addition of ammonium nitrite when necessary (e.g., in etching solutions with a low HNO_3 concentration).

The induction period is followed by the cathodic reduction of HNO_3, resulting in a source of holes which flow to the anode and enter into the oxidation reaction. The oxidation products are reacted in hydrofluoric acid to form the soluble H_2SiF_6. In practice all of these processes occur within a single etch mixture; the result is in the over-all reaction

$$3Si + 4HNO_3 + 18HF \rightarrow 3H_2SiF_6 + 8H_2O + 4NO + 3xh^+ + 3xe^-. \tag{7.10}$$

7.2. FACTORS AFFECTING THE ETCH RATE [3]

The etching process consists of transporting etchant and reaction products to and from the silicon surface respectively, and of the surface reaction itself. If the first of these is the limiting process, the reaction is said to be diffusion limited. Otherwise, it is reaction-rate limited.

A reaction may be diffusion limited if it is so fast that the concentration of etchant cannot be maintained at the surface. For this case a stagnant layer of thickness δ is considered to exist at the surface and the etchant is assumed to diffuse through this layer.

As shown in (7.10) the etching of silicon is accompanied by the evolution of nitric oxide bubbles at the surface, which is effective in breaking up the stagnant layer. The result is very small values† of δ ($\simeq 3\ \mu m$). Thus in most cases the HNO_3-HF etching of silicon is reaction-rate controlled.

The surface reaction rate is an exponential function of temperature. Rapid stirring serves to prevent the buildup of hot spots on the slice, and leads to more uniform etching.

In addition to temperature the surface reaction rate is influenced by the following:

Surface Damage. Mechanically prepared silicon surfaces exhibit an increased etched rate over surfaces that are already chemically prepared. The evidence to date indicates that this is almost entirely caused by their larger effective surface area.

Crystal Defects. Crystal defects such as dislocations are usually associated with the segregation of impurities. In many etches these tend to behave preferentially as localized anodes and lead to an increased etch rate in their vicinity.

Impurities in the Etchant. These often alter the adsorption properties of the semiconductor surface, and hence the surface reaction rate.

†For comparison the spinning of a cylindrical specimen at 1500 rpm results in a value of δ of about 5 μm.

Surface Films. These are formed if the reaction products are insoluble and can lead to a drastic alteration in the etch rate. In some instances adherent surface films can completely terminate the reaction.This sometimes occurs during the etching of silicon in the presence of a high nitric acid concentration. Here it is possible for the formation of the oxidation products to occur at a faster rate than their dissolution in HF, resulting in the buildup of a tough, impenetrable film.

7.3. THE HF-HNO_3 SYSTEM

The HF-HNO_3 system is commonly used for the etching of silicon during microcircuit fabrication. Extensive studies [4, 5] of this system have been made by Robbins and Schwartz, using both water and acetic acid as the diluents. Rapid stirring was used to prevent the formation of localized hot spots and to present continually the silicon surface with fresh etchant. The uncertain induction period (for certain compositions) was avoided by the addition of nitrite ions in the form of $NaNO_2$. Figure 7.1 shows their results in the form of isoetch curves for the various constituents by weight. It should be noted here that normally available concentrated acids are 49.2% HF and 69.5 % HNO_3 by weight respectively.

Either water or acetic acid may be used as the diluent for the system. Qualitatively, both show similar behavior. Common to both systems are the following characteristics:

1. At low HNO_3 concentration, corresponding to the region near the upper vertex of Figure 7.1, the etching contours run parallel to the lines of constant HNO_3. Thus the etch rate is controlled by the HNO_3 concentration in this region. This is due to the fact that there is an excess of HF to dissolve the SiO_2 formed during the reaction.
2. In the region of the lower right vertex (low HF concentration), the etch-rate contours are parallel to the lines of constant HF. Here there is an excess of HNO_3 and the etch rate is governed by the ability of the HF to remove the SiO_2 as it is formed.
3. The etch rates are initially insensitive to the addition of diluent in both the high-HF and high-HNO_3 regions. Eventually, they fall off very sharply until the system becomes critical with respect to the diluent.

The principal advantage of using acetic acid instead of water lies in the greater tolerance of the system for this diluent. This is caused by its lower dielectric constant (6.15 compared with 81 for water). Thus its use results in less dissociation of the nitric acid, and hence in a higher concentration of the undissociated species. This preserves the oxidizing

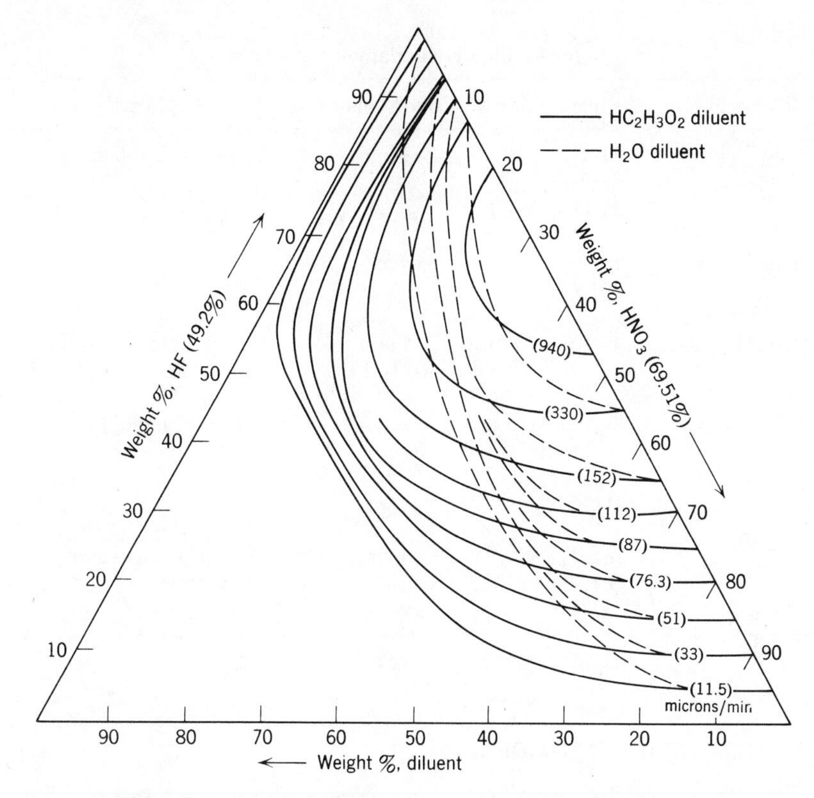

Figure 7.1 Isoetch curves for silicon (Robbins and Schwartz [5]).

power of the HNO_3 for a wider range of acetic acid dilution than for water (see Section 7.1.1).

7.4. CRYSTALLOGRAPHIC ETCHES

A large variety of chemical formulations have been developed for the purpose of delineating various crystal defects and also for exhibiting various crystallographic orientations. Often, they are composed of polishing etches (such as those described in Section 7.3) with the addition of one or more reagents. A few different formulations and their specific applications are shown in Table 7.1 for illustrative purposes.

The etch rate of crystallographic etches is different for different impurity concentrations, as well as for different impurity types. Consequently, they may be used for *p-n* junction delineation.

Table 7.1
Some Etching Formulations for Silicon

Name	Formulation†	Conditions	Application
CP4	3HF 5HNO_3 3CH_3COOH	2–3 min	Chemical polishing
White etch	1HF 3HNO_3	15 sec	Chemical polishing
NaOH or KOH	1–30% solution	1–5 min 50–100°C	Developing of structural details
Silver etch	4 ml HF 2 ml HNO_3 4 ml H_2O 200 mg $AgNO_3$	10 sec	Developing of faults in epitaxial layers
	Conc. HF with 0.1–0.5% conc. HNO_3	One drop on a freshly lapped surface	Delineation of *p-n* junctions

†Constituents are by volume of concentrated commercial reagents.

7.5. THE ETCHING OF OXIDES

The etching of SiO_2 with HF is a complexing reaction and has been described in Section 7.1.3. In practice this is performed in a dilute solution of HF, buffered with NH_4F to avoid depletion of the fluoride ion. A commonly used formulation consists of one part by volume of concentrated (49%) HF, to 10 parts by volume of an NH_4F solution. The NH_4F solution, in turn, consists of 1 lb of NH_4F in 680 ml of water. For this formulation, typical etch rates for SiO_2 are about 7 to 9 Å/sec.

In general, a loosely structured oxide (such as one grown in steam) presents a larger surface area than a more dense one (e.g., one grown in dry oxygen). Consequently steam-grown oxides etch more rapidly than those grown in dry oxygen. In addition, these rates are enhanced in regions of high mechanical stress. Undercutting of SiO_2 near the oxide-silicon interface has been attributed to this fact. In extreme situations this may result in uncovering the oxide protection over a junction edge, as shown in Figure 7.2*b*. This problem is sometimes encountered in the fabrication of shallow-diffused high-speed transistors in which the emitter-metallization window is as large as the emitter-diffusion window (the so-called *washout* emitter).

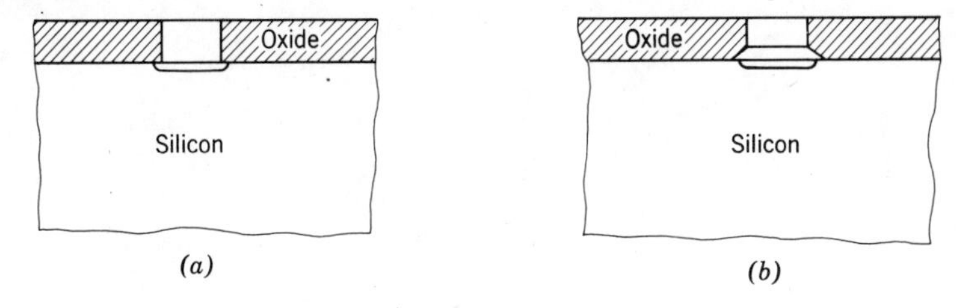

Figure 7.2 Undercutting effects: (*a*) ideal, (*b*) actual.

The surface layers of a diffused slice are usually covered with a thin layer of a mixed-oxide glass. As may be expected, the etch rate of glasses of this type is considerably faster than that for SiO_2. Experimental values of 100 to 200 Å have been quoted[6] for the buffered HF etch formulation described above.

In addition to the buffered HF etch, a number of special etch formulations may be prepared for specific situations. A description of these is beyond the scope of this chapter.

7.6. REFERENCES

[1] H. C. Gatos, and M. C. Lavine, "Chemical Behavior of Semiconductors: Etching Characteristics," in *Progress in Semiconductors*, Vol. 9, Temple, London, 1965 1–46.

[2] D. R. Turner "On the Mechanism of Chemically Etching Ge and Si," *J. Electrochem. Soc.*, **107** (10) 810–816 (1960).

[3] P. J. Holmes (Ed), *The Electrochemistry of Semiconductors*, Academic, London, New York, 1962.

[4] H. Robbins, and B. Schwartz, "Chemical Etching of Silicon, I. The System HF, HNO_3, and H_2O," *J. Electrochem Soc.*, **106** (6), 505–508 (1959).

[5] H. Robbins, and B. Schwartz, "Chemical Etching of Silicon II. The System HF, HNO_3, H_2O and $HC_2H_3O_2$," *J. Electrochem. Soc.*, **107** (2), 108–111 (1960).

[6] W. A. Pliskin, and R. P. Gnall, "Evidence for Oxidation Growth at the Oxide-Silicon Interface from Controlled Etch Studies," *J. Electrochem. Soc.*, 111, 872–873 (1964).

Chapter 8

Photo-aided Processes

A microcircuit is fabricated by accurately positioning a number of appropriately diffused semiconductor regions to form both active and passive components, and subsequently interconnecting these by metallization paths. During its fabrication holes must be cut in the oxide passivation layer to provide windows through which these diffused regions are made. Each diffusion step requires its own series of oxide cut, impurity diffusion, and oxide regrowth steps. Finally, holes must be cut in the oxide to make connections to these regions. The rest of the oxide layer acts as an insulating film over which the interconnections are placed.

Phototechniques play an important part in these fabrication steps. Many of these techniques are proprietary in nature and are described only briefly. On the other hand, problems arising from defective processing have a bearing on the long-term reliability of the final microcircuits and are emphasized in this chapter.

8.1. OXIDE CUTS

Photoengraving is the basic technique by which windows may be cut in the oxide layer. Figure 8.1 shows a flow chart for the many steps encountered in the formation of these windows. It is assumed that the starting point is a silicon slice covered with an oxide layer. For the purpose of illustration only a single window is considered. In an actual microcircuit, however, a large number of identical circuits are made on a single slice; thus a correspondingly large number of windows are "opened" in any single photoengraving step.

Steps in the opening of this window in the oxide are now detailed, with reference to Figure 8.1. Some of the results of these steps are illustrated in Figure 8.2.

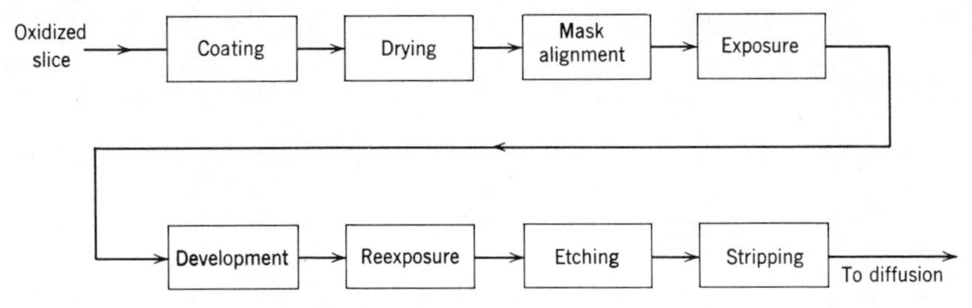

Figure 8.1 Steps in the prediffusion process.

Coating. This consists of laying a film of a photoresist material on the surface of the oxide (see Figures 8.2*a* and *b*). Ideally, such a film should be thin, highly adherent, uniform, and completely free from dust or pinholes.

The coating process consists of spinning the silicon slice at high speed after a small quantity of photoresist is placed on it. Typically, spinning speeds range from 1000 to 5000 rpm and result in films that are about 0.5 μm thick.

Extreme care must be taken to use clean, dry slices to obtain good adhesion of the photoresist. Freshly oxidized slices may be coated directly; however, slices that have been stored on oxidation must be subjected to elaborate cleaning and drying procedures before coating.

To obtain a uniform film it is necessary to bring the spinner rapidly up to full speed. In addition, two thin layers of photoresist are sometimes used in preference to a single thick one, since this minimizes the chances of pinhole formation.

Drying. On coating, the slice is heated for about 10 to 20 min at 100°C to drive out all traces of solvent from the photoresist.

Mask Alignment. A mask is placed over the slice (see Figure 8.2*c*), and manipulated into its desired position by micrometer adjustment. This alignment process is performed with the aid of a microscope.

Masks (see Figure 8.2*d*) are made on high-resolution photographic plates by reduction from a master drawing. This is usually a two-step reduction process and involves a step-and-repeat technique to obtain the required multiple-image pattern from a single master drawing.

An over-all reduction factor of about 100 to 1000 is common practice. Tolerances in the drawing† of the original artwork are about 0.001 to 0.002 in; upon reduction, these become vanishingly small.

A metal mask is sometimes used instead of the high-resolution photographic plate. This consists of a photolithographically etched,

†This is done on a mechanically controlled drafting table known as a coordinatograph.

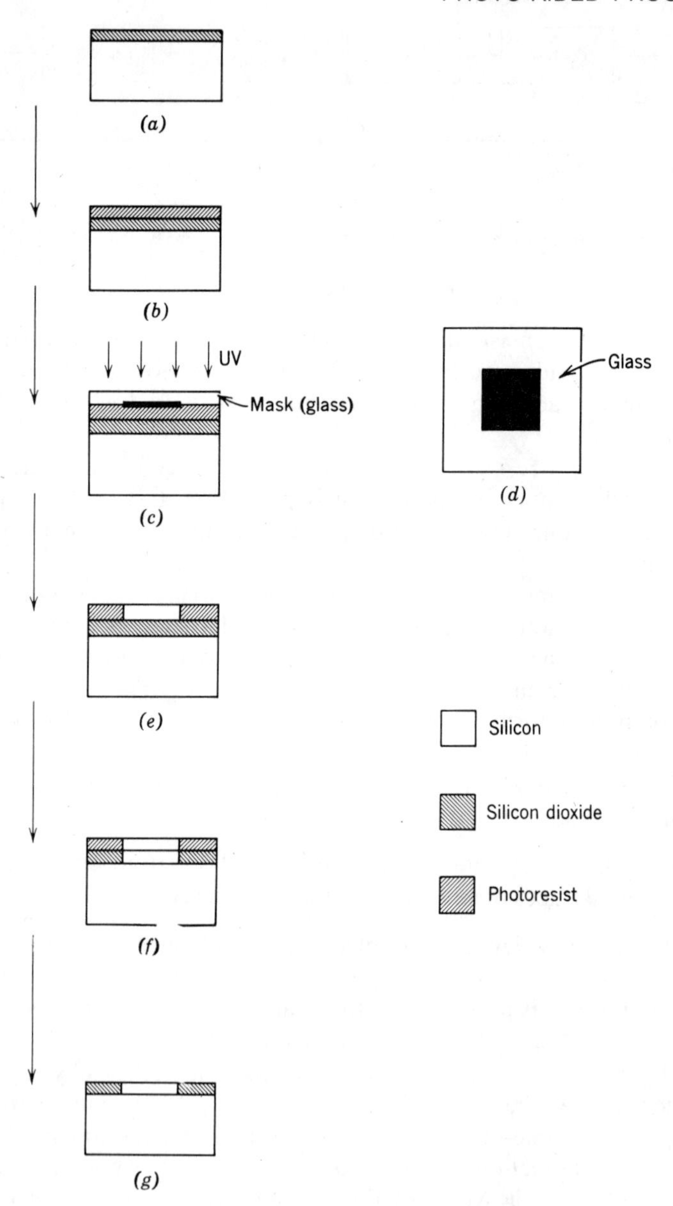

Figure 8.2 Steps in opening a window in the oxide.

chromium-based hard-metal film on a glass substrate and results in sharper edge definition. In addition, it is not as easily damaged as its gelatin counter-part during the mask alignment operation.

Exposure. Photoresists are exposed by means of ultraviolet light, preferably from a collimated source. These materials are usually organic polymers in a solvent base. On exposure, they become polymerized; this toughens them, and makes them impervious to developing agents.

Development. The slice is now rinsed in trichlorethylene or in a proprietary developer. This dissolves the unexposed photoresist but does not effect the exposed regions (see Figure 8.2*e*). A 30-min bakeout period follows development at a temperature specified by the manufacturer (about 150 to 250°C).

Reexposure. After bakeout slices are sometimes flooded with ultraviolet light to toughen further the photoresist that is left. At this stage the oxide layer on the slice is covered with a tough film in those regions where the mask was clear. In regions in which the mask was opaque the oxide layer is not covered (see Figure 8.2*e*).

Etching. The slice is now rinsed in a buffered HF etch of the type described in Section 7.5 This serves to remove the oxide that is not masked, resulting in the desired window. To avoid unnecessary undercutting, this process is visually monitored and arrested as soon as full etching is accomplished (see Figure 8.2*f*).

Stripping. The final step consists of removal of the exposed photoresist. This is usually done by boiling the slices in concentrated sulfuric acid for about 20 min, followed by mechanical agitation. Strong oxidizing agents such as hot chromic acid have also been used for this purpose. However, they leave behind a surface layer of Cr^{3+} ions, which is extremely hard to remove. These ions give rise to long-term instability problems of the type to be described in Section 17.4.5. Once the slice is free of photoresist (see Figure 8.2g), it is ready for the diffusion step.

8.1.1. Photoresists

Most of the photoresists available to the semiconductor industry are of the negative-acting type, that is, they are soluble in their developer until exposed to ultraviolet light. The main difference between them is their solid content. The thinner photoresists are capable of higher resolution; on the other hand, they have a higher pinhole density, often necessitating the use of a double coating procedure. By way of example, KPR (Kodak Photoresist) has a solid content as low as 7%, whereas KMER (Kodak metal-etch resist) has a solid content of 21%. KTFR (Kodak thin-film resist) is somewhere between these extremes in solid content as well as in resistance to etching and in pinhole density. It is perhaps the most popular general-purpose photoresist for microcircuit applications today.

Positive-acting photoresists are also available. These provide a surface coating that is easily dissolved after exposure but insoluble in developer when not exposed. Consequently, they must be used with negatives that are clear in areas in which a window is to be opened and opaque in remaining areas. An advantage of this resist is that its final removal (once the windows are opened) is done by flooding the slice with ultraviolet light. This renders the resist soluble in its developer and simplifies the stripping process.†

A photoresist of this type is manufactured by Shipley under the trade name AZ-111. This product is usually developed in a proprietary developer in which KOH is a constituent. Consequently, the problems of contamination with K^+ ions are a distinct possibility.

8.1.2. Oxide Cut Defects

A number of problems may be encountered during the opening of windows in the oxide. Some result in defective circuits that are rejected. More serious, however, are those problems that result in an inadvertent spread of device parameters which may impair the reliability of the microcircuit and degrade its performance. Some of these defects are now considered.

Undercutting of the Photoresist. If the photoresist adheres poorly to the oxide layer, undercutting may occur during the etching process. This increases the effective area of the diffusion pocket (see Figure 8.3*a*), resulting in a change in the associated device parameters. In extreme cases a defective circuit can result by short-circuiting out adjacent regions.

Undercutting of the Oxide. This effect has been described in Section 7.5. It is of particular importance in the fabrication of extremely small, shallow-diffused structures and results in degraded diode characteristics by removing the protective oxide over a junction edge.

Dimensional Variations. Dimensional variations may occur in the width of oxide cuts (as in Figure 8.3*b*) because of limitations of the photographic process. These, too, result in changing the effective area of the diffusion pocket. The problem is especially severe when long, very narrow ($\simeq 0.0001$ in) cuts are required and may lead to breaks in the window (see Figure 8.3*c*).

†As may be expected, the reexposure step must be omitted with these photoresists.

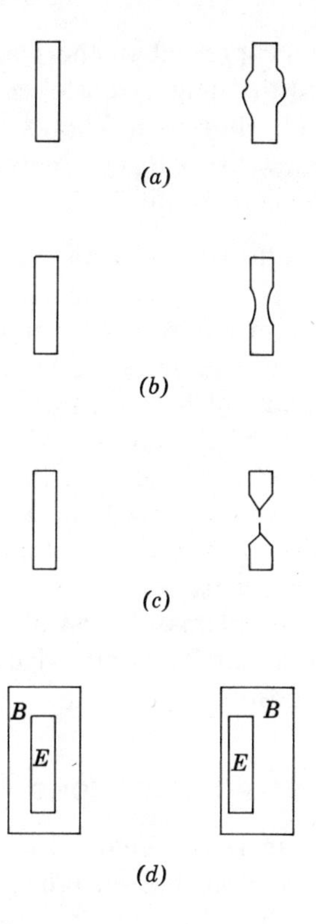

Figure 8.3 Oxide window defects.

In passing, it should be noted that an error of 1 mil (0.001 in) in the artwork results in a line-width error of about 0.01 mil or less in the mask. Thus dimensional variations due to drafting errors can usually be neglected.

Lack of Registration. Misregistration may occur between successive cuts in the oxide. These can lead to altered device configurations, as shown in Figure 8.3*d* for the emitter and base regions of a transistor. Here, too, defective circuits can result in extreme cases. In some situations, misregistration may result in exposing a protected junction edge, thus degrading its characteristics.

Registration problems have become increasingly important with the advent of high-density structures that are required for large-scale integration. At the present time a misregistration of about 0.125 mil aggregate for all processing steps over a 1 in. diameter slice is generally considered the best that can be achieved.

Pinholes. These may be present in the photoresist or in the underlying oxide and result in the inadvertent opening of a small window in the protective layer. Their effect is a function of their specific location. As may be expected, the severity of this problem increases with the component packing density in a microcircuit.

Dust Particles. Every effort is made to carry out the photoengraving process in the total absence of dust particles, with the entire process carried out in a clean-room environment. Nevertheless they present a serious problem during the making of oxide cuts. Dust particles on the clear parts of a photographic plate mask these areas from the ultraviolet light and prevent exposure of the underlying photoresist. With negative acting materials the final result is the same as that caused by pinholes in the oxide. The inverse effect occurs with positive-acting resists, i.e., dust on the photomask results in inadvertently masked spots on the silicon surface. This latter effect is usually considered less serious than the former. In either case, the problems of dust are minimized by the use of good clean-room technique and by careful visual inspection of the masks prior to alignment and exposure.

8.2. CONTACTS AND INTERCONNECTIONS

Aluminum is used for making contacts to the semiconductor regions and also for the interconnections between them. The choice of this metal is made for the following reasons:

1. It can be readily deposited by well-developed vacuum-evaporation techniques.
2. It forms a eutectic with silicon at 577°C (see Figure 3.5). Thus on heating to about 550°C for a short period of time, the aluminum alloys to the silicon to form a strong bond.
3. At the alloying temperature aluminum reacts with the oxide surface on which it is evaporated and bonds strongly to it. It contrast, it is very difficult to obtain adherent gold films directly on SiO_2.
4. It has a low electrical resistivity. A typical aluminum interconnection is about 0.5 μm thick and has a sheet resistance of about 0.025 Ω/square.
5. It welds readily to gold attachment leads by thermocompression bonding. Extremely strong low-resistance connections can be made in this manner. In addition, aluminum leads can be easily welded to this film by ultrasonic techniques.

Contacts and interconnections are made to the microcircuit in the following manner:

1. By use of a mask with a contact pattern, windows are cut in the oxide over the appropriate regions, as shown in Figure 8.4*a* (for a p-n^+ diode).

2. A film of aluminum is evaporated over the slice and, in addition to contacting the silicon, as in Figure 8.4*b*, covers the oxide layer.

3. The slice is now covered with photoresist and exposed through a mask carrying the interconnection pattern (see Figure 8.4*c*). After development the aluminum that is not protected by the mask is removed by means of a suitable† etch. The resulting aluminum connects to the contact regions and projects over the oxide layer, which insulates it from the rest of the silicon surface (see Figure 8.4*d*). Since this oxide is relatively thick (typically 0.5 μm), the interconnection lines may be

†One etch formulation consists of phosphoric acid, nitric acid, and water in a 70:3:27 ratio by volume.

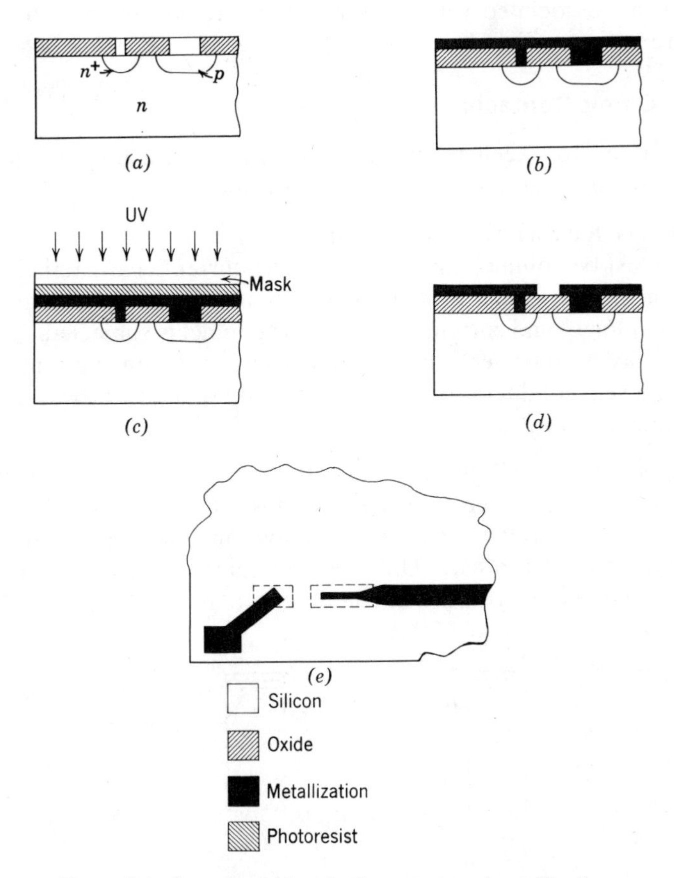

Figure 8.4 Steps in making ohmic contacts and metallization.

widened once contact is made to the device. In addition, it is possible to widen these lines further to form large bonding pads to which leads may be attached. Both situations are shown in the plan view of Figure 8.4*e*.

4. Without removing the photoresist, the slice is now heated in an inert gas ambient at a temperature slightly below the eutectic point (at about 550°C). This serves to alloy the aluminum to the silicon and to bond the metallization to the oxide. This step also removes the few monolayers of SiO_2 that are present at the aluminum-silicon interface, thus lowering the contact resistance.

5. The wrinkled and cracked photoresist resulting from the preceding operation is next removed by softening in organic solvents and by mechanical scrubbing. Strong acids or oxidizing agents cannot be used at this point since the aluminum is attacked by them.

Problems associated with the formation of contacts and interconnections are now considered.

8.2.1. Ohmic Contacts

In addition to excellent mechanical properties, a good contact to a semiconductor must have the following attributes:

1. It must have a low electric resistance.
2. It must be "ohmic," that is, its voltage-current characteristic should be a straight line going through the origin and extending over the entire range of voltages and currents to which the contact is subjected.
3. Finally, it must serve purely as a means for getting current into and out of the semiconductor but play no part in the active processes occurring within the device.

Consider the metallurgical processes that occur when an evaporated aluminum film is alloyed on a silicon substrate in the neighborhood of the eutectic temperature. Figure 8.5*a* shows an aluminum-silicon system of the type considered here. On raising its temperature, a fraction of the aluminum combines with the silicon to form a melt (see Figure 8.5*b*). On

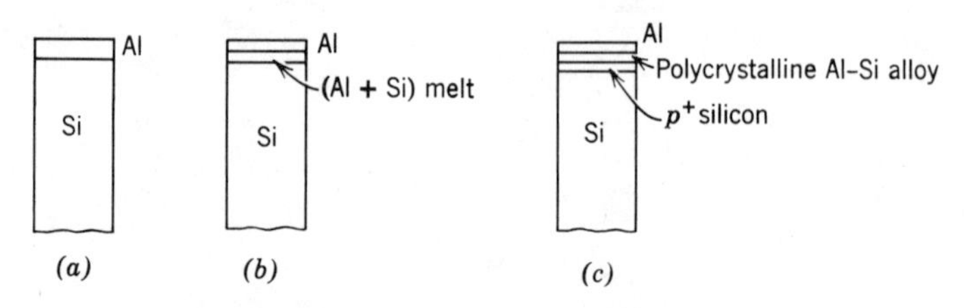

Figure 8.5 The ohmic contact.

cooling, a solid solution of aluminum in silicon freezes out epitaxially as an extension of the single-crystal substrate. This is followed by a polycrystalline Al-Si alloy and, finally, by pure aluminum, as shown in Figure 8.5*c*. Thus a highly doped p^+ region is formed on the silicon; connection is made to it by a polycrystalline Al-Si alloy.

It would at first appear necessary to perform this alloying operating above the eutectic temperature. However, this is not the case because localized melting at the points of contact between the aluminum and the silicon occurs below this temperature. An advantage over eutectic melting is that it results in a very shallow penetration depth ($\simeq 0.1\ \mu m$) and minimizes the chance of mechanical strain to the silicon lattice.

The behavior of the p^+ contact on various types of semiconductor regions is now considered.

p^+-n^+ Contact. A contact of this type is made between highly damaged, low-lifetime semiconductor regions. Thus it is a very inefficient injector of minority carriers into the semiconductor. In addition, its built-in contact potential† is sufficient to cause breakdown at zero volts, resulting in a low-resistance "ohmic" characteristic.

A contact of this type is normally made to the emitter region of an *n-p-n* transistor.

p^+-p Contact. This also results in "ohmic" behavior. In effect, the junction so formed by this combination has an extremely high-leakage current which completely masks the usual diodelike characteristic. Thus, the voltage-current characteristic approximates a straight line going through the origin and may be considered "ohmic." The connection to the base region of an *n-p-n* transistor is of this type.

p^+-n Contact. This is encountered if an aluminum contact is made to the collector of an *n-p-n* transistor. It results in a well-defined *p-n* junction diode characteristic and is quite nonohmic in nature. Consequently an ohmic contact to the collector is made by first diffusing an n^+ region‡ into it and making the p^+ contact to this region. The resulting sandwich of p^+-n^+ and n^+-n is ohmic in character.

Figure 8.6*a* shows the manner in which these contacts are made to an *n-p-n* transistor. In Figure 8.6*b* is shown a *p-n-p* transistor together with its contacting arrangement. In this case it is necessary to make an n^+ diffusion on the base region, which requires an additional diffusion step and results in an enlarged base pocket. Emitter and collector contacts are quite straightforward.

†The electrical properties of *p-n* junctions are more fully treated in Chapter 12.
‡This n^+ region is usually made at the same time as the emitter diffusion and requires no additional diffusion steps.

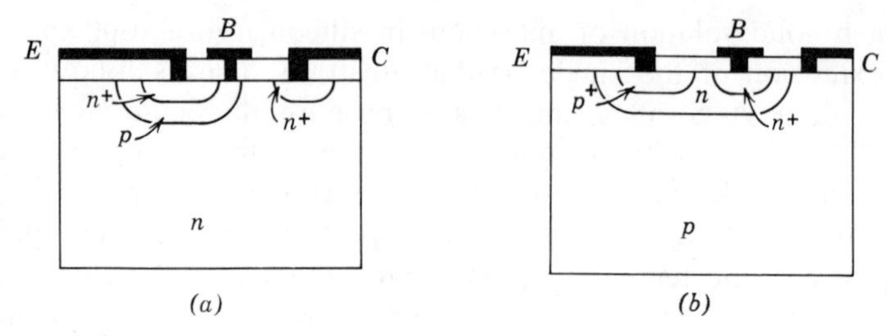

Figure 8.6 Contacts to different transistor types.

Problems with contacts can generally be traced to insufficient oxide removal from the window before sintering or to the sintering step itself. In either case traces of SiO_2 may still be present at the aluminum-silicon interface, resulting in weak, high-resistance, nonohmic contacts. Overheating can result in a high-resistance contact and in the formation of mechanical strains in the silicon lattice. In addition it can lead to embrittlement of the aluminum, thus affecting the long-term reliability of the microcircuit.

8.2.2. Interconnection Defects

Defects in the metallization steps can result in a number of electrical problems. Some of these are considered in the following.

Scratches. These are by far the most common type of defect and are caused by rough handling during the fabrication process. Scratches greatly increase the current density in their vicinity, giving rise to the formation of hot spots. Thus, although the circuit operation is unchanged, its reliability may be greatly impaired in this manner. In extreme cases, scratches result in defective circuits by breaking the metallization path.

Tears. Tearing of the metallization may occur during the removal of the photoresist. This is often encountered when the alloying step is inadequate or if the scrubbing operation is too violent. This defect also results in reducing the effective area of the current path or in an open circuit in extreme cases.

Shadowing. Shadowing may occur when an excessively thin layer of metal goes over a step in the oxide, as seen in Figure 8.7. This can give rise to a high-resistance region and result in a hot spot. Alternately the metal film may break open at this edge.

Pinholes. The vacuum evaporation of aluminum on a silicon dioxide layer with pinholes can result in short circuits to points in the semi-

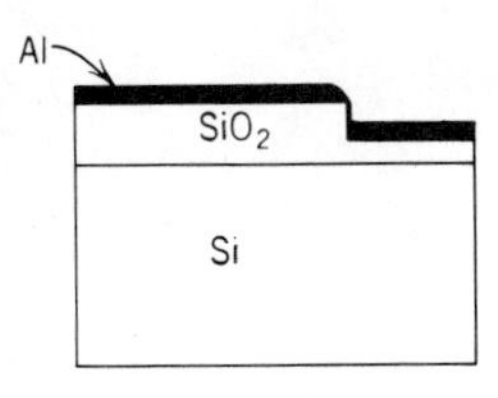

Figure 8.7 Shadowing effects.

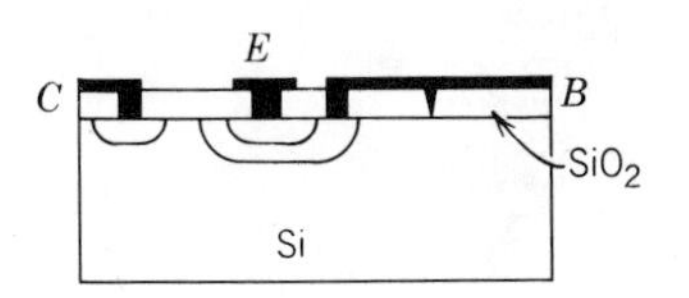

Figure 8.8 Pinholes.

conductor. An example is shown in Figure 8.8, where a collector-base short circuit results from a spike of this type. Often such spikes are extremely small in diameter ($\approx 1\ \mu m$) and cause a resistive shunt path rather than a perfect short circuit.

Sharp Corners. A poorly designed metallization path with sharp corners can result in the localized concentration of current in their vicinity. This, in turn, gives rise to hot spots. The problem can be eliminated by proper design.

8.3. SOME ADDITIONAL COMMENTS

In this chapter, a number of photo-aided processes have been outlined together with the many problems to which they are prone. Those problems resulting in defective circuits are of importance to the process engineer because they reduce the production yield. More serious, however, are the problems which result in workable circuits whose long-term reliability has been impaired.

Many of these problems can be diagnosed by suitable measurements and inspection during fabrication. Nevertheless it is common practice for users of microcircuits to subject them to further testing before acceptance. This testing normally consists of x-ray inspection, electric cycling over a range of temperatures, and burn-in procedures.

8.4. REFERENCES

[1] "Photosensitive Resists for Industry," Eastman Kodak Company, Rochester, New York, (1966).

[2] "Techniques for Photolithography," Eastman Kodak Company, Rochester, New York, (1966).

[3] "Integrated Device Technology," Vol. III, *Photoengraving,* Research Triangle Institute Document *AD603714,* Durham, N. C. (1964).

[4] R. M. Warner, Jr., and J. N. Fordemwalt, *Motorola Integrated Circuits—Design Principles and Fabrication,* McGraw-Hill, New York, (1965).

[5] Integrated Circuit Engineering Corp., "Integrated Circuits Course," *Electron. Eng.,* 86–138 (Aug. 1966).

[6] J. B. Brauer, "Poor QC Yields Bad IC's," *Electron. Eng.,* 78–82 (Aug. 1966).

Chapter 9

Packaging

This chapter describes briefly the basic principles underlying the various bonding processes required in the fabrication of microcircuits. These include the bonding of the semiconductor chip (or die) to the package, attachment of leads between the aluminum bonding pads on the silicon chip to the terminal posts on the package, and the final sealing of the package itself.

As with photo-aided processes, the semiconductor chip is subjected to considerable handling during these operations. Consequently, many of the problems of both yield and reliability may be traced to inadequate control during these processes.

9.1. DIE BONDING

Microcircuit chips range in size from as small as 20×20 mils to as large as 80×200 mils. These chips are bonded to a gold-plated header by means of a gold-germanium eutectic (88% Au-12% Ge by weight) preform. As shown in Figure 9.1, this preform is placed between the header and the silicon. The temperature of the combination is raised to 390 to 400°C and pressure is applied to the die in conjunction with a vibratory scrubbing motion. Eutectic melting occurs and a Au-Ge-Si bond results upon cooling.

In some applications, in which a good ohmic contact is required, doped preforms may be used with 0.1% gallium or antimony for p^+ and n^+ contacts respectively. In many microcircuits, however, this contact serves no electrical function.

A gold-silicon system of the type shown in Figure 3.6 may also be used. However, the gold-germanium system has been found to have better wetting characteristics and is preferred.

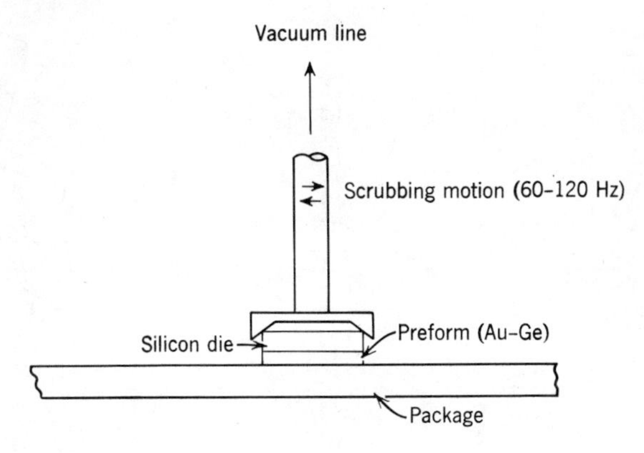

Figure 9.1 Die bonding.

9.2. LEAD ATTACHMENT

Both gold and aluminum wires, ranging from 0.5 to 2 mils in diameter, are used for connecting the bonding pads to the terminal posts. Gold wires are usually drawn from 99.999% pure gold and are work-hardened to provide enough stiffness for handling purposes. With aluminum, however, it is necessary to use a 99% Al-1% Si alloy to provide this stiffness.

9.2.1. Thermocompression Bonding

Gold welds readily to silicon by the simultaneous application of heat and pressure. Of the many types of bonding equipment available, the nail-head bonder is perhaps the most common. Figure 9.2 shows the various steps in its operation. A gold wire with a sphere at one end is fed through a capillary from a spool. The sphere is aligned over the aluminum bonding pad and then lowered. During this process the semiconductor die, mounted on its header, is maintained at 280 to 300°C in a nitrogen gas ambient. The gold sphere is pressed against this pad for a few seconds until a weld is formed. This also results in plastic flow of the sphere into the form of a nail head. On raising the capillary, gold wire is drawn out from the spool, resulting in a lead. The capillary is now aligned over the terminal post and the procedure repeated. Finally, a hydrogen torch is used to break the wire, resulting in the formation of spheres at its ends. Thus the wire is ready for the next bonding operation.

Figure 9.3 shows the typical dimensions of the lead, the nail head, and the sphere formed by this process. A disadvantage of this type of bonding is the presence of the cantilever "tail" formed with each bond. Under

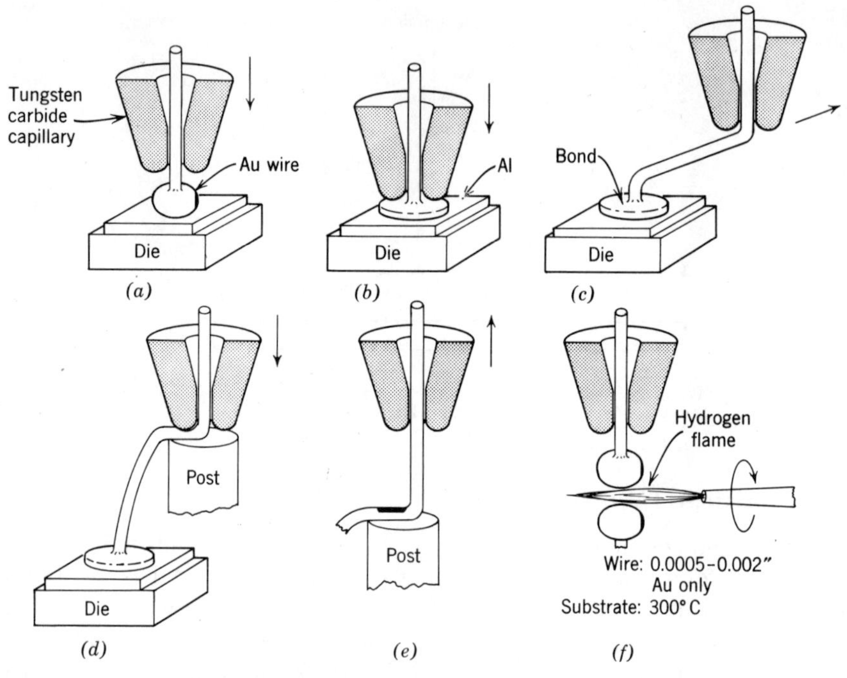

Figure 9.2 Nail-head bonding (adapted from [3]).

conditions of extreme vibration this tail can be broken off within the package, and is thus an incipient cause for circuit failure.

A stitch bonder may be used to avoid this problem. Here steps in the bonding operation are as shown in Figure 9.4, and cutters are used to prevent the formation of the dangling lead.

In practice, the nail-head bonder has been found to be considerably

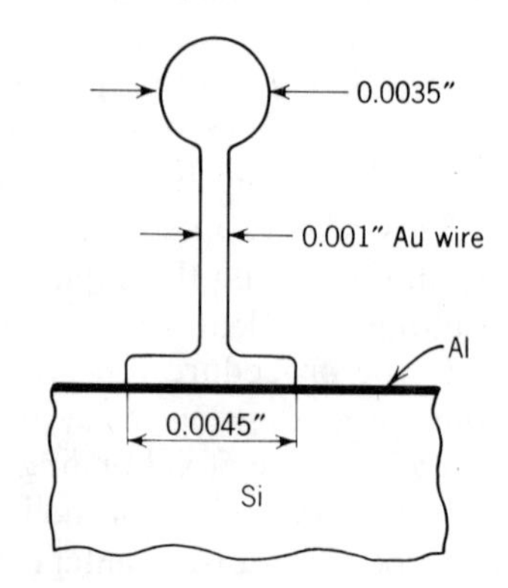

Figure 9.3 Details of a nail-head bond.

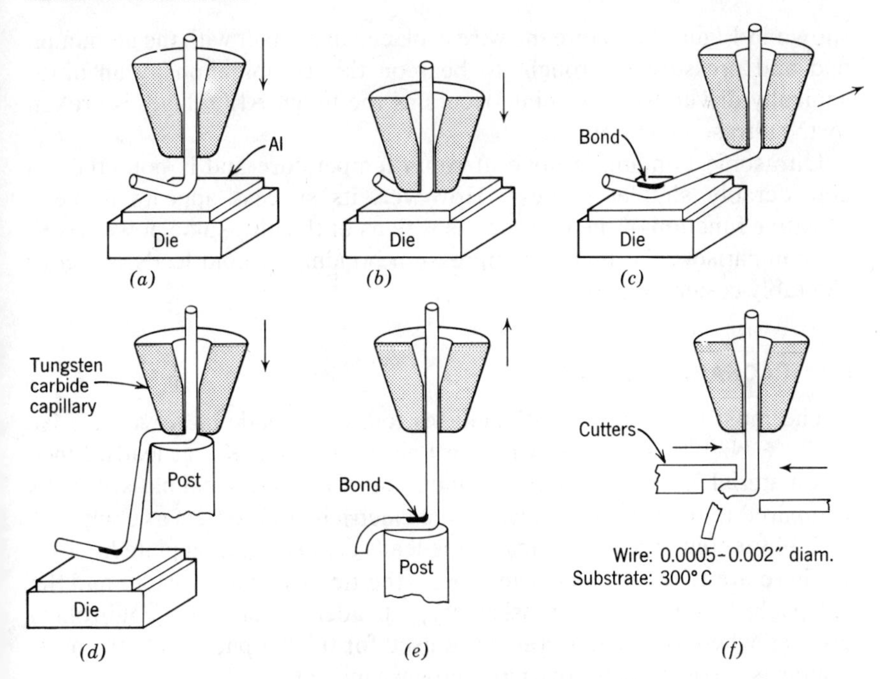

Figure 9.4 Stitch bonding (adapted from [3]).

simpler and faster to operate and results in a stronger bond. However, the stitch bonder is potentially superior and should eventually surpass it in popularity.

The problems associated with the purple plague have been described in section 3.2.4. Of the many systems that have been tried to avoid this problem, the simplest and best appears to be one in which aluminum leads are bonded to aluminum pads. In this way the binary Au-Al system is avoided at the lead-chip interface. The aluminum lead, however, must eventually be connected to the gold-plated terminal post. Thus it would appear as if this only serves to postpone the problem rather than to eliminate it.

As described in Section 3.2.4, the rate of formation of intermetallic gold-aluminum compounds is greatly reduced in the absence of the silicon. Hence the problems of making a reliable, plague-free contact at the post are not as severe as on the silicon die.

9.2.2. Ultrasonic Bonding

Ultrasonic methods have been found suitable for bonding aluminum leads to microcircuit pads. The bonding operation is accomplished as shown in Figure 9.5. Here the wire is placed in contact with the aluminum pad and pressure is brought to bear on this combination by an ultrasonically driven tool. Welding occurs as the tough Al_2O_3 layer is broken by the ultrasonic vibrations.

Ultrasonic bonding is done at room temperature and is potentially a considerably simpler process. However, its success appears to be a sensitive function of the pressure as well as of the ultrasonic power level. By comparison, the thermocompression bonding of gold leads is a considerably easier technique.

9.3. PACKAGE SEALING

The most commonly used packages today are made of Kovar (a 54% Fe-29% Ni-17% Co alloy) with hard-glass seals and Kovar leads. Since this material is a relatively poor conductor of heat (0.04 cal/cm-sec-°C as compared to 0.7 for gold) as well as of electricity (33 $\mu\Omega$-cm as compared to 2.44 for gold), the substrate and the leads are usually gold plated.

There are two basic package types – the transistor-like header and the flat pack. Sealing of the transistor-type header is done by a cold-rolling process whereas a solder preform is used for the flat pack. In both cases, sealing is carried out in a dry nitrogen-gas ambient.

Recent advances in surface passivation and plastics technology have resulted in the introduction of plastic packages. These packages have found ready acceptances for many applications by virtue of their low cost.

9.4. ADDITIONAL COMMENTS

It is at the packaging stage that the concept of batch processing breaks down; once the slice is separated into individual chips, further handling is on a one-at-a-time basis. It is not surprising, therefore, that packaging represents a major fraction of the cost of the microcircuit.

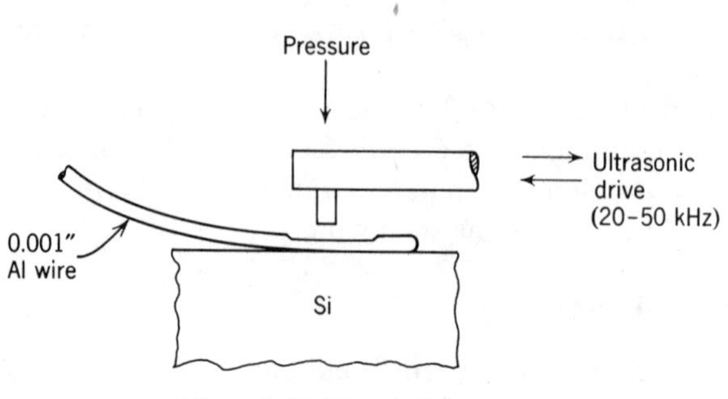

Figure 9.5 Ultrasonic bonding.

Many approaches are currently being undertaken to reduce these costs. These include the development of sophisticated chip-handling equipment, the adoption of a large-scale integration approach, and the use of beam lead and nitride technology to provide surface passivation and lead attachment directly on the slice before separating it into individual microcircuits.

The large-scale integration approach appears to be most attractive to the computer designer. However, the use of beam lead technology is very promising in the majority of applications where individual microcircuits are required. A further discussion of these approaches is beyond the scope of this chapter. The interested reader is referred to References 5 to 10.

9.5. GENERAL REFERENCES

[1] E. Keonjian (Ed.), *Microelectronics,* McGraw-Hill, New York, (1963).

[2] R. M. Warner, Jr., and J. N. Fordemwalt, *Integrated Circuits-Design Principles and Fabrication,* McGraw-Hill, New York, (1965).

[3] Integrated Circuit Engineering Corp. "Integrated Circuits Course", *Electron. Eng.* 63–102 (1966).

[4] D. Nixen, "Materials for Packaging Microelectronic Devices," *Electron. Industries,* 66–75 (1965).

[5] R. L. Petritz, "Technological Foundations and Future Directions of Large Scale Integration," *Proc., 1966 Fall Joint Computer Conference,* 65–87 (1966).

[6] R. D. Lohman, "LSI – The Fabrication Viewpoint," *Digest, 1967 International Solid State Circuits Conference,* 30–31 (1967).

[7] T. R. Finch, "LSI – Digital Electronics," *Digest, 1967 International Solid State Circuits Conference,* 32–33 (1967).

[8] V. Y. Doo et. al., "Preparation and Properties of Pyrolytic Silicon Nitride," *J. Electrochem. Soc.* **113**, 1279 (1966).

[9] M. P. Lepselter, "Beam-Lead Technology," *Bell Systems Tech. J.,* **45**, 233–253 (1966).

[10] J. H. Forster and J. B. Singleton, "Beam-Lead Sealed-Junction Integrated Circuits," *Bell Lab. Record,* **44**, 312–317, (Oct.–Nov. 1966).

Chapter 10

Monolithic Processes

The preceding chapters have described a number of basic processes that are used for fabricating semiconductor components. It remains now to describe those special steps that allow the fabrication and interconnection of a number of these components into a single "monolithic" piece of semiconductor. These processes are called *monolithic processes* [1–3]. They are not unique; rather, they are specific ways in which the basic processes may be applied.

Monolithic processes are described in this chapter together with their unique features. The detailed electrical characteristics of the components so formed are described in later chapters. No attempt has been made to be all-inclusive here; since industry is engaged in a substantial research effort to develop new processes, such an attempt would be short-lived.

10.1. ISOLATION

The ability to isolate components from one another is a fundamental requirement of the monolithic process. For example, it would be necessary for all transistors to share a common collector node (as shown in Figure 10.1), if this were not possible. Although this type of circuit is useful in switching applications such as direct-coupled transistor logic, this requirement is far too restrictive from the point of view of the circuit designer. The various techniques that follow have all been developed to remove this restriction.

10.2. OXIDE ISOLATION

Conceptually the most straightforward approach is to use an insulating layer [4] to isolate a number of pockets of single-crystal semiconductor

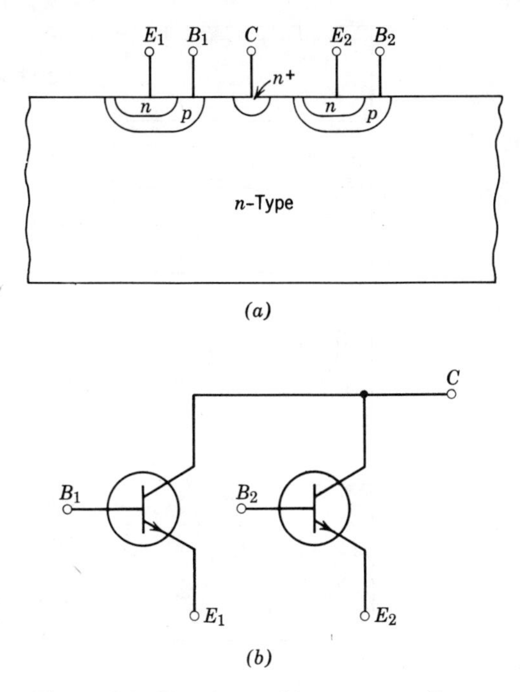

Figure 10.1 Transistors with a common collector.

material in a slice of silicon. It is only quite recently, however, that schemes of this type have been put into commercial practice. Silicon dioxide is the usual choice of insulator, although other materials are feasible.

One method for obtaining oxide isolation is described here to illustrate the approach to the problem. Most practical techniques used are based on this process, with some modifications:

1. A silicon slice is masked as shown in Figure 10.2*a* and etched so as to result in the structure shown in Figure 10.2*b*.
2. A thick oxide layer is grown on this slice (see Figure 10.2*c*).
3. Next, polycrystalline silicon is grown on top of this oxide layer (see Figure 10.2*d*). This is often accomplished by the pyrolitic decomposition of silane. The thickness of the polycrystalline layer is about 200 to 250 μm.
4. The single-crystal side of the slice is now lapped down, resulting in the structure shown in Figure 10.2*e*. The resulting slice consists of the desired pockets of single-crystal silicon, isolated from each other by a layer of silicon dioxide. Various types of components may be fabricated within these pockets as desired.

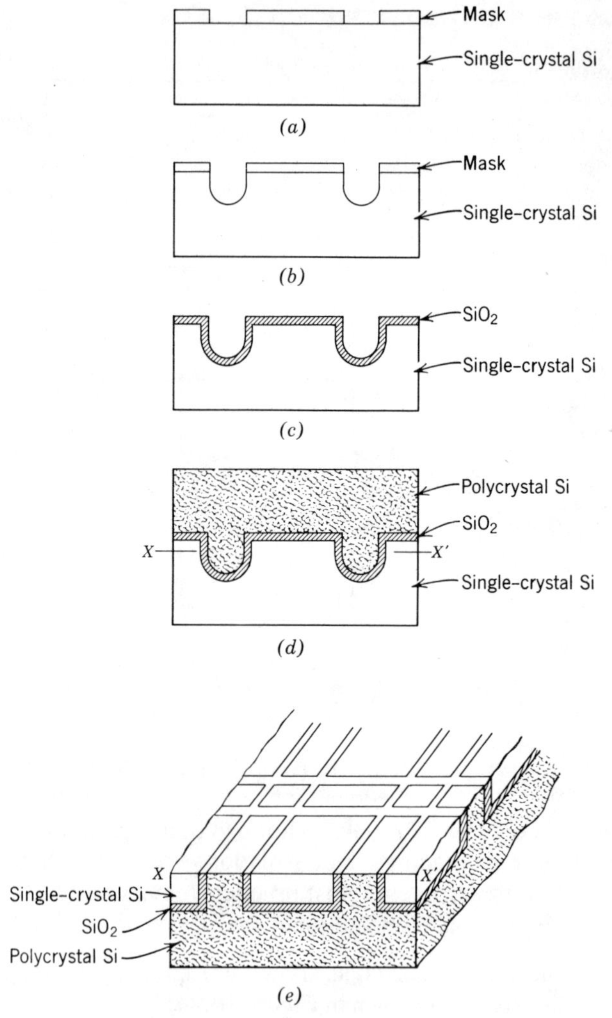

Figure 10.2 Oxide layer isolation.

This process results in near-perfect isolation between the regions. Although the heat conductivity of the SiO_2 is $\frac{1}{50}$ times that of silicon, this does not present a serious problem since the layer is thin (0.5 to 1 μm).

The main disadvantage of oxide isolation is that it requires extremely careful mechanical alignment of the slice during the lapping step. Since the depth of the isolation pocket is only 10 to 12 μm, any misalignment results in single-crystal pockets of variable depth. As a consequence, these processes are only used in special situations. The most common approach is one of *p-n* junction isolation, which is next described.

10.3. *P-N* JUNCTION ISOLATION

This isolation process makes use of the fact that a reverse-biased *p-n* junction silicon diode has an extremely low leakage current (in the nanoampere range at room temperature). Thus two regions of a semiconductor are effectively isolated to direct current if they are of opposite conductivity type and are suitably reverse-biased. The significant coupling between regions is primarily capacitive in nature; its effect on the microcircuit must be considered at high frequencies.

10.3.1. The Two-sided Monolithic Process

As its name implies, this process is carried out from both sides of the silicon slice. Although not in use today, it is the forerunner of the modern monolithic process.

Figure 10.3 illustrates this process. Starting with *n*-type material, *p*-type isolation diffusions are conducted from both sides until the regions meet. The *p* diffusion is made through the entire reverse side and through the top side in the form of moats. These diffusions result in pockets of *n*-type material which are isolated from each other, provided that the *p* regions are biased negatively with respect to the *n* regions. In practice, this is done by tying them to the most negative point in the microcircuit.

One or more components may now be fabricated in each of these isolated regions. Thus, if it is required to fabricate two transistors with separate collectors, two such regions are used. Alternately, both transistors may be placed in a single isolated *n*-region if they share a common-collector node.

An isolation process of this type has a number of disadvantages, as follows:

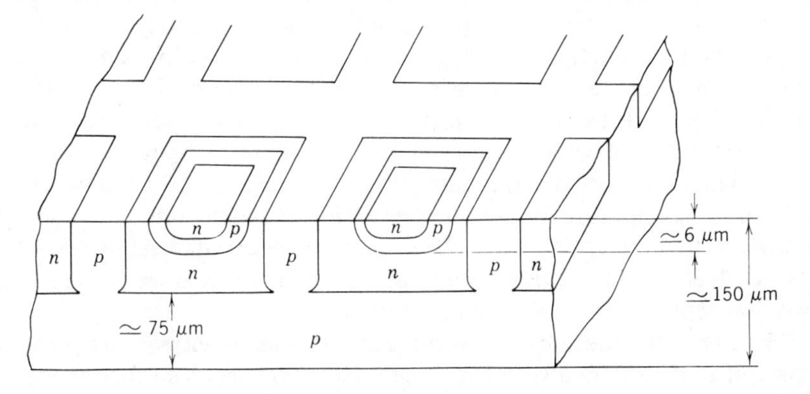

Figure 10.3 The two-sided transistor.

1. The silicon slice must be made as thin as possible to minimize the diffusion time for the isolation wall. Typically, slices used in this process are 125 to 150 μm (5 to 6 mils) thick and present severe breakage problems during fabrication.

2. Even with thin slices it is necessary to conduct the p diffusions at 1250°C for durations in excess of 40 to 50 hr. Thus a thick oxide mask must be used with one or two intermediate reoxidation steps during the diffusion run. These long diffusion times greatly increase the possibility of slice contamination and result in reverse diodes with high-leakage and soft-breakdown characteristics.

Transistors are diffused into these isolated pockets by the methods outlined in Section 4.6. Typically, the collector-base junction of these is located at 5 to 6 μm from the surface, although the effect of the isolation diffusion is only apparent some 75 μm beyond this junction. Thus the isolation diffusion is too far from the transistor to affect its doping profile. As a consequence, transistors fabricated with this process are of the double-diffused nonepitaxial type shown in Figure 4.17.

10.3.2. The Triple-diffused Process

In this process isolation is obtained by conducting n-type diffusions through the top of the silicon slice, as shown in Figure 10.4*a*. Again, the pockets so formed are isolated from one another by eventually connecting the substrate to the most negative point in the microcircuit.

Two additional diffusions are required to fabricate a transistor in this pocket; it is for this reason that the technique is referred to as the triple-diffused (TD) process. Two such transistors are shown here; the use of the n^+ region for the collector contact has been described in Section 8.2.1.

The doping profile for a typical transistor of this types is shown in Figure 10.4*b*. A number of points may be noted with reference to this profile:

1. The base diffusion is made into an n-type-graded background. This results in a doping profile that is different from that of the usual double-diffused structures. In particular, it results in a low-impurity concentration for the collector, hence, a high collector body resistance.

2. The collector-substrate junction must be placed far from the collector-base junction to avoid the necessity of subjecting the collected carriers to a retarding field. This, in turn, necessitates the use of an excessively high-resistivity substrate material to achieve a high penetration depth for the isolation diffusion.

3. Typically, the substrate is p-type, with a resistivity of 10 Ω-cm. The presence of a thermally grown oxide tends to invert its surface characteristics to n-type, resulting in short circuits between isolated areas.

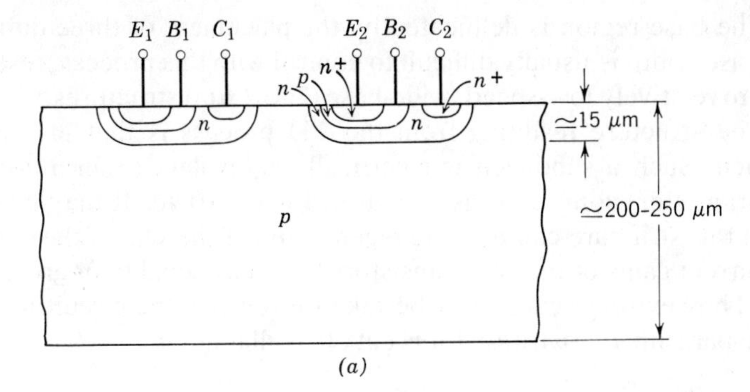

(a)

Fig. 10.4a

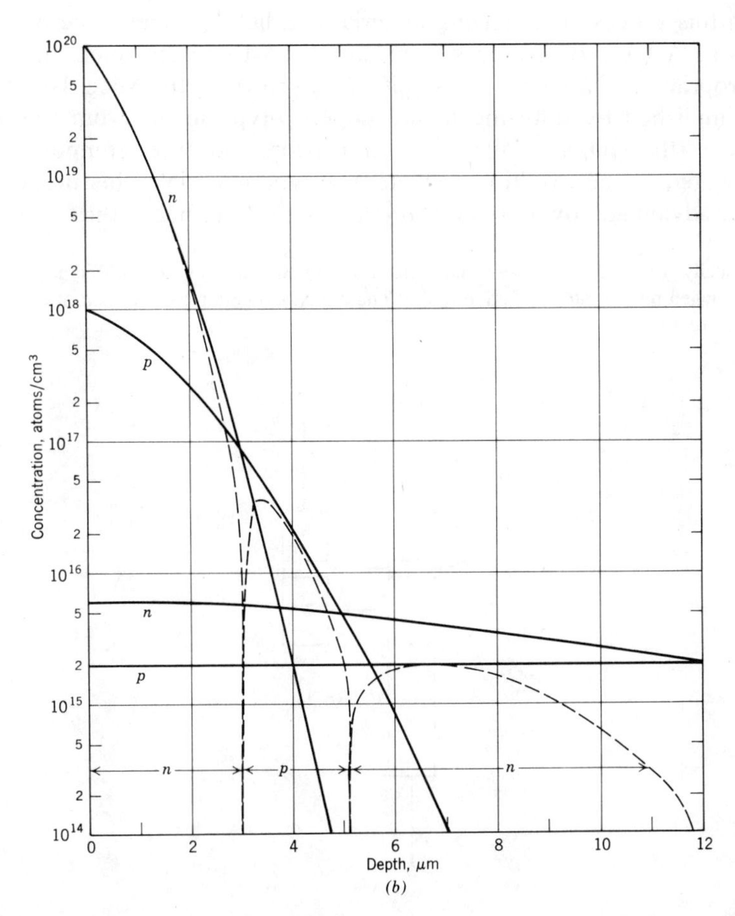

(b)

Figure 10.4 The triple-diffused process: (a) cross section; (b) doping profile;

4. The base region is delineated by the placement of three diffusions. Thus base width is usually difficult to control with this process, restricting its use to relatively low-speed, wide-base (2 to 3 μm) structures.

5. The structure resulting from the TD process is that of a *p-n-p-n* sandwich. Such a sandwich is electrically equivalent to an *n-p-n* and a *p-n-p* transistor, connected as shown in Figure 10.4*c*. It may be shown [5] that this structure can become regenerative if the sum of the common-base current gains of the two transistors becomes equal to or greater than unity. Thus extreme care must be taken in biasing the circuit to ensure that the parasitic *p-n-p* transistor is cut off at all times.†

10.3.3. The Double-diffused Epitaxial (DDE) Process

In this process the starting material is a lightly doped slice of *p*-type silicon. A layer of *n*-type silicon, usually 5 to 15 μm thick and of the appropriate resistivity type, is epitaxially grown on this slice. Isolation is accomplished by diffusing highly doped *p*-type moats from the upper surface (the epitaxial layer side); transistors and other components are fabricated in the resulting pockets of *n*-type material. This process has many advantages over previous processes, as described in the following:

†Typically, the wide-base *p-n-p* parasitic transistor has a common-emitter gain of about 2 to 3 when biased into the active range. The common-emitter current gain of the *n-p-n* is on the order of 50.

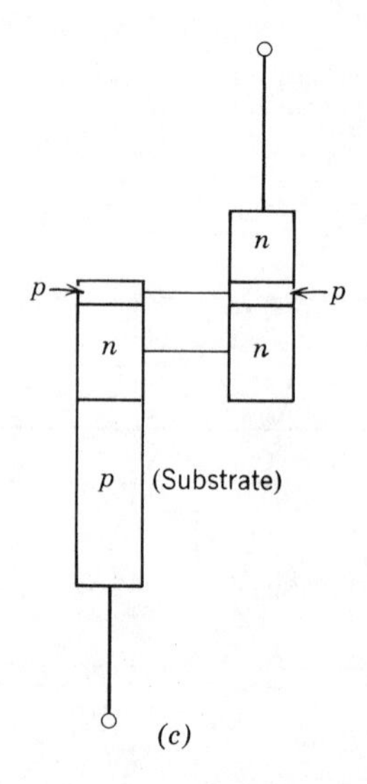

Figure 10.4(*c*) The triple-diffused process: the *p-n-p-n* structure.

1. The epitaxial layer needs only to be thin enough to fabricate a device. Thus the time taken for the isolation diffusion assumes manageable proportions and contamination during this process is minimized.

2. The substrate may be of arbitrary thickness, thus reducing handling problems during microcircuit fabrication. Thicknesses from 200 to 250 μm are typical of modern practice.

3. Since epitaxy results in a relatively abrupt doping profile, the choice of layer resistivity may be based solely on device-design considerations.

4. The lateral penetration of the isolation diffusion is approximately equal to its penetration normal to the silicon surface. Consequently, the relatively shallow diffusions required by the DDE process result in a minimum of waste area on the silicon chip and lead to greatly improved yields in processing.

Figure 10.5*a* shows a cross section of isolated transistors made by the DDE process. These are of the double-diffused non-epitaxial type; a typical doping profile for a narrow-base high-speed device of this type is shown in Figure 10.5*b*.

The DDE process may be readily extended to allow the fabrication of transistors with properties similar to those of epitaxial devices. This is done with the aid of buried layers.

10.4. BURIED LAYERS

Figure 10.6*a* shows a cross section of an isolated transistor made by the DDE process. The path of current flow from the edge of the emitter to the n^+ collector contact is also shown. This path represents a high parasitic collector resistance.

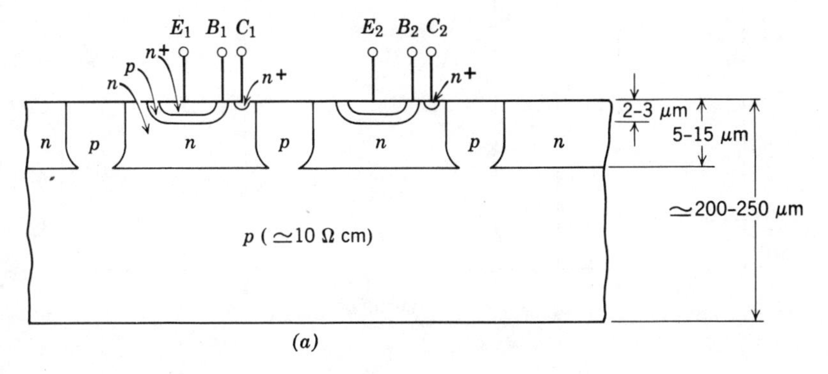

Figure 10.5(*a*) The double-diffused epitaxial process: cross section.

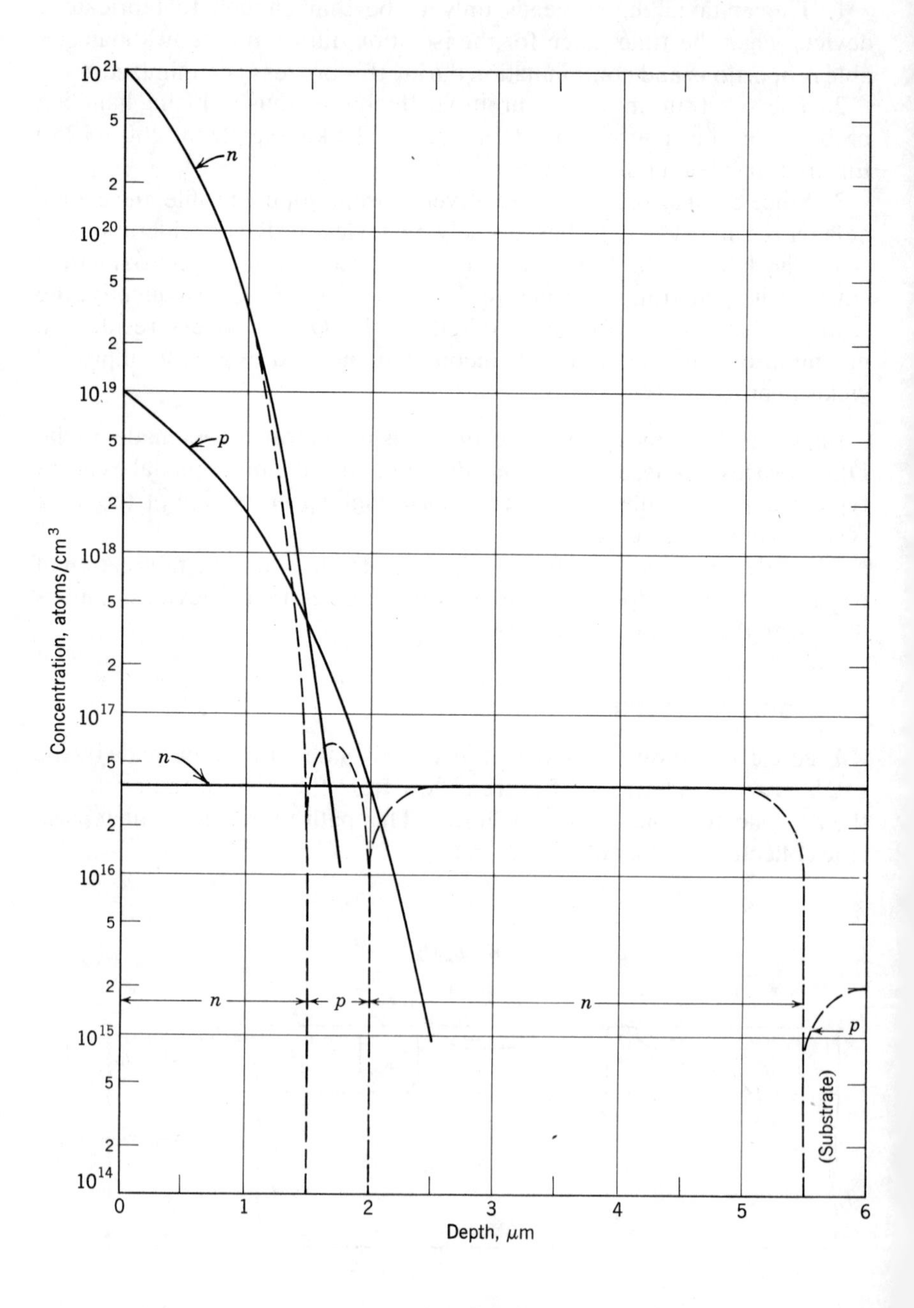

Figure 10.5(*b*) The double-diffused epitaxial process: doping profile.

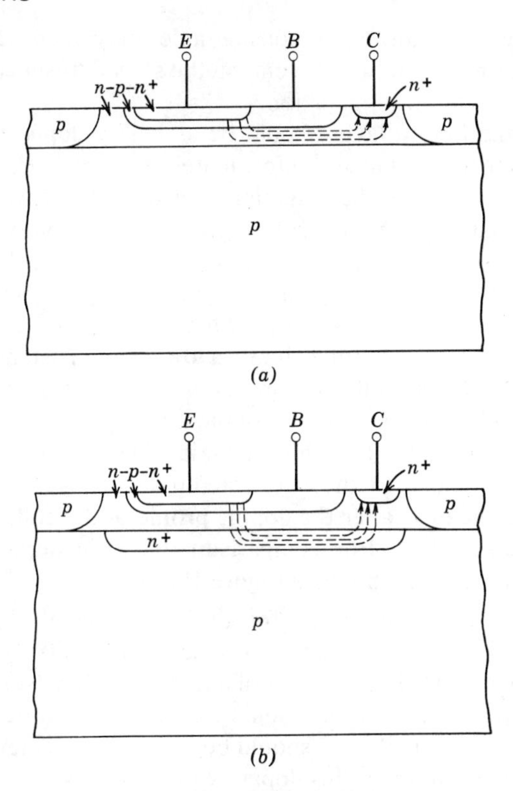

Figure 10.6 The buried layer.

Figure 10.6*b* shows the same device with a buried layer. Prior to the growth of the *n*-type epitaxial layer the substrate is masked and a region of high-concentration *n*-type material is diffused into it. The epitaxial layer is grown over this "buried layer" and the microcircuit is fabricated from the resulting slice.

The collector-current flow path is shown for this transistor in Figure 10.6*b*. In this case much of this path is traversed through the low-resistivity buried layer, resulting in a reduction of the collector parasitic resistance. For a typical high-speed transistor of small geometry, the use of the buried layer may reduce the collector resistance from 200 Ω to as low as 20 Ω, a significant improvement in many switching applications. In this manner performance comparable to that of epitaxial transistors may be obtained.

The buried layer is the first region to be introduced into the silicon slice. In addition, it is a relatively highly doped region. This causes three problems:

1. The layer is subject to considerable movement during further processing. To minimize this movement, the slow diffusers (antimony and arsenic) are used for this application.

2. Ideally, the buried layer should be as highly doped as possible, to minimize its sheet resistance. Unfortunately this leads to strain-induced damage, on top of which the epitaxial layer must be grown. In addition, severe redistribution takes place during subsequent epitaxial growth and diffusion steps. Partially to offset these problems, the buried layer is diffused by a two-step process of the type outlined in Section 4.5.3. A long drive-in step is used, typically 10 to 15 hr at 1200°C, to reduce the surface concentration and still maintain a low sheet resistance.

3. As a result of out-diffusion and etch-back effects, the ideal abrupt junction is not obtained and profiles of the type shown in Figures 5.7*a* and 5.8*a* are more common. For thin epitaxial layers, where the collector-base junction is close to the substrate-epitaxial layer interface, these effects can significantly alter the doping profile of the collector region, as shown in Figure 10.7*a*. For comparison the ideal doping profile of a buried layer transistor is shown in Figure 10.7*b*.

Antimony is commonly used for the buried layer. Its large misfit factor ($\epsilon = 0.153$) and consequent low solid solubility limits its use to buried layers with a sheet resistance of about 15 to 20 Ω/square.

In view of the perfect fit of arsenic to the silicon lattice and its consequent high solid solubility, it should be possible to achieve a substantially higher concentration of this dopant without excessive lattice damage. Assuming that solid solubility is the only limitation here, a sheet resistance of less than 5 Ω/square can be achieved. Unfortunately, arsenic has a high vapor pressure and is subject to serious etch-back effects during the growth of the layer. This fact, combined with the highly toxic nature of arsenic, has limited its use at the present time.

10.5. CHOICE OF TRANSISTOR TYPE

All of the monolithic processes result in the possibility of only one type of transistor for a specific type of substrate. Thus only *n-p-n* transistors can be fabricated in the examples shown in Figures 10.3 to 10.5. Conversely, if an *n*-type starting material were chosen, only *p-n-p* devices would be possible.

In general, the use of *n-p-n* devices is preferred over *p-n-p* devices in microcircuit fabrication for the following reasons:

1. Electron mobility is higher than hole mobility in silicon (by a factor of about 3 to 1). Thus minority carrier motion is inherently faster in the *n-p-n* structure.

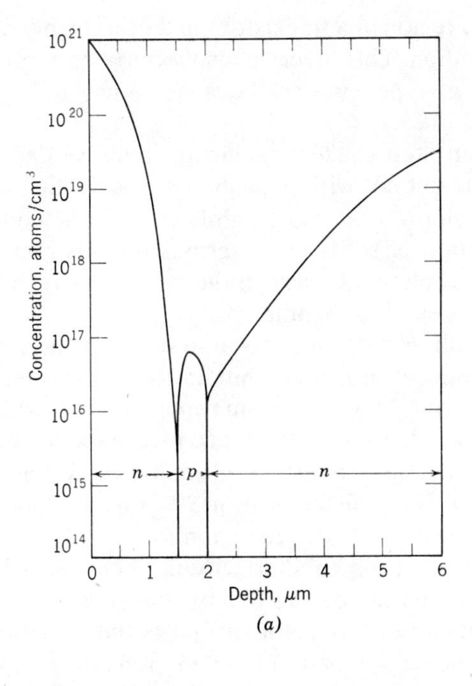

(a)

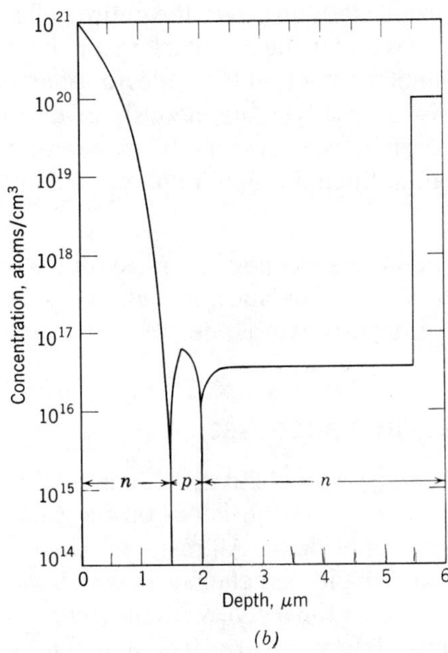

(b)

Figure 10.7 Effects of out-diffusion.

2. The emitter region of a transistor must be as highly doped as possible for efficient operation. This favors phosphorus over boron (i.e., an n^+-type emitter over a p^+-type emitter) because it has a lower misfit factor (0.068 versus 0.254).

3. The drive-in step causes significant impurity depletion in regions doped with boron but not with phosphorus (see Section 4.8.1). In a *p-n-p* transistor the region that is most lightly doped (the collector) is *p* type; hence the depletion effect is most serious for this region. For an *n-p-n* transistor boron depletion occurs in the more heavily doped base region, and the problem is easier to handle.

4. It has been observed that the charge state of a thermally grown oxide layer tends to induce an *n*-type shift in the surface regions beneath it; that is, *n*-type surface layers become more *n* type while *p*-type layers become less *p* type. Thus a lightly doped *p*-type region is most likely to be affected and can actually invert to *n* type under certain conditions. The collector of a *p-n-p* transistor is prone to this surface inversion effect because it is the most lightly doped region.

Surface inversion leads to short circuits between adjacent base and isolation regions. This can be avoided by diffusing a narrow p^+ guard ring through the middle of the collector surface at the same time as the emitter diffusion. Being heavily doped, this region remains *p* type, and prevents the inverted layer from extending over the entire collector surface. However, its presence necessitates the use of a larger collector pocket.

5. Finally, the ohmic contact on the collector of an *n-p-n* is made to an n^+ region which is diffused simultaneously with the emitter. For a *p-n-p* device, however, it is necessary to make contact to the base by means of an n^+ region. Such a region requires an additional masking and diffusion step.

As a consequence of the foregoing, almost all microcircuits are made with *n-p-n* transistors. Thus the starting materials shown in Figures 10.3 to 10.6 are indicative of current practice.

10.6. COMPLEMENTARY PROCESSES

Many attempts have been made to develop monolithic processes which provide both *p-n-p* and *n-p-n* transistors on the same substrate. A few of these are reviewed here briefly. All result in relatively poor transistors, and find their main use in circuit schemes in which the high-performance demands are only made on the *n-p-n*, with the *p-n-p* often serving merely as a voltage translation device. Figure 10.8*a* illustrates one such approach. Here, of necessity, the following compromises are made:

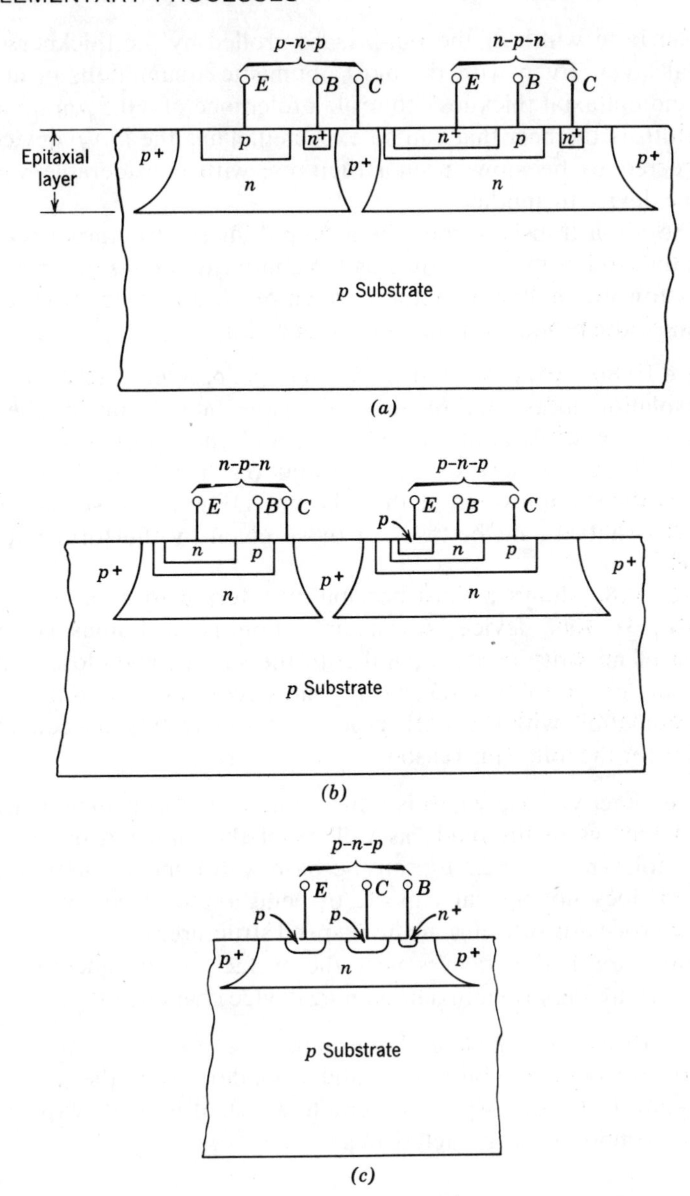

Figure 10.8 Complementary processes.

1. All *p-n-p* devices share a common collector. In addition, the resistivity of this substrate collector is far too high to be compatible with good transistor design practice.

2. The base width of the *p-n-p* is controlled by the thickness of the epitaxial layer. Even with the most optimistic assumptions of diffusion depth and epitaxial thickness control, a tolerance of $\pm 0.5\ \mu m$ on a 4-μm base width is the best that can be expected. Thus the *p-n-p* devices can be expected to be slow-speed structures, with considerable variation from one device to another.

3. The *n-p-n* transistor must be a deep-diffused structure, because its base is diffused at the same time as the emitter of the *p-n-p*. The present trend is toward shallow structures which result in devices with a higher gain bandwidth product for the same base width.

Figure 10.8*b* shows a second approach to the problem. Here, in addition to the isolation diffusion, three more diffusions must be made. The resulting *n-p-n* is a triple-diffused structure, and the *p-n-p* is a deep-base structure. This is a more satisfactory solution than the first; however, it combines the complexities of the TD and DDE processes and results in devices that are no better than those given by the former (poorer) approach.

Figure 10.8*c* shows a third possibility, referred to as a lateral *p-n-p* transistor. In this device, transistor action is accomplished by the diffusion of minority carriers parallel to the surface and close to it. The biggest advantage of this scheme is that it requires no additional steps and is compatible with the DDE process. It, too, results in a compromise transistor for the following reasons:

1. The effective base width is a function of the mechanical placement of two windows in the oxide as well as of the diffusion process. This places a tolerance of $\pm 2.5\ \mu m$ on the base width under optimum conditions and does not appear capable of being reduced below this point. Thus the process results in a medium-speed structure.

2. Transistor action occurs near the surface. Consequently surface phenomena are important in determining device characteristics.

At the present time the lateral transistor appears to be the most promising means for obtaining both *p-n-p* and *n-p-n* devices on the same microcircuit chip. Consequently, considerable effort [6] is being expended by industry to improve its characteristics.

10.7. MONOLITHIC DIODES

A monolithic transistor may be used to perform the function of a diode in a number of different ways. For example, the emitter-base junction, with the collector left unconnected, may be used for this function.

Alternately, the emitter-base junction, with the collector shorted to the base, is yet another possibility and has quite different electrical characteristics. Figure 10.9 shows this latter structure, which is commonly used because it results in a high-conductance diode with low-minority-carrier storage. The details of this and other combinations will be discussed in a later chapter.

10.8. MONOLITHIC RESISTORS

Figure 10.10*a* shows a monolithic resistor which consists of a base diffusion into the *n*-type collector background. A number of such resistors may be placed within the same pocket because they are isolated from each other. The collector region is eventually connected to the most positive point in the microcircuit.

Because the sheet resistance of an emitter diffusion is considerably lower than that of a base diffusion (typically 2 to 10 Ω/square as compared to 150 Ω/square), it is common practice to use the emitter diffusion for low-value resistors. This is done by first conducting a base diffusion in order to form a *p*-type pocket in which the *n*-type resistor region is diffused (see Figure 10.10*b*). This pocket must be connected to the most negative point in the microcircuit in order to reverse-bias it with respect to the resistor.

10.9. PRECISION RESISTORS

In addition to variations imposed by limitations of the photolithographic process, the diffused resistor is subject to variation caused by diffusion tolerances. As a result, a ±5% tolerance on the desired value at room temperature is the best that can be accomplished at the present state of

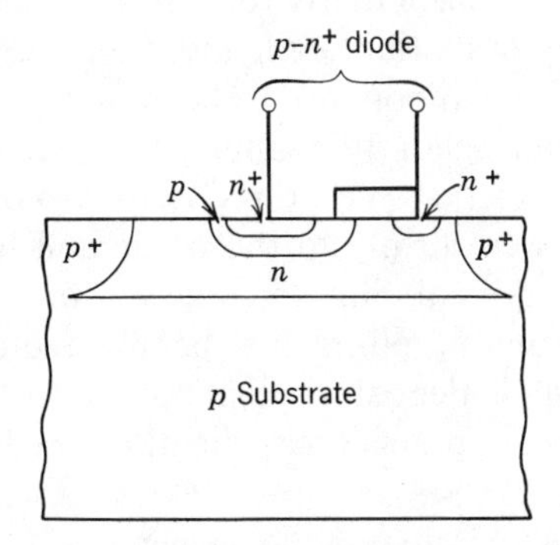

Figure 10.9 A monolithic diode.

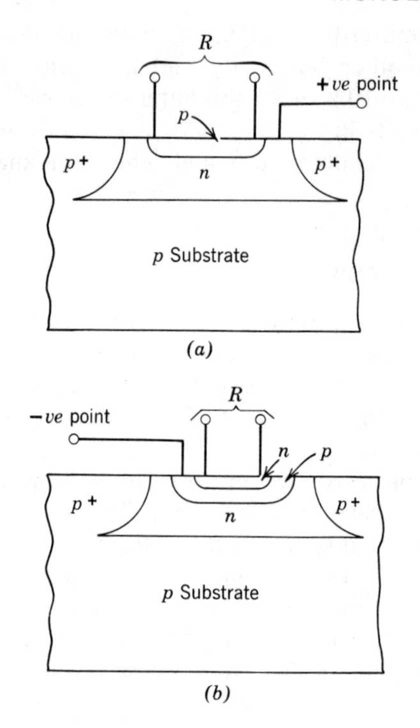

Figure 10.10 Monolithic resistors.

the art. In addition, the value of the diffused resistance changes over the operating temperature range of the microcircuit owing to changes in the carrier mobility and concentration. Temperature variations on the order of 2000 ppm/°C are typical for these resistors.

In many applications, particularly for linear circuits, resistor variations of this magnitude cannot be tolerated, and it is necessary to provide ultra-stable, high-precision components. These are made by the vacuum evaporation or sputtering of thin films of resistive materials, such as nichrome, chromium, or tantalum, directly on top of the oxide layer. All of these materials adhere firmly to the oxide and have usable values of sheet resistance for typical film thicknesses of about 0.1 to 0.5 μm. Finally, a vast amount of effort has been devoted to studying their characteristics and their deposition technology to the point where they are quite satisfactory† for precision applications.

†By way of example values have been quoted with a temperature variation of under 100 ppm/°C.

Means are available for hand trimming of these films after fabrication. Methods for trimming have included the use of anodic etching, laser-beam etching, and partial oxidation by transient thermal overloads.

Thin-film technology is a subject in itself and will not be covered here. The interested reader is referred to the References 7 to 10 at the end of this chapter.

10.10. CAPACITORS

The capacitive reactance of a reverse-biased *p-n* diode may be used for this function. With the use of monolithic processes various different diode configurations may be considered; all of these result in relatively lossy voltage-variable capacitors, only suitable for bypass and speedup applications.

It is customary to make *p-n* capacitors by using the emitter for one region because of its low sheet resistance. If this is done, the aluminum metallization may be used as the second electrode, with the thermally grown oxide as the dielectric,† resulting in a metal-oxide-semiconductor (MOS) capacitor. In this manner, as shown in Figure 10.11, a reasonably stable high-*Q* capacitor can be made.

†The relative permittivity of grown oxide layers is about 3.4 to 3.8.

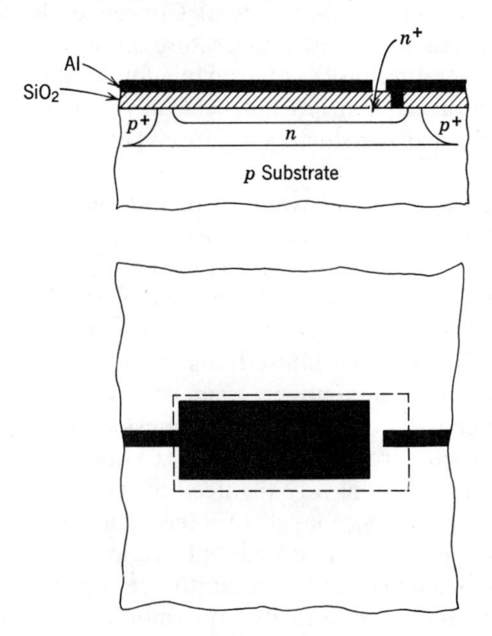

Figure 10.11 MOS capacitor.

10.11. INDUCTORS

Inductors do not lend themselves to monolithic process techniques. At very best the aluminum metallization can be formed in the shape of a spiral inductor. Values of such inductors are typically in the 0 to 0.5-μH range and are useful in very limited situations.

A considerable amount of research is currently being undertaken to simulate the electrical function of inductance in microcircuits. Much of this work is based on active network techniques and on the utilization of electromechanical resonant structures. There does not appear to be any generally satisfactory solution to this problem, even though solutions to specific situations have been achieved.

10.12. CROSSOVERS

In many circuits it is impossible to lay out the component locations without crossovers in some of the interconnections. The most direct solution to this problem is to deposit an insulating film over the first layer of interconnections and place the crossovers on top of this film. Unfortunately this imposes a number of metallurgical requirements which are often incompatible. Thus once the first metallization layer is placed, the circuit cannot be subsequently heated in excess of 550°C (about 25° below the eutectic point of the Al-Si system). Consequently the insulating film must be deposited by a low-temperature process. A completely satisfactory low-temperature process for laying down a thin, pinhole-free, adherent, stable, inorganic film has not been developed at the present time although some partly satisfactory solutions [11] have recently been obtained.

An alternate approach takes advantage of the fact that microcircuit resistors are covered by a grown oxide layer, except at their extremities where ohmic contacts are made to them. Thus the space above a resistor is available to make a crossover, as shown in Figure 10.12.

Sometimes no resistor is conveniently located to effect this crossover. In this case a low-value (emitter-diffused) resistor is deliberately inserted into one of the leads. This can only be done, however, if the insertion of this resistance (typically 2 to 3 Ω) does not deteriorate the performance of the microcircuit. Such a crossover is known as a *tunnel* or a *via*.

With increasing circuit complexity the need for crossovers will doubtless increase. Thus two or more levels of interconnection are already in demand by the proponents of large-scale integration schemes. There is no doubt that these demands will be met in the very near future; in fact, a number of manufacturers are already using multilevel interconnection schemes in their laboratory prototypes.

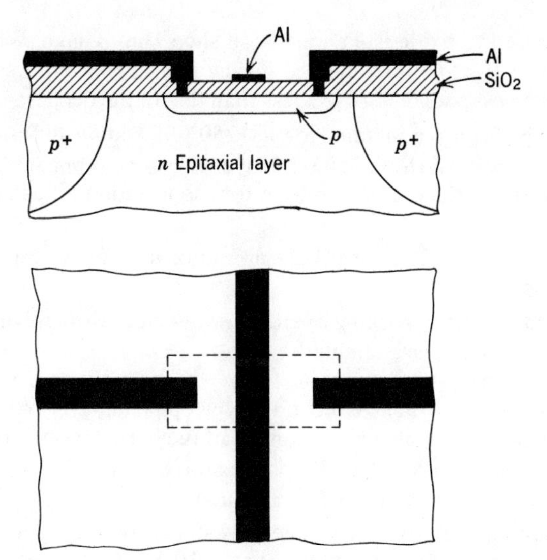

Figure 10.12 A crossover.

10.13. REFERENCES

[1] A. J. Khambatta, *Introduction to Integrated Circuits,* Wiley, New York, 1963.

[2] R. M. Warner, Jr., and J. N. Fordemvalt, *Integrated Circuits – Design Principles and Fabrication,* McGraw-Hill, New York, 1965.

[3] E. Keonjian (Ed.), *Microelectronics*, McGraw-Hill, New York, 1963.

[4] W. Steinmaier, and J. Bloem, "Successive Growth of Si and SiO_2 in Epitaxial Apparatus," *J. Electrochem. Soc.*, **111**, 206–209 (1964).

[5] J. L. Moll *et al.,* "*P-N-P-N* Transistor Switches," *Proc. IRE,* **44**, 1174–1182 (Sept., 1956).

[6] J. Lindmayer, and W. Schneider, "Theory of Lateral Transistors," *Solid State Electron.* **10**. 225–234 (1967).

[7] L. Holland, *The Vacuum Deposition of Thin Films*, Chapman and Hall, London 1963.

[8] L. Holland (Ed.), *Thin Film Electronics*, Wiley, New York, 1965.

[9] R. W. Berry, "Tantalum Thin Film Circuitry and Components," *Bell Lab. Record,* **41**, 46 (Feb., 1963).

[10] T. V. Sikina, "High Density Tantalum Film Microcircuits," *Proc. 1962 Electronic Components Conference* **24**, (1962).

[11] N. Goldsmith, and W. Kern, "The Deposition of Vitreous Silicon Dioxide Films from Silane," *R.C.A. Review*, **28** (1), 153–165 (1967).

10.14. PROBLEMS

It is suggested that Figure 11.20 be used in conjunction with the following problems.

1. It is desired to oxide-isolate a silicon slice 1-in in diam. What is the degree of misalignment that may be allowed during the lapping process so that the depth of the shallowest pocket is 10 μm less than that of the deepest.

2. Compute the drive-in time for a 1250°C isolation diffusion with the DDE process. A 6-μm 0.5 Ω-cm epitaxial layer is used in conjunction with a 10 Ω-cm substrate. The surface concentration for the isolation diffusion is required to be 10^{19} atoms/cm^3.

3. Repeat Problem 2 for a 0.1 Ω-cm epitaxial layer. Compare the result with Problem 2.

4. To avoid the extra masking operations involved with growing the buried layer, the following process is attempted:

A 0.05 Ω-cm n^+-type epitaxial layer, 1 μm thick, is grown over the entire 10 Ω-cm p-type substrate. A second n-type epitaxial layer of 0.5 Ω-cm resistivity and 5 μm thickness is next grown on top of the buried layer. Compute the drive-in time for isolation with the DDE process of Problem 2, assuming that all other impurities are immobile during this diffusion.

5. Calculate the approximate sheet resistance of the collector region for the narrow-base transistor shown in Figure 10.5*b*. Compare this with the sheet resistance of a buried layer.

6. A 0.25 mil-wide aluminum interconnection is laid over a 0.5 μm oxide layer on a silicon substrate. Assuming that the substrate is a ground plane, determine the equivalent circuit for the resulting *RC* transmission line. What is its characteristic impedance?

7. Determine an approximate equivalent circuit for a 1×1-mil MOS capacitor. An oxide thickness of 0.5 μm may be assumed in conjunction with the doping profile of Figure 10.5*b*. At what frequency does this capacitor have a Q of unity?

Chapter 11

Current Transport

The current that flows in a semiconductor is given by the time rate at which charge is transported through any given surface in a direction normal to it. The movement of this charge is associated with the transport properties of mobile carriers; in a semiconductor these carriers are electrons and holes in the conductance and valence bands respectively, and are free to wander around in the crystal lattice.

Immobile carriers are also present and are associated with the electrons that constitute the covalent bonding in a semiconductor. These electrons are also in motion; however, the time average for their transport velocity is zero because they are confined to the neighborhood of their respective lattice sites. Consequently, aside from fluctuation noise (Johnson noise), they do not contribute to the flow of current in the semiconductor.

The movement of mobile carriers may take place under the influence of an electric field by a drift process. Alternately, movement may occur because of a concentration gradient of mobile carriers, to the exclusion of an electric field. Here carriers are considered to move by diffusion. Finally, both processes may occur at one time.

In this chapter the current transport properties of a semiconductor are described by first developing the carrier statistics to determine the mobile carrier concentration. Next we consider those factors that determine the manner in which these carriers are transported under conditions of drift and diffusion resulting in current flow.

11.1. CARRIER STATISTICS

It is reasonable to expect some statistical distribution of energies for the various constituent electrons of a crystal lattice which is maintained at a temperature in excess of 0°K. The nature of this distribution may be found by considering the principle of detailed balance.

11.1.1. Principle of Detailed Balance

This principle applies to any system in thermal equilibrium and states that, for any process, there is an inverse process that occurs at the same rate. Consider, for example, a system of particles in thermal equilibrium. Let this system consist of N_1 particles in state ① and N_2 particles in state ②. Let this two-level system be immersed in a heat bath at a temperature T, as shown in Figure 11.1. Associated with N_1 and N_2 are energy levels of E_1 and E_2 respectively.

Let W_{12} be the probability that a single particle in state ① makes a transition to state ② in unit time. In like manner, we define W_{21} for ② →① .

Pauli's exclusion principle is considered to apply to the system. Then the number of particles which make a transition from ① to ② in unit time is equal to the number originally in ① *times* (W_{12}) *times* the number of available states in ② . The number of particles which make a transition from ② → ① can be written in a similar manner.

Let N_{s1} be the *total* number of states at E_1, and N_{s2} the number of states at E_2. Then the principle of detailed balance leads directly to the equality,

$$N_1 W_{12}(N_{s2} - N_2) = N_2 W_{21}(N_{s1} - N_1). \tag{11.1}$$

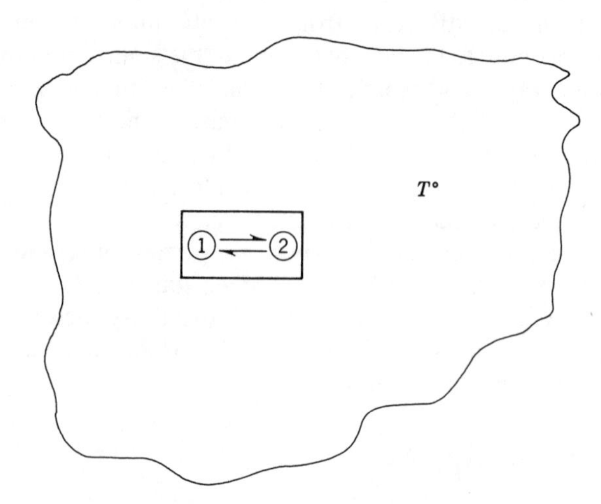

Figure 11.1 The transition process.

11.1.2. Transition Probability Ratio

The transition probability ratio is given by W_{12}/W_{21}. Thus (11.1) may be rewritten as

$$\frac{W_{12}}{W_{21}} = \frac{(N_{s1}-N_1)/N_1}{(N_{s2}-N_2)/N_2} \tag{11.2}$$

The right-hand side of this expression may be evaluated by considering the various configurations in which indistinguishable particles may be arranged in the available states.

The number of ways that N_1 indistinguishable particles can be made to fit in N_{s1} states is

$$\frac{N_{s1}!}{(N_{s1}-N_1)!N_1!} = C_{N_1}^{N_{s1}} \tag{11.3}$$

The number of ways that N_2 indistinguishable particles can be made to fit in N_{s2} states is

$$\frac{N_{s2}!}{(N_{s2}-N_2)!N_2!} = C_{N_2}^{N_{s2}} \tag{11.4}$$

The joint number of ways in which N_1 indistinguishable particles can fit in N_{s1} states and N_2 indistinguishable particles in N_{s2} states is given by the product of these two combinations.

The entropy associated with this system[1] is given by $S = k \ln$ (number of ways):

$$S = k \ln C_{N_1}^{N_{s1}} + k \ln C_{N_2}^{N_{s2}}, \tag{11.5}$$

where k is Boltzmann's constant $(8.62 \times 10^{-5}\ \mathrm{eV/^\circ K})$. The internal energy of the system is $E = N_1E_1 + N_2E_2$. Therefore the Helmholtz free energy of the system is given by

$$F = E - TS, \tag{11.6}$$

$$= N_1E_1 + N_2E_2 - kT \ln C_{N_1}^{N_{s1}} - kT \ln C_{N_2}^{N_{s2}} \tag{11.7}$$

The most probable situation for equilibrium is the one in which the total free energy of the system is minimized for particles at any energy level. This situation occurs when

$$\left(\frac{\partial F}{\partial N_1}\right)_{T=\text{const}} = 0 \tag{11.8}$$

and

$$\left(\frac{\partial F}{\partial N_2}\right)_{T=\text{const}} = 0. \tag{11.9}$$

Applying the first condition gives

$$0 = E_1 - kT\frac{\partial}{\partial N_1}\left(\ln \mathrm{C}_{N_1}^{N_{s1}}\right)$$

$$= E_1 - kT\frac{\partial}{\partial N_1}[\ln N_{s1}! - \ln (N_{s1} - N_1)! - \ln N_1!]$$

$$\simeq E_1 - kT \ln\frac{N_{s1} - N_1}{N_1}. \tag{11.10}$$

Thus†,

$$\frac{N_{s1} - N_1}{N_1} \simeq e^{E_1/kT}. \tag{11.11}$$

Similarly, application of the second condition gives

$$\frac{N_{s2} - N_2}{N_1} \simeq e^{E_2/kT}, \tag{11.12}$$

so that

$$\frac{W_{12}}{W_{21}} \simeq \frac{e^{E_1/kT}}{e^{E_2/kT}}. \tag{11.13}$$

Although this equation has been derived only for those systems in which Pauli's principle holds, it is actually considerably more general in nature. It is now used to determine the Fermi-Dirac distribution for electron occupancy in a solid.

11.1.3. Fermi-Dirac Distribution Function

Consider a system of M particles (in this case, electrons). Let

f_i = probability of having an electron in the ith state at an energy level E_i,

f_j = probability of having an electron in the jth state at an energy level E_j,

W_{ij} = probability of an individual electron making a transition from the ith state to the jth state, in unit time,

M = total number of electrons considered.

†Note that Stirling's approximation for the factorial of a large number is $\ln x! \simeq x \ln x - x$. Thus $(\partial/\partial x)(\ln x!) \simeq \ln x$.

Then, in unit time, the number of electrons available for making the transition from the ith state to the jth state is Mf_i. However, the number of unoccupied states is $M(1-f_j)$ if we assign a single state to each electron.† In like manner, similar expressions may be written for electrons making the transition from the jth state to the ith state.

Applying the principle of detailed balance, we obtain

$$Mf_i \,.\, W_{ij} \,.\, M(1-f_j) = Mf_j \,.\, W_{ji} \,.\, M(1-f_i). \tag{11.14}$$

Thus

$$\frac{Mf_i}{M(1-f_i)} e^{E_i/kT} = \frac{Mf_j}{M(1-f_j)} e^{E_j/kT} \tag{11.15}$$

A similar procedure, applied to transitions between the ith and any other state in the system (for example, the hth state), gives

$$\frac{Mf_i}{M(1-f_i)} e^{E_i/kT} = \frac{Mf_h}{M(1-f_h)} e^{E_h/kT} \tag{11.16}$$

Thus for this system,

$$\frac{Mf_h}{M(1-f_h)} e^{E_h/kT} = \frac{Mf_i}{M(1-f_i)} e^{E_i/kT} = \frac{Mf_j}{M(1-f_j)} e^{E_j/kT} = \dots . \tag{11.17}$$

The only way this equality can hold for *all* states is for each term to be equal to some constant C, which is independent of the individual probabilities $f_h, f_i, f_j, \dots$. Consequently,

$$\frac{Mf_j}{M(1-f_j)} e^{E_j/kT} = C. \tag{11.18}$$

Solving,

$$f_j = \frac{1}{1+(1/C)e^{E_j/kT}} \tag{11.19}$$

This is the probability of an electron being in the jth state.

By definition of E_f such that

$$C \equiv e^{E_f/kT} \tag{11.20}$$

(11.19) reduces to

$$f_j = \frac{1}{1+e^{(E_j-E_f)/kT}} \tag{11.21}$$

†The effects of spin degeneracy [2] are to double this number. This will be considered at a later point.

This is the Fermi-Dirac distribution function. Since the jth state is associated with electrons at an energy level E_j, we can replace f_j with $f(E_j)$. Thus

$$f(E_j) = \frac{1}{1 + e^{(E_j - E_f)/kT}}. \tag{11.22}$$

The probability of finding an electron with an energy of E_f is $\frac{1}{2}$. This defines the Fermi level for a given system.

Figure 11.2 shows a sketch of the Fermi-Dirac function for values of T. At $T \simeq 0°\text{K}$, the function shows an abrupt discontinuity between the values of 0 and 1, and becomes more and more gradual at elevated temperatures. Sketching in the valence and conduction bands for the semiconductor, it is seen for this situation that, with increasing temperature, there is an increasing probability of finding electrons with energies within the conduction band.

The actual carrier concentration at any energy level is directly proportional to the probability of finding electrons at that energy level and to the density of available states† at this level. Thus $dn(E)$, the number of electrons per unit volume in an energy range between E and $E + dE$, is given by

$$dn(E) = f(E)g(E)\,dE. \tag{11.23}$$

†This is a direct outcome of the exclusion principle.

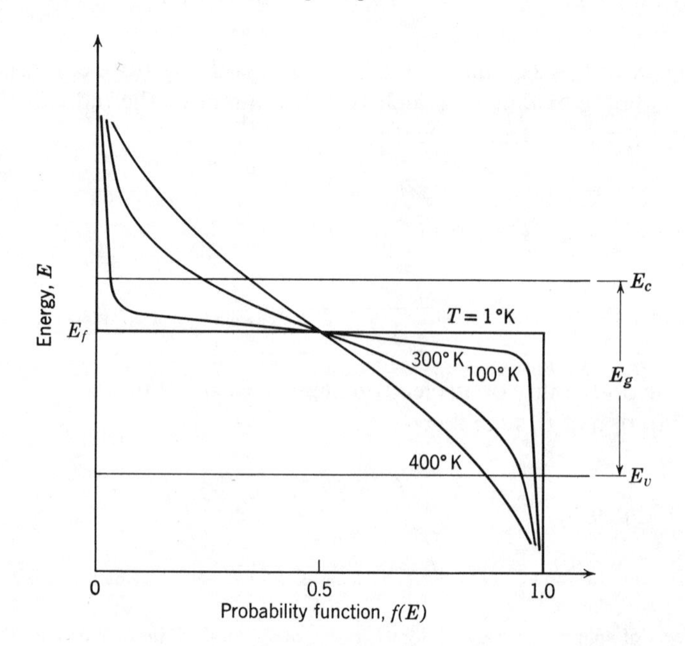

Figure 11.2 The Fermi-Dirac function.

Here $g(E)\,dE$ is the number of available states per unit volume in the energy range between E and $E+dE$. The function $g(E)$ is defined as the density-of-states function.

11.1.4. Density-of-States Function

The number of available states in a semiconductor is finite only in the valence and conduction bands and zero in the forbidden gap. To determine their density in an energy range between E and $E+dE$, it is first necessary to determine the number of different momenta values that can be acquired by electrons in this energy range. In a quantum-mechanical system this number is finite.

The motion of carriers in a solid is governed by the applied field and by the internal field associated with the crystal-binding energy. The analysis of this problem is quantum-mechanical in nature; however, the results of a treatment of this type lead to the somewhat remarkable conclusion that classical methods may be applied to the problem if its quantum-mechanical nature is taken into consideration by assuming that the carriers have an "effective" mass. It must be emphasized that this effective mass is an artifice. Consequently, an appropriate value must be used for the situation in question. Thus a "density-of-states" effective mass[3] is used to determine the motion of carriers in a situation from which the density-of-states function is to be determined. In like manner, a "conductivity" effective mass is more suited for a situation where the movement of carriers gives rise to current flow. The symbol m^* is used for both cases, since it is usually clear which effective mass is appropriate for the situation being considered.

Figure 11.3 shows the energy-band picture for a semiconductor. For electrons in the energy range between E and $E+\Delta E$, classical mechanics gives

$$E-E_c=\tfrac{1}{2}m_n^*v^2=\frac{P_r^2}{2m_n^*}, \tag{11.24}$$

where m_n^* is the electron density-of-states mass for a semiconductor and P_r is the momentum.

In a three-coordinate system, all momentum vectors terminating on a spherical surface correspond to electrons with constant kinetic energy. In "momentum-space" therefore, all electrons within a ΔE energy interval must have momentum vectors of length between P_r and $P_r+\Delta P_r$; i.e., vectors terminating within a differential volume between spheres of radius P_r and $P_r+\Delta P_r$. This volume ΔV_p is given by

$$\Delta V_p=4\pi P_r^2\,\Delta P_r. \tag{11.25}$$

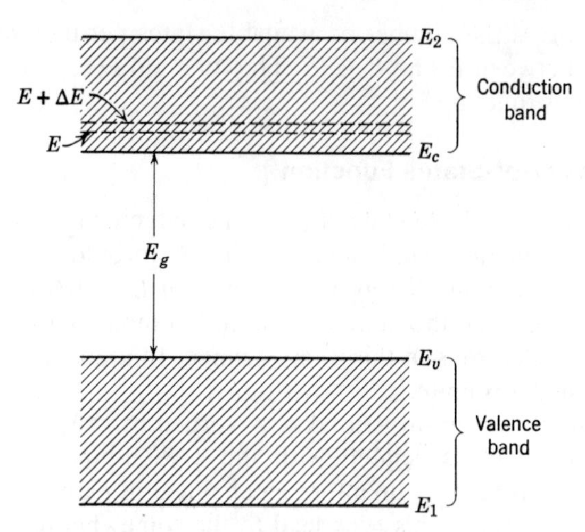

Figure 11.3 The energy-band picture for silicon.

Combining (11.24) and (11.25) gives

$$\Delta V_p = \frac{2}{\sqrt{\pi}}(2\pi m_n^*)^{3/2}\sqrt{E-E_c}\,\Delta E. \tag{11.26}$$

To take into account the discrete nature of the momenta values, each energy level in an incremental volume $\Delta x\,\Delta y\,\Delta z$ of the crystal is associated with an incremental volume $\Delta P_x\,\Delta P_y\,\Delta P_z$ in momentum space (see Figure 11.4). The uncertainty principle states that, at best,

$$(\Delta x\,\Delta P_x)(\Delta y\,\Delta P_y)(\Delta z\,\Delta P_z) = h^3, \tag{11.27}$$

where h is Planck's constant (6.63×10^{-27} erg-sec). It follows that all that can be said for the magnitude of each momentum vector is that it terminates in a cube of momentum space of volume $h^3/\Delta x\,\Delta y\,\Delta z$. Thus in a crystal of unit "space-volume", the number of such cubes that can be contained in a momentum volume ΔV_p is given by $\Delta V_p/h^3$. This is the number of momenta values that can be accommodated in the energy interval ΔE in a unit volume of the semiconductor. Since each momentum value is associated with a distinct energy level, it follows that

$$\Delta\,(\text{energy levels}) = \frac{2}{\sqrt{\pi}}\left(\frac{2\pi m_n^*}{h^2}\right)^{3/2}\sqrt{E-E_c}\,\Delta E. \tag{11.28}$$

There are twice as many available states as energy levels, because two spin states (↑ and ↓) can be accommodated at each level. Thus the density-of-states function may be written directly as

$$g(E) = \frac{4}{\sqrt{\pi}}\left(\frac{2\pi m_n^*}{h^2}\right)^{3/2}\sqrt{E-E_c}. \tag{11.29}$$

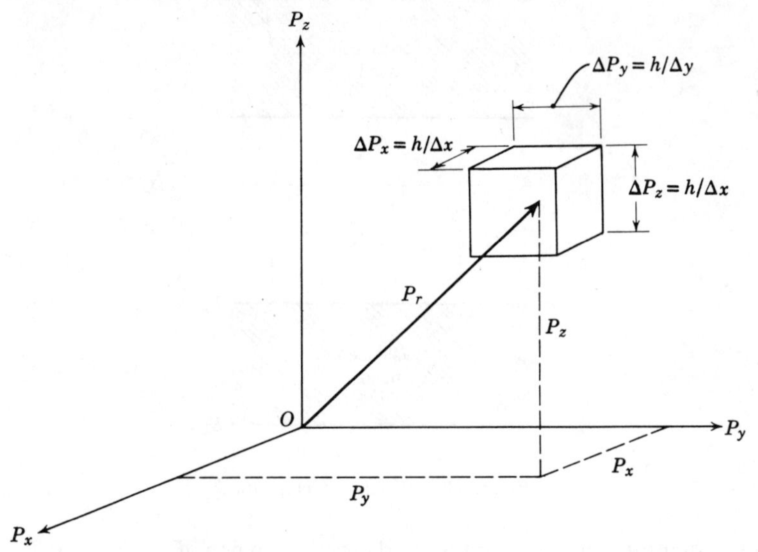

Figure 11.4 A volume in momentum space.

By similar reasoning the density-of-states function for holes in the valence band is

$$g(E) = \frac{4}{\sqrt{\pi}}\left(\frac{2\pi m_p^*}{h^2}\right)^{3/2}\sqrt{E_v - E} \tag{11.30}$$

where m_p^* is the density-of-states mass for holes in the semiconductor. For silicon [4],

$$m_n^* = 1.10\, m_0, \tag{11.31}$$

$$m_p^* = 0.59\, m_0, \tag{11.32}$$

where m_0 is the mass of an electron in free space.

11.1.5. The Effective Density of States

The equilibrium carrier concentration in a semiconductor may be conveniently specified in terms of an effective density of states. Consider the energy-band picture of Figure 11.5, shown with the conduction band extending from E_c to E_2. For this semiconductor

$$dn(E) = f(E)g(E)\, dE, \tag{11.33}$$

$$= \left(\frac{1}{1 + e^{(E - E_f)/kT}}\right)\frac{4}{\sqrt{\pi}}\left(\frac{2\pi m_n^*}{h^2}\right)^{3/2}\sqrt{E - E_c}\, dE. \tag{11.34}$$

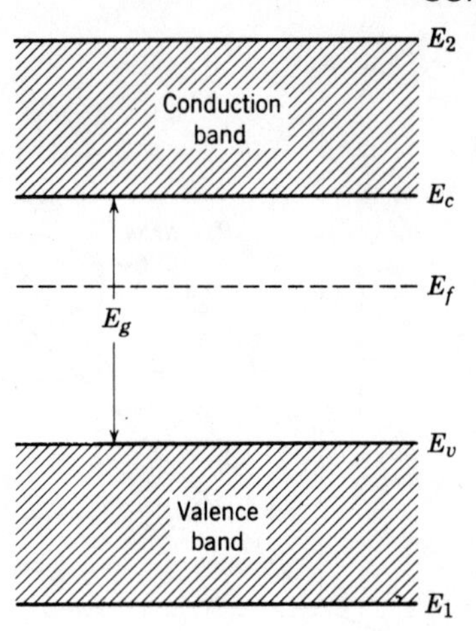

Figure 11.5 Energy-band picture.

This equation may be integrated readily when $E_2 - E_c > 3kT$ and $E_c - E_f > 3kT$. For these cases the contribution of electrons in the energy range $3kT$ beyond E_c may be ignored, and E_2 replaced by infinity. In addition, the Boltzmann function may be used in place of the Fermi-Dirac function. Thus the equilibrium concentration for electrons $\bar{n}$ is given by

$$\bar{n} \simeq \frac{4}{\sqrt{\pi}}\left(\frac{2\pi m_n^*}{h^2}\right)^{3/2} \int_{E_c}^{\infty} e^{-(E-E_f)/kT} \sqrt{E-E_c}\, dE. \tag{11.35}$$

Setting $x = (E - E_c)/kT$,

$$\bar{n} \simeq \frac{4}{\sqrt{\pi}}\left(\frac{2\pi kTm_n^*}{h^2}\right)^{3/2} e^{-(E_c - E_f)/kT} \int_0^{\infty} \sqrt{x}e^{-x}\, dx. \tag{11.36}$$

But,

$$\int_0^{\infty} \sqrt{x}e^{-x}\, dx = \frac{\sqrt{\pi}}{2} \tag{11.37}$$

so that

$$\bar{n} = 2\left(\frac{2\pi kTm_n^*}{h^2}\right)^{3/2} e^{-(E_c-E_f)/kT} \tag{11.38}$$

In like manner, the equilibrium hole concentration is given by

$$\bar{p} = 2\left(\frac{2\pi kTm_p^*}{h^2}\right)^{3/2} e^{-(E_f - E_v)/kT} \tag{11.39}$$

These expressions may be written more compactly as

$$\bar{n} = N_c\, e^{-(E_c - E_f)/kT} \tag{11.40a}$$

$$\bar{p} = N_v\, e^{-(E_f - E_v)/kT}. \tag{11.40b}$$

Here N_c and N_v are defined as the *effective density of states* at the conduction- and valence-band edges respectively. This is because transitions to the band edges require less energy than transitions to points within the band, and are statistically favored.

For silicon

$$N_c = 2.8 \times 10^{19}\left(\frac{T}{300}\right)^{3/2} \text{ cm}^{-3}$$

$$N_v = 1.02 \times 10^{19}\left(\frac{T}{300}\right)^{3/2} \text{ cm}^{-3}.$$

Note that, regardless of the doping level,

$$\bar{p}\bar{n} = N_c N_v\, e^{-(E_c - E_v)/kT}, \tag{11.41a}$$

$$= N_c N_v\, e^{-E_g/kT}. \tag{11.41b}$$

Thus the $\bar{p}\bar{n}$ product is independent of the doping level and is related to the energy gap E_g. For silicon $E_g = 1.11$ eV and $\bar{p}\bar{n} = 1.9 \times 10^{20}$ electrons/cm^6 at 300°K.

11.2. INTRINSIC SEMICONDUCTORS

In intrinsic, or undoped, semiconductors in thermal equilibrium the only mobile carriers are thermally generated electrons and holes. These are present in equal number.

Substituting $\bar{p} = \bar{n} = n_i$, the intrinsic carrier concentration

$$\bar{p}\bar{n} = n_i^2 = N_c N_v\, e^{-E_g/kT}. \tag{11.42}$$

The position of the Fermi level in an intrinsic semiconductor may be determined by setting $\bar{p} = \bar{n}$ in (11.40*a*) and (11.40*b*). When this is done,

$$E_f - E_v = \frac{E_c - E_v}{2} + \frac{kT}{2}\ln\frac{N_v}{N_c}$$

$$= \frac{E_c - E_v}{2} + \frac{3kT}{4}\ln\frac{m_p^*}{m_n^*}. \tag{11.43}$$

At 300°K the second term is approximately −10 mV for silicon. Thus the Fermi level in intrinsic silcon is 10 mV below the center of the energy gap. For all practical purposes this correction may be ignored, and the intrinsic level considered as being midway in the energy gap.

Writing the intrinsic Fermi level as E_i, it follows that

$$n_i = N_c e^{-(E_c - E_i)/kT}, \tag{11.44a}$$

$$= N_v e^{-(E_i - E_v)/kT}. \tag{11.44b}$$

11.3. IMPURITY-DOPED SEMICONDUCTORS

Although the $\bar{n}\bar{p}$ product is constant, the separate values of $\bar{n}$ and $\bar{p}$ (hence the $\bar{n}+\bar{p}$ sum which controls the conductivity) can be varied by orders of magnitude. This is done by doping with impurities from columns III and V of the periodic table. These impurities enter the silicon lattice in substitutional sites and behave as acceptors and donors respectively. Typical doping levels in microcircuits range from 10^{15} atoms/cm^3 to 10^{20} atoms/cm^3. Since silicon has 5×10^{22} atoms/cm^3, these concentrations are very dilute. The impurities may thus be considered as a system of noninteracting particles in a dielectric medium (for silicon the relative permittivity is 12). Consequently they exhibit discrete energy levels.

An estimate of the position of the energy levels may be made by assuming that such impurities are hydrogen-like in character. Elementary atomic theory shows that the energy levels of the hydrogen atom are given by

$$E = -\frac{m_0 q^4}{8a^2 h^2 \epsilon_0{}^2}, \tag{11.45}$$

where ϵ_0 is the permittivity of free space and a takes on the values 1, 2,... An equivalent situation can be considered for a crystal by replacing the mass of the electron with its effective mass and the permittivity of free space by that of the crystal. Thus,

$$E = -\frac{m_n^* q^4}{8a^2 h^2 (\epsilon\epsilon_0)^2}, \tag{11.46}$$

where ϵ is the relative permittivity of the crystal.

The energy required to remove an electron from the hydrogen atom is 13.6 eV. The comparable number in a crystal would thus be

$$E_{\text{ionization}} = \frac{13.6}{\epsilon^2}\frac{m_n^*}{m_0}. \tag{11.47}$$

Table 11.1
Ionization Energies for Column III and V Impurities in Silicon

Dopant	P	As	Sb	B	Al	Ga	In
Acceptor level—distance from valence band, eV	—	—	—	0.045	0.057	0.065	0.16
Donor level—distance from conduction band eV	0.045	0.049	0.039	—	—	—	—
Type of dopant		*n*-type			*p*-type		

Assuming an effective mass ratio† of about 0.6 and a permittivity of 12 for silicon gives an ionization energy of about 0.06 eV.

Table 11.1 lists the usual donors and acceptors for silicon together with their ionization energies. Since most of these are on the order of kT eV, they may be assumed fully ionized at room temperature. Impurities of this type are referred to as *shallow*. It is worth noting that indium is not generally considered a shallow impurity in silicon.

In principle an impurity may exhibit more than one energy level for each of its charge states. These additional levels may be observed if they are within the forbidden gap, as is the case for many deep-lying impurities. Only one impurity level, however, is present within the forbidden gap for the shallow donors and acceptors in silicon.

The hydrogen-like model is an extremely elementary one; indeed, it is remarkable that it can predict the energy levels of the shallow impurities with any degree of accuracy. Needless to say, the model cannot be extended too far. Thus it cannot explain the ionization energy of indium or that for the deep-lying impurities, such as those listed in Table 11.2.

†This is roughly the average of the conductivity effective mass ratio and the density-of-states effective mass ratio.

Table 11.2
Ionization Energies of Deep-Lying Impurities in Silicon

Dopant	Zn	Cu	Au	Ag	Ni	Tl	Mn
Acceptor levels—distance from valence band, eV	0.31 0.56	0.49	0.57	0.89	0.21 0.76	0.26	—
Donor levels—distance from conduction band, eV	—	0.87	0·76	0.79	—	—	0.53

11.3.1. Electron Occupation at an Impurity Level

To describe the statistics of electron occupation at an impurity level, we use the same line of reasoning that was applied to the derivation of the Fermi-Dirac distribution function for a multilevel system.

Combining (11.18) and (11.20) gives

$$\frac{Mf_j}{M(1-f_j)} e^{E_j/kT} = e^{E_f/kT}. \tag{11.48}$$

At this point it is necessary to consider the physical significance of this relation. Equation 11.48 states that, in unit time, for transitions from the jth energy level,

$$\left(\frac{\text{number of electrons}}{\text{number of available states}}\right) e^{E_j/kT} = e^{E_f/kT}. \tag{11.49}$$

Since unit volume is considered, concentrations may be used instead of actual numbers.

The transition process is, in essence, one by which an electron is removed from an impurity atom, thus promoting this atom to the next higher charge state. Once the electron has been removed, it leaves behind an atom that is now available to accept an electron; that is, an available state. Using the symbol M^{s+} to denote an impurity atom having a net charge of $+s$, the transition process at the jth energy level may be described by

$$(M^{s+} - e^- \rightleftharpoons M^{s+1})_j. \tag{11.50}$$

If the equilibrium concentration of impurity atoms in the charge state s is $\bar{N}^s$, the statistics of the transition process at the jth level are given by [see (11.49)]

$$\left(\frac{\bar{N}^s}{g_j \bar{N}^{s+1}}\right)_j e^{E_j/kT} = e^{E_f/kT}. \tag{11.51}$$

Here g_j is a multiplying constant, because the removal of an electron from the atom may result in more than one available state. This constant is known as the *degeneracy factor*. Specific cases of the application of (11.51) now follow.

11.3.2. *n*-Type Semiconductors

An *n*-type semiconductor is formed by doping with impurities from column V of the periodic table. Figure 11.6 shows the energy diagram for this semiconductor, with the donor level E_d close to the conduction

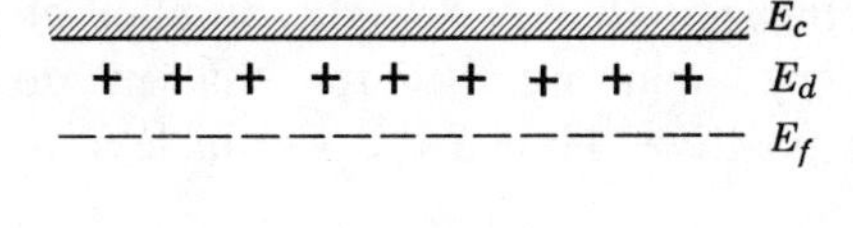

Figure 11.6 Extrinsic n-type material.

band (see Section 11.3). The unit volume of the semiconductor is considered as before.

The transition process at the donor level may be written as

$$(M^0 - e^- \rightleftharpoons M^+)_d, \tag{11.52}$$

where M^0 refers to the unionized donor impurity and M^+ refers to the impurity in the + charge state. Writing the equilibrium concentration of ionized donors as $\bar{N}_d^+$ and the concentration of unionized donors as $\bar{N}_d^0$ (11.51) reduces to

$$\frac{\bar{N}_d{}^0}{g_d \bar{N}_d^+} e^{E_d/kT} = e^{E_f/kT}, \tag{11.53}$$

where g_d is the degeneracy factor for the donor level.

The degeneracy factor is ascribed a value by noting that an electron can be placed in a specific level in one of two ways—spin up or spin down ($\uparrow$ or $\downarrow$). Thus each electron that makes a transition from the impurity level leaves behind *two* available states. Only *one* of these states may be filled by a recaptured electron; nevertheless this recapture can take place in *two* distinctly different ways! The resulting spin degeneracy factor is 2. Equation 11.53 may thus be rewritten as

$$\frac{\bar{N}_d{}^0}{2\bar{N}_d^+} e^{E_d/kT} = e^{E_f/kT}. \tag{11.54}$$

The total number of donors (in a unit volume) is given by $N_d = \bar{N}_d{}^0 + \bar{N}_d^+$. Hence (11.54) results in

$$\frac{\bar{N}_d{}^0}{N_d} = \frac{1}{1 + \frac{1}{2} e^{(E_d - E_f)/kT}}. \tag{11.55}$$

Since ionized atoms are those which have an electron removed from them, the probability of electron occupation of the donor level is given by

$$f(E_d) = \frac{\bar{N}_d{}^0}{N_d} = \frac{1}{1 + \frac{1}{2} e^{(E_d - E_f)/kT}}. \tag{11.56}$$

It should be emphasized that $\bar{N}_d{}^0$ is the number of *unionized* donors. Since this is a very small number for shallow donors, we can expect $E_d - E_f$ to be positive. Hence the Fermi level is below the donor level.

11.3.3. *p*-Type Semiconductors

A similar reasoning may be applied to *p*-type semiconductors. Here elements from column III are utilized. Figure 11.7 shows the energy gap for such a semiconductor together with the Fermi level and the acceptor level.

For this situation the transition process at the acceptor level is written as

$$(M^- - e^- \rightleftharpoons M^0)_a, \tag{11.57}$$

hence

$$\frac{\bar{N}_a^-}{g_a \bar{N}_a{}^0} e^{E_a/kT} = e^{E_f/kT}, \tag{11.58}$$

where g_a is the degeneracy factor for the acceptor level. Here, $\bar{N}_a^-$ and $\bar{N}_a{}^0$ are the equilibrium concentrations of ionized and unionized acceptors respectively.

In this situation an electron can leave an ionized acceptor in either of two ways ($\uparrow$ or $\downarrow$), but only one electron of the correct spin can return to the neutral acceptor. Thus $g_a = \frac{1}{2}$ and

$$f(E_a) = \frac{\bar{N}_a^-}{N_a} = \frac{1}{1 + 2e^{(E_a - E_f)/kT}}. \tag{11.59}$$

Note here that $\bar{N}_a^-$ is the number of *ionized* acceptors. Since nearly all shallow acceptors are ionized, we can expect $E_a - E_f$ to be negative. Thus the Fermi level is above the acceptor level.

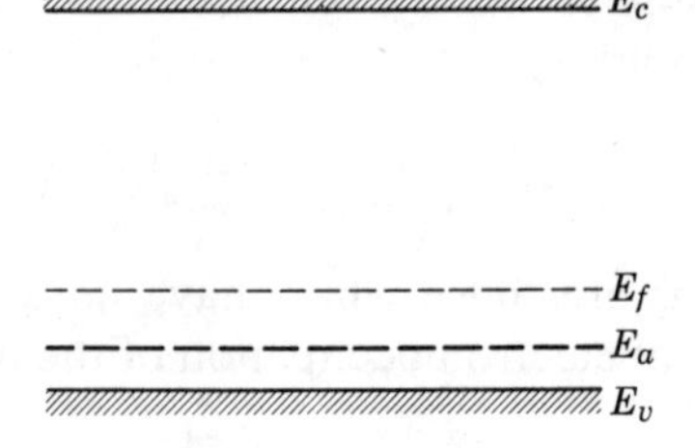

Figure 11.7 Extrinsic *p*-type material.

11.3.4. Comments on Degeneracy

The following comments should be made concerning the use of degeneracy factors:

1. Whereas electrons bound to impurity atoms have a specific identity (i.e., are either ↑ or ↓), this is not the case for electrons in the conduction band or for absence of electrons (holes) in the valence band. Thus the problem of specific spin states only arises in connection with impurity levels.

2. The spin degeneracy factor can only be computed if there is specific knowledge of the electronic states of the impurity atom. This is often a problem when elements from columns other than III and V are considered.

3. Other degeneracy effects, such as orbital degeneracies, are ignored. Thus the true degeneracy factor is often different from that attributed to spin alone.

4. Errors resulting from ignoring the spin degeneracy factor are relatively slight. For example, since $e^{\pm 0.018/kT} = 2$ or $\frac{1}{2}$ at 300°K, an error of 18 mV in the value of the energy level often totally masks the degeneracy factor correction. This is especially true for deep-lying impurities the energy levels of which are not accurately known. In addition, degeneracy is often implicitly accounted for in the experimental determination of deep-lying impurity levels.

11.3.5. Equilibrium Carrier Concentrations

The equilibrium carrier concentrations for an arbitrarily doped semiconductor have been derived in Section 11.1.5. These concentrations were given by

$$\bar{n} = N_c e^{-(E_c - E_f)/kT}, \tag{11.40a}$$

$$\bar{p} = N_v e^{-(E_f - E_v)/kT}. \tag{11.40b}$$

Note that these equations are applicable to both intrinsic as well as extrinsic situations. For the intrinsic case the Fermi level lies approximately half way in the band gap and $\bar{p} = \bar{n} = n_i$. For the extrinsic case the Fermi level moves much closer to the impurity level. Thus for an n-type semiconductor, $E_c - E_f \ll E_f - E_v$, hence $\bar{n} \gg \bar{p}$. Similarly, for a p-type semiconductor, the Fermi level shifts toward the acceptor level, resulting in $\bar{p} \gg \bar{n}$. Note that

$$\bar{p}\bar{n} = N_c N_v e^{-E_g/kT}, \tag{11.60}$$

for all levels of doping.

Equilibrium concentrations may also be obtained as a function of the displacement of the Fermi level from the intrinsic level. Thus combining (11.40*a*) and (11.44*a*) gives

$$\bar{n} = n_i e^{(E_f - E_i)/kT}. \tag{11.61a}$$

In addition, combining (11.40*b*) and (11.44*b*) gives

$$\bar{p} = n_i e^{(E_i - E_f)/kT}. \tag{11.61b}$$

These relations serve to emphasize the symmetric nature of the equilibrium concentration of carriers with respect to the departure of the Fermi level from the intrinsic level.

11.3.6. Determination of the Fermi Level

The central problem of calculating the equilibrium concentration of free carriers in a semiconductor lies in the determination of the Fermi level. This is done by setting up a charge balance between positive and negative charges in a semiconductor in thermal equilibrium. Let

$\bar{p}$ = concentration of mobile positive charges (holes),
$\bar{N}_d^+$ = concentration of bound positive charges,
$\bar{n}$ = concentration of mobile negative charges (electrons),
$\bar{N}_a^-$ = concentration of bound negative charges.

For charge balance,

$$\bar{p} + \bar{N}_d^+ = \bar{n} + \bar{N}_a^- \tag{11.62}$$

A graphical solution is shown for this equation in Figure 11.8 for the case of a semiconductor doped with a donor impurity ($\bar{N}_a^- = 0$). Expressions for the various terms of the charge-balance equation are given by (11.40*a*), (11.40*b*), (11.55), and (11.59).

11.3.6.1. The Shallow-Impurity Approximation.

From (11.61*a*),

$$E_f - E_i = kT \ln \frac{\bar{n}}{n_i}. \tag{11.63}$$

It may be assumed that shallow impurities are fully ionized at room temperature (see Problem 4). Consequently, $\bar{n} = \bar{N}_d^+ \simeq N_d$ for an *n*-type semiconductor. Since $\bar{n} \gg \bar{p}$ for the extrinsic case, (11.63) may be written as

$$E_f - E_i \simeq kT \ln \frac{N_d}{n_i}. \tag{11.64a}$$

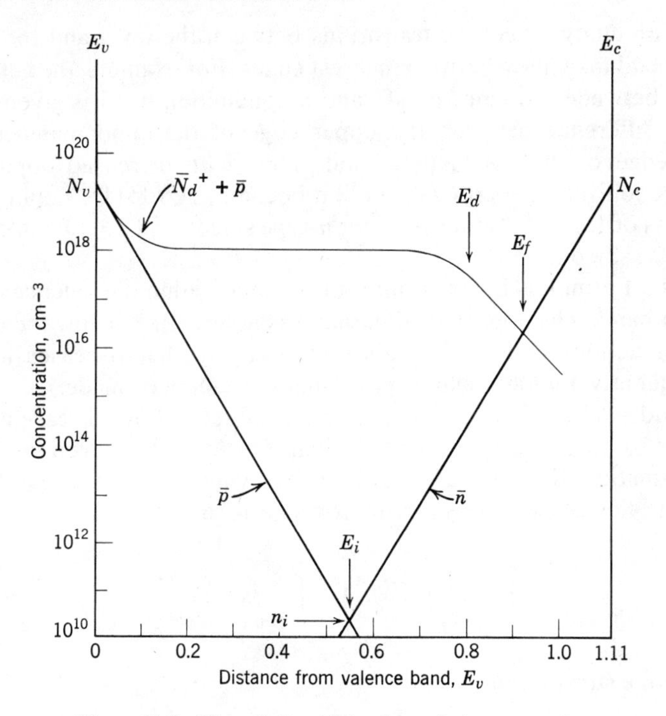

Figure 11.8 The solution of the charge-balance equation.

In like manner, a shallow acceptor the concentration of which is sufficiently high so that $\bar{p} \gg \bar{n}$ leads to

$$E_i - E_f \simeq kT \ln \frac{N_a}{n_i}. \tag{11.64b}$$

Equations 11.64*a* and 11.64*b* give the position of the Fermi level for reasonably extrinsic semiconductors. Note that in both cases the Fermi level is located between the intrinsic level and the impurity level.

11.3.6.2. Heavily Doped Semiconductors. Often doping concentrations are sufficiently high so that the approximate statistics do not apply. The emitter region of a transistor is typical of this situation. In such cases there are two important differences that must be considered to determine the carrier concentrations and the position of the Fermi level. These are as follows:

1. The impurity concentration is so large that it is not possible to ignore the interaction of impurity atoms with one another. Thus the impurity levels broaden out into bands. This results in an effective decrease in the

activation energy, because transitions between the level and the appropriate band take place between nearest edges. For example, the activation energy between a donor "band" and a conduction band is given by the energy difference between the upper edge of the donor band and the lower edge of the conduction band. Thus with increased doping, the effective ionization energy falls until it becomes zero [5] for doping levels in excess of 1.8×10^{18} atoms/cm^3 for n-type silicon and 7×10^{18} atoms/cm^3 for p-type silicon.

2. The Fermi level moves until it is located within the valence or conduction band. The approximation that it is at least $3kT$ below the edge of the conduction band (or $3kT$ above the valence band) ceases to hold. Consequently, the Boltzmann approximation cannot be made.

Consider a heavily doped n-type semiconductor. For this case it is still possible to make the approximation that $E_2 - E_c > 3kT$, even though the approximation that $E_c - E_f > 3kT$ is not valid. Then the equilibrium concentration of electrons is given from (11.34) as

$$\bar{n} = \frac{4}{\sqrt{\pi}}\left(\frac{2\pi m_n^*}{h^2}\right)^{3/2} \int_{E_c}^{\infty} \frac{\sqrt{E-E_c}\, dE}{1+e^{(E-E_f)/kT}}. \tag{11.65}$$

Making the substitutions,

$$\eta = \frac{E-E_c}{kT}, \tag{11.66a}$$

$$\eta_f = \frac{E_f - E_c}{kT}, \tag{11.66b}$$

(11.65) reduces to

$$\bar{n} = \frac{4}{\sqrt{\pi}}\left(\frac{2\pi kT m_n^*}{h^2}\right)^{3/2} \int_0^{\infty} \frac{\eta^{1/2}\, d\eta}{1+e^{\eta-\eta_f}}, \tag{11.67}$$

$$= \frac{2}{\sqrt{\pi}} N_c F_{1/2}(\eta_f), \tag{11.68}$$

where N_c is the density of states at the edge of the conduction band and

$$F_{1/2}(\eta_f) = \int_0^{\infty} \frac{\eta^{1/2}\, d\eta}{1+e^{\eta-\eta_f}}. \tag{11.69}$$

Figure 11.9 shows a plot of $(2/\sqrt{\pi})\, F_{1/2}(\eta_f)$ as a function of η_f.

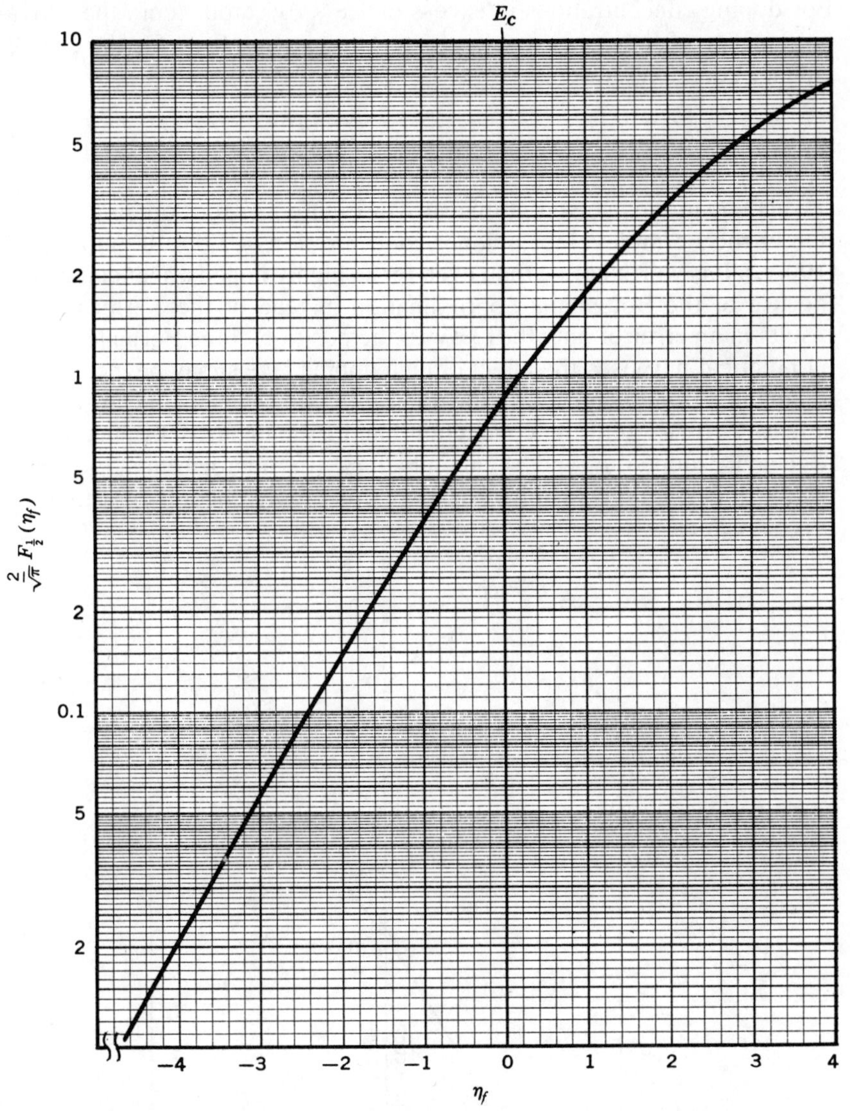

Figure 11.9 The Fermi integral (adapted from Blakemore[2]).

The position of the Fermi level may now be determined for a highly doped n-type semiconductor. For such a material the charge-balance equation is given by $\bar{n} = \bar{N}_d^+$. Combining (11.54) and (11.68) gives

$$\frac{2}{\sqrt{\pi}} N_c F_{1/2}(\eta_f) = \frac{N_d}{1 + 2e^{(E_f - E_d)/kT}}. \tag{11.70}$$

For doping concentrations in excess of 1.8×10^{18} atoms/cm^3, the activation energy of the donor level is zero. Thus setting $E_d = E_c$, the charge-balance equation reduces to

$$\frac{2}{\sqrt{\pi}} N_c F_{1/2}(\eta_f) = \frac{N_d}{1 + 2e^{\eta_f}}. \tag{11.71}$$

This equation is solved graphically in Figure 11.10 to obtain the position of the Fermi level for a donor concentration of 2.8×10^{21} atoms/cm^3 ($100N_c$). It is worth noting that, even for so highly doped a semiconductor, the use of the Boltzmann approximation instead of the Fermi integral results in an error of only 0.013 eV in the position of the Fermi level. For all practical purposes the effect of high doping levels may be included by assuming a Boltzmann approximation for the free carrier concentration to at least 2 to $3kT$ into the conduction band. Analogous reasoning applies to heavily doped p-type semiconductors. In microcircuits, however, highly doped n-type regions are more commonly encountered than are p-type regions.

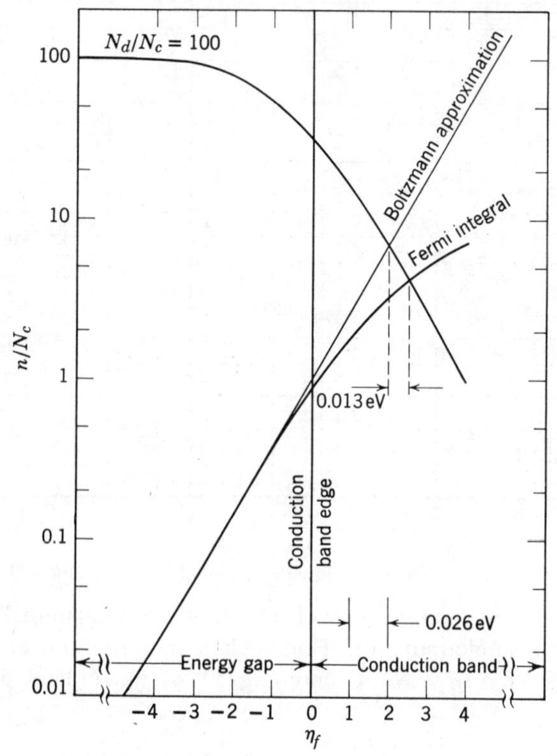

Figure 11.10 A degenerately doped situation.

The term *degenerate* may be applied to highly doped semiconductors to indicate the extent to which the Fermi level penetrates the conduction (or valence) band. The amount of the degeneracy is then given by $E_f - E_c$ or by $E_v - E_f$ depending on the type of semiconductor.

11.4. MULTIPLE-CHARGE STATES[6]

The doping of a semiconductor with elements other than those of columns III and V of the periodic table often gives rise to a somewhat complex energy-level structure. In general these impurities exhibit more than one energy level, often of more than one type (i.e., both donor and acceptor levels). Furthermore, these levels are usually found quite deep in the forbidden gap. It is this last characteristic which leads to the use of these elements in reducing minority carrier lifetime in high-speed microcircuits.

The most complex energy-level structure arises for a monovalent impurity atom. Such an atom, in the neutral state, has only one attached electron which provides covalent binding with its neighboring silicon lattice atoms. When additional electrons are attached to it, it is successively transferred to a more and more negatively charged state, as follows:

$$M^0 + e^- \rightleftharpoons M^- \tag{11.72a}$$

$$M^- + e^- \rightleftharpoons M^{2-}, \tag{11.72b}$$

$$M^{2-} + e^- \rightleftharpoons M^{3-}. \tag{11.72c}$$

Each additional electron gives rise to a possible new energy level. Since these electrons are attached sucessively to a more and more negatively charged atom, it is probable that the value of the associated energy level will continually increase in sequence until it goes beyond the conduction-band edge. At this point the atom will lose this electron to the conduction band, resulting in no further identifiable energy levels.

In addition, a monovalent impurity atom may lose an electron as follows:

$$M^0 - e^- \rightleftharpoons M^+. \tag{11.72d}$$

Thus it is reasonable to assume that the energy-level structure for a monovalent impurity atom may consist of as many as one donor level and three acceptor levels, progressively spaced in order of increasing negative charge. In most instances only a few of these various levels are identifiable within the energy gap. A notable exception is gold in germanium which exhibits all four energy levels.

11.4.1. Carrier Statistics for Multilevel Impurities

The equilibrium carrier statistics can be derived by recognizing that the process of adding or removing an electron transfers the atom to its adjacent charge state. Since this transition process may be described at the jth energy level by

$$(M^s - e^- \rightleftharpoons M^{s+1})_j, \tag{11.73}$$

it follows that

$$\left(\frac{\bar{N}^s}{g_j \bar{N}^{s+1}}\right)_j e^{E_j/kT} = e^{E_f/kT}, \tag{11.74}$$

where E_j is the energy level at which the transition occurs and g_j is the appropriate degeneracy factor. Each transition may be described by an equation of this type.

11.4.2. Application to Gold in Silicon [7]

Gold is commonly used for the reduction of minority carrier lifetime in silicon microcircuits. It has a maximum solid solubility of about 10^{17} atoms/cm^3 in silicon (at 1280°C). Of this, more than 90% is located in active sites in the silicon lattice.

Gold exhibits one donor and one acceptor state within the forbidden gap, as shown in Figure 11.11. As suggested in the earlier argument, the donor state is below the acceptor state. For this reason, gold is often referred to as an *amphoteric impurity* in silicon.

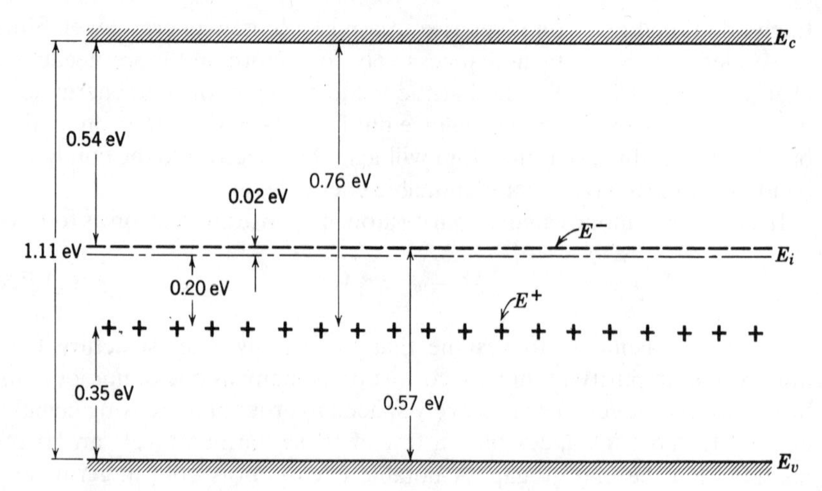

Figure 11.11 Energy levels of gold in silicon.

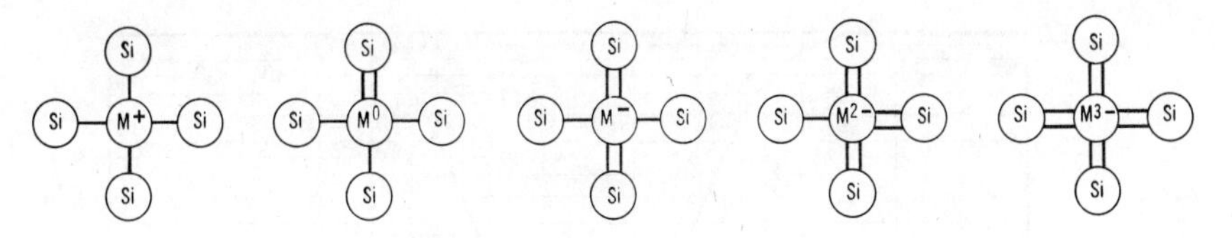

Figure 11.12 Bond arrangments for a monovalent impurity.

Figure 11.12 shows the covalent bonding configuration for the various charge states of gold in silicon. Writing N^+ and N^- as the concentrations of gold in positive (donor) - and negative (acceptor)-charge states respectively, and E^+ and E^- as their corresponding energy levels†, the transition process at the donor level is given by

$$\frac{\bar{N}^0}{g_d \bar{N}^+} e^{E^+/kT} = e^{E_f/kT}, \tag{11.75}$$

while that at the acceptor level is given by

$$\frac{\bar{N}^-}{g_a \bar{N}^0} e^{E^-/kT} = e^{E_f/kT} \tag{11.76}$$

where $\bar{N}^0$ is the equilibrium concentration of unionized gold. Assuming the bonding configurations of Figure 11.12, it is seen that the spin degeneracy factor for the donor state g_d is given by 4. In like manner g_a, the degeneracy factor for the acceptor state, is $\frac{3}{2}$. The total gold concentration is given by

$$N = \bar{N}^+ + \bar{N}^0 + \bar{N}^-. \tag{11.77}$$

Combining with (11.75) and (11.76), the net charge on the gold is given by

$$\frac{\bar{N}^+ - \bar{N}^-}{N} = \frac{(1/g_d) e^{(E^+ - E_f)/kT} - g_a e^{(E_f - E^-)/kT}}{1 + (1/g_d) e^{(E^+ - E_f)/kT} + g_a e^{(E_f - E^-)/kT}} \tag{11.78}$$

This function is shown in Figure 11.13. It is seen that all the gold is in the positive-charge state when the Fermi level is close to the valence band. Thus gold behaves as a donor in strongly *p*-type silicon. As the silicon is made less *p*-type (possibly by reducing its shallow acceptor concentration), the net charge on the gold gets less and less positive. Eventually, when the Fermi level is at 0.436 eV from the valence band,

†N^{2-} and N^{3-} give rise to energy levels within the conduction band. They are not considered for the reasons already stated.

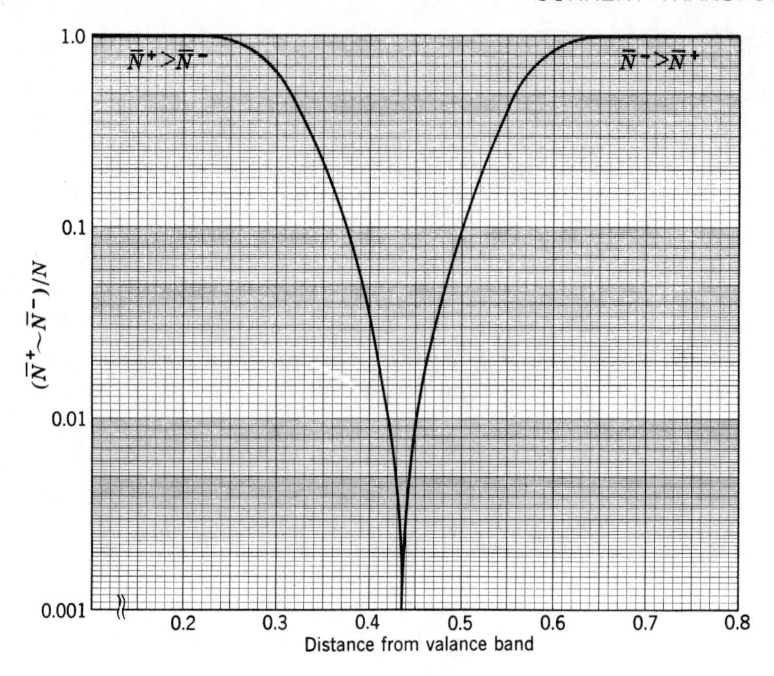

Figure 11.13 Net charge on the gold.

the gold is entirely unionized. As the Fermi level moves still further from the valence band, the sign of the net charge on the gold becomes negative. Thus the gold behaves like an acceptor in n-type silicon.

The energy levels of gold are not symmetric about the intrinisic level, but are displaced toward the valence band. Consequently gold is weakly p-type in silicon; that is, its compensation effect is stronger in n-type silicon than p-type silicon. In fact, with sufficient gold, it is possible to convert n-type silicon to p-type!

The position of the Fermi level may be found by setting up the charge balance conditions. Since gold is normally introduced into a semiconductor which is already doped with a shallow impurity from either columns III or V of the periodic table, the charge on this impurity must also be considered in the equation. Let

$\bar{N}_{sd}^{+}, \bar{N}_{sa}^{-}$ = concentrations of positively charged shallow donors and negatively charged shallow acceptors respectively

$\bar{N}^{+}, \bar{N}^{-}$ = concentrations of gold in the positive- and negative-charge states,

$\bar{n}, \bar{p}$ = equilibrium concentrations of electrons and holes repectively.

Then charge balance requires that

$$\bar{p}+\bar{N}^{+}+\bar{N}_{sd}^{+}=\bar{n}+\bar{N}^{-}+\bar{N}_{sa}^{-} \tag{11.79}$$

This equation may be solved graphically with the aid of Figure 11.13 to obtain the free carrier concentration and the position of the Fermi level. A common approximation that may be made is that all shallow impurities are ionized.

Curves of resistivity versus gold concentration, based on this solution, are shown in Figures 11.14 and 11.15. The following can be seen from these figures:

1. Both p- and n-type silicon become p type at high values of gold concentration. The actual limiting value is not precisely known and depends on the choice† of degeneracy factors.

2. For small gold concentrations there is a negligible resistivity change in the silicon. Ultimately a point is reached when this change becomes

†To be more correct the degeneracy factors should be picked to result in the experimentally found limiting value.

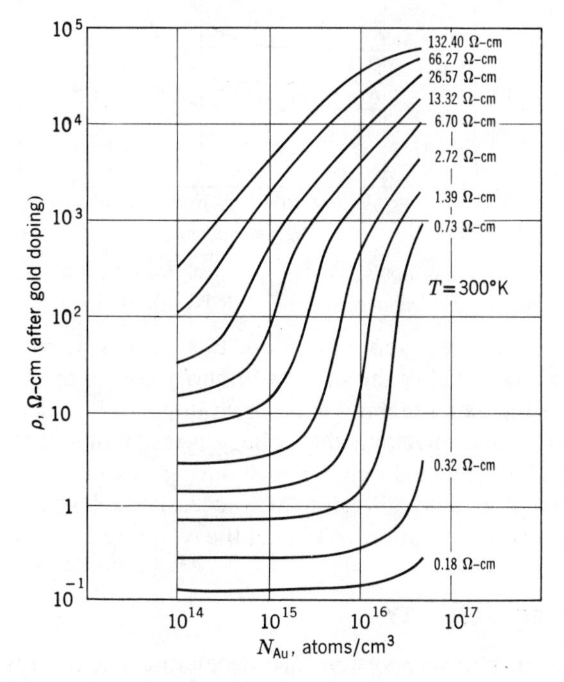

Figure 11.14 Compensation of p-type silicon by gold [From W. Runyan, *Silicon Semiconductor Technology*, McGraw Hill, New York (1965)].

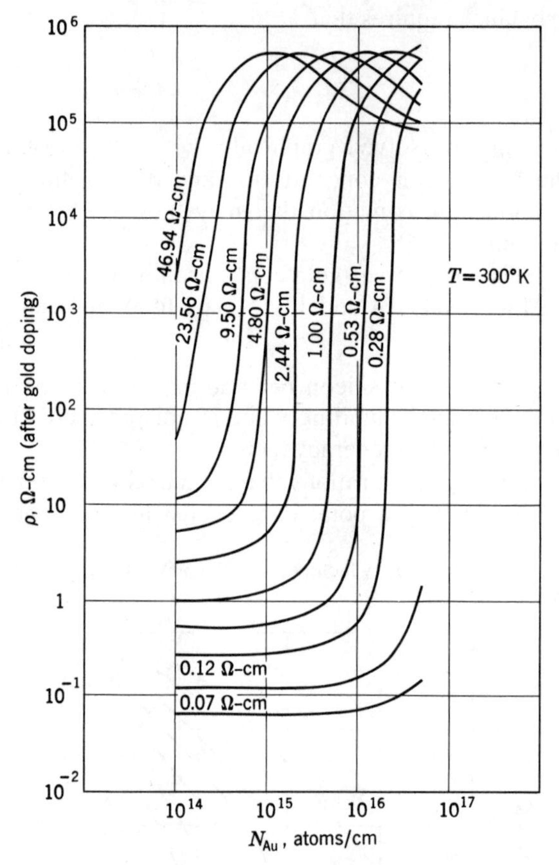

Figure 11.15 Compensation of *n*-type silicon by gold [From W. Runyan, *Silicon Semiconductor Technology*, McGraw Hill, New York (1965)].

very rapid. To a rough approximation, this occurs when the gold concentration exceeds 30% of the shallow impurity concentration.

In practice the usual technique for gold doping is to increase the gold level in steps while checking the collector resistivity of the devices to ensure that it has changed only slightly during this process. This is done by conducting the gold diffusion at successively higher temperatures and monitoring the saturation voltage of the resulting devices.

11.5. CARRIER VELOCITY

Mobile carriers in any system are accelerated in the presence of an electric field. In free space this acceleration proceeds until the carriers approach the speed of light. In a solid, however, they are continually subjected to collisions and deflections, and thus experience a restraining force in addition to an accelerating force. Consequently, they accelerate until they reach a steady-state velocity when these forces are equal in magnitude. This steady-state velocity is known as the *drift velocity* v_d.

Consider the semiconductor bar of unit cross-section, with $dn(E)$ electrons in an energy interval between E and $E+dE$. For these electrons let the restraining force be directly proportional to their velocity $v(E)$ in a direction parallel to that of the electric field. The equation of motion for these electrons is given by

$$F = m_n^* \frac{dv(E)}{dt} + \alpha(E)v(E), \tag{11.80}$$

where F is the force on the electrons, m_n^* is their effective mass, and $\alpha(E)$ is a factor of proportionality.

Since carrier flow is being considered, it is appropriate to use a conductivity effective mass. For silicon this effective mass is

$$m_n^* = 0.26\, m_0, \tag{11.81a}$$

$$m_p^* = 0.38\, m_0, \tag{11.81b}$$

where m_0 is the mass of the electron in free space.

For electrons $F = -q\mathscr{E}$, where $\mathscr{E}$ is the electric field. Substituting into (11.80) and solving for steady-state conditions results in the drift velocity

$$v_d(E) = -\frac{q}{\alpha(E)}\mathscr{E}. \tag{11.82a}$$

Defining a collision time $\tau_c(E) \equiv m_n^* \alpha(E)$, (11.82$a$) may be rewritten as

$$v_d(E) = -\frac{q\tau_c(E)}{m_n^*}\mathscr{E}. \tag{11.82b}$$

The minus sign indicates that the direction of electron drift velocity is opposite to that of the electric field.

The current density associated with these carriers is

$$dJ(E) = qv_d(E)\, dn(E). \tag{11.83}$$

the total current density is given by

$$\int_{\text{all } E} dJ(E) = -qv_d n = -q \int_{\text{all } E} v_d(E)\, dn(E) \tag{11.84}$$

where v_d is the average drift velocity of the electrons and n is their concentration. Combining with (11.82b) gives

$$v_d = \frac{\mathscr{E}}{n} \int_{\text{all } E} \frac{q\tau_c(E)}{m_n^*} dn(E). \tag{11.85}$$

Writing the average collision time as $\langle \tau_c \rangle$,

$$\langle \tau_c \rangle = \frac{\int_{\text{all } E} \tau_c(E)\, dn(E)}{\int_{\text{all } E} dn(E)} \tag{11.86}$$

Equation 11.85 reduces to

$$v_d = -\frac{q\langle \tau_c \rangle}{m_n^*} \mathscr{E}. \tag{11.87}$$

The quantity $q\langle \tau_c \rangle / m_n^*$ is denoted by μ_n, the electron mobility.† A similar line of reasoning may be applied in developing the relation for μ_p, the hole mobility. Note that, for this case,

$$v_d = +\mu_p \mathscr{E}, \tag{11.88}$$

where

$$\mu_p = q\langle \tau_c \rangle / m_p^* \tag{11.89}$$

11.5.1. Lattice Scattering

Consider an ideal lattice, free from chemical or mechanical defects. In such a lattice, the atoms have a vibrational energy associated with the temperature of their surroundings, giving rise to standing waves (or phonon waves) at a frequency ν_{phonon}. If mobile electrons are present in this lattice, together with their associated photon waves of frequency ν_{photon}, an electromagnetic wave interaction will occur. The magnitude of this interaction depends on their relative frequencies. Typically, the wavelength λ is on the order of a lattice spacing for phonons (about 1 Å), but about 10^4 Å for electrons having an energy of 1 eV. In addition, the velocity of propagation v of phonons is about 10^5 cm/sec, and that of photons is about 3×10^{10} cm/sec. Since $\nu = v/\lambda$, it is seen that

$$\frac{\nu_{\text{photon}}}{\nu_{\text{phonon}}} \simeq 30. \tag{11.90}$$

†This is also called the *drift mobility* to distinguish it from the Hall mobility. Often, this distinction is not made, and the word drift is omitted.

Because of this large difference between frequencies, there is a relatively weak interaction between these waves. Thus the effect of lattice scattering is a gradual distortion of the motion of the mobile carriers in the lattice. Shockley and Bardeen[8] have studied the quantum-mechanical nature of this electron-phonon wave interaction, and shown that

$$l = \frac{h^4 C_\mu}{16\pi^3 (m^* E_{1n})^2 kT}, \tag{11.91}$$

where

l = mean free path of the carriers,
C_μ = elastic constant for longitudinal waves.
E_{1n} = change in the position of the conduction band edge per unit dilation of the crystal lattice,
m^* = conductivity effective mass of the carrier under consideration.

If v_t is the average thermal velocity of these carriers, then

$$\tfrac{1}{2} m^* v_t^2 = \tfrac{3}{2} kT \tag{11.92}$$

or

$$v_t = \sqrt{\frac{3kT}{m^*}} \tag{11.93}$$

The mean time between collisions is given by

$$\langle \tau_c \rangle = l/v_t = \frac{l}{\sqrt{3kT/m^*}} \tag{11.94}$$

Combining (11.91) and (11.94), the lattice mobility μ_L is given by

$$\mu_L = \frac{q\langle \tau_c \rangle}{m^*}, \tag{11.95}$$

$$\propto T^{-3/2}. \tag{11.96}$$

Thus lattice mobility is *inversely* proportional to the three halves power of the temperature.

11.5.2. Ionized Impurity Scattering

This type of scattering is coulombic in nature. Depending on the sign of the mobile carrier charge and of the charge on the ionized impurity atom, it takes the form of either attractive or repulsive deflection of the carriers.

An ionized impurity is more effective in scattering low-velocity carriers; thus its effects become more important with decreasing lattice temperature. As a result the mobility associated with impurity scattering increases with increasing temperature. In addition, the scattering effect increases with the concentration of ionized impurities.

Conwell and Weisskopf have carried out a detailed analysis of this process, based on the Rutherford scattering of electrons by nuclei. Based on their analysis the mobility in silicon due to ionized impurity scattering μ_I is given[9] by

$$\mu_I = \frac{4.7 \times 10^{17} \sqrt{m_0/m^*}\, T^{3/2}}{N_I \ln\left(1 + 4.5 \times 10^{8} T^2 / N_I^{2/3}\right)}, \tag{11.97}$$

where μ_I is the mobility in cm^2/volt-sec, and N_I is the concentration of ionized impurities in atoms/cm^3. It is seen that, to an approximation (i.e., ignoring the correction due to the term involving the natural logarithm), the ionized impurity mobility is *inversely* proportional to the concentration of ionized impurities and *directly* proportional to the three halves power of the lattice temperature.

11.5.3. Neutral Particle Scattering

It is necessary for a mobile carrier to get extremely close to the core of a neutral atom in order to be scattered by it. In addition, a neutral particle may lead to a localized strain field which distorts the path of mobile carriers in its vicinity. Both effects are usually ignored because they are quite secondary in nature.

11.5.4. Mobile Carrier Scattering

Scattering by mobile carriers of opposite charge type is somewhat similar to impurity scattering, except that both particles deflect about a common center of mass. In extrinsic semiconductors, in which one species of carrier is in considerably larger concentrations than the other, the probability of encounters of this type is small. Note, however, that the minority carrier is exposed to a large number of majority carriers in the semiconductor. Hence scattering effects of this type are more apparent for the minority carrier than for the majority carrier.

Scattering between carriers of the same charge type is also possible. Since the total momentum is conserved during this process, however, the net effect on conductivity is zero. Hence this type of scattering may be ignored.

11.5.5. Effective Mobility

It has been shown that, in general, lattice and impurity scattering are the two mechanisms that determine the mobility of carriers in a semiconductor. Thus in lightly doped materials, lattice scattering is predominant. With increasing doping concentration, the impurity-scattering term becomes more important.

To take these two scattering terms into consideration, an effective mobility μ may be defined such that

$$\frac{1}{\mu} \equiv \frac{1}{\mu_L} + \frac{1}{\mu_I} \tag{11.98}$$

In like manner additional scattering terms may be included whenever necessary.

Figure 11.16 illustrates the mobility versus temperature curve for a number of boron-doped silicon samples, and serves to illustrate the nature of both μ_L and μ_I. In all samples the effect of impurity scattering predominates at low temperatures (but more so in the highly doped samples). This is seen in the fact that the mobility rises with increasing temperature.

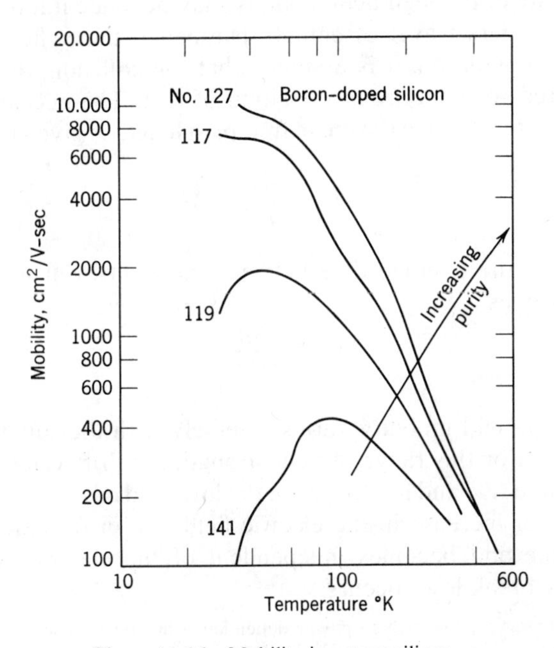

Figure 11.16 Mobility in p-type silicon.

At higher temperatures lattice scattering predominates and mobility falls† with increasing temperature.

The behavior of mobility with temperature is shown in Figure 11.17 for various doping concentrations. These curves cover the range of values over which microcircuits are usually operated.

Figure 11.18 shows the variation of mobility with impurity concentration. As predicted by the theory, the mobility falls inversely with concentration, once impurity scattering effects become important.

A slight difference exists between the mobilities of majority and minority carriers at high-doping concentration. This has been accounted for by the fact that the scattering effect of minority carriers by majority carriers is slightly greater than the converse situation. Thus the majority carrier mobility is seen to be slightly higher than that for minority carriers.

11.5.6. High-field Conditions

The velocity of mobile carriers has been seen to be directly proportional to the electric field, for low values of the electric field. As this field increases, the electrons gain energy, thus increasing the frequency of collisions with the lattice as well as the amount of energy transferred to it.

An estimate of the high-field mobility may be made if it is assumed that all mobile carriers travel essentially parallel to the $\mathscr{E}$ field under these conditions. In addition, it is assumed that, on collision, the carriers can be considered to move from a position of rest. The acceleration of the carriers is $q\mathscr{E}/m^*$. Hence the mean free-path length is given by

$$l = (\tfrac{1}{2}) \frac{q\mathscr{E}}{m^*} \langle \tau_c \rangle^2, \tag{11.99}$$

where $\langle \tau_c \rangle$ is the average time between collisions. Substituting (11.99) into (11.89) gives

$$\mu = \left(\frac{2ql}{m^*\mathscr{E}} \right)^{1/2} \tag{11.100}$$

Thus the high-field mobility varies inversely with the square root of the electric field. For this range of field strength the drift velocity is directly proportional to $\sqrt{\mathscr{E}}$ and not to $\mathscr{E}$ as in the low-field case.

With further increase in the electric field, the drift velocity reaches a limiting value and becomes independent of the electric field. At even higher fields, breakdown occurs.

†Contrary to theory, a five halves power dependence has been observed with silicon. It is believed that this is caused by the presence of optical phonon scattering effects.

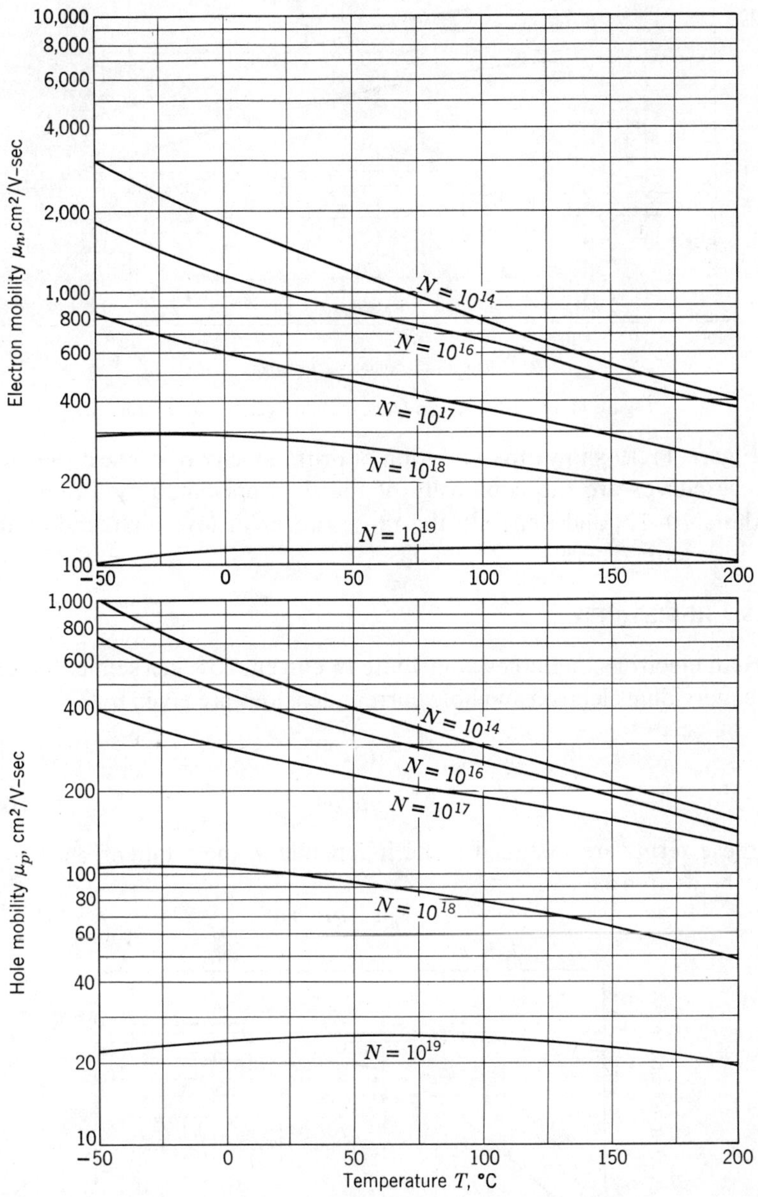

Figure 11.17 Mobility versus temperature [From A. Phillips, *Transistor Engineering*. McGraw Hill, New York (1962)].

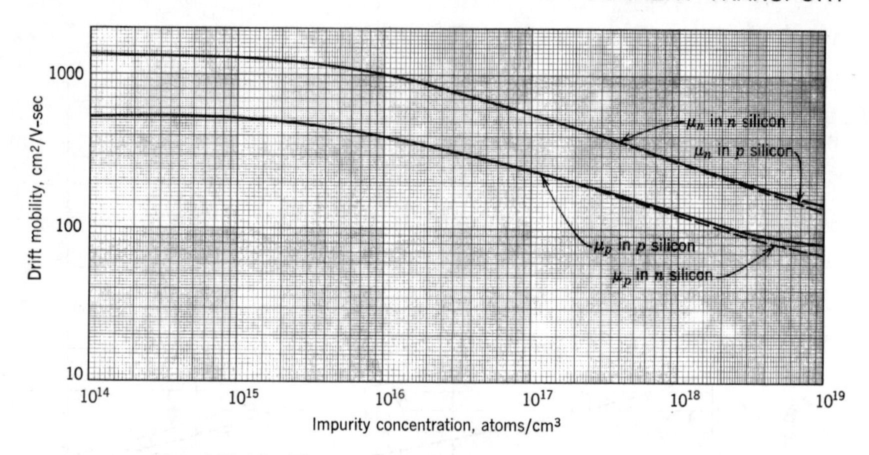

Figure 11.18 Electron and hole mobility versus concentration.

Figure 11.19 shows the variation of drift velocity of carriers in silicon. These curves are the composite of the data presented by a number of authors[10–12] and illustrate the three ranges of low, high, and limiting fields described here.

11.6. RESISTIVITY

Both electrons and holes contribute to current flow in a semiconductor. The individual electron and hole current densities are given by

$$J_n = q\mu_n n\mathscr{E}, \tag{11.101a}$$

$$J_p = q\mu_p p\mathscr{E}. \tag{11.101b}$$

If these terms are assumed to be independent, the total current density may be written as

$$J = (q\mu_n n + q\mu_p p)\mathscr{E}, \tag{11.102}$$

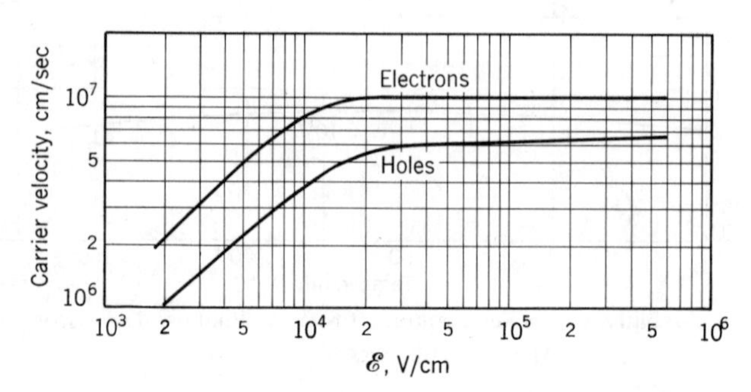

Figure 11.19 Carrier velocity versus electric field.

where J = current density, in amperes per square centimeter,
n,p = electron and hole concentration, per cubic centimeter,
μ_n, μ_p = drift mobilities, in square centimeters per volt-second,
$\mathscr{E}$ = electric field, in volts per centimeter,
q = the magnitude of the electronic charge (1.6×10^{-19} coulomb).

Note that, whereas holes and electrons drift in opposite directions under the influence of the electric field, the sign of their charges is opposite. Thus their contributions to the current flow are additive.

The resistivity ρ is thus given by

$$\rho = \frac{1}{q(\mu_n n + \mu_p p)}, \tag{11.103}$$

and has the dimensions of ohm-centimeters. It must be emphasized that the electron and hole concentrations in this expression are not thermal-equilibrium values since current is flowing. As long as the material, however, exhibits a linear voltage-current characteristic (i.e., as long as Ohm's law holds), these concentrations are not far different from their equilibrium values and may be replaced by $\bar{n}$ and $\bar{p}$ respectively.

Experimental[13] values of resistivity as a function of doping concentration are given in Figure 11.20 for a wide range of impurity concentrations at room temperature. They may be used in conjunction with the curves of Figure 11.17 to determine the resistivity of extrinsic silicon as a function of temperature.

11.7. CARRIER DIFFUSION

Diffusion processes have been considered for chemical impurities in Chapter 4. At this point an analogous line of reasoning is applicable to the behavior of mobile carriers. Thus defining the flux density j_n as the time rate of change of number of mobile electrons per unit area, and n as the electron concentration in a one-dimensional situation, it follows directly from Fick's first law that [see (4.10)]

$$j_n = -D_n \frac{\partial n}{\partial x}, \tag{11.104}$$

where D_n is the diffusion coefficient† for electrons.

†Values of D_n range from 3 to 30 cm^2/sec for electrons at room temperature. In comparison, the diffusion constant of boron and phosphorus is 10^{-12} cm^2/sec at 1150°C.

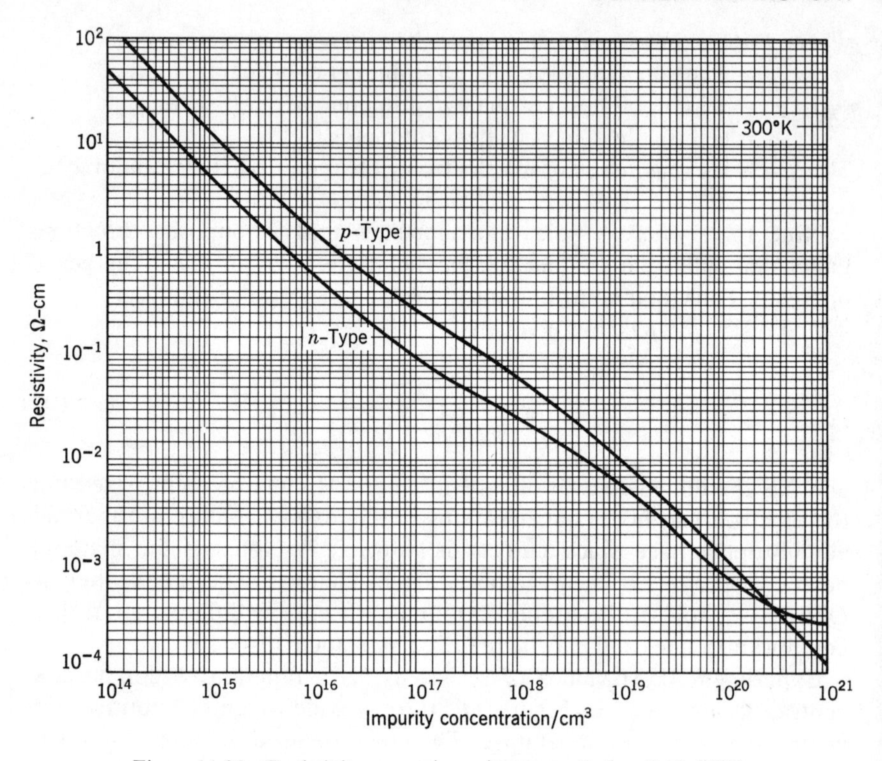

Figure 11.20 Resistivity versus impurity concentration (Irvin [13]).

A charge of $-q$ is associated with each electron. Thus the current density due to the diffusion of electrons J_n is given by

$$J_n = +qD_n \frac{\partial n}{\partial x}\,. \tag{11.105a}$$

A similar reasoning may be applied to holes, noting that the charge associated with each hole is $+q$. Thus,

$$J_p = -qD_p \frac{\partial p}{\partial x}\,, \tag{11.105b}$$

where J_p = hole current density, in amperes per square centimeter,
D_p = hole diffusion constant, in square centimeter per second,
p = hole concentration, per cubic centimeter.

11.8. THE TOTAL CURRENT

If diffusion and drift are considered to be independent events, (11.101*a*) and (11.105*a*) may be combined to write the total electron current density in a semiconductor as

$$J_n = q\mu_n n\mathscr{E} + qD_n\frac{\partial n}{\partial x}, \tag{11.106a}$$

The hole current density is given by

$$J_p = q\mu_p p\mathscr{E} - qD_p\frac{\partial p}{\partial x}. \tag{11.106b}$$

11.8.1. The Diffusion Constants

It is reasonable to expect that the constants that characterize drift and diffusion are interrelated. This is because both phenomena involve the same random wandering of mobile carriers through the crystal lattice, even though an additional directed motion is present in each case. This inter-relationship is obtained by considering thermal-equilibrium conditions. For electrons, $J_n = 0$, and (11.106*a*) reduces to

$$\frac{\partial n}{\partial x} + \frac{\mu_n\mathscr{E}}{D_n}n = 0, \tag{11.107}$$

hence

$$n = \bar{n} = Ae^{-(\mu_n/D_n)\int\mathscr{E}dx} \tag{11.108}$$

Noting that

$$-\int\mathscr{E}\,dx = V = -\frac{E}{q}, \tag{11.109}$$

where V is the electrostatic potential, gives

$$\bar{n} = Ae^{(\mu_n/D_n)(E_2-E_1)/q}. \tag{11.110}$$

Here E_1 and E_2 are the energy levels associated with the integration. But

$$\bar{n} = n_i e^{(E_f-E_i)/kT}. \tag{11.111}$$

If we compare (11.111) with (11.110), it will follow that

$$D_n = \frac{kT}{q}\mu_n. \tag{11.112a}$$

In like manner,

$$D_p = \frac{kT}{q}\mu_p. \tag{11.112b}$$

These are the well-known Einstein relations. Note that they are applicable only to materials that are not heavily doped, that is, to nondegenerate materials.

11.9. REFERENCES

[1] C. Kittel, *Elementary Statistical Physics*, Wiley, New York, 1958.
[2] J. S. Blakemore, *Semiconductor Statistics*, Pergamon, New York, 1962.
[3] B. Lax and J. G. Mavroides, "Statistics and Galvanomagnetic Effects in Germanium and Silicon with Warped Energy Surfaces," *Phys. Rev.*, **100** (6), 1650–1657 (Dec. 15, 1953).
[4] R. A. Smith, *Semiconductors*, The University Press, Cambridge, 1961.
[5] H. Brooks, *Advances in Electronics and Electron Physics*, Vol. 7, 120, Academic, New York, 1955.
[6] W. Shockley and J. T. Last, "Statistics of the Charge Distribution for a Localized Flaw in a Semiconductor," *Phys. Rev.*, vol. **107** (no. 2), 392–396, (July 15, 1957).
[7] W. M. Bullis, "Properties of Gold in Silicon," *Solid State Electron.*, **9**, 143–168 (1966).
[8] W. Shockley and J. Bardeen, "Energy Bands and Mobility in Monatomic Semiconductors," *Phys. Rev.*, **77**(3), 407–408 (Feb. 1950).
[9] E. M. Conwell, "Properties of Silicon and Germanium," *Proc. IRE*, **40**(11), 1327–1342 (1952).
[10] C. B. Norris and J. F. Gibbons, "Measurement of High Field Carrier Drift Velocities in Silicon by a Time of Flight Technique," *IEEE Trans. on Electron Devices*, **ED-14**(1), 38–43 (1967).
[11] V. Rodriguez *et al.*, "Measurement of the Drift Velocity of Holes in Silicon at High Field Strengths," *IEEE Trans. on Electron Devices*, **ED-14**(1), 44–46 (1967).
[12] C. Y. Duh and J. L. Moll, "Electron Drift Velocity in Avalanching Silicon Diodes," *IEEE Trans. on Electron Devices*, **ED-14**(1), 46–49 (1967).
[13] J. C. Irvin, "Resistivity of Bulk Silicon and of Diffused Layers," *Bell System Tech. J.*, **41**, 387–410 (1962).

11.10. PROBLEMS

1. Using the principle of detailed balance [i.e., the results of (11.13)], determine the equilibrium concentration of Frenkel defects (1.23) in a crystal.

2. Using the principle of detailed balance, compute the $\bar{n}\bar{p}$ product of an extrinsic semiconductor.

3. Nickel has two acceptor states in silicon, N^- and N^{2-}. With the aid of Table 11.2, sketch the energy-band picture for nickel-doped silicon. Sketch also the net charge on the nickel as the Fermi level sweeps through the energy gap.

4. A piece of n-type silicon having a phosphorus concentration of 10^{16} atoms/cm^3 is doped with 10^{16} atoms/cm^3 of gold. Determine the position of the Fermi level and the free-carrier concentration.

5. Repeat Problem 4 for a p-type starting material.

6. Compute the carrier concentration for silicon doped with a phosphorus concentration of: (a) 10^{17} atoms/cm^3; (b) 10^{19} atoms/cm^3; (c) 10^{21} atoms/cm^3.

Explain physically why the fraction of ionized impurities is so large in (a), even though the activation energy of phosphorus is on the order of kT.

7. Compute the carrier concentration for silicon doped with 10^{17} atoms/cm^3 of indium and determine the position of the Fermi level.

8. Assuming phosphorus-doped silicon, determine $E_f - E_i$ as a function of impurity concentration over the range from 10^{15} atoms/cm^3 to 10^{21} atoms/cm^3. Compare the result with (11.64a).

9. A microcircuit resistor is fabricated by conducting a boron diffusion with a surface concentration of 10^{18} atoms/cm^3 into silicon with 10^{16} atoms/cm^3 background concentration. The junction depth is 2 μm, and the dimensions of the resistor are 10×100 μm. Neglecting end effects, compute the value of the resistance so formed at −50, 25, and 150°C. (*Hint:* Assume a gaussian profile and use a piecewise linear approximation.)

10. A resistor consists of an n-type silicon bar, 20×20 μm long. Assuming a resistivity of 2000 Ω-cm (near-intrinsic), compute its voltage-current characteristic over the range from 0 to 200 volts.

Chapter 12

Nonequilibrium Processes

The condition of thermal equilibrium postulates a reciprocal transfer of energy between a system and its environment. In a semiconductor at any given temperature this results in the continual generation and recombination of hole-electron pairs with identical generation and recombination rates. The end result of this process is an equilibrium concentration of carriers, the statistics of which have been discussed in Chapter 11. If such a system is subjected to a stimulus (by illuminating it with steady penetrating light, for example), both carrier generation and recombination rates are altered until a new steady-state condition is reached. It is important to note that this is *not* a thermal-equilibrium condition because there is no reciprocal transfer of energy between the light source and the semiconductor.

Such a situation is referred to as a *steady-state non-equilibrium condition*. On removal of this additional stimulus a *transient* non-equilibrium phase occurs until the system returns to thermal equilibrium. This is brought about by the process of recombination between the excess electrons and holes. The mechanisms involved in the recombination process, together with the manner in which they may be characterized and controlled, are considered in this chapter.

12.1. RECOMBINATION AND GENERATION RATES

Recombination takes place when holes and electrons encounter each other in the crystal lattice. Since their separate movements are statistically independent, this process occurs at a rate that is directly proportional to their density at any given time.

Consider an n-type semiconductor under conditions of illumination.

At any time the carrier concentration in this semiconductor may be written as

$$n = \bar{n} + n' \tag{12.1a}$$

$$p = \bar{p} + p' \tag{12.1b}$$

where $\bar{n}$, $\bar{p}$ are thermal equilibrium concentrations and n', p' are the time dependent concentrations of excess electrons and holes respectively. The recombination rate is then given by

$$R\,(n, p) = C(\bar{n} + n')(\bar{p} + p') \tag{12.2}$$

where C is a constant.

Consider now the commonly encountered case of an extrinsic, n-type semiconductor, with $\bar{n} \gg \bar{p}$. For this material consider also the situation where the excess majority carrier concentration is small compared to the equilibrium concentration, that is, where $n' \ll \bar{n}$. Note that it does not necessarily follow that $p' \ll \bar{p}$. In fact, the more common situation is for the excess minority carrier concentration to be much greater than the equilibrium concentration of these carriers, that is, for $p' \gg \bar{p}$.

An example serves to illustrate this point. Assume n-type silicon with $\bar{n} = 10^{16}$ electrons/cm^3 and $\bar{p} = 1.9 \times 10^4$ holes/cm^3. For this material an excess concentration of 1.9×10^{10} hole-electron pairs/cm^3 corresponds to 10^6 times the equilibrium concentration of minority carriers, but only 1.9×10^{-6} times the equilibrium concentration of majority carriers. For this situation (12.2) reduces to

$$R\,(n, p) \simeq C\bar{n}p. \tag{12.3}$$

In like manner, for a p-type semiconductor, it follows that

$$R(n, p) \simeq C\bar{p}n. \tag{12.4}$$

Note that in each case the carrier recombination rate is directly proportional to the concentration of carriers which are in the minority, that is, *minority carriers*.

The recombination rate for thermal-equilibrium conditions is equal to the generation rate and is given by

$$R\,(\bar{n}, \bar{p}) = C\bar{n}\bar{p}. \tag{12.5}$$

Thus the recombination rate for excess carriers is given by subtracting (12.5) from (12.2), resulting in

$$R\,(n', p') = C\,(\bar{n}p' + \bar{p}n' + p'n'), \tag{12.6}$$

$$\simeq C\bar{n}p'. \tag{12.7}$$

The generation rate is given at all times by the sum of the generation rates for thermal-equilibrium carriers and for excess carriers. Thus

$$G(n,p) = G(\bar{n},\bar{p}) + G(n',p'). \tag{12.8}$$

For thermal-equilibrium conditions,

$$G(\bar{n},\bar{p}) = R(\bar{n},\bar{p}). \tag{12.9}$$

12.2. CARRIER LIFETIME

Suppose that the n-type semiconductor is subjected to a uniform penetrating illumination of constant intensity. Let this give rise to G_0 hole-electron pairs/sec in unit volume of the material. The change in the excess minority carrier concentration at any given time may be written as

$$\frac{dp'}{dt} = G(n',p') - R(n',p'), \tag{12.10}$$

$$= G_0 - C\bar{n}p'. \tag{12.11}$$

When steady-state conditions prevail, (12.11) reduces to

$$p'_{\text{steady state}} = \frac{G_0}{C\bar{n}}. \tag{12.12}$$

The minority carrier lifetime may be determined by removing the light stimulus and observing the manner in which the excess carriers decay to zero. For this condition (12.11) reduces to

$$\frac{dp'}{dt} + C\bar{n}p' = 0. \tag{12.13}$$

This equation may be solved, subject to the boundary condition that the initial value of the excess carrier concentration be [see (12.12)] equal to $G_0/C\bar{n}$. Solving,

$$p' = \frac{G_0}{C\bar{n}} e^{-tC\bar{n}}. \tag{12.14}$$

We now define a minority carrier lifetime τ_p such that

$$\tau_p \equiv \frac{1}{C\bar{n}} = -\frac{p'}{dp'/dt}. \tag{12.15}$$

Substituting in (12.14), gives

$$p' = G_0\tau_p e^{-t/\tau_p} \tag{12.16}$$

From (12.15) it is seen that the minority carrier lifetime is given by the negative of the ratio of the excess carrier concentration to the rate of change of excess carriers. It is important to note that this lifetime is independent of the actual excess carrier concentration for the low level situation where $n' \ll \bar{n}$. In general, the minority carrier lifetime becomes concentration dependent when this condition is violated.

The majority carrier lifetime may be defined in an analogous manner. However, it is the recombination rate of minority carriers that is of interest in studying non-equilibrium processes.

12.3. RECOMBINATION CENTERS

The physical process of recombination may be considered in terms of the energy-band picture for silicon. In Figure 12.1*a* the momentum-energy relations for silicon are shown for the various crystallographic directions. The more commonly used picture of the band structure is shown in Figure 12.1*b*, which depicts the energy gap as the minimum between the valence and conduction surfaces.

The physical process of recombination of electrons and holes is one in which momentum must be conserved. Since this results in the annihilation of these particles, direct recombination can only occur when they have exactly equal and opposite momenta before the encounter. This

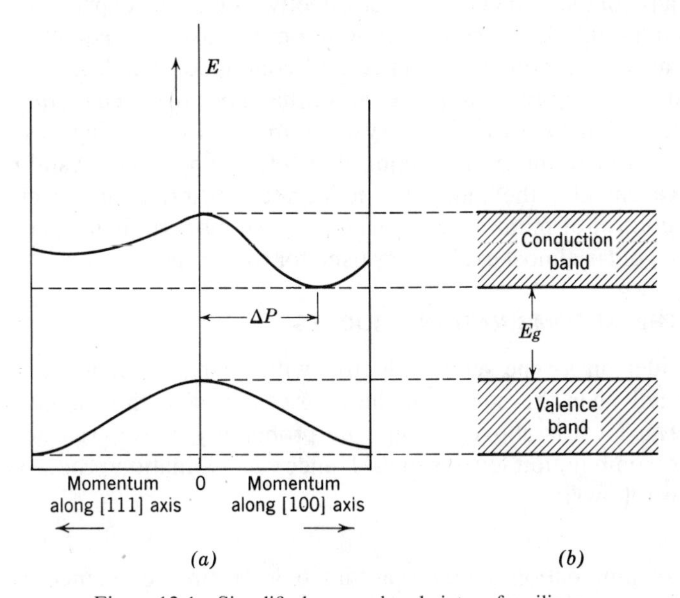

Figure 12.1 Simplified energy-band picture for silicon.

fact, combined with the requirement that they must be in the vicinity of each other, leads to the conclusion that direct recombination in silicon is highly improbable. Indeed, computations based on this possibility result in values of lifetime that are two or three orders of magnitude larger than those found in even the best grades of semiconductor silicon!

From Figure 12.1*a* it is also seen that a relatively large momentum change (ΔP) is involved when an electron and a hole combine to re-establish a valence bond. This energy must either be given off as photons or transferred to the lattice as phonons.

The momentum of a wave quantum is h/λ. For phonons in a crystal lattice, $\lambda \simeq 1$ Å. In contrast, photons in an energy range of 1 eV have a wavelength of 10^4 Å. Thus the momentum associated with a phonon is about 10^4 times larger than that associated with a photon. Consequently, it is highly probable that phonons are involved in the recombination process in silicon (or for that matter, in any "indirect gap" semiconductor). Following this argument it is reasonable to postulate [1] the presence in the forbidden gap of one or more energy levels, which are only partially occupied by electrons. Recombination occurs when a mobile carrier of the appropriate charge species is captured at these levels and recombines with a carrier of the opposite species, which is subsequently captured. The excess momentum is transferred to the crystal lattice in the form of phonon vibrations.

For a recombination center to be effective it must be capable of existing in two different charge states. In addition, it must have equal (or almost equal) access to both the valence and conduction band. Such a role is fulfilled by a deep-lying level. Levels of this type are called *recombination centers*, and may arise from crystal imperfections during growth and subsequent handling or from impurities left behind after crystal purification. Alternately, they may be deliberately inserted in the crystal to obtain control of minority carrier lifetime. In modern, high-speed microcircuits, gold is almost exclusively used for this purpose.

12.4. THE RECOMBINATION PROCESS

Consider an *n*-type semiconductor, with a single recombination level E_r located near the center of the energy gap. For this semiconductor the Fermi level is above E_r, and the probability of electron occupancy of the recombination level is high. Hence recombination proceeds in two steps, as follows:

STEP 1

The recombination centers capture holes from the valence band and are thus promoted to the next higher charge state. This is actually a reversible process, where the capture of holes from the valence band is simultaneously accompanied by the emission of holes back to the valence band. Indeed, in thermal equilibrium, the principle of detailed balance states that these two processes must occur at the same rate. Defining the rate of capture of holes from the valence band as R_{cp}, and the rate of emission of holes as R_{ep}, this process can be described by

$$\left(M^s + h^+ \underset{R_{ep}}{\overset{R_{cp}}{\rightleftharpoons}} M^{s+1}\right)_r. \tag{12.17}$$

The net rate of hole capture is equal to $R_{cp} - R_{ep}$. Under thermal-equilibrium conditions,

$$\bar{R}_{cp} = \bar{R}_{ep}, \tag{12.18}$$

where the bar refers to thermal equilibrium values.

STEP 2

The recombination centers are now available to capture electrons from the conduction band, and are thus returned to their original charge state. If R_{cn} is the rate of capture of electrons from the conduction band and R_{en} is the rate of emission of electrons to the conduction band,

$$\left(M^{s+1} + e^- \underset{R_{en}}{\overset{R_{cn}}{\rightleftharpoons}} M^{s}\right)_r. \tag{12.19}$$

The net rate of electron capture is thus given by $R_{cn} - R_{en}$. For thermal-equilibrium conditions,

$$\bar{R}_{cn} = \bar{R}_{en}. \tag{12.20}$$

Figure 12.2*a* illustrates this process for an *n*-type semiconductor in which the recombination center is a singly ionizable acceptor. The situation for a semiconductor with a singly ionizable donor-like center is shown in Figure 12.2*b*. Needless to say, the same line of reasoning applies to multiple-charge states as well.

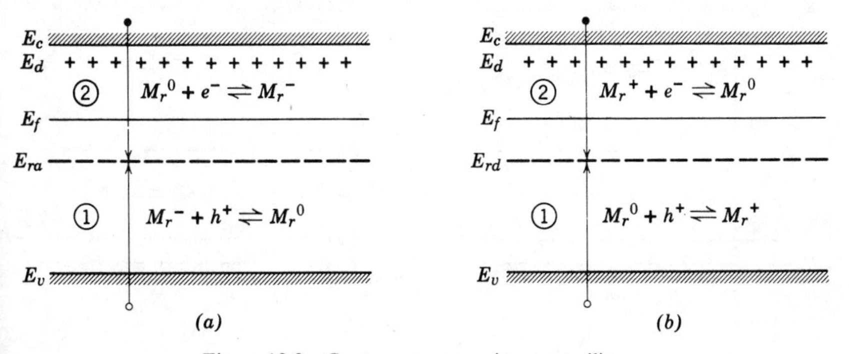

Figure 12.2 Capture processes in *n*-type silicon.

In a p-type semiconductor the steps in the recombination process are reversed. First, electrons are captured at the centers, the process being described by (12.19). Recombination occurs when the centers subsequently capture holes [as described by (12.17)] and return to their original charge state. This process is illustrated in Figure 12.3 for singly ionizable recombination centers which are both donor- and acceptor-like in character.

It is interesting to note that in all of these processes recombination is initiated by the capture of the minority carrier. This is a consequence of the relative locations of the Fermi level and the recombination level.

12.5. CAPTURE CROSS SECTION

The ability of a recombination center to capture mobile carriers may be described by its capture cross section. Consider that an electron of momentum P is captured when it comes within a circular neighborhood of the recombination center. Let $A_n\ (P)$ be the capture cross section for this center.

Since the electron velocity is P/m_n^*, it sweeps out an effective volume $(P/m_n^*)A_n\ (P)$ per second. This is the rate of electron capture. Considering electrons in all energy ranges, the average rate of electron capture is given by $\langle (P/m_n^*)A_n(P)\rangle$. It is more convenient to write this average in terms of experimentally derived quantities. Thus let

$$\langle P/m_n^*\ A_n(P)\rangle = v_{tn}A_n = \alpha_n, \tag{12.21a}$$

where A_n is the average electron capture cross section of the recombination center and v_{tn} is the average electron thermal velocity.

In like manner the average rate of hole capture is given by

$$\langle P/m_p^* A_p(P)\rangle = v_{tp}A_p = \alpha_p. \tag{12.21b}$$

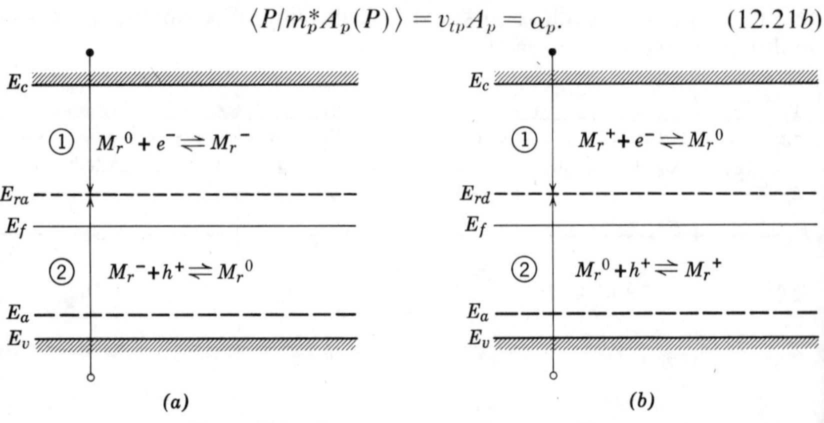

Figure 12.3 Capture processes in p-type silicon.

Here A_p is the average hole capture cross section and v_{tp} is the average hole thermal velocity.

12.6. RECOMBINATION THROUGH A SINGLE LEVEL[2, 3]

Consider an n-type semiconductor, with a density of recombination centers N_r at an energy level E_r. For the purposes of discussion it is assumed that these centers are donors. In thermal equilibrium some of these centers will be neutral and the rest ionized. Thus

$$N_r = \bar{N}_r^{+} + \bar{N}_r^{0}, \tag{12.22}$$

where the bar serves to distinguish the thermal equilibrium values from the nonequilibrium values. Under steady-state non-equilibrium conditions, the concentrations of neutral and ionized recombination centers will be altered, so that

$$N_r = N_r^{+} + N_r^{0}. \tag{12.23}$$

The capture rate of electrons from the conduction band is proportional to the electron concentration and to the concentration of positively ionized centers. Thus,

$$R_{cn} = \alpha_n n N_r^{+}. \tag{12.24a}$$

The rate of emission of electrons to the conduction band is proportional to the concentration of neutral centers and to the density of available states at the conduction-band edge. Thus

$$\begin{aligned} R_{en} &= C_1 N_r^{0}(N_c - n), \\ &\simeq C_1 N_r^{0} N_c\,, \end{aligned} \tag{12.24b}$$

where C_1 is a proportionality factor.

The rate of capture of holes from the valence band is proportional to the hole concentration and to the concentration of neutral centers. Thus

$$R_{cp} = \alpha_p p N_r^{0}. \tag{12.25a}$$

Finally, the rate of emission of holes to the valence band is proportional to the concentration of ionized centers and to the density of available states at the valence-band edge. Thus

$$\begin{aligned} R_{ep} &= C_2 N_r^{+} (N_v - p) \\ &\simeq C_2 N_r^{+} N_v\,, \end{aligned} \tag{12.25b}$$

where C_2 is a proportionality factor.

C_1 and C_2 may be evaluated by using the conditions for thermal equilibrium. For example, since $\bar{R}_{cn} = \bar{R}_{en}$,

$$C_1 = \frac{\alpha_n \bar{n}}{N_c} \frac{\bar{N}_r^+}{\bar{N}_r{}^0}. \tag{12.26}$$

Combining (12.24*a*), (12.24*b*) and (12.26), the rate of change of the excess electron concentration is given by

$$\begin{aligned} -\frac{dn'}{dt} &= R_{cn} - R_{en} \\ &= \alpha_n \left(nN_r^+ - \bar{n}N_r{}^0 \frac{\bar{N}_r^+}{\bar{N}_r{}^0} \right). \end{aligned} \tag{12.27}$$

In thermal equilibrium,

$$\bar{n} = N_c e^{-(E_c - E_f)/kT} \tag{12.28}$$

and (ignoring degeneracy factors)

$$\frac{\bar{N}_r{}^0}{\bar{N}_r^+} e^{E_r/kT} = e^{E_f/kT}. \tag{12.29}$$

Let n_1 be the equilibrium concentration of electrons that would be obtained if the Fermi level were at the recombination level. Then

$$n_1 = N_c \, e^{-(E_c - E_r)/kT} \tag{12.30}$$

by definition. Substituting (12.28), (12.29), and (12.30) into (12.27) gives

$$-\frac{dn'}{dt} = \alpha_n (nN_r^+ - n_1 N_r{}^0). \tag{12.31a}$$

In like manner the rate of change of the excess hole concentration is given by

$$-\frac{dp'}{dt} = \alpha_p \, (pN_r{}^0 - p_1 N_r^+), \tag{12.31b}$$

where p_1 is the equilibrium value of the hole concentration if the Fermi level were at the recombination level.

Equations 12.31*a* and 12.31*b* must be solved to obtain the non-equilibrium concentrations of ionized and unionized donors. In steady state, electrons and holes are simultaneously annihilated. Thus

$$\frac{dn'}{dt} = \frac{dp'}{dt}. \tag{12.32}$$

In addition,

$$N_r = N_r^+ + N_r^0. \tag{12.33}$$

Combining (12.32) and (12.33) with (12.31*a*) and (12.31*b*) gives

$$N_r^+ = N_r \frac{\alpha_n n_1 + \alpha_p p}{(n+n_1)\alpha_n + (p+p_1)\alpha_p} \tag{12.34a}$$

and

$$N_r^0 = N_r \frac{\alpha_n n + \alpha_p p_1}{(n+n_1)\alpha_n + (p+p_1)\alpha_p} . \tag{12.34b}$$

Substituting back into (12.31*b*) and noting that $n_1 p_1 = n_i^2$,

$$-\frac{dp'}{dt} = \frac{np - n_i^2}{(n+n_1)/\alpha_p N_r + (p+p_1)/\alpha_n N_r} . \tag{12.35}$$

Writing

$$\tau_{p0} = (\alpha_p N_r)^{-1} \tag{12.36a}$$

$$\tau_{n0} = (\alpha_n N_r)^{-1} \tag{12.36b}$$

the minority carrier lifetime is given by

$$\tau_p = -\frac{p'}{dp'/dt} \tag{12.37}$$

$$= p' \frac{(n+n_1)\,\tau_{p0} + (p+p_1)\,\tau_{n0}}{np - n_i^2} \tag{12.38a}$$

A similar analysis applies to *p*-type semiconductors. Here

$$\tau_n = n' \frac{(n+n_1)\tau_{p0} + (p+p_1)\tau_{n0}}{np - n_i^2} . \tag{12.38b}$$

12.6.1. The Low-level Injection Case

The minority carrier lifetime for an *n*-type semiconductor is obtained by substituting (12.1*a*) and (12.1*b*) into (12.38*a*). In addition, it is noted that $n' = p'$, since the process of recombination consists of the simultaneous annihilation of excess electrons and holes. Under low-level injection conditions, it is possible to make the additional assumptions that $\bar{n} \gg n'$. Making these substitutions, (12.38*a*) reduces to

$$\tau_p = \frac{(\bar{n} + n_1)\,\tau_{p0} + (\bar{p} + p_1 + p)\,\tau_{n0}}{\bar{n} + \bar{p}} \tag{12.39}$$

This equation may be used to determine the minority carrier lifetime for a number of situations.

Consider the case of an extrinsic n-type semiconductor, with recombination centers situated approximately midway in the energy gap. Setting $n_1 \simeq p_1 \simeq n_i$, and $\bar{n} \gg \bar{p}$, (12.39) simplifies to

$$\tau_p \rightarrow \tau_{p0} = (\alpha_p N_r)^{-1}. \tag{12.40}$$

Physically this represents the case where nearly all the recombination centers are occupied by electrons (i.e., $N_r \simeq N_r{}^0$) prior to injection, since the Fermi level is well above the recombination level. Consequently, the capture of holes by the centers is the rate-limiting process, because there are many electrons to combine with them. Furthermore, this condition persists as long as the Fermi level is above the recombination level. For this situation, the hole capture rate is given by $\alpha_p p N_r{}^0 \simeq \alpha_p p N_r$. The hole emission rate is $C_2 N_r^+ N_v \simeq \alpha_p \bar{p} N_r$. Thus the net rate of change of excess hole concentration is

$$-\frac{dp'}{dt} = \alpha_n p' N_r. \tag{12.41}$$

It follows that the minority carrier lifetime is thus approximately equal to $1/\alpha_p N_r$ for this situation.

For a highly doped p-type semiconductor, in like manner,

$$\tau_n \rightarrow \tau_{n0} = (\alpha_n N_r)^{-1}. \tag{12.42}$$

In general it is found that the cross section for holes is larger than that for electrons, so that $\tau_{n0} > \tau_{p0}$.

As less extrinsic semiconductors are considered, the Fermi level approaches the recombination level; less of the recombination centers are occupied by electrons, and their efficiency of capturing holes is thus impaired. Consequently the hole lifetime increases with increasing resistivity. For intrinsic material (12.39) reduces to

$$\tau = \frac{\tau_{p0}}{2}\left(1 + \frac{n_1}{n_i}\right) + \frac{\tau_{n0}}{2}\left(1 + \frac{p_1}{p_i}\right). \tag{12.43}$$

If the recombination level is at the intrinsic level, i.e., if it is in the center of the band gap, then

$$\tau = \tau_{p0} + \tau_{n0}. \tag{12.44}$$

The usual case is one in which the level is not at the center. For this case

$$\tau = \frac{\tau_{p0}}{2}(1 + e^{\Delta E/kT}) + \frac{\tau_{n0}}{2}(1 + e^{-\Delta E/kT}), \tag{12.45}$$

where ΔE is the displacement above† the center of the band gap.

†Both positive and negative values of ΔE occur in practice.

Figure 12.4 shows a sketch of the appropriate minority carrier lifetime for *n*- and *p*-type semiconductors for various positions of the Fermi level. The precise nature of this curve depends upon the location of the recombination level with respect to the intrinsic level and on the values of τ_{n0} and τ_{p0}. The typical case, in which there is a finite separation between the recombination level and the intrinsic level (i.e., for $\Delta E \neq 0$), is shown together with the special case of $\Delta E = 0$, for which the minority carrier lifetime varies nearly monotonically from τ_{n0} to τ_{p0}. A value of $\tau_{n0} = 10\tau_{p0}$ was chosen in this example because this represents a practical situation.

12.6.2. The High-level Injection Case

Consider an *n*-type semiconductor in thermal equilibrium. Here the Fermi level is above E_r, and the majority of the recombination centers have electrons attached to them. As a result, they are ideally situated to capture holes. Under low-level injection conditions, therefore, the hole lifetime is relatively short. With increasing injection, the concentration of centers having electrons attached to them falls as more holes are captured. Consequently they become less effective at capturing holes and more effective at capturing electrons. Thus the hole lifetime rises with increasing injection level; conversely the electron lifetime falls (see Problem 12.7).

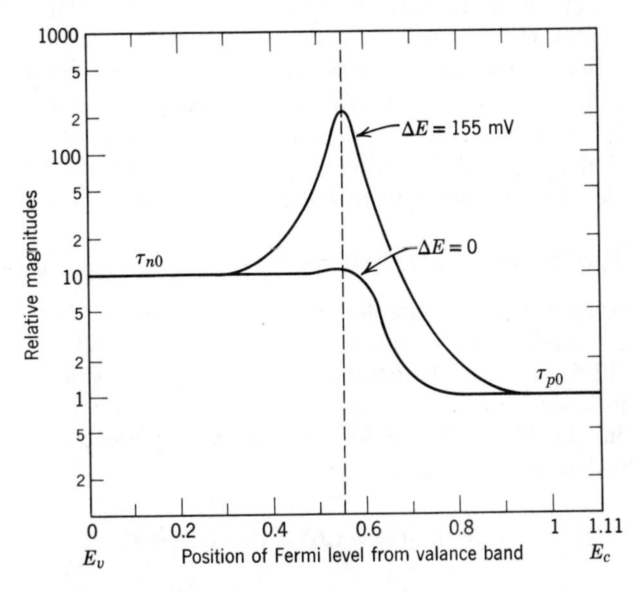

Figure 12.4 Minority carrier life-time versus position of the recombination center.

In the limit the injected electrons and holes far exceed the thermal equilibrium value, so that $n \gg \bar{n}$ and $p \gg \bar{p}$. In addition, $n \gg n_1$ and $p \gg p_1$. Making these substitutions in (12.38*a*) gives

$$\tau_{p,\,\text{high}} \rightarrow \tau_{p0} + \tau_{n0}\,. \tag{12.46a}$$

In like manner,

$$\tau_{n,\,\text{high}} \rightarrow \tau_{p0} + \tau_{n0}\,. \tag{12.46b}$$

Thus the majority and minority carrier lifetimes become both equal to $\tau_{p0} + \tau_{n0}$ at high injection levels. Normal operation of devices is somewhere between the extremes of low- and high-level injection.

12.7. RECOMBINATION THROUGH MULTIPLE LEVELS

Recombination centers are generally associated with more than one level. Thus it has been noted in Section 1.5.1 that vacancies and interstitials often exhibit two acceptor and donor levels respectively. Most deep-lying impurities that are inadvertently present in silicon also exhibit more than one level (see Table 11.2). Finally, gold, which is deliberately introduced for the control of minority carrier lifetime, exhibits two impurity levels in silicon.

The analysis of non-equilibrium processes with two recombination levels is considerably more complex than for a single recombination level. Here four capture processes are involved (with four different capture cross sections), corresponding to the capture of electrons and holes at each of the two recombination levels. In addition, four emission processes, corresponding to the emission of electrons and holes from each recombination level, must also be considered. The actual procedure for obtaining the minority carrier lifetime proceeds along lines similar to that for the single-level case, and has been detailed by Moll[4].

12.7.1. Application to Gold in Silicon

Gold exhibits both an acceptor and a donor level in silicon; the location of these levels in the energy gap is given in Figure 11.11.

Figure 12.5 shows a schematic of the energy gap in silicon, indicating the various capture and emission processes associated with these levels. In addition, the nomenclature for the various capture cross sections is also established. For gold, at the acceptor level[5],

$$\alpha_n = v_{tn} A_n^{\,0} = 1.65 \times 10^{-9}\ \text{cm}^3/\text{sec}, \tag{12.47a}$$

$$\alpha_p = v_{tp} A_p^{\,-} = 1.15 \times 10^{-7}\ \text{cm}^3/\text{sec}. \tag{12.47b}$$

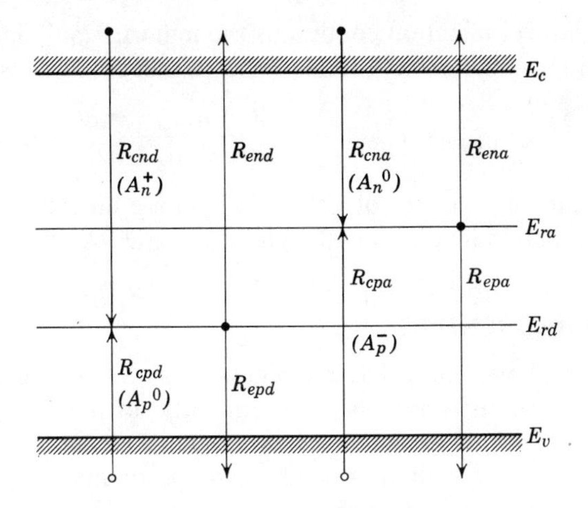

Figure 12.5 Recombination at two levels.

At the donor level,

$$\beta_n = v_{tn}A_n^+ = 6.3 \times 10^{-8}\ \text{cm}^3/\text{sec}, \tag{12.47c}$$

$$\beta_p = v_{tp}A_p^0 = 2.4 \times 10^{-8}\ \text{cm}^3/\text{sec}. \tag{12.47d}$$

The minority carrier lifetime in extrinsic, gold-doped silicon may be roughly estimated as follows: For an n-type semiconductor much of the gold concentration is in the negative-charge state and is acceptor-like in character. The actual concentration in this charge state, $\bar{N}_r^-$, is given by (11.78b). It is assumed that the capture of excess holes at this level is the limiting process, and occurs at a rate given by $v_{tp}A_p^-p'N_r^-$. It follows, then that

$$\tau_p \simeq (v_{tp}A_p^-N_r^-)^{-1} \simeq (\alpha_p\bar{N}_r^-)^{-1}. \tag{12.48}$$

For the highly doped case a further simplification may be made by setting $\bar{N}_r^- \simeq N_r$. Thus

$$\tau_p \to (\alpha_p N_r)^{-1}. \tag{12.49}$$

For extrinsic p-type silicon recombination processes are dominated by electron capture at the donor level of gold. Consequently

$$\tau_n \simeq (v_{tn}A_n^+N_r^+)^{-1} \simeq (\beta_n\bar{N}_r^+)^{-1}. \tag{12.50}$$

For highly doped p-type silicon this reduces to

$$\tau_n \to (\beta_n N_r)^{-1}. \tag{12.51}$$

Under high-level injection conditions the minority and majority carrier lifetimes again become equal. Moll has shown that for this case

$$\tau_{n,\text{high}} = \tau_{p,\text{high}} \rightarrow \frac{\alpha_n\beta_n + \alpha_p\beta_p + \alpha_p\beta_n}{N_r\alpha_p\beta_n(\alpha_n + \beta_p)}. \tag{12.52}$$

Thus the details of behavior of gold in silicon are different to those for a single recombination level, even though the general features are similar.

12.8. TRAPPING CENTERS

It has been shown that a recombination center is one which, upon capturing a carrier of one type, subsequently attracts a carrier of the opposite type, resulting in its annihilation. Such a center has ready access to both valence and conduction bands, and approximately equal capture rate constants for electrons and holes. If either of these rate constants becomes very small with respect to the other, the center becomes a trapping site. Figure 12.6 shows a site of this type, which captures electrons from the conduction band, holds them for some mean time τ_g, and then releases them to the conduction band. If the rate constant for this trap is α_{nt}, the time rate of change of trapped electrons is given[7] by

$$\frac{\partial n_t}{\partial t} = \alpha_{nt} n(N_t - n_t) - \frac{n_t}{\tau_g} \tag{12.53}$$

where n is the free-electron concentration, and N_t is the total density of traps located at an energy level E_t. Since these trapped electrons are not available for recombination, their presence tends to complicate the relatively simple recombination theory that has been outlined; for example, they lead to such phenomena as abrupt jumps superimposed on the otherwise exponential decay of carriers during recombination.

The existence of at least four different trapping centers have been established[8] in silicon, with times as large as 1000 s observed for τ_g. A

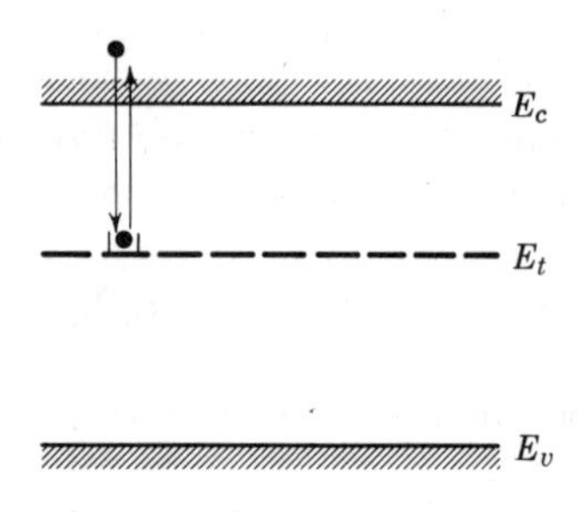

Figure 12.6 Trapping.

rough correlation has been shown between their concentration and the rate of crystal rotation during Czochralski growth. Thus it is currently believed[9] that they are associated with the presence of oxygen in the crystal. In present-day silicon crystals, which are grown at relatively slow spin rates, the effect of traps in the bulk may be neglected.

12.9. REFERENCES

[1] J. S. Blakemore, *Semiconductor Statistics*, Pergamon, New York, 1962.

[2] R. N. Hall, "Electron-Hole Recombination in Germanium," *Phys. Rev.*, **87**, 387 (1952).

[3] W. Shockley and W. T. Read, "Statistics of the Recombination of Holes and Electrons," *Phys. Rev.*, **87**, 835–842 (1952).

[4] J. L. Moll, *Physics of Semiconductors*, McGraw–Hill, New York, 1964.

[5] W. M. Bullis, "Properties of Gold in Silicon," *Solid State Electron.* **9**, (2), 143–168 (1966).

[6] J. M. Fairfield and B. V. Gokhale, "Gold as a Recombination Center in Silicon," *Solid State Electron.*, **8**, 685–691 (1965).

[7] N. B. Hannay (Ed.), *Semiconductors*, Reinhold, New York, 1960, 482–507.

[8] J. R. Haynes and J. A. Hornbeck, "Trapping of Minority Carriers in Silicon," *Phys. Rev.*, **100**, 606–615 (1955).

[9] W. Kaiser *et al.*, "Infrared Absorption and Oxygen Content in Silicon and Germanium," *Phys. Rev.*, **101**, 1264–1268 (1956).

12.10. PROBLEMS

1. A bar of n-type silicon is illuminated with F photons/sec of penetrating light. Develop the expression for the growth of excess electrons and holes in the bar, assuming that each photon gives rise to an electron-hole pair.

2. An extrinsic n-type semiconductor has acceptor-like recombination centers. On the basis of physical reasoning, sketch the net charge on these centers as the injection level is increased from zero to a high value. Assume that $\alpha_p = 2\alpha_n$.

3. Draw a graph of the minority carrier lifetime in highly doped n-type silicon as a function of the gold-doping temperature. Plot this graph in the form of log τ versus $1/T°$.

4. Repeat Problem 3 for p-type silicon.

5. A piece of silicon is doped with 10^{17} atoms/cm^3 of a deep-lying impurity which is located at the intrinsic level. The semiconductor is next doped with phosphorus at 10^{16}, 10^{17}, and 10^{18} atoms/cm^3. Compute the minority carrier lifetime for each case, given that $\alpha_n = 10^{-15}$ cm^3/sec, and $\alpha_p = 10^{-14}$ cm^3/sec.

6. Show that a shallow impurity level, such as arsenic, is ineffective in reducing minority carrier lifetime in either p-type or n-type silicon.

7. Show that, for a lightly doped semiconductor, it is possible for the minority carrier lifetime to fall with injection level. Assuming $\tau_{n0} = 10\tau_{p0}$, and a recombination center located at 155 mV from the center of the energy gap, what is the resistivity of an n-type semiconductor in which the minority carrier lifetime is the same at low and high injection levels?

Chapter 13

The P–N Junction

Two types of p-n junctions are commonly encountered in a microcircuit. The first of these, the emitter-base junction, is delineated by the intersection of an n-type complementary error function (erfc) diffusion and a p-type gaussian diffusion. The second type corresponds to the intersection of the p-type gaussian doping profile and the n-type background concentration, and represents the collector-base junction.

Figure 13.1 shows the doping profile for the narrow base transistor of Figure 10.5*b*, plotted on linear axes to exhibit the relative character of the junctions. For this device, the emitter-base junction has an impurity gradient of 1.86×10^{22} atoms/cm^4 while the collector-base junction gradient is only 1.08×10^{21} atoms/cm^4.

The emitter-base junction is often considered as an abrupt structure in which the n-type region is infinitely highly doped, whereas the p-type region is doped to a uniform concentration given by the peak value of the impurity concentration in the base (7.0×10^{16} atoms/cm^3 for this device). The collector-base junction, on the other hand, more nearly approaches a linearly graded structure, and is often considered as such. Both of these approximations are shown in Figure 13.1.

In this chapter an analysis is made of the behavior of both abrupt and linearly graded junctions. In addition, the properties of diffused junctions are also considered, especially as applicable to both the emitter-base structure and the collector-base structure.

13.1. THE ABRUPT JUNCTION

An abrupt p-n junction is formed when the doping level changes suddenly from p type to n type at some point within the single crystal. Figure 13.2*a* shows the energy-band structure of both n- and p-type

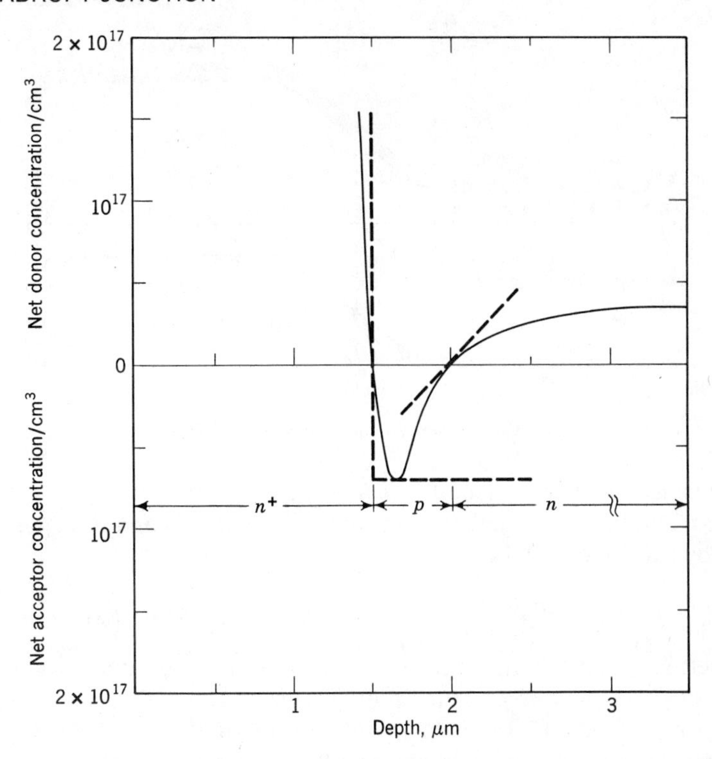

Figure 13.1 Transistor doping profile.

semiconductor regions, with the abscissa denoting distances along these regions. Figure 13.2*b* shows this diagram when these regions are combined in a single-crystal semiconductor bar. In thermal equilibrium the Fermi level must be reestablished as a straight line throughout both regions. This may be shown as follows:

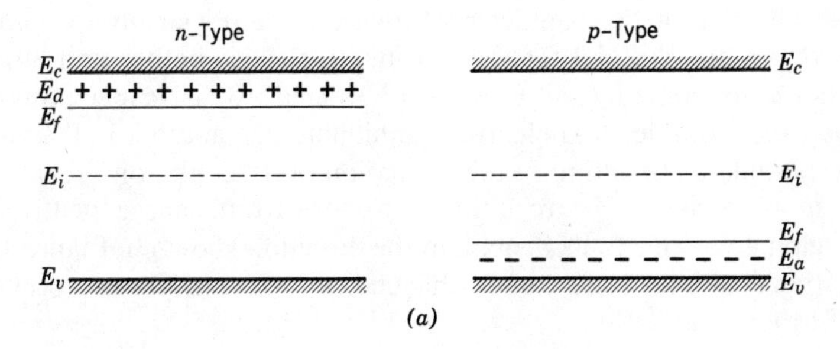

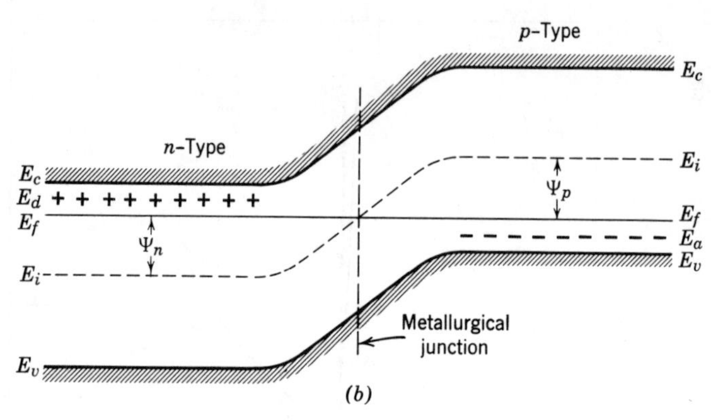

Figure 13.2 Energy-band picture of a *p-n* junction.

Consider two points A and B in a semiconductor. Writing f_{jA} and f_{jB} as the probabilities of having electrons in the jth state at A and B respectively, it follows that [see (11.14)]

$$Mf_{jA} \cdot W_{AB} \cdot M(1-f_{jB}) = Mf_{jB} \cdot W_{BA} \cdot M(1-f_{jA}) \tag{13.1}$$

In thermal equilibrium, electrons with the same energy are completely free to move from one point to another without expending energy. Hence $W_{AB} = W_{BA}$, and (13.1) leads to the condition that $f_{jA} = f_{jB}$. Defining E_{fA} and E_{fB} as the Fermi level at A and B respectively, it follows that

$$\frac{1}{1+e^{(E_{fA}-E_j)/kT}} = \frac{1}{1+e^{(E_{fB}-E_j)/kT}}, \tag{13.2}$$

hence

$$E_{fA} = E_{fB} \tag{13.3}$$

Thus for any arbitrarily doped semiconductor in thermal equilibrium the Fermi level is a straight line. As seen from Figure 13.2, this leads to a bending of the valence and conduction levels, and establishes a built-in contact potential across the junction.

The presence of the contact potential may be physically explained in terms of a simple model for the *p-n* junction, both before and after the two sides are joined (see Figures 13.3*a* and 13.3*b* respectively). On contact, mobile holes and electrons annihilate one another in the vicinity of the boundary, resulting in the formation of a "depletion" region, free from such carriers. The resulting departure from charge neutrality in this region gives rise to an $\mathscr{E}$ field, in the direction shown in Figure 13.3*b*. Once the depletion region is established, mobile carriers are prevented from further annihilation.

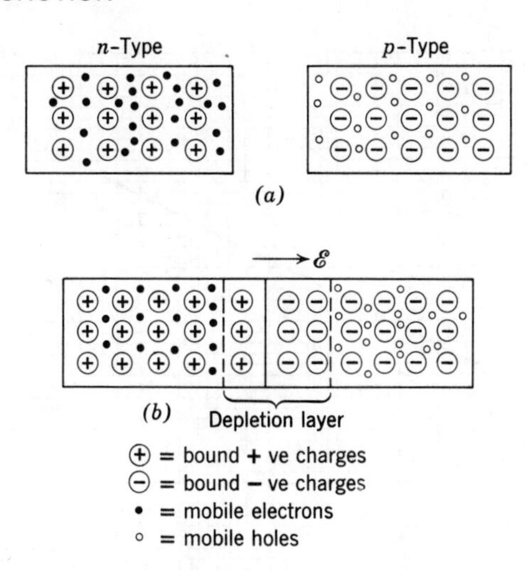

Figure 13.3 Establishment of the depletion layer.

The contact potential may be computed with the aid of Figure 13.2*b*. Assuming a nondegenerately doped *p* region with fully ionized acceptors, it has been shown that [see (11.64*b*)]

$$E_i - E_f = kT \ln \frac{N_a}{n_i} \tag{13.4}$$

The potential Ψ_p, measured from the Fermi level, is thus given by

$$\Psi_p = \frac{kT}{q} \ln \frac{N_a}{n_i} \tag{13.5}$$

Figure 13.4 shows values of Ψ_p for various concentrations of N_a. An identical expression holds for Ψ_n for the case of the moderately doped *n*-type region, shown in Figure 13.2.

The *n*-type region of an emitter-base diode is highly doped ($\simeq 10^{21}$ atoms/cm^3). Thus for a structure of this type, degenerate Fermi statistics must be used to determine the value of Ψ_n. The solution may be obtained graphically in the manner outlined in Section 11.3.6.2. However, the approximation that the Fermi level is $2kT$ within the conduction band may be conveniently made for this case. Thus $\Psi_n \simeq 0.61$ V for an *n*-type emitter region.

The total contact potential is given by $\Psi_n + \Psi_p$. This is a built-in reverse potential across a *p-n* junction in thermal equilibrium, and constitutes a barrier to current flow across it. Under applied reverse-bias conditions, this barrier height is increased, and the depletion region widens in order

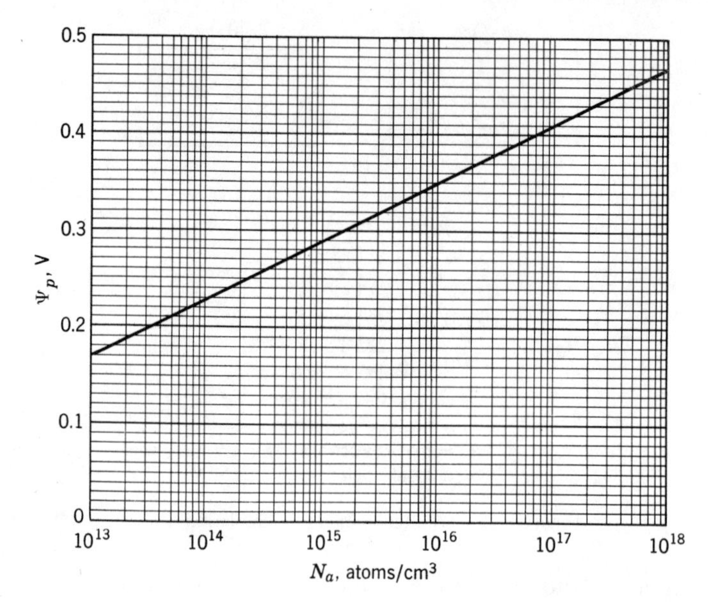

Figure 13.4 Contact potential.

to accommodate this extra voltage. With forward-applied voltage, the barrier height is reduced, resulting in the more ready flow of carriers across the junction.

The presence of a finite barrier at all times, under both forward- and reverse-bias conditions, cannot be too highly emphasized. That this is the case is clear if the consequences of zero barrier height are considered; for such a situation, carriers may be transported between the regions without any externally applied force, resulting in infinite current flow.†

13.1.1. Reverse-bias Conditions

With reverse bias the depletion layer widens, providing the additional space-charge region required to support this voltage. For a junction with a highly doped n^+ region, the widening of the depletion layer is confined to the p side, as shown in Figure 13.5*a*.

Figure 13.5*b* shows the space-charge density $\rho(x)$ associated with the depletion layer. In order that charge neutrality be applicable over the depletion layer as a whole, the total space charge must be equal on both sides of the junction. On the p side this space charge is given by $-qN_a x_p$ per unit area, where x_p is the penetration of the depletion layer into the

†Provided that ohmic drops in the p and n region are neglected.

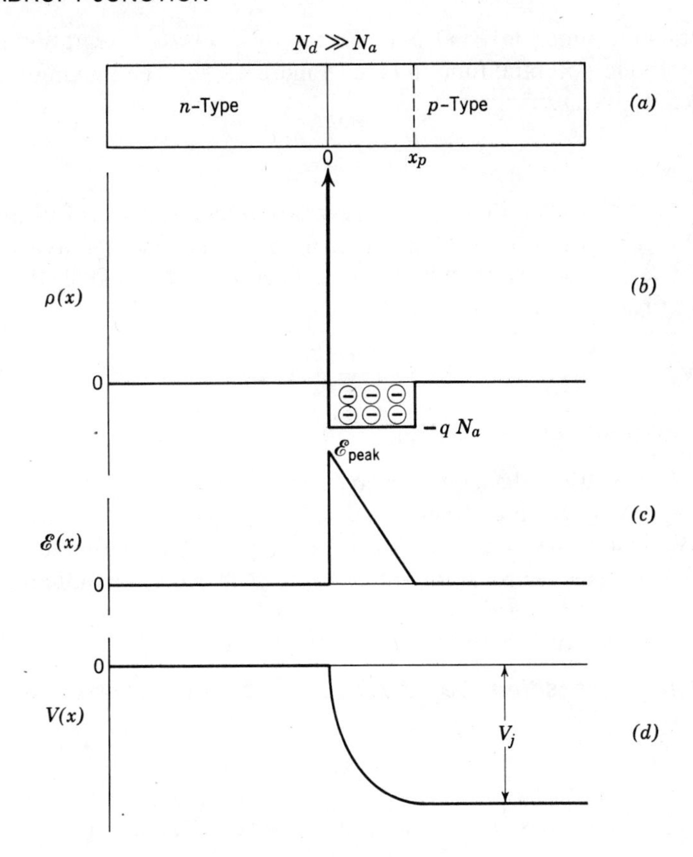

Figure 13.5 The reverse-biased n^+-p junction.

p region and N_a is the acceptor concentration. It is assumed that these acceptors arc fully ionized.

The n region is assumed infinitely highly doped. It follows then that $x_n = 0$.

13.1.1.1. *The Depletion-Layer Width.* Poisson's equation for a one-dimensional situation states that

$$\nabla^2 V = -\frac{\partial \mathscr{E}}{\partial x} = -\frac{\rho(x)}{\epsilon\epsilon_0}, \tag{13.6}$$

where ϵ is the relative permittivity and ϵ_0 is the dielectric constant of free space. Integrating, the resulting $\mathscr{E}$ field is shown in Figure 13.5c, and has a peak value given by

$$\mathscr{E}_{\text{peak}} = +\frac{qN_a x_p}{\epsilon\epsilon_0} \tag{13.7}$$

The potential function $V(x)$ is equal to $-\int \mathscr{E}(x)\,dx$. Integrating results in a parabolic potential function (see Figure 13.5*d*), the maximum value of which is given by

$$V_j = \frac{qN_a x_p^2}{2\epsilon\epsilon_0}, \tag{13.8}$$

where V_j is the magnitude of the reverse voltage supported by the depletion layer. (It is assumed that ohmic drops in the bulk regions are negligible.) If V_R is the magnitude of the applied reverse voltage, then $V_j = V_R + \Psi$. From (13.8) it is seen that

$$x_p = \left(\frac{2\epsilon\epsilon_0 V_j}{qN_a}\right)^{1/2}. \tag{13.9}$$

This is the total width of the depletion layer, since $x_n = 0$. Here

$x_0 = x_p$ = width of the depletion layer, in centimeters,
V_j = voltage across the depletion layer, in volts,
N_a = acceptor concentration, in atoms per cubic centimeters,
ϵ_0 = dielectric constant of free space (8.85×10^{-14} farad/cm),
ϵ = relative permittivity of silicon is 12,
q = electron charge (1.6×10^{-19} coulomb).

13.1.1.2. Transition Capacitance. The space charge associated with the depletion layer is given by

$$Q = qN_a A x_0, \tag{13.10}$$

where A is the cross-section area of the diode. Combining with (13.9) gives

$$Q = A\sqrt{2\epsilon\epsilon_0 q N_a V_j} \tag{13.11}$$

Thus the space charge is a function of the voltage across the depletion layer. The capacitance associated with this space-charge layer is known as the depletion layer or *transition capacitance*, and is given by C_t, where

$$C_t = \frac{dQ}{dV_j} \tag{13.12a}$$

$$= A\left(\frac{\epsilon\epsilon_0 q N_a}{2V_j}\right)^{1/2}. \tag{13.12b}$$

An alternate approach is to recognize that a small change in the reverse voltage does not appreciably change the shape of the space-charge layer, but only alters the charge distribution at its edges. Thus in order to compute the differential capacitance C_t, it is only necessary to consider the depletion layer as a dielectric of thickness x_0, bounded by low-resistance "plates" having surface charges on them. For such a capacitor,

$$C_t = \frac{A\epsilon\epsilon_0}{x_0}, \qquad (13.13a)$$

$$= A\left(\frac{\epsilon\epsilon_0 q N_a}{2V_j}\right)^{1/2}, \qquad (13.13b)$$

as before.

Figure 13.6 shows the depletion-layer width and the transition capacitance of an abrupt p-n^+ diode as a function of the reverse voltage across the junction.

13.1.1.3. The Peak Field. From (13.7) and (13.9) the magnitude of the peak field is given by

$$\mathscr{E}_{\text{peak}} = \left(\frac{2qN_aV_j}{\epsilon\epsilon_0}\right)^{1/2}, \qquad (13.14)$$

$$= \frac{2V_j}{x_0} \qquad (13.15)$$

For a typical silicon diode with $N_a = 10^{16}$ atoms/cm³, the depletion-layer width is approximately 1.0 μm for an applied reverse voltage of

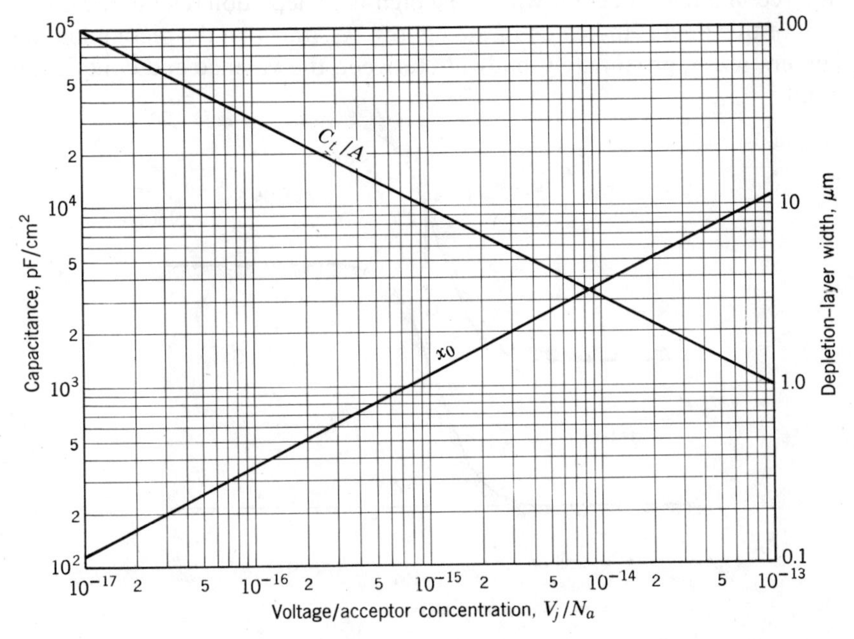

Figure 13.6 Capacitance and depletion-layer width for the abrupt junction.

8 V, resulting in a peak field of 1.6×10^5 V/cm. Thus it is seen that a high field is present in the depletion layer of a diode, even for modest values of reverse voltage. The various consequences of this high electric field are discussed in succeeding sections.

13.1.1.4. The Reverse Current. The reverse current of a diode consists of three components, as follows:

1. Leakage current due to surface effects.
2. Reverse current due to diffusion.
3. Reverse current due to charge generation in the depletion layer.

In oxide-passivated silicon devices, surface effects are minimized and the first of these terms may be neglected. Of the latter two terms it will be shown that, for silicon, the diffusion-limited reverse current is about three orders of magnitude less than that due to charge generation in the depletion layer. (Such a comparison is made in the discussion of the diffusion model for an *p-n* junction. See Section 13.1.2.2). As a consequence, it is reasonable to consider the reverse leakage of a *p-n* silicon diode to be entirely due to the charge generation component.

Figure 13.7 shows the energy-band picture for a *p-n* junction with a reverse voltage of magnitude V_j. For this condition consider the behavior at a recombination center within the high-field depletion region (the direction of this field is indicated in the figure). Assume that the recombination centers are approximately midway between the valence and conduction band.

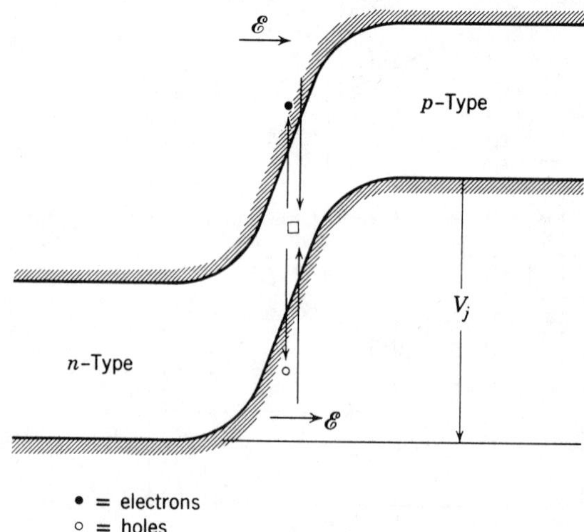

Figure 13.7 Charge generation in the depletion layer.

The generation rate for electrons (or holes) is directly proportional to the number of appropriate charged centers at the recombination level and to the number of available states in the conduction (or valence) bands. Thus as far as these rates are concerned, the situation inside the depletion layer is essentially the same as that outside it.

The capture rate, on the other hand, is proportional to the number of mobile carriers in the valence and conduction bands, and to the number of recombination centers of the appropriate charge state. Since the number of mobile carriers in the depletion layer is very small (indeed, it was originally assumed that the region had *no* mobile carriers), the capture rates are considerably lower that those encountered outside the depletion layer.

Thus the presence of deep-lying levels in the depletion layer gives rise to the generation of charge, with each center emitting electrons and holes in succession. Upon generation these charges are swept out of the depletion layer under the influence of the high electric field associated with this region. Note that, for every hole-electron pair generated in this manner, only one mobile carrier is transported past any given reference plane. Thus only one carrier is delivered to the external circuit for each hole-electron pair that is generated.

The magnitude of the leakage current may be determined for steady-state conditions. Thus (12.35) may be rewritten as

$$\frac{dp'}{dt}=\frac{dn'}{dt}=\frac{n_i^2-np}{(n+n_1)/\alpha_p N_r+(p+p_1)/\alpha_n N_r} \tag{13.16}$$

Combining with (12.36*a*) and (12.36*b*), and noting that both n and p are very small within the depletion layer, (13.16) reduces to

$$\frac{dn'}{dt}=\frac{dp'}{dt}=\frac{dn}{dt}=\frac{dp}{dt}\simeq\frac{n_i^2}{n_1\tau_{p0}+p_1\tau_{n0}} \tag{13.17}$$

The reverse current due to recombination is given by

$$I_{\text{rec}}\simeq\frac{A}{2}\left(\int qv\,dn+\int qv\,dp\right) \tag{13.18}$$

$$\simeq\frac{Aq}{2}\left(\int\frac{dn}{dt}dx+\int\frac{dp}{dt}dx\right) \tag{13.19}$$

$$\simeq\frac{qn_i^2x_0A}{n_1\tau_{p0}+p_1\tau_{n0}} \tag{13.20}$$

Note that the leakage current is a volume effect; that is, it is directly proportional to the volume of the depletion layer. From (13.9) it is seen that this reverse current varies as the square root of V_j for an abrupt junction. In addition, (13.20) indicates one of the important consequences of deliberately introducing gold for the purpose of reducing minority carrier lifetime. The introduction of gold increases the reverse leakage current in almost direct proportion to the reduction of lifetime. Figure 13.8 shows the magnitude of this leakage current as a function of the temperature at which gold is introduced into the silicon microcircuit [1].

For recombination centers located midway in the energy gap the charge-generation current is directly proportional to n_i, and thus doubles every 11°C approximately. Figure 13.9 shows a plot of the charge-generation current for this type of recombination center.

13.1.1.5. *Zener Breakdown.* Figure 13.10 shows the energy-band picture of a reverse-biased *p-n* junction. Also shown is an energy level E_1 and the triangular energy barrier which separates valence and conduction bands at this level.

Electrons at E_1 in the valence band may make transitions to available states at the same energy level in the conduction band by a quantum-mechanical process known as *tunneling*, provided that the barrier is

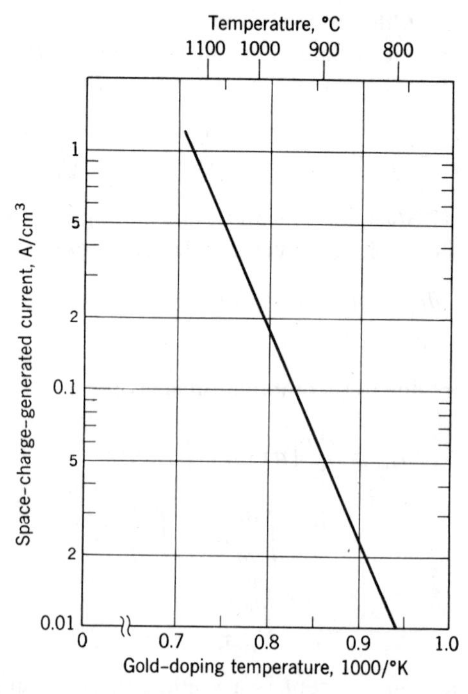

Figure 13.8 Leakage current for gold-doped devices (Kressel et al. [1]).

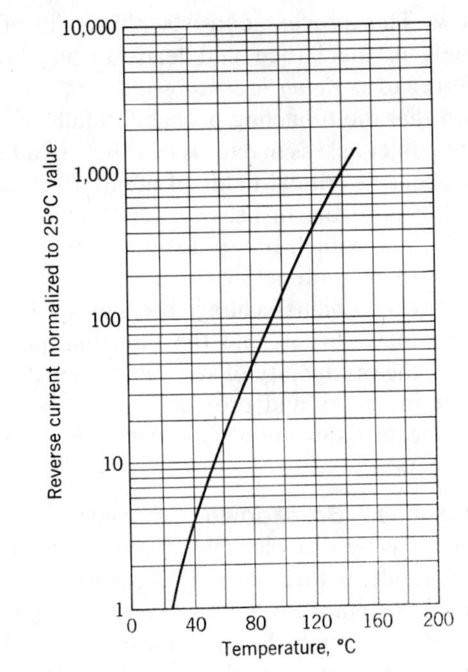

Figure 13.9 Leakage current versus temperature (Phillips[7]).

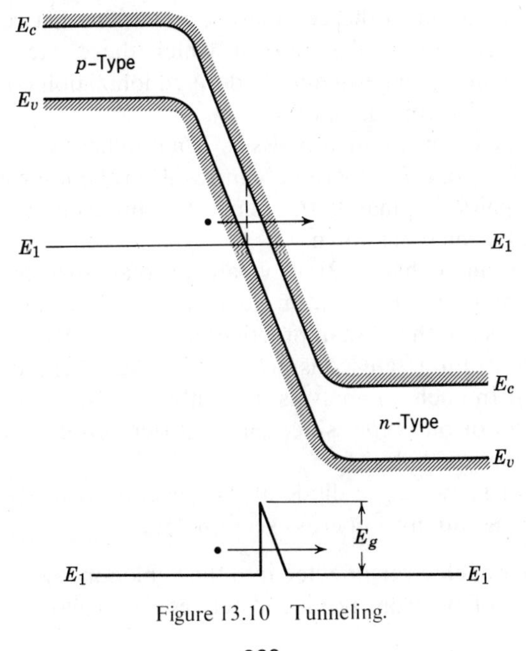

Figure 13.10 Tunneling.

sufficiently narrow. This process consists physically of the rupture of the covalent bonds in the lattice and leads to an abrupt increase in conductivity, referred to as *Zener breakdown.*

It can be shown that the tunneling probability falls off very rapidly for energy barriers with thicknesses in excess of 100 Å. In addition, tunneling is only found to occur at critical fields of about 10^6 V/cm. Thus Zener breakdown is highly probable in heavily doped low-voltage structures. In general, *p-n* junctions which break down at 5 V or lower are considered to have done so by pure tunneling.

With junctions having a higher voltage breakdown, the depletion-layer width is correspondingly larger, and the tunneling probability falls off rapidly. In addition, the probability of secondary ionization effects within the depletion layer increases and provides an alternate mechanism for breakdown. This mechanism, known as *avalanche breakdown,* is described in the next section.

13.1.1.6. Avalanche Breakdown. Consider a semiconductor subjected to an increasing electric field. Eventually a point is reached when mobile carriers attain their terminal velocity, as described in Section 11.5.6. Beyond this point further increase in the field does not alter the drift velocity, even though the velocities of the individual carriers exceed their thermal velocity. At high fields, these "hot" carriers collide with atoms and impart enough energy to valence-band electrons so as to allow their promotion to the conduction band. The net result is a hole-electron-pair generation. The rate at which these excess carriers are generated is given by experimentally derived ionization constants, which are different for electrons and holes.

The secondary ionization process is a multiplicative one. Thus each electron and hole that is generated can result in the ionization of further electron-hole pairs. Ultimately this generation proceeds at an infinite rate, leading to avalanche breakdown.

A relatively simple theory [2] of avalanche breakdown may be presented if the ionization rate constants for electrons and holes are considered to be equal. This is not the case in practice; it is possible, however, to define an average ionization rate constant which takes these differences into consideration. In such an analysis it is only necessary to follow the history of carriers of one type, since each carrier results in the generation of both electrons and holes.

Figure 13.11 shows a *p-n* diode at the onset of avalanche breakdown For this diode, assuming unit cross section, let

n_1 = number of electron injected into the depletion layer,
n = number of electrons produced between zero and x,

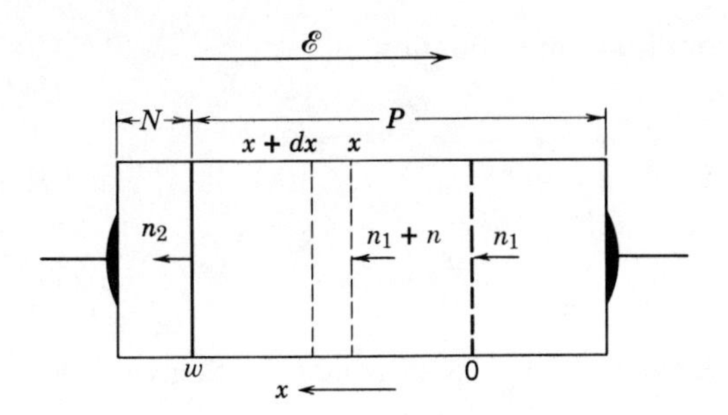

Figure 13.11 Avalanche processes.

n_2 = number of electrons leaving the depletion layer at $x = w$,
w = width of the depletion layer at the onset of breakdown,
α_i = number of electron-hole pairs produced per electron, per centimeter traveled in the direction of the $\mathscr{E}$ field.

Then the number of electrons generated between x and w is $n_2 - n_1 - n$. Hence the number of electrons produced in the interval between x and $x + dx$ is given by

$$dn = (n_1 + n)\alpha_i\, dx + (n_2 - n_1 - n)\alpha_i\, dx \tag{13.21}$$

$$= n_2\alpha_i\, dx. \tag{13.22}$$

Integrating over the depletion layer gives

$$1 - \frac{n_1}{n_2} = \int_0^w \alpha_i\, dx. \tag{13.23}$$

Avalanche breakdown occurs when n_2/n_1 becomes infinite, that is, when

$$\int_0^w \alpha_i dx = 1. \tag{13.24}$$

To solve this equation an empirically derived ionization rate constant is used. A suitable approximation† is given by

$$\alpha_i = A\mathscr{E}^5 \tag{13.25}$$

where $A = 1.65 \times 10^{-24}$.

Writing BV as the breakdown voltage and neglecting the contact potential,

$$w = \left(\frac{2\epsilon\epsilon_0 BV}{qN_a}\right)^{1/2}. \tag{13.26}$$

†See also the approximation given in Problem 13.4.

In addition, from Poisson's equation,

$$\frac{d\mathscr{E}}{dx} = \frac{qN_a}{\epsilon\epsilon_0}. \tag{13.27}$$

Finally,

$$\mathscr{E}_{\text{peak}} = \frac{2BV}{w}. \tag{13.28}$$

Combining (13.25), (13.26), (13.27), and (13.28) gives

$$BV = \left[\frac{3}{4A}\left(\frac{\epsilon\epsilon_0}{qN_a}\right)^2\right]^{1/3}, \tag{13.29}$$

$$= 2.72 \times 10^{12}\, N_a^{-2/3} \text{ volts}, \tag{13.30}$$

where N_a is the acceptor concentration in atoms/cm³.

Figure 13.12 shows the actual breakdown voltage for an abruptly doped junction of this type, together with the curve of (13.30). It is seen that there is an excellent fit between theory and experiment, except at low voltages (for heavily doped structures) where breakdown is controlled by tunneling.

In general, *p-n* junctions which break down at 8 V or higher are considered to do so by secondary ionization processes. Since the upper limit

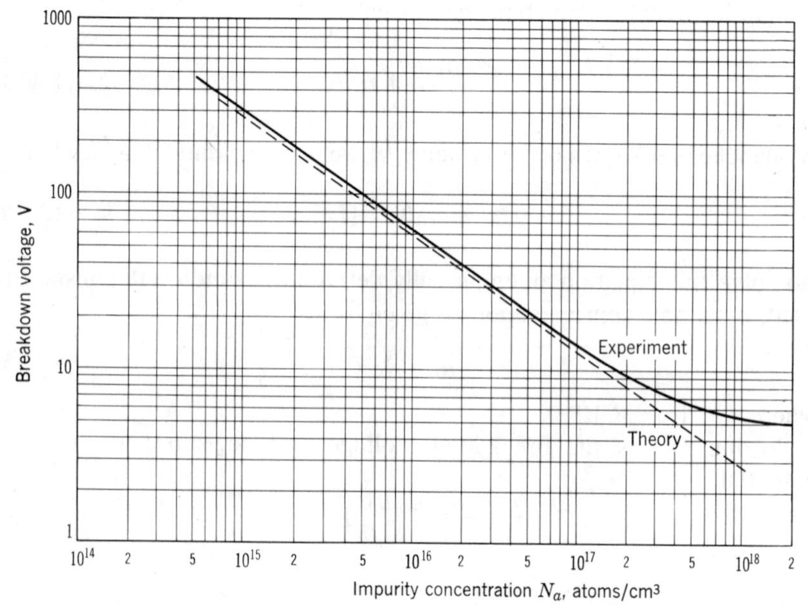

Figure 13.12 Avalanche breakdown voltage versus concentration.

for tunneling is about 5 V, both phenomena are considered to occur in devices with breakdown between 5 and 8 V.

It has been experimentally observed that the avalanche breakdown voltage increases at a rate of about 0.1%/°C over the range of temperature from -55 to $+125$°C. This increase may be attributed to changes in α_i with temperature.

The quantity n_2/n_1 is sometimes called the multiplication factor. In practice this is found to be empirically given by

$$M = \frac{1}{1-(V_j/BV)^m}, \tag{13.31}$$

where $m = 4$ for n-type silicon and 2 for p-type silicon.

13.1.2. Forward-bias Conditions

It has been shown that a contact potential is developed across a p-n junction in thermal equilibrium and constitutes a barrier to current flow across it. For this situation [see (11.106a)] the current density due to electron flow is given at any point at

$$J_n = 0 = q\mu_n \bar{n}\mathscr{E} + qD_n\frac{d\bar{n}}{dx}, \tag{13.32}$$

where $\bar{n}$ is the equilibrium concentration of majority carriers at that point. An analogous equation applies for the current density due to holes.

Equation 13.32 shows that the electron current flow across a junction consists of two components. These are the drift and diffusion currents respectively; in thermal equilibrium they are equal in magnitude but opposite in direction.

It is instructive to obtain an estimate of the magnitude of these currents in thermal equilibrium. Consider an n-p junction of unit area, with $N_d = 10^{20}$ atoms/cm^3, and $N_a = 10^{16}$ atoms/cm^3. Assuming that all impurities are fully ionized, the electron concentration in the n and p regions are 10^{20}/cm^3 and 1.9×10^4/cm^3 respectively. From Figure 13.4, $\Psi = 0.96$ V, resulting in a depletion layer that is 0.35 μm thick. Hence $(d\bar{n}/dx)_{av} \simeq 2.86 \times 10^{24}$/cm^4. From Figure 11.18, $\mu_n = 1000$ cm^2/V-sec. Thus the diffusion component of the electron current density is given by

$$J_{n,\text{diff}} \simeq qD_n\left(\frac{d\bar{n}}{dx}\right)_{av} \tag{13.33}$$

$$\simeq kT\mu_n\left(\frac{d\bar{n}}{dx}\right)_{av} \tag{13.34}$$

$$= 1.18 \times 10^7 \text{ A/cm}^2. \tag{13.35}$$

In like manner, the drift component is also 1.18×10^7 A/cm^2. These components are orders of magnitude in excess of the current density that is actually present under conditions of heavy current flow. Thus for forward-bias conditions, it is still quite reasonable to describe the current density due to electron flow by

$$J_n = q\mu_n n\mathscr{E} + qD_n\frac{dn}{dx} \simeq 0. \tag{13.36}$$

A similar relation applies to the hole current density.

To obtain the carrier concentrations on either side of the depletion layer, it is necessary to integrate (13.36), subject to the appropriate boundary conditions. Using the intrinsic level as the reference and noting that $\mathscr{E} = -dV/dx$, (13.36) reduces to

$$\frac{dV}{dx} - \frac{kT}{q}\frac{1}{n}\frac{dn}{dx} \simeq 0. \tag{13.37}$$

Figure 13.13 shows the *p-n* junction in thermal equilibrium and under forward bias. In thermal equilibrium, let the electron concentrations in the *p* and *n* regions be denoted by $\bar{n}_P$ and $\bar{n}_N$ respectively, and let the voltage supported by the depletion layer be Ψ. When a forward voltage V_A is applied to the diode, the voltage across the depletion layer is $\Psi - V_A$

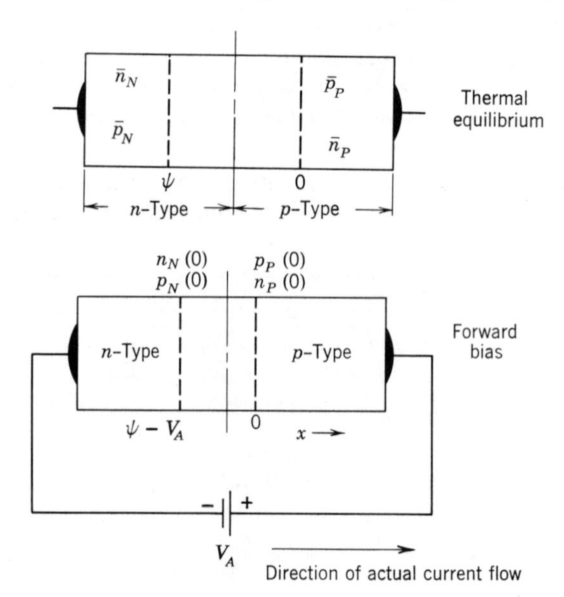

Figure 13.13 The forward-biased junction.

and gives rise to non-equilibrium electron concentrations. At the edge of the depletion layer these electron concentrations are denoted by $n_P(0)$ and $n_N(0)$ in the p and n regions respectively. If we assume low-level injection conditions, where the majority carrier concentration is not significantly altered from its equilibrium value under forward-bias conditions, then

$$n_N(0) \simeq \bar{n}_N \tag{13.38}$$

Equation (13.37) may be written as

$$\int_{\Psi - V_A}^{0} dV - \frac{kT}{q} \int_{\bar{n}_N(0)}^{n_P(0)} \frac{1}{n} dn = 0. \tag{13.39}$$

Solving,

$$V_A - \Psi - \frac{kT}{q} \ln \frac{n_P(0)}{\bar{n}_N} = 0. \tag{13.40}$$

However,

$$\Psi = \frac{kT}{q} \ln \frac{\bar{n}_N \bar{p}_P}{n_i^2}. \tag{13.41}$$

Equation 13.41 holds strictly for non-degenerate semiconductor regions. However, it provides a reasonably good approximation, even when one of the regions is heavily doped. Combining this equation with (13.40) gives

$$n_P(0) = \bar{n}_P e^{qV_A/kT}. \tag{13.42a}$$

This is known as the *law of the junction*. It relates the minority carrier concentration at the edge of the depletion layer to the equilibrium concentration of minority carriers and to the applied voltage. For the n region, in like manner,

$$p_N(0) = \bar{p}_N e^{qV_A/kT}, \tag{13.42b}$$

where $p_N(0)$ is the minority carrier concentration at the edge of the depletion layer in the n region.

Figure 13.14a shows the minority carrier concentrations in a p-n diode under conditions of forward bias (V_A positive). For this example it is assumed that $N_d > N_a$. In addition, the p and n regions of the diode are considered long, so that the excess carrier concentrations eventually decay to their equilibrium values. The origin for these curves is taken at the edges of the depletion layer. In practice, however, it is customary to ignore the thickness of this region and set the origin at the metallurgical junction.

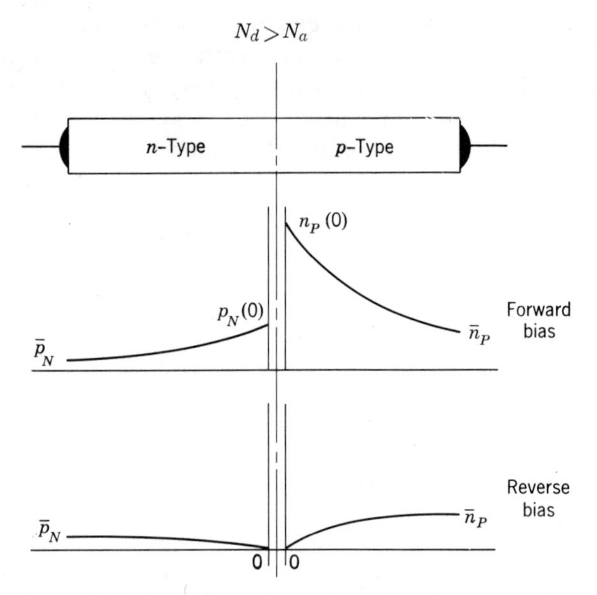

Figure 13.14 Minority carrier concentrations in a *p-n* junction.

Minority carrier concentrations under conditions of reverse bias (V_A negative) are shown in Figure 13.14*b*. Note that $n_P(0) \simeq p_N(0) \simeq 0$ for negative values of V_A. Finally, $\bar{p}_N \simeq 0$ if the *n* region is heavily doped, and current flow is almost entirely caused by the electron current. Thus the current contribution of minority carriers in the n^+ region of a p-n^+ diode may be ignored.

It is seen that the application of a forward voltage across the diode results in establishing, at the edge of the depletion layer, a minority carrier concentration that is in excess of the equilibrium concentration. For the low-level injection case considered here, the majority carrier concentration is unchanged from its equilibrium value. Thus forward bias results in the injection of minority carriers into a space-charge neutral region beyond the depletion layer.

13.1.2.1. The Continuity Equation. The continuity equation describes the behaviour of electrons and holes in any elemental volume of semiconductor. For electrons in a *p*-type material, for example, this equation states that

$$\begin{bmatrix}\text{The time rate of increase of electrons}\\ \text{in a volume of semiconductor}\end{bmatrix} = \begin{bmatrix}\text{The rate of increase due to generation} -\\ \text{rate of decrease due to recombination}\end{bmatrix} - \begin{bmatrix}\text{The divergence of the electron flow in}\\ \text{the volume}\end{bmatrix} \quad (13.43)$$

The rate of increase of electrons due to generation processes may be written as U_n. The rate of decrease due to recombination is dn'_P/dt, where n'_P is the excess electron concentration. If $\bar{n}_P$ is the equilibrium concentration of electrons, n_P is the total electron concentration, and τ_n is the electron lifetime, then [see (12.15)]

$$\frac{dn'_P}{dt} = -\frac{n'_P}{\tau_n} = -\frac{n_P - \bar{n}_P}{\tau_n}. \tag{13.44}$$

The divergence of the electron flow is $-(1/q)\nabla\cdot J_n$, where J_n is the current density due to electron flow. For the one-dimensional case this reduces to $-(1/q)(\partial J_n/\partial x)$. Hence the continuity equation for electrons in a p-type semiconductor is

$$\frac{\partial n_P}{\partial t} = U_n - \frac{n_P - \bar{n}_P}{\tau_n} + \frac{1}{q}\frac{\partial J_n}{\partial x} \tag{13.45}$$

But

$$J_n = q\mu_n n_P \mathscr{E} + qD_n \frac{\partial n_P}{\partial x} \tag{13.46}$$

Substituting (13.46) into (13.45) gives

$$\frac{\partial n_P}{\partial t} = U_n - \frac{n_P - \bar{n}_P}{\tau_n} + D_n \frac{\partial^2 n_P}{\partial x^2} + \mu_n \frac{\partial}{\partial x}(n_P \mathscr{E}). \tag{13.47a}$$

A similar reasoning for holes in an n-type semiconductor gives

$$\frac{\partial p_N}{\partial t} = U_p - \frac{p_N - \bar{p}_N}{\tau_p} + D_p \frac{\partial^2 p_N}{\partial x^2} - \mu_p \frac{\partial}{\partial x}(p_N \mathscr{E}). \tag{13.47b}$$

For minority carriers in a field-free region in which there are no generation processes, these equations reduce to their commonly used form,

$$\frac{\partial n_P}{\partial t} = -\frac{n_P - \bar{n}_P}{\tau_n} + D_n \frac{\partial^2 n_P}{\partial x^2} \tag{13.48a}$$

$$\frac{\partial p_N}{\partial t} = -\frac{p_N - \bar{p}_N}{\tau_p} + D_p \frac{\partial^2 p_N}{\partial x^2} \tag{13.48b}$$

Solutions of these equations, subject to the appropriate boundary values, are applicable to the majority of situations encountered in practice.

13.1.2.2. The Diode Equation. Consider a p-n^+ diode, shown in Figure 13.15, with a uniformly doped p region. A voltage V_A is impressed across the diode; neglecting ohmic drops, this entire voltage appears across the junction. The thickness of the depletion layer as well as charge generation processes within it are neglected in the analysis.

For electrons in the p region under steady-state conditions,

$$\frac{\partial n_P}{\partial t} = 0 = -\frac{n_P - \bar{n}_P}{\tau_n} + D_n \frac{\partial^2 n_P}{\partial x^2} \tag{13.49}$$

Defining a diffusion length for electrons L_n such that $L_n = (\tau_n D_n)^{1/2}$ (13.49) may be written as

$$\frac{d^2 n_P}{dx^2} - \frac{n_P - \bar{n}_P}{L_n^2} = 0. \tag{13.50}$$

Therefore the excess electron concentration is given by

$$n_P - \bar{n}_P = Ae^{-x/L_n} + Be^{+x/L_n}. \tag{13.51}$$

where A and B are constants.

Let the base region be of finite width W. At $x = W$, an ohmic contact is made to the device. This contact presents an extremely low lifetime termination to the p region; consequently, it is reasonable to expect both majority and minority carriers to maintain their equilibrium values at this contact. Thus the boundary values are

$$n_P = \bar{n}_P e^{qV_A/kT}, \qquad \text{at } x = 0, \tag{13.52a}$$

$$n_P = \bar{n}_P, \qquad \text{at } x = W. \tag{13.52b}$$

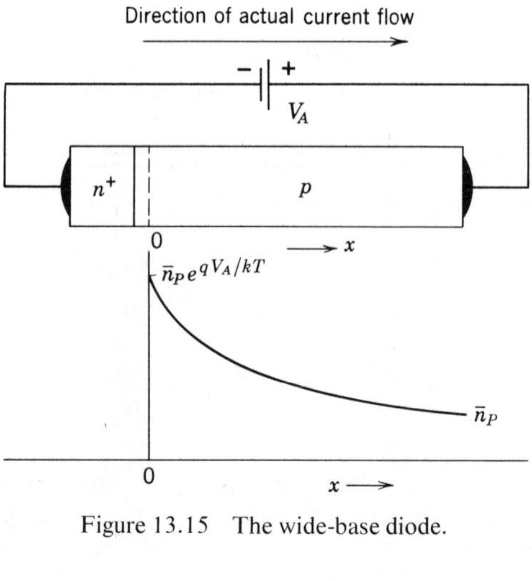

Figure 13.15 The wide-base diode.

Solving (13.51) subject to these boundary values,

$$n_P - \bar{n}_P = \bar{n}_P(e^{qV_A/kT} - 1)\,\frac{\sinh (W-x)/L_n}{\sinh W/L_n}. \tag{13.53}$$

The total junction current due to the flow of electrons is given by

$$I_n = qAD_n\left(\frac{dn_P}{dx}\right)_{x=0} \tag{13.54a}$$

$$= -\frac{qAD_n\bar{n}_P}{L_n \tanh W/L_n}(e^{qV_A/kT} - 1). \tag{13.54b}$$

This is the current due to the flow of electrons in a p-n^+ diode of arbitrary base width. The current due to hole flow may be found in an analogous manner.

We now consider the special case of a diode in which the p region is wide compared to L_n. Then this region may be treated as if it were infinitely wide, so that

$$n_P = \bar{n}_P e^{qV_A/kT}, \qquad \text{at } x = 0, \tag{13.55a}$$

$$n_P = \bar{n}_P, \qquad \text{at } x \simeq \infty. \tag{13.55b}$$

Solving (13.51) subject to these conditions,

$$n_P - \bar{n}_P = \bar{n}_P(e^{qV_A/kT} - 1)e^{-x/L_n} \tag{13.56}$$

Combining (13.54a) and 13.56) and evaluating at $x = 0$ result in the current due to the flow of electrons,

$$I_n = -\frac{qAD_n\bar{n}_P}{L_n}(e^{qV_A/kT} - 1), \tag{13.57}$$

where A = area of the junction in square centimeters,
D_n = diffusion constant of electrons in the p region in square centimeters per second,
$\bar{n}_P$ = equilibrium electron concentration in the p region, per cubic centimeter,
L_n = diffusion length of electrons in the p region, in centimeters,
V_A = voltage applied across the junction in volts.

In like manner, the current due to hole flow is

$$I_P = -\frac{qAD_p\bar{p}_N}{L_p}(e^{qV_A/kT} - 1). \tag{13.58}$$

The total diode current is the sum of these terms. For a n^+ region, however, $\bar{p}_N \simeq 0$; the total current flowing through a wide p-n^+ diode is thus

$$I = -\frac{qAD_n\bar{n}_P}{L_n}(e^{qV_A/kT}-1). \tag{13.59}$$

Note that, for forward-bias conditions, I is negative, i.e., current flows from p to n through the junction. For reverse bias (V_A negative),

$$I = I_0 \simeq +\frac{qAD_n\bar{n}_P}{L_n}. \tag{13.60}$$

This is the diffusion component of the reverse leakage current. It is instructive to compare its magnitude with the charge-generation current given by (13.20). Assuming recombination centers located midway in the energy gap, $n_1 = p_1 = n_i$, and

$$I_{\text{rec}} \simeq \frac{qAn_ix_0}{\tau_{p0}+\tau_{n0}}. \tag{13.61}$$

The diffusion component of the reverse current is given by

$$I_{\text{diff}} \simeq \frac{qAD_n\bar{n}_P}{\sqrt{(D_n\tau_{n0})}}, \tag{13.62}$$

since the p region is extrinsic. Choosing $N_a = 10^{16}$ atoms/cm^3, $\tau_{p0} = \tau_{n0} = 10^{-6}$ sec, $\mu_n = 1000$ cm^2/V-sec, and $x_0 = 1\ \mu$m,

$$I_{\text{rec}}/I_{\text{diff}} \simeq 7250. \tag{13.63}$$

Thus the diffusion component of the reverse current can be ignored in comparison to the charge-generation component. This is generally true for wide-gap semiconductors such as silicon.

13.1.2.3. The Narrow-base Diode. From (13.56) it is seen that for an infinitely wide region the excess minority carriers decay exponentially with distance beyond the edge of the depletion layer. An important case is that of the narrow-base diode, for which this situation does not hold. This case is commonly encountered in practice and describes the emitter-base diode of a transistor. Here $W \ll L_n$, and (13.53) reduces to

$$n_P - \bar{n}_P = \bar{n}_P(e^{qV_A/kT}-1)(1-x/W). \tag{13.64}$$

Thus the excess carrier concentration in a narrow-base diode falls off linearly with distance from the edge of the depletion layer. Consequently,

$$I_n = -\frac{qAD_n\bar{n}_P}{W}(e^{qV_A/kT} - 1). \tag{13.65}$$

This is the total current, because the n region is heavily doped. Note that the basic form of the narrow- and wide-base-diode equations is identical, the only difference being that associated with the coefficient preceding the exponential.

Figure 13.16 shows the minority carrier concentration in a narrow-base diode under forward-bias conditions. In comparison to the wide-base diode, the falloff in excess carrier concentration is faster, since $W \ll L_n$. Thus the narrow-base diode has a higher conductance than its equivalent wide-base counterpart.

13.1.2.4. The Forward Current. The equations for the forward current of a diode were derived by neglecting the effects of charge generation in the depletion layer. In addition, it was shown that charge generation in the depletion layer of silicon devices gives rise to a reverse-current component which is much higher than the diffusion-limited component.

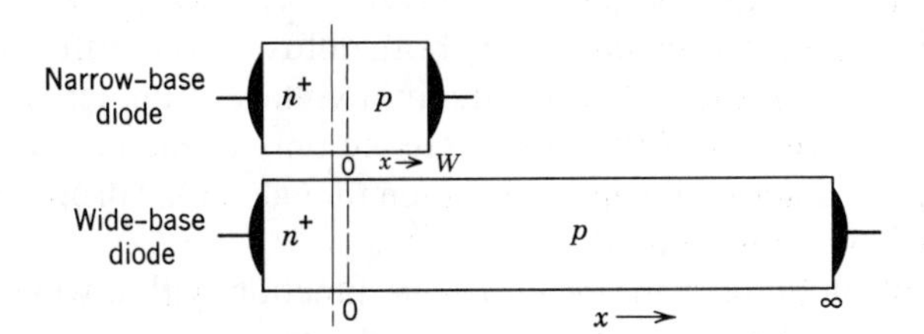

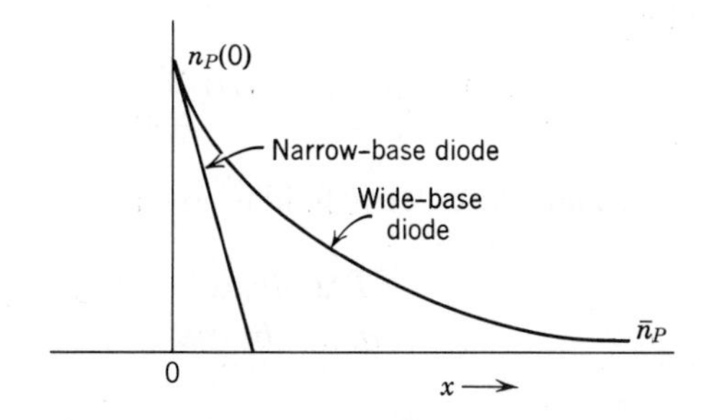

Figure 13.16 The narrow-base diode.

At low forward bias it can be shown that charge generation-recombination processes also dominate the forward-current characteristic. The magnitude of this forward current has been given by Sah *et al.*[3] as

$$I_F \simeq \frac{qAn_ix_0}{\tau_{p0}+\tau_{n0}} e^{qV_A/2kT}, \tag{13.66}$$

and assumes recombination centers in the middle of the forbidden gap. This forward-current relation holds at low voltages ($\simeq$ 0 to 0.5 V) until the diffusion-current term becomes significant. For a typical diode having an active area of 1×1 mil, this corresponds to a forward current of about 1 to 5 μA.

Beyond about 0.5 V the diffusion-current term becomes dominant. Here the magnitude of the forward current is given by

$$I_F = \frac{qAD_n\bar{n}_P}{L_n}(e^{qV_A/kT}-1). \tag{13.67}$$

Equation 13.67 is an accurate description of the forward current in a diode in which minority carrier transport is by diffusion alone. This condition is satisfied in an abrupt junction at low injection levels. At high levels of injection there is a significant alteration of the majority carrier concentration outside the depletion layer, which gives rise to an $\mathscr{E}$ field. Thus carrier motion is influenced by both diffusion and drift in this region.

The presence of the $\mathscr{E}$ field results in a voltage drop across the space-charge neutral region of the diode. Hence only a fraction of the applied voltage appears across the junction, even though ohmic drops due to body resistance have been ignored.

Consider, as before, an abrupt $p-n^+$ junction with a wide, uniformly doped p region. The minority and majority carrier concentrations in this region are shown in Figure 13.17 for high-level injection conditions. For approximate charge neutrality, it is assumed that $dp_P/dx \simeq dn_P/dx$.

For the p region

$$J_p \simeq 0 = q\mu_p p_P \mathscr{E} - qD_p\frac{dp_P}{dx} \tag{13.68}$$

But $D_p = (kT/q)\mu_p$. Thus the electric field is given† by

$$\mathscr{E} = \frac{kT}{q}\frac{1}{p_P}\frac{dp_P}{dx} \tag{13.69}$$

†Note that there is no $\mathscr{E}$ field in a region of uniform majority carrier concentration.

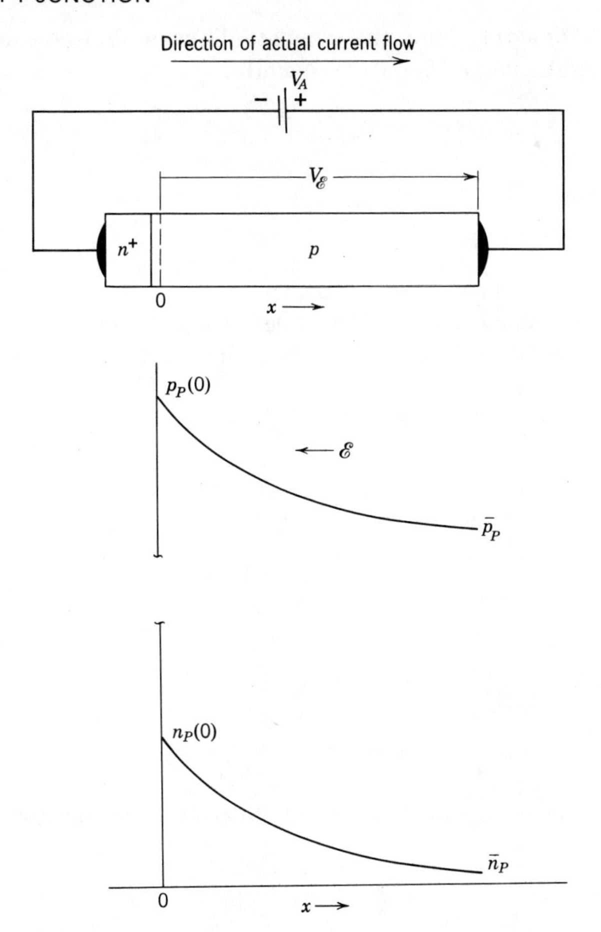

Figure 13.17 High-level injection.

The direction of this electric field is negative (since dp_P/dx is negative), and aids the motion of minority carriers (electrons) in this region. Thus the electron current is given by

$$J_n = q\mu_n n_P \mathscr{E} + qD_n \frac{dn_P}{dx} \tag{13.70}$$

$$= qD_n\left(1 + \frac{n_P}{p_P}\right)\frac{dn_P}{dx} \tag{13.71}$$

The limiting case for high-level occurs when $n_P \simeq p_P \to \infty$. For this case

$J_n \to 2qD_n(dn_P/dx)$. Thus the effective diffusion constant for minority carriers doubles in the limit. Consequently,

$$I_n \simeq -\frac{2qAD_n\bar{n}_P}{L_n}(e^{qV_A'/kT}-1), \tag{13.72}$$

$$\simeq -\frac{2qAD_n\bar{n}_P}{L_n}e^{qV_A'/kT}, \tag{13.73}$$

where V_A' is the voltage across the depletion layer.

The applied voltage across the diode is equal to the sum of the voltage across the depletion layer V_A', and the voltage required to support the $\mathscr{E}$ field ($V_{\mathscr{E}}$). Hence

$$V_A' = V_A - V_{\mathscr{E}}, \tag{13.74}$$

$$= V_A + \int_0^W \mathscr{E}(x)\,dx, \tag{13.75}$$

$$= V_A + \int_{p_P(0)}^{\bar{p}_P} \frac{kT}{q}\frac{dp_P}{p_P} \tag{13.76}$$

For the limiting case of high-level injection,

$$p_P(0) \simeq n_P(0) = \bar{n}_P e^{qV_j/kT}. \tag{13.77}$$

In addition, $\bar{p}_P\bar{n}_P = n_i^2$. Solving (13.76) and making these substitutions,

$$V_A' \simeq \frac{V_A}{2} + \frac{kT}{q}\ln\frac{n_i}{\bar{n}_P} \tag{13.78}$$

Combining with (13.73) gives

$$I_n \simeq -\frac{2qAD_n n_i}{L_n}e^{qV_A/2kT}. \tag{13.79}$$

Thus under high-level conditions the diffusion constant is effectively doubled, the voltage across the junction is halved, and the minority carrier concentration is replaced by the intrinsic carrier concentration.

The transition point between high-level and low-level injection may be found by combining (13.57) and (13.79). The transition voltage is thus given by

$$V_0 = \frac{2kT}{q}\ln\frac{2\bar{p}_P}{n_i} \tag{13.80}$$

For a diode with $N_a = 10^{16}$ atoms/cm^3, the transition voltage occurs at 0.74 V.

Most forward-biased diodes are usually operated at some point between high- and low-level injection. Thus the exponent in the diode equation is usually expressed as $e^{qV_A/mkT}$, where m is taken somewhere between 1 and 2.

Figure 13.18 shows the forward characteristic of a typical p-n^+ diode, indicating these important regions of operation. Although the ohmic drop has been ignored in this analysis, it is this term that eventually dominates the forward characteristic.

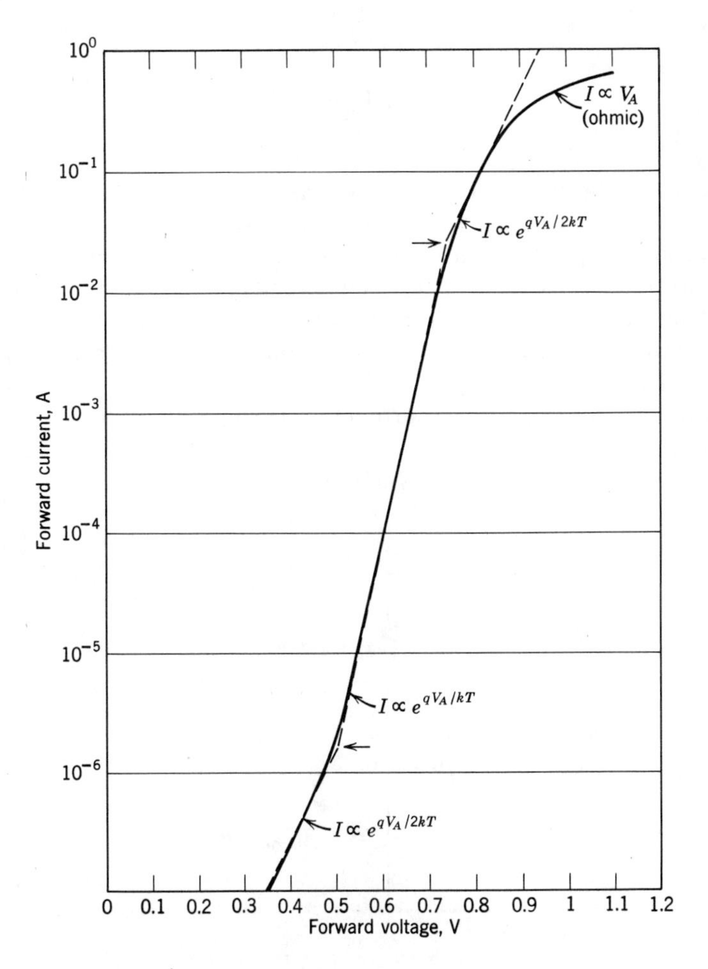

Figure 13.18 Forward-diode characteristics.

13.1.3. Small-signal Behavior

At very low frequencies the dynamic resistance of a diode at any operating point is given by the slope of its $V-I$ characteristic at that point. In the reverse direction this $V-I$ characteristic is dominated by charge-generation effects, and the current is directly proportional to the width of the depletion layer. This current may be written for an abrupt junction diode as

$$I_R \simeq K(V_j)^{1/2}. \tag{13.81}$$

Thus the dynamic resistance is

$$r_d = \left(\frac{dV_j}{dI}\right)_{V_j=V_R+\Psi} \tag{13.82}$$

$$= \frac{2(V_R+\Psi)}{I_R} \tag{13.83}$$

where V_R is the magnitude of the reverse voltage across the junction. Since I_R is in the nanoampere range, this resistance is usually masked by the transition capacitance at all but the very lowest frequencies.

Figure 13.19*a* shows the equivalent circuit of a reverse-biased diode. If the ohmic series resistance of the neutral region r_s is included, the equivalent circuit of Figure. 13.19*b* results. In this circuit the term r_d has been neglected, as shown.

In the forward direction the current through a p-n^+ diode is

$$I \simeq -\frac{qAD_n\bar{n}_p}{L_n}e^{qV_j/kT}, \tag{13.84}$$

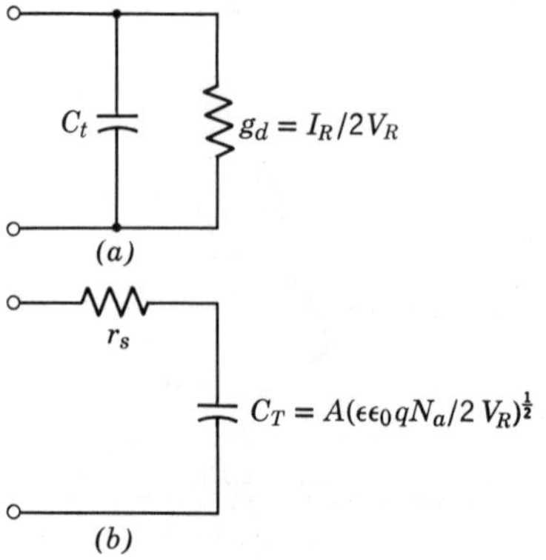

Figure 13.19 Equivalent circuit of a reverse-biased diode.

where V_j is a positive voltage. Hence the dynamic resistance at low frequencies is given by

$$r_d = \left(\frac{dV_j}{dI}\right)_{I=I_F} \simeq \frac{kT}{qI_F} \tag{13.85}$$

where I_F is the direct forward current through the junction.

There are two capacitances associated with the forward-biased diode. The first of these, the depletion layer capacitance C_t is a function of the width of the depletion layer, as described in Section 13.1.1.2. The second capacitive effect is that created by the motion of excess minority carriers in the device when the voltage across it is altered. Thus, if dQ is the change in the charge stored in a diode for a voltage change of dV, the effective diffusion capacitance presented by the diode is proportional to dQ/dV. To a first-order approximation the dynamic admittance of a forward-biased diode is thus given by

$$y_d \simeq \frac{1}{r_d} + j\omega C_t + j\omega C_d \tag{13.86}$$

where C_d is the diffusion capacitance. This capacitance may be determined[4] by assuming a voltage $V = V_F + v(t)$ applied to the device, resulting in a current $I = I_F + i(t)$. Here $v(t)$ and $i(t)$ are the time-varying components of voltage and current respectively. A uniformly doped p-n^+ diode is considered.

For minority carriers in the p region.

$$\frac{\partial n_p}{\partial t} = -\frac{n_p - \bar{n}_p}{\tau_n} + D_n \frac{\partial^2 n_p}{\partial x^2} \tag{13.87}$$

Writing the excess electron concentration as n', so that

$$n' = n_p - \bar{n}_p \tag{13.88}$$

this equation may be rewritten as

$$\frac{\partial n'}{\partial t} = -\frac{n'}{\tau_n} + D_n \frac{\partial^2 n'}{\partial x^2} \tag{13.89}$$

Assume a solution of the form

$$n' = n'_{\mathrm{dc}} e^{-x/L_n} + n'_{\mathrm{ac}} e^{j\omega t}, \tag{13.90}$$

where n'_{dc} and n'_{ac} are the amplitudes of the steady-state and time-varying components of the excess electron concentration respectively. Here only n'_{ac} is a function of distance.

Now
$$L_n = (D_n \tau_n)^{1/2}. \tag{13.91}$$

Substituting this equation and (13.90) into (13.89) gives

$$\frac{d^2 n'_{ac}}{dx^2} - \frac{n'_{ac}(1 + j\omega\tau_n)}{L_n{}^2} = 0. \tag{13.92}$$

Note that the equation governing the a-c behaviour of minority carriers in a semiconductor is of the same form as that for the d-c case, except that the diffusion length L_n is now replaced by L'_n, where

$$L'_n = \frac{L_n}{(1 + j\omega\tau_n)^{1/2}} \tag{13.93}$$

Solving (13.92),
$$n'_{ac} = Ae^{-x/L'_n} + Be^{+x/L'_n} \tag{13.94}$$

The constants A and B may be evaluated for the appropriate boundary conditions. Since the contact is made to a region of very low lifetime,

$$n'_{ac} = 0 \qquad \text{at } x = W, \tag{13.95}$$

where W is the width of the p region.

The magnitude of n'_{ac} at the edge of the depletion layer is found by setting

$$n'(0) = \bar{n}_p e^{q[V_F + v(t)]/kT}, \tag{13.96}$$

$$= \bar{n}_p e^{qV_F/kT}\left[1 + \frac{qv(t)}{kT} + \ldots\right]. \tag{13.97}$$

The time-varying component of excess electron concentration is thus

$$n'_{ac}(0) \simeq \frac{q\bar{n}_p v(t)}{kT} e^{qV_F/kT} \qquad \text{at } x = 0, \tag{13.98}$$

for the small-signal case. This is the second boundary condition. Equation 13.94 may be solved for these conditions to obtain n'_{ac}. The time-varying component of the electron current is next given by

$$i(t) = qAD_n\left(\frac{dn'_{ac}}{dx}\right)_{x=0} \tag{13.99}$$

Combining (13.94), (13.95), (13.98), and (13.99), we obtain the magnitude of $i(t)$ as

$$i(t) = \frac{qAD_n n'_{ac}(0)}{L'_n} \coth \frac{W}{L'_n} \tag{13.100}$$

Up to this point we have considered the general case of a diode of finite width W. Let us now consider the approximations for the narrow-base structure, where $W \ll L_n$. Here

$$W\left(\frac{1+j\omega\tau_n}{L_n^2}\right)^{1/2} \simeq \left(\frac{j\omega\tau_n W^2}{L_n^2}\right)^{1/2}, \tag{13.101}$$

$$\simeq \left(\frac{j\omega W^2}{D_n}\right)^{1/2}. \tag{13.102}$$

Also,

$$\frac{1}{r_d} = \frac{q}{kT} I_F, \tag{13.103}$$

$$= \frac{q}{kT} \frac{qAD_n \bar{n}_P}{W} e^{qV_F/kT}. \tag{13.104}$$

Combining (13.98), (13.102), and (13.104) with (13.100) gives

$$i(t) = \frac{v(t)}{r_d} \left(\frac{j\omega W^2}{D_n}\right)^{1/2} \coth \left(\frac{j\omega W^2}{D_n}\right)^{1/2}. \tag{13.105}$$

Hence the dynamic admittance of the narrow-base diode is given by

$$y_d = \frac{1}{r_d} \left(\frac{j\omega W^2}{D_n}\right)^{1/2} \coth \left(\frac{j\omega W^2}{D_n}\right)^{1/2} \tag{13.106}$$

At low frequencies this admittance can be approximated by

$$y_d \simeq \frac{1}{r_d} + \frac{j\omega W^2}{3D_n r_d} \tag{13.107}$$

Thus the diffusion capacitance of a narrow-base diode is given by

$$C_d \simeq \frac{W^2}{3D_n r_d} \tag{13.108}$$

where

$$r_d = \frac{kT}{qI_F} \tag{13.109}$$

Figure 13.20*a* shows the equivalent circuit for a forward-biased narrow-base diode at low frequencies. The impedance plot of (13.107) is also shown in Figure 13.20*b* as a function of frequency.

The a-c impedance of a long-base diode may be obtained in a similar manner by setting

$$n'_{\mathrm{ac}} = 0 \qquad \text{at } x = \infty. \tag{13.110}$$

Thus

$$i(t) = \frac{v(t)}{r_d}(1 + j\omega\tau_n)^{1/2}. \tag{13.111}$$

From this equation it is seen that the high-frequency dynamic admittance of the diode varies directly as the square root of frequency. At low to

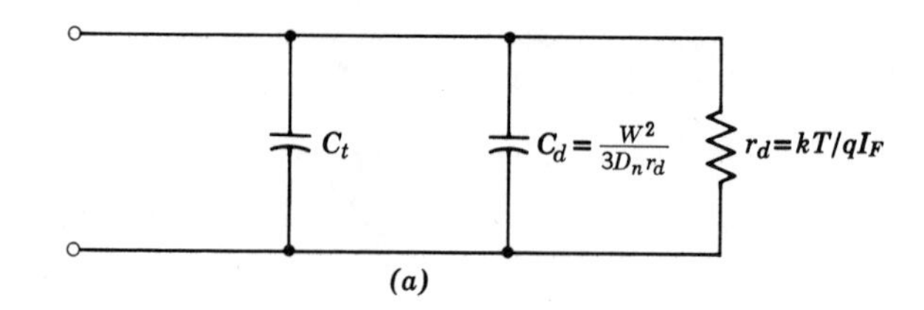

(*a*)

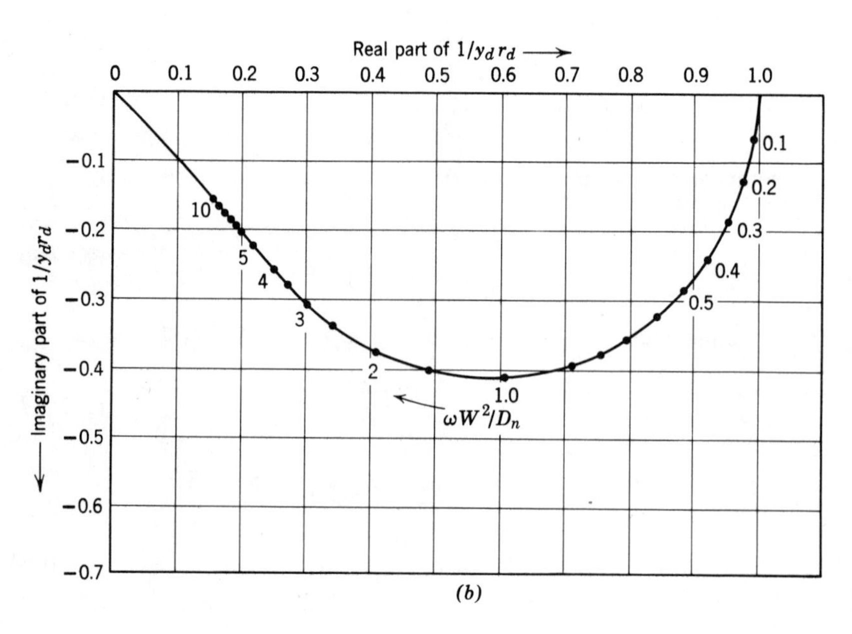

(*b*)

Figure 13.20 Equivalent circuit of a forward-biased narrow-base diode (Lindmayer and Wrigley [4]).

medium frequencies, expansion in Taylor series form allows the admittance to be written as

$$y_d \simeq \frac{1}{r_d} + j\omega C_d \tag{13.112a}$$

Here the diffusion capacitance is C_d, where

$$C_d \simeq \frac{L_n^2}{2D_n r_d} \tag{13.112b}$$

13.1.4. Transit Time in the Narrow-base Diode

Under forward-bias conditions, minority carriers are injected into a space-charge neutral (i.e., a field-free) region. Here the movement of these carriers comes about as a result of diffusion in a concentration gradient. As described in Section 4.2, carriers move at their thermal velocities, and with a random-jump motion. In the presence of a concentration gradient this movement gives rise to a flux density, the magnitude of which is given by Fick's first law. Thus the net result of this jump motion is that carriers appear to traverse a distance Δx in a time Δt. Thus it is possible to speak of an apparent velocity $v = \Delta x/\Delta t$.

Negligible minority carrier recombination is assumed to occur in the space-charge neutral region of a narrow-base diode. Hence the flux density of minority carriers may be assumed constant throughout this region. Since, however, the excess minority carrier concentration falls linearly with distance, the apparent carrier velocity must increase correspondingly to maintain this flux density. This velocity cannot increase beyond the thermal velocity; thus the excess carrier concentration at the ohmic contact cannot be zero, as assumed in our model, and is actually small but finite.

Consider a n^+-p diode with an abrupt junction. If v is the apparent velocity of excess minority carriers (electrons) at any point x in the narrow-base region and n_P' is the excess carrier concentration, the magnitude of the forward current is given by $qAn_P'v$, where

$$n_P' = [n_P(0) - \bar{n}_P]\left(1 - \frac{x}{W}\right). \tag{13.113}$$

But

$$I_F = qAD_n \frac{n_P(0) - \bar{n}_P}{W} \tag{13.114}$$

Combining these equations gives

$$v = \frac{D_n}{W - x} \tag{13.115}$$

Thus the apparent velocity of the minority carriers is D_n/W at the edge of the depletion layer, and tends to infinity as the ohmic contact is approached. The transit time is thus given by

$$t_B = \int_0^W \frac{1}{v}\,dx, \tag{13.116}$$

$$= \frac{W^2}{2D_n}. \tag{13.117}$$

This is an important relation, because it can be used to compute the transit time of minority carriers through the base of a transistor.

Again it must be emphasized that individual carriers do not take this time to traverse the space-charge neutral region of the diode. All we can say is that electrons are injected at the edge of the depletion layer, that electrons arrive at the ohmic contact, and that the time between these two events is t_B. In distinction, for conditions of drift, we can speak of the actual motion of electrons from one point to another.

It is instructive to repeat this computation on the basis of the charge-control principle. Figure 13.21 shows the minority carrier concentration in this diode. If Q_B is the magnitude of the excess charge in the narrow-base region at any instant of time and I_F is the forward current, it follows that $Q_B = I_F t_B$. For a narrow-base diode

$$Q_B = \left(\frac{W}{2}\right) qA[n_P(0) - \bar{n}_P]. \tag{13.118}$$

Combining these relations with (13.114) gives the base transit time as $W^2/2D_n$, as before.

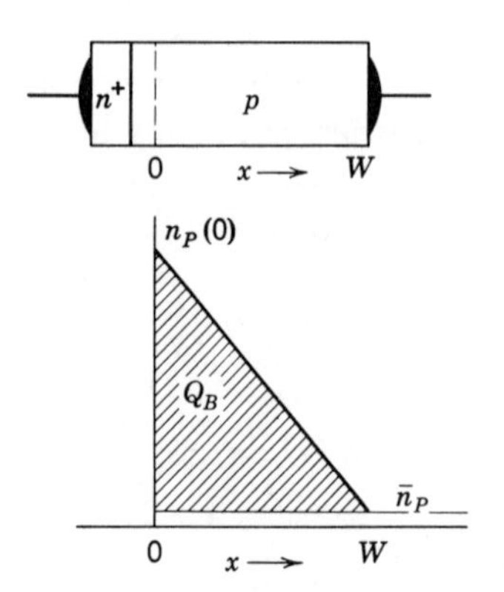

Figure 13.21 Stored charge in a narrow-base diode.

13.1.5. Large-signal Behavior

The response of a *p-n* diode to a step of voltage provides one method of determining those device parameters which are of importance in digital applications. These applications are by far the most commonly encountered with microcircuits at the present time.

Consider a p-n^+ diode in the circuit configuration of Figure 13.22*a*. Initially the switch is in position *A*. Steady-state conditions prevail at this point, and the magnitude of the forward diode current is given by $I_F \simeq V_R/I_R$, if R is assumed to be large compared to the forward impedance. For this condition let the forward drop across the diode be V_D, where $V_D \ll V_F$.

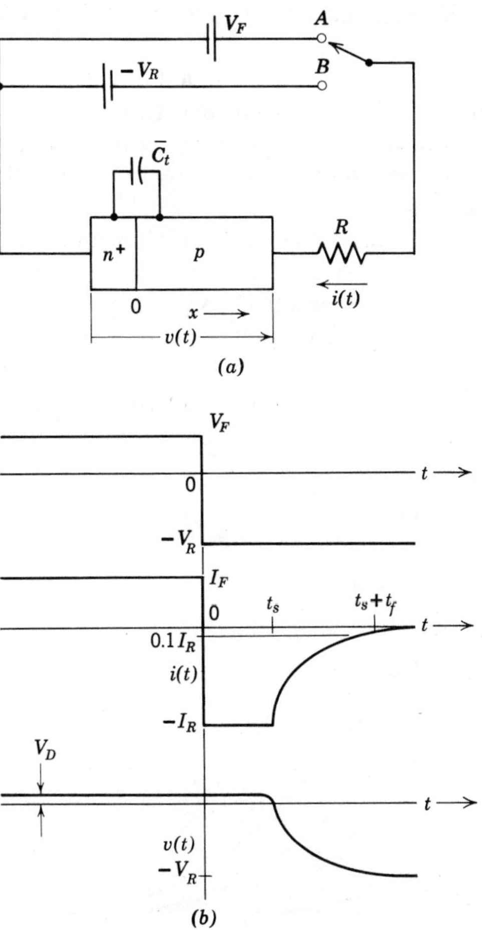

Figure 13.22 Large-signal characteristics.

At time $t = 0$ the switch is moved over to position B and a reverse voltage of magnitude V_R is applied to the diode. This initiates the reverse recovery phase. Carriers moving by diffusion in the space-charge neutral region are swept out of the diode in the reverse direction upon approaching the edge of the depletion layer. The velocity with which these carriers are moved through the depletion layer is on the order of the limiting velocity (about 10^7 cm/sec for electrons). Thus the current is limited only by the resistance in the external circuit. It has a magnitude of $I_R \simeq V_R/R$ and flows during the recovery phase. This reverse current is maintained as long as there is sufficient stored charge in the base.

At time $t = t_s$, this constant-current phase is terminated. From here on the current to the external circuit falls off until it eventually becomes zero.†

Figure 13.23. illustrates the distribution of excess minority carriers in a wide-base diode for $t \geqslant 0$. Note that, from $t = 0$ to $t = t_s$, the slope $(dn_P/dx)_{x=0}$ is constant, corresponding to the delivery of a constant reverse current to the circuit. From $t = t_s$ to infinity, this slope falls, as does the reverse current.

Waveforms of voltage and current are shown in Figure 13.22*b*. Note that the diode remains forward-biased until $t = t_s$. Beyond this point the voltage across it becomes negative, eventually reaching a magnitude of V_R (since the reverse leakage current is ignored).

†The leakage current has been neglected in this discussion. This is a good assumption for silicon devices.

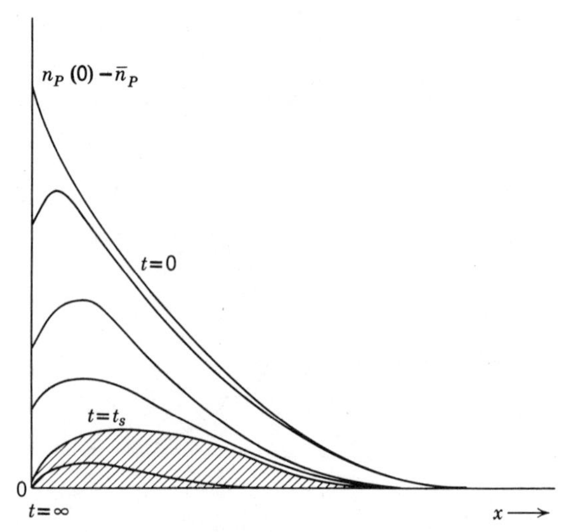

Figure 13.23 Decay of stored charge.

A charge-control method of analysis[5] may be used to describe this behavior. This method relates directly to the physical behavior of the diode under large-signal operation conditions.

Consider a p-n^+ diode of unit cross-sectional area. The continuity equation for electrons in the p region may be written as

$$\frac{\partial(n_P - \bar{n}_P)}{\partial t} = -\frac{n_P - \bar{n}_P}{\tau_n} - \frac{1}{q}\frac{\partial i_n(t)}{\partial x}, \tag{13.119}$$

where $i_n(t)$ is the current due to electron flow, defined positive in the direction shown in Figure 13.22a.

The charge on the electron is $-q$ (where q is a positive number). Multiplying by $-q$ and integrating over the p region.

$$\frac{\partial}{\partial t}\int_0^W -q(n_P - \bar{n}_P)\,dx = \int_0^W \frac{\partial i_n(t)}{\partial x}\,dx - \int_0^W \frac{-q}{\tau_n}(n_P - \bar{n}_P)\,dx. \tag{13.120}$$

Writing $Q(t)$ as the excess stored charge in the p region at any given time, (13.120) reduces to

$$\frac{dQ(t)}{dt} + \frac{Q(t)}{\tau_n} - i_n(W,t) = -i_n(0,t). \tag{13.121}$$

$Q(t)$ is associated with electrons in the base and is thus a negative quantity.

Let us define an effective lifetime in the forward direction τ_F such that

$$\frac{Q(t)}{\tau_F} \equiv \frac{Q(t)}{\tau_n} - i_n(W,t). \tag{13.122}$$

For a wide-base diode, $i_n(W,t) = 0$. Thus τ_F is equal to the minority carrier lifetime. For the narrow-base diode, however,

$$i_n(W,t) \simeq -\frac{Q(t)}{t_B}, \tag{13.123}$$

where t_B is the transit time of carriers through the base. Thus a contact close to the depletion layer results in a sink for diffusing electrons and reduces the effective lifetime in the narrow-base region. Combining (13.122), (13.123), and (13.118),

$$\frac{1}{\tau_F} = \frac{1}{\tau_n} + \frac{2D_n}{W^2} \tag{13.124}$$

Substitution of (13.122) into (13.121*a*) gives

$$\frac{dQ(t)}{dt}+\frac{Q(t)}{\tau_F}=-i_n(0,t). \tag{13.121b}$$

The total forward current of a p-n^+ diode consists of the current through the depletion-layer capacitance as well as of the current due to electron flow. If we write the average depletion-layer capacitance as $\bar{C}_t$, (13.121*b*) may be modified to

$$\frac{dQ(t)}{dt}+\frac{Q(t)}{\tau_F}-\bar{C}_t\frac{dv(t)}{dt}=-i(t). \tag{13.121c}$$

This equation is a statement of the *charge-control principle*. Since it does not involve the detailed behavior of minority carriers in the semiconductor, it can be used with different impurity configurations.

Before reversal of the voltage, that is, at $t \leqslant 0$, $i(t)=I_F$. Thus the steady-state excess stored charge is $-I_F\tau_F$. This is the boundary condition at $t=0$. Upon reversal of the voltage, a constant current $-I_R$ flows during the time interval from 0 to t_s. During this interval the change in the voltage drop across the diode is small (see Figure 13.22*b*). Consequently the current through the depletion-layer capacitance may be ignored. Hence (13.121*c*) reduces to

$$\frac{dQ(t)}{dt}+\frac{Q(t)}{\tau_F}=I_R. \tag{13.125}$$

Solving this equation, subject to the initial value condition, gives

$$Q(t)=I_R\tau_F-(I_R+I_F)\tau_F e^{-t/\tau_F}. \tag{13.126}$$

It is necessary to make some assumptions concerning the charge remaining in the p region at $t > t_s$. Let us intuitively write this charge as

$$Q(t)=-i(t)\tau_R \tag{13.127}$$

where τ_R is an effective reverse time constant. Then the excess stored charge at $t=t_s$ is given by $-I_R\tau_R$. Substituting this condition into (13.127) gives

$$t_s=\tau_F\left[\ln\left(1+\frac{I_F}{I_R}\right)-\ln\left(1+\frac{\tau_R}{\tau_F}\right)\right]. \tag{13.128}$$

The ratio τ_F/τ_R is equal to the ratio of the excess charge stored in the base under forward bias to the excess charge stored at $t = t_s$. For a narrow-base diode, this ratio may be estimated from a highly idealized sketch of the minority carrier concentration, as shown in Figure 13.24. Here the shaded area representing the stored charge is 25% of the total, so that $\tau_F/\tau_R \simeq 4$. In practice, this ratio is experimentally determined and is usually found to be anywhere from 2 to 4 for various types of diodes.

The current through $\bar{C}_t$ cannot be ignored during the fall period $(t \geqslant t_s)$. From the loop equation for the network of Figure 13.22*a* it is seen that $dv(t) = -R\,di(t)$. Substituting this relationship and (13.127) into (13.121*c*), the equation determining the fall period is

$$\frac{di(t)}{dt}(\tau_R + R\bar{C}_t) + i(t)\left(1 + \frac{\tau_R}{\tau_F}\right) = 0. \tag{13.129}$$

At $t = t_s$, $i(t) = -I_R$. Thus

$$i(t) = -I_R\, e^{-(t-t_s)/\tau} \tag{13.130}$$

where

$$\tau = \frac{\tau_R + R\bar{C}_t}{1 + \tau_R/\tau_F} \tag{13.131}$$

The fall time, measured to 10% of the initial value of reverse current, is thus

$$t_f \simeq 2.3\,\frac{\tau_R + R\bar{C}_t}{1 + \tau_R/\tau_F} \tag{13.132}$$

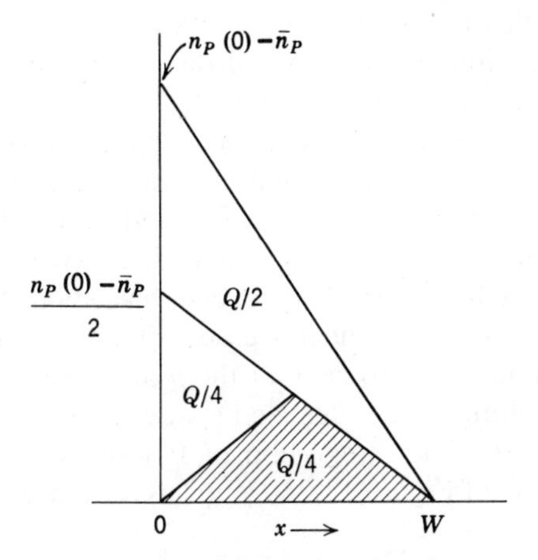

Figure 13.24 Stored-charge configuration.

It is thus seen that the recovery characteristics of a diode are related to the minority carrier lifetime, to the transit time (for a narrow-base diode), to the forward and reverse currents, and to circuit parameters. In addition, the manner in which the device is fabricated plays a role in determining τ_R.

In (13.132), $\bar{C}_t$ is the average capacitance value as the reverse voltage across the diode varies from zero to $-V_R$. Since this capacitance is voltage dependent, we may write its average value as

$$\bar{C}_t = \frac{1}{V_R} \int_{-V_R}^{0} C_t(v)\, dv. \tag{13.133}$$

13.2. THE LINEARLY GRADED JUNCTION

As its name implies, this type of junction is formed when the impurity concentration in a semiconductor changes linearly from p-type to n-type over a finite length of crystal. Characteristic of this junction is a grade constant $\mathscr{A}$, which defines the slope of the impurity concentration and has the dimensions of cm^{-4}.

In linearly graded junctions the impurity grading and, hence, the space charge extends beyond the depletion layer and into the base region. Morgan and Smits[6] have considered the potential distribution of a junction of this type in detail and obtained numerical solutions for a number of different values of grading. In general their findings indicate that, for values of impurity grading in excess of $10^{17}/\text{cm}^4$, it is possible to make the assumption that the space-charge region is confined to the depletion layer, with the base regions considered field-free and quasi-neutral. As a consequence of this approximation it follows that the behavior of the linearly graded junction in the forward direction approximates that of the abrupt junction. This is because injected carriers travel through the quasi-neutral regions to the ohmic contact by diffusion alone, as was the case for the abrupt junction. In addition, the minority carrier concentration at the edge of the depletion layer is independent of its impurity grading. The essential difference between these two structures, then, lies in the quantitative behavior of the reverse-biased junction.

Consider a reverse-biased, linearly graded junction. Let the depletion layer extend an amount x_0 into each of the p and n regions, as shown in Figure 13.25a, resulting in the idealized space-charge-density configuration of Figure 13.25b. The integration of Poisson's equation results in the field configuration of Figure 13.25c, and has a value given by

$$\mathscr{E} = \mathscr{E}_{\text{peak}}\left[1 - \left(\frac{x}{x_0}\right)^2\right], \tag{13.134}$$

where

$$\mathscr{E}_{\text{peak}} = \frac{q\mathscr{A}x_0^2}{2\epsilon\epsilon_0} \tag{13.135}$$

In like manner a second integration yields a potential function $V(x)$ of the form shown in Figure 13.25d, with the n region taken as the reference. Noting that the area under a parabola is $\frac{2}{3}$ base times height, the voltage across the junction is given by

$$V_j = \frac{2q\mathscr{A}x_0^3}{3\epsilon\epsilon_0} \tag{13.136}$$

This is the voltage supported by the depletion layer.

13.2.1. The Contact Potential

Let 2δ be the width of the depletion layer (δ in each of the p and n regions) with no reverse voltage. Then the contact potential associated with this depletion layer is given by

$$\Psi = \frac{2q\mathscr{A}\delta^3}{3\epsilon\epsilon_0}. \tag{13.137}$$

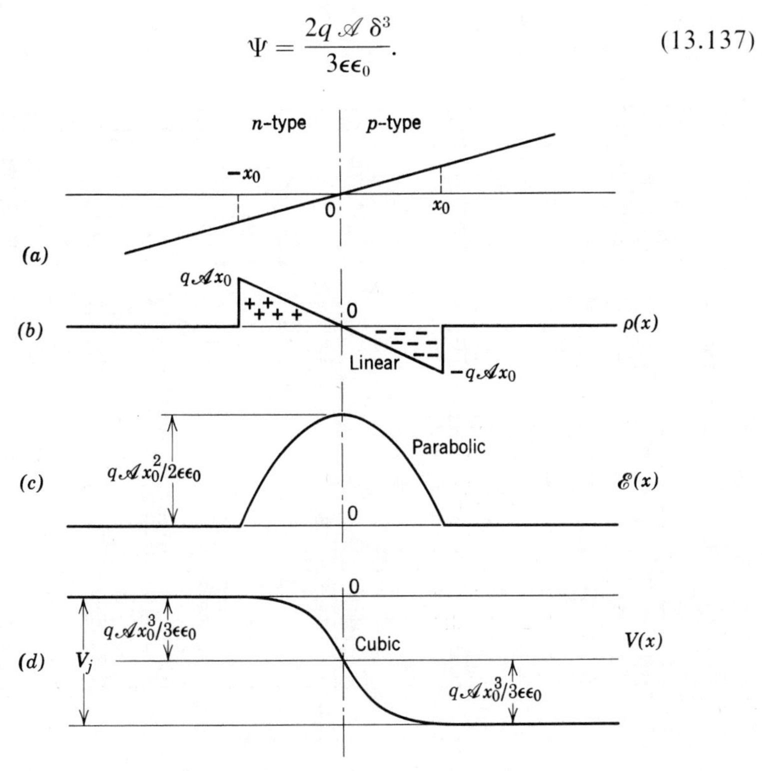

Figure 13.25 The linearly-graded junction.

But

$$\Psi = \Psi_n + \Psi_p = \frac{kT}{q} \ln \frac{N_a N_d}{n_i^2}, \tag{13.138a}$$

$$= \frac{kT}{q} \ln \frac{\mathscr{A}^2 \delta^2}{n_i^2}. \tag{13.138b}$$

Solving (13.137) and (13.138*b*) yields

$$\Psi = \frac{2kT}{3q} \ln \frac{3\epsilon\epsilon_0 \mathscr{A}^2 \Psi}{2qn_i^3}. \tag{13.139}$$

Figure 13.26 shows the solution to this transcendental equation. It is seen that a reasonably close fit to this curve is given over commonly encountered values of the grade constant by

$$\Psi \simeq \frac{4kT}{3q} \ln \mathscr{A} - 1 \tag{13.140}$$

where $\mathscr{A}$ is expressed in cm^{-4}.

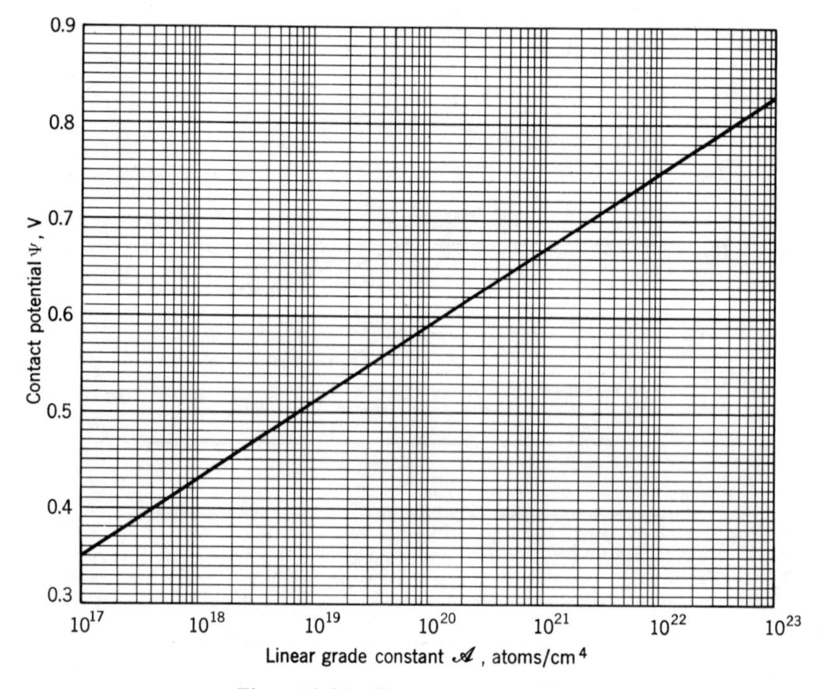

Figure 13.26 The contact potential.

13.2.2 The Transition Capacitance

The transition capacitance of a linearly graded junction may be calculated by considering separately the capacitances associated with the p and n sides of the depletion layer. Each of these capacitances is denoted by $2C_t$ so as to result in an overall series capacitance of C_t.

Let Q be the charge associated with one-half of the depletion layer. Since the voltage across this half is $V_j/2$, the capacitance is given by

$$2C_t = \frac{dQ}{dV_j/2} \tag{13.141a}$$

or

$$C_t = \frac{dQ}{dV_j}. \tag{13.141b}$$

Assuming a cross-sectional area A,

$$Q = \frac{qA\,\mathscr{A}\,x_0^2}{2} \tag{13.142}$$

By combining (13.136) and (13.142), and differentiating we have

$$C_t = \frac{dQ}{dV_j} = A\left(\frac{\epsilon^2\epsilon_0^2 q\mathscr{A}}{12V_j}\right)^{1/3}. \tag{13.143}$$

Thus the capacitance† of the reverse-biased, linearly graded junction varies inversely as the cube root of the voltage, as shown in Figure 13.27. Substituting (13.135) into (13.136), the peak field is given by

$$\mathscr{E}_{\text{peak}} = \left(\frac{9q\,\mathscr{A}\,V_j^2}{32\epsilon\epsilon_0}\right)^{1/3}, \tag{13.144a}$$

$$= \frac{3V_j}{4x_0}. \tag{13.144b}$$

Thus for equal depletion layer thicknesses the graded junction has a lower peak field than the abrupt junction, and can be expected to have a higher breakdown voltage.

†Note that the same value of C_t may be obtained by considering the depletion layer as a dielectric of thickness $2x_0$.

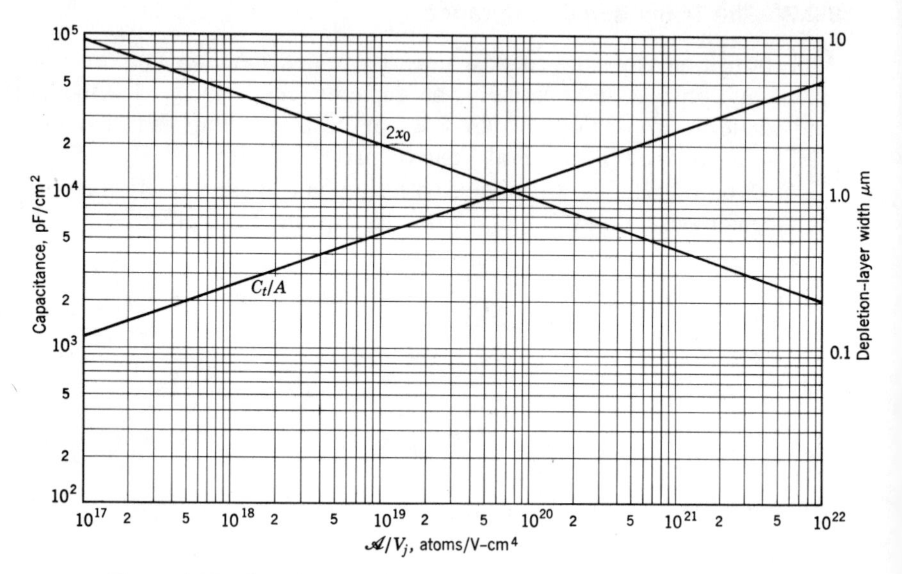

Figure 13.27 Capacitance and depletion-layer width for the linear junction.

13.2.3. Breakdown Voltage

It is not possible to obtain breakdown by Zener tunneling in linearly graded junctions. This is due to the relatively large depletion layer width that exists, even in the most highly graded structures.

The conditions for avalanche breakdown may be established in a manner similar to that for abrupt junctions, with breakdown occurring when

$$\int_0^w \alpha_i \, dx = 1. \tag{13.145}$$

Let $\mathscr{E}'$ be the peak field associated with the breakdown voltage, and w the width of the depletion layer on each side of the junction, at breakdown. The $\mathscr{E}$ field is given by

$$\mathscr{E} = \mathscr{E}'\left[1 - \left(\frac{x}{w}\right)^2\right]. \tag{13.146}$$

From (13.135)

$$w = \left(\frac{2\epsilon\epsilon_0 \mathscr{E}'}{q\mathscr{A}}\right)^{1/2}, \tag{13.147}$$

hence the ionization integral reduces to

$$\left(\frac{\epsilon\epsilon_0}{2q\mathscr{A}}\right)^{1/2}\int_0^{\mathscr{E}'}\frac{\alpha_i\,d\mathscr{E}}{(\mathscr{E}'-\mathscr{E})^{1/2}}=1, \tag{13.148}$$

where $\alpha_i = 1.65 \times 10^{-24}\mathscr{E}^5$ as in (13.25).

Setting $\mathscr{E} = \mathscr{E}'\sin^2\theta$, and integrating after making this change of variable (13.148) results in

$$\left(\frac{2\epsilon\epsilon_0\mathscr{E}'}{q\mathscr{A}}\right)^{1/2}\alpha_i\frac{\Gamma(6)\Gamma(0.5)}{\Gamma(6.5)}=1, \tag{13.149}$$

where $\Gamma(y)$ is the gamma function of y. But [see (13.144a)]

$$\mathscr{E}'=\left(\frac{9q\mathscr{A}\,BV^2}{32\epsilon\epsilon_0}\right)^{1/3}, \tag{13.150}$$

where BV is the breakdown voltage. Substituting (13.150) into (13.149), and solving for the various constants,

$$BV = 1.71 \times 10^9\mathscr{A}^{-4/11} \qquad \text{for silicon.} \tag{13.151}$$

Here BV is the breakdown voltage in volts and $\mathscr{A}$ is the linear grade constant in cm^{-4}. Figure 13.28 shows this relationship for a linearly graded junction. As may be expected, the breakdown voltage falls as the junction becomes more and more abrupt in character.

13.3. THE DIFFUSED JUNCTION

The diffused junction is a special class of inhomogeneous structure that is commonly encountered in microcircuits. The salient features of this type of structure will now be described.

Let N be the net ionized impurity concentration (donors minus acceptors) at any point, and let $\bar{n}$ and $\bar{p}$ be the equilibrium concentrations of carriers. For the one-dimensional case, Poisson's equation may be written as

$$\frac{\partial^2 V}{\partial x^2}=-\frac{q}{\epsilon\epsilon_0}(N+\bar{p}-\bar{n}), \tag{13.152}$$

where V is the potential. In addition,

$$\bar{n} = n_i\,e^{qV/kT} \tag{13.153a}$$

$$\bar{p} = n_i\,e^{-qV/kT}. \tag{13.153b}$$

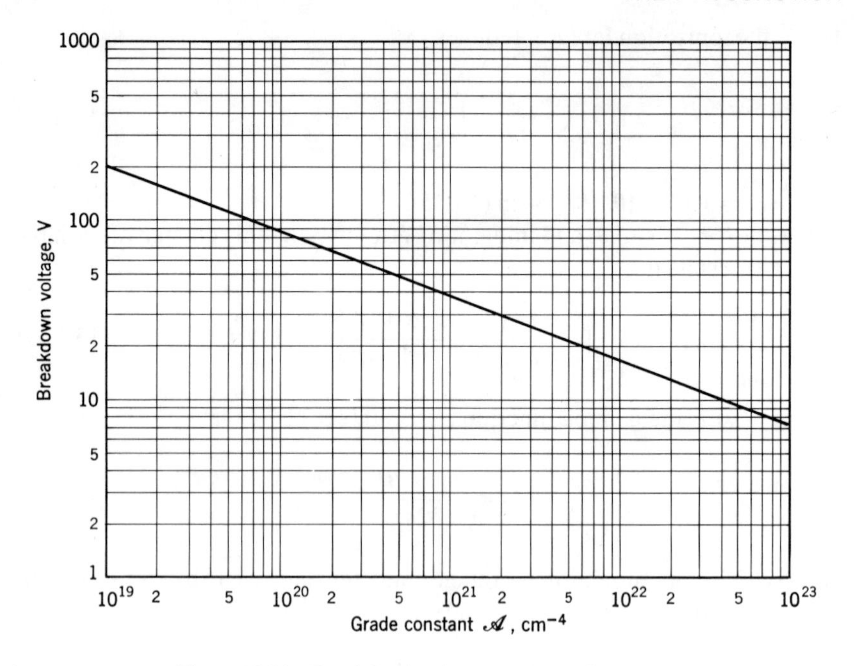

Figure 13.28 Breakdown voltage versus grade constant.

Therefore

$$\frac{\partial^2 V}{\partial x^2} = -\frac{q}{\epsilon\epsilon_0}\left(N - 2n_i \sinh\frac{qV}{kT}\right). \tag{13.154}$$

Exact solution† of this equation is only possible by numerical methods.

Junctions made by microcircuit fabrication techniques are of the diffused type. Thus the collector-base junction of a transistor made by the DDE process involves a p-type gaussian diffusion, whereas the emitter-base junction utilizes both a p-type gaussian and an n-type erfc diffusion. Both these diffusions can be closely approximated by exponential impurity gradings for a considerable distance on either side of the junction. The net impurity concentration may therefore be approximated by

$$N(x) \simeq N_0 e^{-kx} - N_C, \tag{13.155}$$

as shown in Figure 13.29a. This impurity concentration profile is essentially exponential on one side of the depletion layer and constant on the other.

†Within a depletion layer this equation reduces to (13.6), since $\bar{p}$ and $\bar{n}$ are negligible.

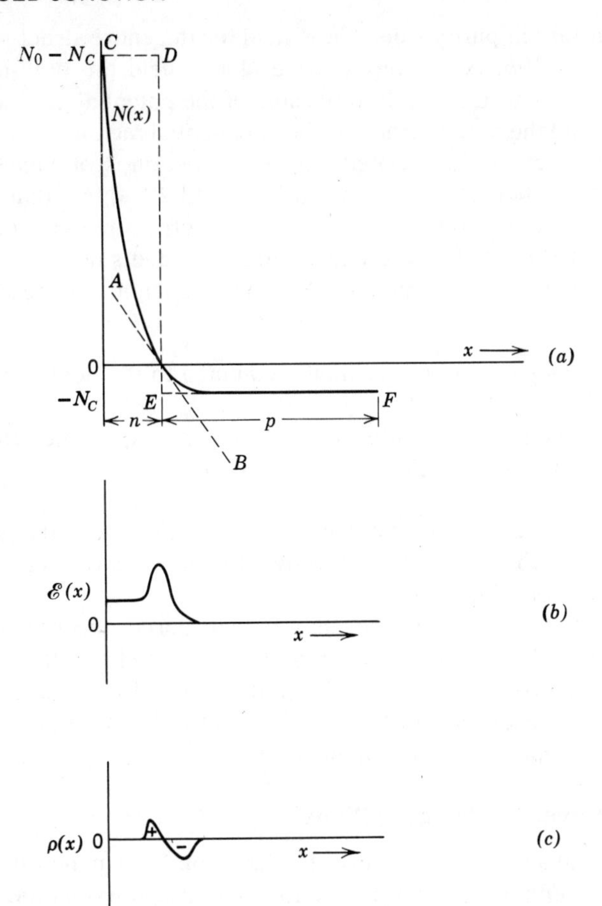

Figure 13.29 The diffused junction.

The electric field at any point in a semiconductor is given by

$$\mathscr{E} = +\frac{kT}{qp_P}\frac{dp_P}{dx} \tag{13.156a}$$

for a p-type semiconductor [see (13.69)] and

$$\mathscr{E} = -\frac{kT}{qn_N}\frac{dn_N}{dx} \tag{13.156b}$$

for an n-type semiconductor. It follows, then, that there is a constant electric field in the exponentially graded side of the junction and no field in the constant impurity side. The $\mathscr{E}$ field for the entire structure is shown in Figure 13.29*b*. As a consequence of this field the built-in potential across the structure is equal to the sum of the potential across the depletion layer and the potential across the exponential region.

The space charge associated with this junction is obtained by differentiating the $\mathscr{E}$ field, and is shown in Figure 13.29*c*. Note that the regions beyond the depletion layer are quasi-neutral, notwithstanding the presence of this field in the exponentially graded side of the structure.

The properties of the exponentially graded junction may be summarized as follows:

1. The space-charge is essentially confined to the depletion layer and to the surface.
2. The regions outside of the depletion layer are quasi-neutral.
3. There is no $\mathscr{E}$ field in the constant concentration side of the junction.
4. There is a constant $\mathscr{E}$ field in the exponential side of the junction. Its direction is such as to oppose the flow of minority carriers in this region, that is, it is a retarding field.
5. The built-in potential consists of two parts—the potential across the depletion layer and that caused by the $\mathscr{E}$ field in the exponential side. In addition, there is a slight shift in the effective position of the junction as a result of this $\mathscr{E}$ field. This is of little consequence in determining the properties of the *p-n* junction.

13.3.1. Reverse-bias Conditions

At low values of reverse bias, the depletion layer penetrates an equal distance on either side of the junction. To an approximation, then, this junction may be considered as a linearly graded one, the grading of which is equal to the slope of the line *AB* in Figure 13.29*a*.

With increasing values of reverse bias, the depletion layer widens, penetrating much deeper into the constant side than into the exponentially graded side. In the limit, the junction may be treated as an abrupt structure, the profile of which is given by the dashed line *CDEF*.

Let V_0 be the transition voltage that divides the operating ranges over which these two approximations hold. Below V_0 the junction capacitance exhibits an inverse cube-root dependence with voltage. An inverse square-root dependence exists above V_0. Combining (13.13*b*) and (13.143) around the transition voltage gives

$$V_0 = \frac{18qN_C{}^3}{\epsilon\epsilon_0\mathscr{A}^2}. \tag{13.157}$$

For the collector-base junction shown in Figure 13.1, $N_C = 3.5 \times 10^{16}$ atoms/cm^3 and $\mathscr{A} = 1.08 \times 10^{21}$ atoms/cm^4, and thus $V_0 = 29.5$ V. For reverse voltages below this value (provided breakdown has not previously occurred), this junction may be treated as a linearly graded structure. The effects of the additional built-in potential due to the $\mathscr{E}$ field are relatively small compared with reverse voltages that are encountered in practice, and may be ignored.

For reverse voltages in the neighborhood of V_0, however, it is necessary to solve the Poisson's equation (13.154) in detail. This has been done in numerical form by Lawrence and Warner[8]; some of their results are shown in Figure 13.30 for a junction depth of x_j, and are useful for computing junction properties for a variety of situations encountered in microcircuits.

The emitter-base junction of Figure 13.1 has a grade constant $\mathscr{A} = 1.86 \times 10^{22}$ atoms/cm^4. For this diode the linear approximation ceases to hold at a reverse bias in excess of 2.5 V. In practice it is often necessary to determine the transition capacitance of an emitter-base junction at low values of both reverse and forward voltage, where the diode exhibits a capacitance with a cube-root dependence on voltage. Thus application of (13.143) results in the desired value of capacitance, provided the appropriate values of V_j are chosen.

An estimate of the built-in potential may be made by recognizing that the two ends of the junction are quasi-neutral. Since the Fermi level is a straight line for thermal-equilibrium conditions, the built-in potential is a function of the impurity concentration at these ends and corresponds to the contact potential for the abrupt structure shown as CDEF in Figure 13.29. Here the n side is heavily doped, and the built-in potential associated with it is approximately 0.61 V. The potential associated with the p side is obtained with the aid of Figure 13.2.

In practice it is customary to circumvent the necessity of computing the actual magnitude of the concentration on the p side and use instead the concentration N_B' at which the emitter and base diffusions intersect. This is a reasonably good approximation for modern shallow structures. Figure 13.31 shows normalized curves which may be used for determining the magnitude of N_B'. In addition, the grade constant at the emitter-base junction is given by

$$\mathscr{A} \simeq \frac{2N_B'}{x_{EB}}\left[\left(\operatorname{erfc}^{-1}\frac{N_B'}{N_{0E}}\right)^2 - \ln\frac{N_{0B}}{N_B'}\right], \qquad (13.158)$$

where the various terms of this equation are defined in Figure 13.31.

The avalanche breakdown voltage of a diffused junction may be computed by solving the ionization integral subject to the appropriate boundary

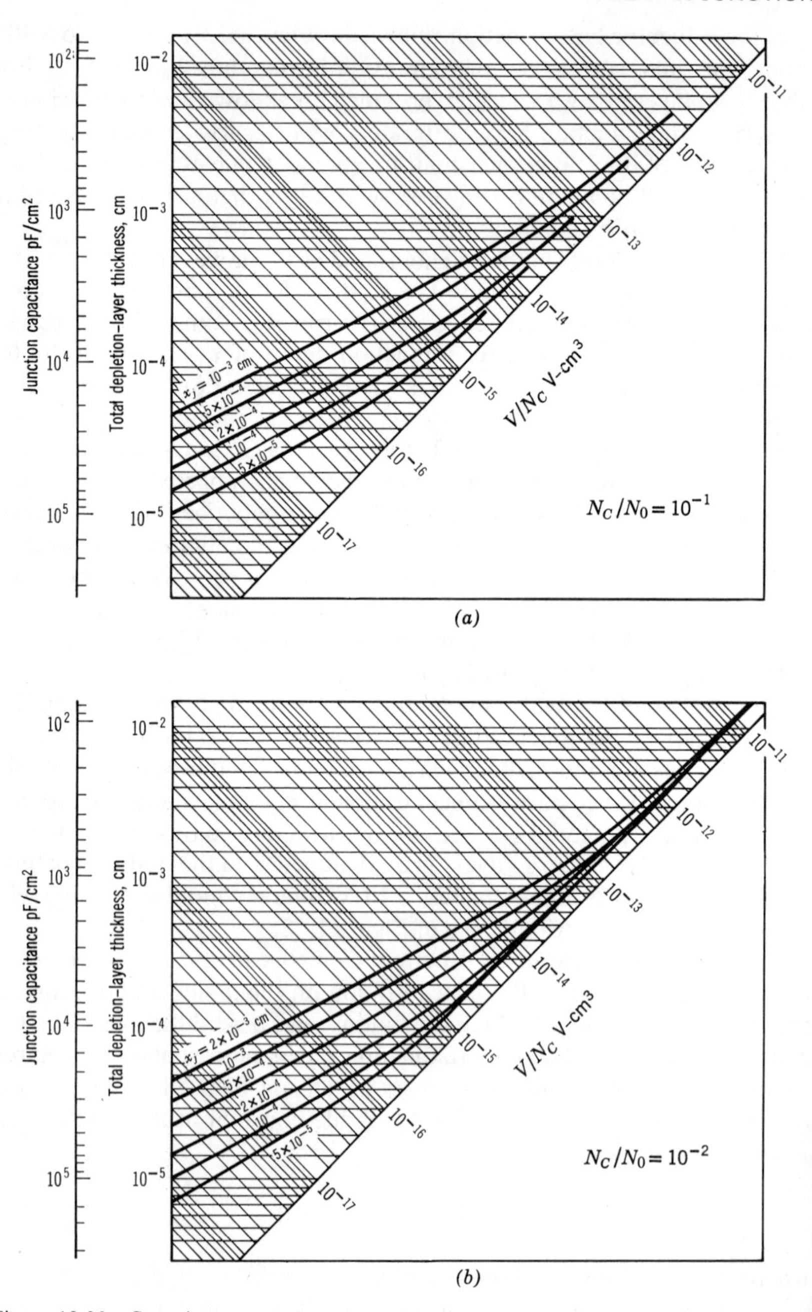

Figure 13.30 Capacitance and depletion-layer width for the diffused junction (Lawrence and Warner [8]).

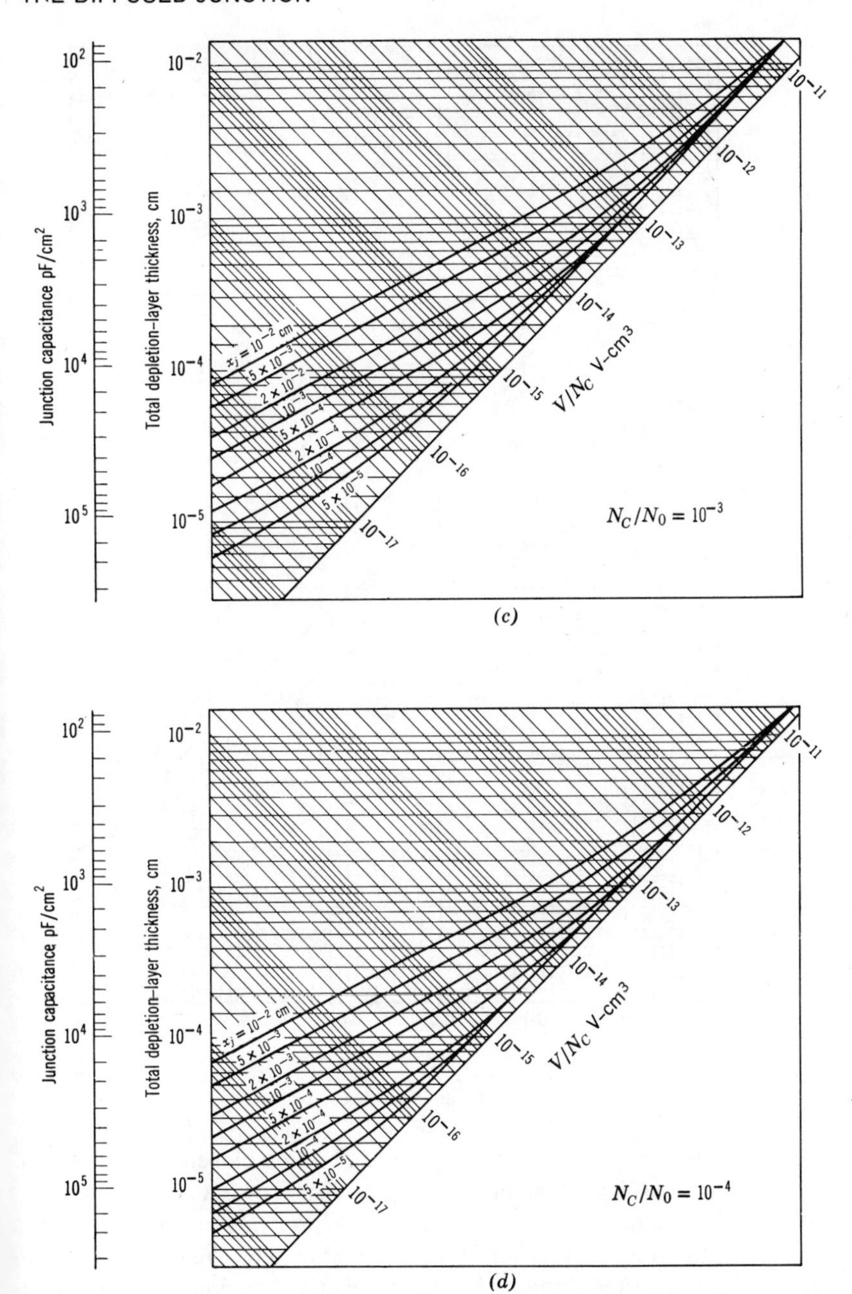

Figure 13.30 (*continued*)

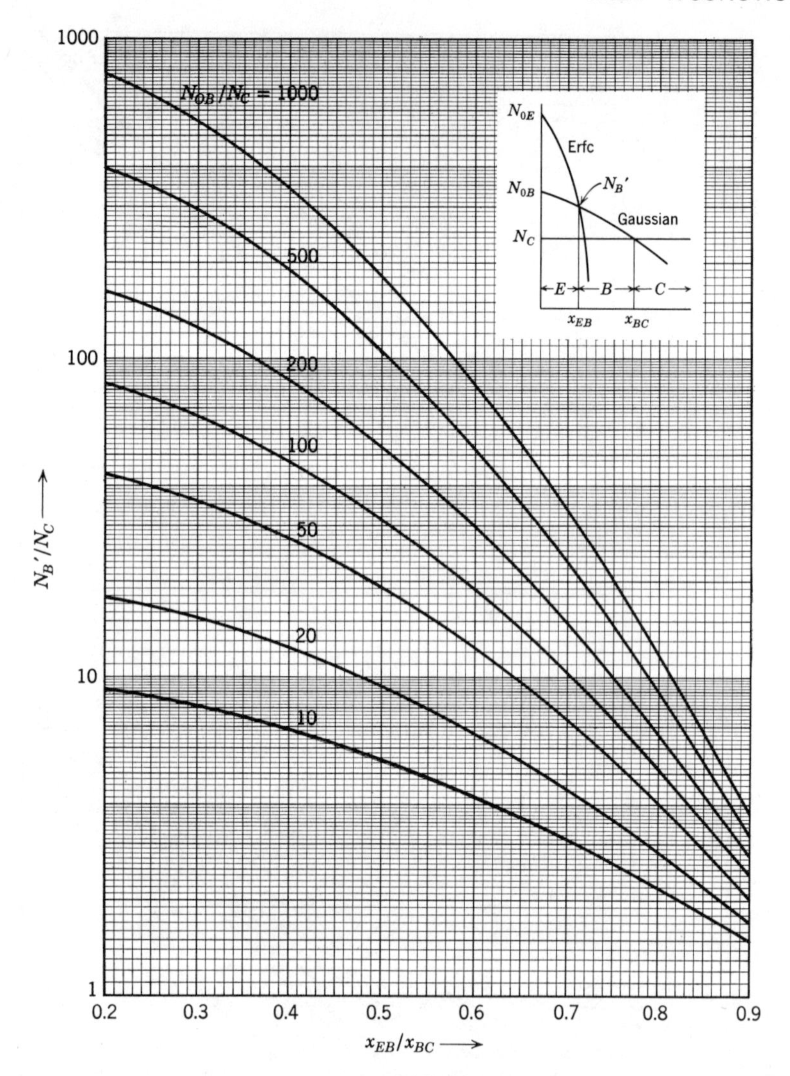

Figure 13.31 Doping-profile calculations.

conditions. In general, it is not possible to obtain a closed-form solution for either the gaussian or the erfc doping profile. However, Kennedy and O'Brien [9] have obtained numerical solutions for this problem with the aid of a digital computer. Their results are shown in Figure 13.32 for a parallel-plane junction obtained by making a erfc diffusion into a constant background. This result may also be used for a gaussian-diffusion profile because of its generally similar nature.

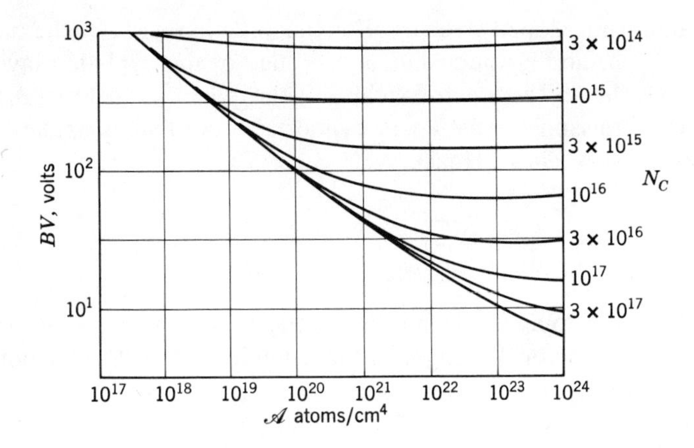

Figure 13.32 Breakdown voltage for diffused junctions (Kennedy and O'Brien[9]).

It is interesting to note that the breakdown voltage for a diffused junction is higher than that for a linear junction having the same grade constant. For high values of background concentration, however, the diffused -junction profile approximates that of a linearly graded structure. In the limiting case, the breakdown voltage of the diffused junction approaches that of its linearly graded counterpart.

13.3.1.1. The Cylindrical Junction[10]. Microcircuit junctions are parallel-plane in character, except at the edge of the window cut in the oxide mask. Here owing to lateral diffusion effects of the type discussed in Section 4.6.1, the junction takes on a cylindrical shape. As a result, the space-charge lines in this region are distorted, resulting in an $\mathscr{E}$ field that is different from that obtained in a parallel-plane structure. To an approximation, the radius of this cylindrical junction is equal to the diffusion depth; thus distortion effects are more severe in shallow structures than in deep structures.

Since avalanche breakdown is a function of the critical field strength attained in a semiconductor, we may expect that the breakdown voltage in this cylindrical region to be different from that in the parallel-plane region. Consider a n^+-p cylindrical junction, as shown in Figure 13.33,

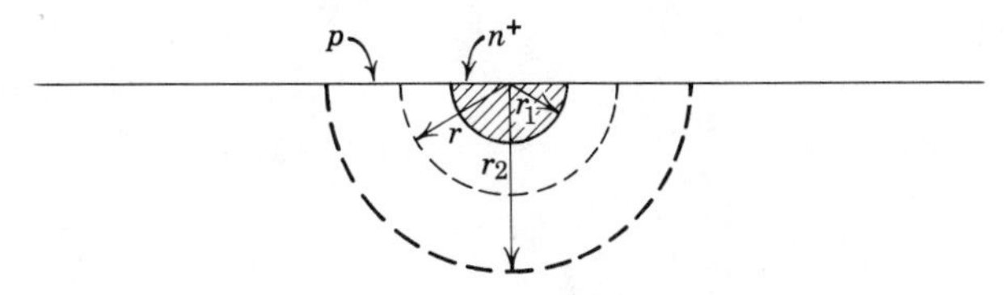

Figure 13.33 The cylindrical junction.

with a uniformly doped p region. For this junction let r_1 be the radius of the n^+ region, and r_2 the radius at the edge of the depletion layer. Let $\mathscr{E}'$ and BV define the electric field and the potential at breakdown respectively. Poisson's equation in cylindrical coordinates applies to this situation for $r_1 \leqslant r \leqslant r_2$. Hence

$$\frac{1}{r}\frac{d}{dr}\left(r\frac{dV}{dr}\right) = -\frac{1}{r}\frac{d}{dr}(r\mathscr{E}) = -\frac{qN_a}{\epsilon\epsilon_0}, \tag{13.159}$$

where N_a is the doping concentration of the p region and V is the potential along a radius vector. Solutions of this equation, subject to the boundary conditions that

$$\mathscr{E} = \mathscr{E}' \qquad \text{at } r = r_1, \tag{13.160a}$$

$$\mathscr{E} = 0 \qquad \text{at } r = r_2, \tag{13.160b}$$

gives the magnitude of the peak field at breakdown as

$$\mathscr{E}' = \frac{qN_a}{2\epsilon\epsilon_0}\frac{r_2^2 - r_1^2}{r_1}. \tag{13.161}$$

A second integration, subject to the boundary conditions that

$$V = 0 \qquad \text{at } r = r_1, \tag{13.162a}$$

$$V = -BV \qquad \text{at } r = r_2, \tag{13.162b}$$

gives the magnitude of the breakdown voltage as

$$BV = \frac{qN_a}{2\epsilon\epsilon_0}\left(r_2^2 \ln\frac{r_2}{r_1} - \frac{r_2^2 - r_1^2}{2}\right). \tag{13.163}$$

Combining (13.161) and (13.163) to eliminate r_2 and writing

$$y = \frac{2\epsilon\epsilon_0\mathscr{E}'}{qN_a r_1} \tag{13.164}$$

results in

$$BV = \frac{\epsilon\epsilon_0\mathscr{E}'^2}{2qN_a}\frac{2}{y^2}\left[(1+y)\ln(1+y) - y\right], \tag{13.165}$$

$$= \frac{\epsilon\epsilon_0\mathscr{E}'^2}{2qN_a}F(y). \tag{13.166}$$

But the breakdown voltage for a parallel-plane p-n^+ junction is given by [see (13.14)]

$$BV_{P\text{-}P} = \frac{\epsilon\epsilon_0 \mathscr{E}'^2}{2qN_a}, \tag{13.167}$$

hence

$$\frac{BV}{BV_{P\text{-}P}} = F(y). \tag{13.168}$$

This relationship is shown in Figure 13.34. It is seen that the breakdown voltage of a cylindrical junction approaches that of a parallel-plane structure for large values of r_1 (small values of y), but falls without limit as the radius is reduced. It follows, then, that the breakdown voltage of a junction made by oxide masking and diffusion processes will be given by the voltage for which breakdown occurs at the edge of the oxide window. Furthermore, with shallow diffusion, r_1 falls, and the breakdown voltage is further reduced.

Kennedy and O'Brien[11] have developed numerical solutions of this problem for a range of background concentrations and junction depths, as shown in Figure 13.35. These curves illustrate the manner in which the breakdown voltage varies as a function of the radius of curvature of the cylindrical junction. Although they have been obtained for erfc diffusions into a constant background concentration, they may also be used for gaussian diffusions with reasonable accuracy.

13.3.2. Forward-bias Conditions[12]

Consider a diffused p-n junction of unit cross section. Let N be the net ionized impurity concentration at any point, and let n and p be the electron and hole concentrations respectively. Note that N, n, and p are all functions of distance.

Under steady-state forward-bias conditions the current density due to electron flow in the p region is given by

$$J_n = q\mu_n n_P \mathscr{E} + qD_n \frac{dn_P}{dx}. \tag{13.169}$$

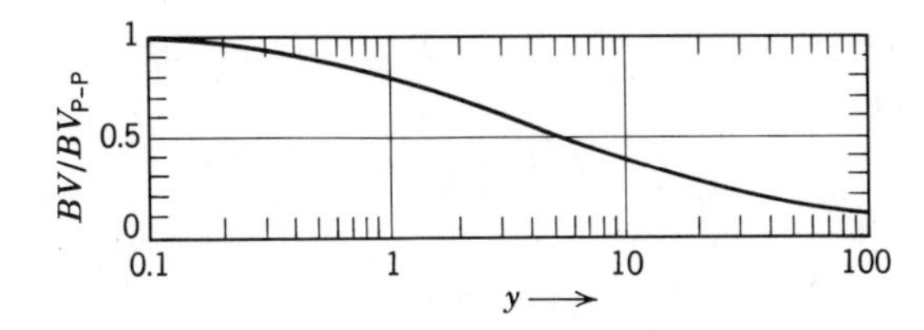

Figure 13.34 Breakdown voltage of cylindrical junctions.

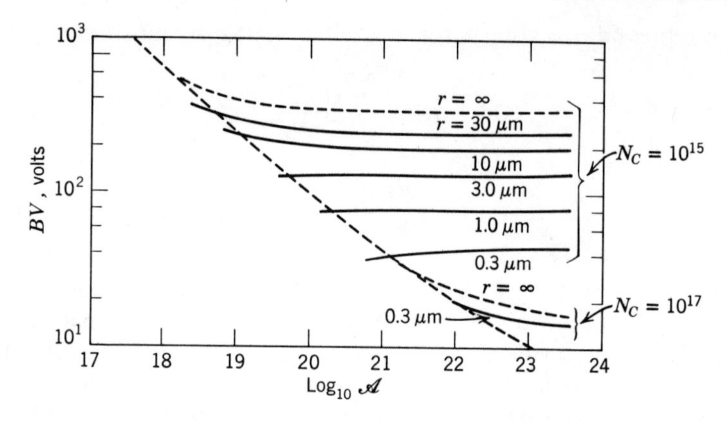

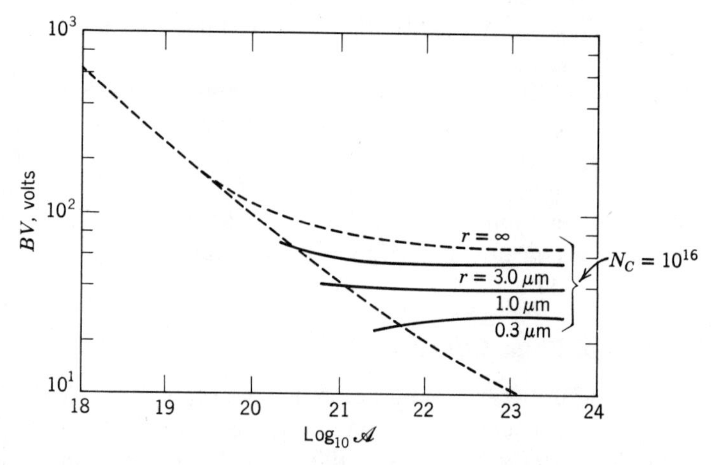

Figure 13.35 Breakdown voltage in diffused cylindrical junctions. (Kennedy and O'Brien [11]).

The $\mathscr{E}$ field is given by

$$\mathscr{E} = \frac{kT}{q}\frac{1}{N}\frac{dN}{dx}. \tag{13.170}$$

Combining these equations with the Einstein relation gives the current due to electron flow as

$$I_n = -\frac{qAD_n}{N}\left(N\frac{dn_P}{dx} + n_P\frac{dN}{dx}\right) \tag{13.171}$$

$$= \frac{qAD_n}{N}\frac{d(Nn_P)}{dx} \tag{13.172}$$

The solution of this equation is

$$Nn_P = \frac{I_n}{qAD_n} \int_C^x N\, dx, \qquad (13.173)$$

where C is a constant of integration.

Let W_P be the width of the p region. If this region is narrow, $n_p \simeq 0$ at $x = W_P$. This condition can be satisfied if the constant of integration is W_P. Making this substitution, (13.173) is solved to obtain

$$n_P = -\frac{I_n}{qAD_n N} \int_x^{W_P} N\, dx. \qquad (13.174)$$

The electron concentration at the edge of the depletion layer (in the p region) is thus given by

$$n_P(0) = -\frac{I_n}{qAD_n N(0)} \int_0^{W_P} N\, dx, \qquad (13.175)$$

where $N(0)$ is the impurity concentration at the edge of the depletion layer (extending into the p region). But from (13.42a)

$$n_P(0) = \bar{n}_P(0)\, e^{qV_A/kT}, \qquad (13.176)$$

where $\bar{n}_P(0)$ is the equilibrium concentration at the edge of the depletion layer (in the p region), and is equal to $n_i^2/N(0)$. Combining these equations,

$$I_n = -\frac{qAD_n n_i^2}{\int_0^{W_P} N\, dx} e^{qV_A/kT}. \qquad (13.177)$$

In like manner, the current due to hole flow may be given as

$$I_p = -\frac{qAD_p n_i^2}{\int_{W_N}^{0} N\, dx} e^{qV_A/kT} \qquad (13.178)$$

where the n region is narrow and has a width of W_N. The total current is thus

$$I = -qAn_i^2 \left(\frac{D_p}{\int_{W_N}^{0} N dx} + \frac{D_n}{\int_0^{W_P} N dx} \right) e^{qV_A/kT} \qquad (13.179)$$

This equation is of the same general form as that for a forward-biased diode in which injection takes place into field-free regions. If either of the regions is long, it must be modified so that the integration is taken over the diffusion length for the minority carrier in question. Note that the negative sign implies that, under conditions of forward bias, the total current flows through the junction from the p region to the n region.

13.4. REFERENCES

[1] H. Kressel et al., "Design Considerations for Double Diffused Silicon Switching Transistors," *RCA Review*, **23**(4), 587–616 (1962).

[2] K. G. McKay, "Avalanche Breakdown in Silicon," *Phys. Rev.* , **94**, 877–884 (1954).

[3] C. T. Sah et al., "Carrier Generation and Recombination in p-n Junctions and *p-n* Junction Characteristics," *Proc. IRE*, **49**, 1228–1243 (1957).

[4] J. Lindmayer, and C. Y. Wrigley, *Fundamentals of Semiconductor Devices*, *Van Nostrand*, New York, 1965.

[5] H. J. Kuno, "Analysis and Characterization of *p-n* Junction Diode Switching," *IEEE Trans. on Electron Devices*, **ED-11**(1), 8–14 (1964).

[6] S. P. Morgan, and F. M. Smits, "Potential Distribution and Capacitance of a Graded *p-n* Junction," *Bell System Tech. J.*, **39**(6), 1573–1602 (1960).

[7] A. B. Phillips, *Transistor Engineering*, McGraw-Hill, New York, (1962).

[8] H. Lawrence and R. M. Warner, Jr., "Diffused Junction Depletion Layer Calculations," *Bell System Tech. J*, **39**(2), 389–403 (1960).

[9] D. P. Kennedy, and R. R. O'Brien, "Avalanche Breakdown Characteristics of a Diffused *p-n* Junction," *IRE Trans. on Electron Devices*, **ED-9**(6), 478–483 (1962).

[10] H. L. Armstrong, "A Theory of Voltage Breakdown of Cylindrical *p-n* Junctions, With Applications," *IRE Trans. on Electron Devices*, **ED-4**(1), 15–16 (1957).

[11] D. P. Kennedy and R. R. O'Brien, "Avalanche Breakdown Calculations for a Planar *p-n* Junction," *IBM J. Res. Dev.*, **10**(3), 213–219 (1966).

[12] J. L. Moll, and I. M. Ross, "The Dependence of Transistor Parameters on Base Resistivity, *Proc. IRE*, **44**(1), 72–78 (1956).

13.5. PROBLEMS

1. An abrupt p-n^+ junction is fabricated so that the peak field in the depletion layer is 10^6 V/cm with no applied reverse voltage. What is the doping concentration of the p region? Sketch the V-I characteristics for this diode in both forward and reverse directions.

2. A p-n^+ diode, 1×1 mil in area, is fabricated from 0.25 Ω-cm silicon. The diode is reverse-biased to 10 V. At time $t = 0$, the diode is turned on by means of a constant current of 1 mA. Determine the time required to bring it to a zero bias condition. Hence determine the average diode capacitance during this turn-on delay period.

3. A transistor is fabricated so that $N_{0E} = 10^{20}$ atoms/cm^3, $N_{0B} = 10^{18}$ atoms/cm^3, $N_C = 10^{16}$ atoms/cm^3, $x_{EB} = 3.0\ \mu$m, and $x_{BC} = 3.5\ \mu$m. Sketch the doping profile for this device and compare it with that shown in Figure 13.1.

4. The relation $\alpha_i = ae^{-b/\mathscr{E}}$, where $a = 9 \times 10^5$ and $b = 1.8 \times 10^6$, provides a better approximation to the average ionization constant than the power series form of (13.25). Using this expression, determine a more accurate relation for the avalanche voltage of an abrupt junction p-n^+ diode.

5. Plot the small-signal impedance of a long-base p-n^+ junction over the full frequency range. Compare your result with that of Figure 13.20.

6. A narrow base p-n^+ diode is turned on from a constant-current source. Determine an expression that describes the growth of stored charge in the diode as a function of time. Assuming that the minority carrier distribution is a linear function of distance at all times, develop an expression for the buildup of voltage across the diode.

7. A p-n^+ diode is switched as shown in Figure 13.22*a*. Develop the expression for the fall time, assuming that the diode has an average capacitance $\bar{C}_t$ during this phase.

8. The collector-base junction of a transistor is made by diffusing boron into a background concentration of 10^{16} atoms/cm^3. The surface concentration of the boron diffusion is 10^{18} atoms/cm^3 and the junction depth is 3.5 μm. Determine the junction capacitance per unit area for a reverse bias of 10 V, assuming a linearly graded structure. Compare your answer with that obtained from Figure 13.30 for a diffused junction.

9. Compute the breakdown voltage of the junction described in Problem 8, assuming a linearly graded structure. Compare you answer with that obtained from Figure 13.32 for a parallel-plane diffused junction and also with that for a junction made by oxide masking techniques.

10. A p-n^+ diode is biased in the forward direction to a constant current. Show that the voltage drop across the diode falls linearly with increasing temperature at a rate of about 1.8 mV/°C.

11. The transistor of Problem 3 is fabricated with an emitter area of 3×0.25 mil. Compute the capacitance of the emitter-base diode at a forward bias of 5 mA. Assume a narrow-base structure and do not neglect the emitter sidewall area.

Chapter 14

The Transistor

The bipolar junction transistor is the most important element in a microcircuit. As a consequence the choice of basic doping profiles for the microcircuit is determined by the requirements of the transistors to be used in it. Other elements, such as resistors and capacitors, are subsequently designed to use these same doping profiles.

This chapter describes the properties of the bipolar junction transistor. The theory is first developed for the uniform-base structure, and then is extended to the double-diffused structure encountered in modern microcircuits fabricated by the DDE process. Throughout this chapter the term *transistor* is used to describe this device for the sake of brevity.

14.1. TRANSISTOR ACTION

The transistor is a charge-controlled device. Fundamental to any device of this type is a source of carriers, an interaction region through which these carriers must pass, and a sink at which they must be collected. The flow of carriers between source and sink is controlled in the interaction region by means of an input signal. If the power level of the controlling signal is lower than that of the controlled carriers, the device is said to have power gain.

In a transistor the base region provides the interacting medium, and the type of carrier utilized is the minority carrier. Thus for proper operation a transistor requires an efficient means by which minority carriers may be injected into this base. Next it is necessary to have an effective interaction mechanism between these carriers and the control. Finally means must be provided by which only minority carriers are collected, resulting in a controlled current in the external circuit.

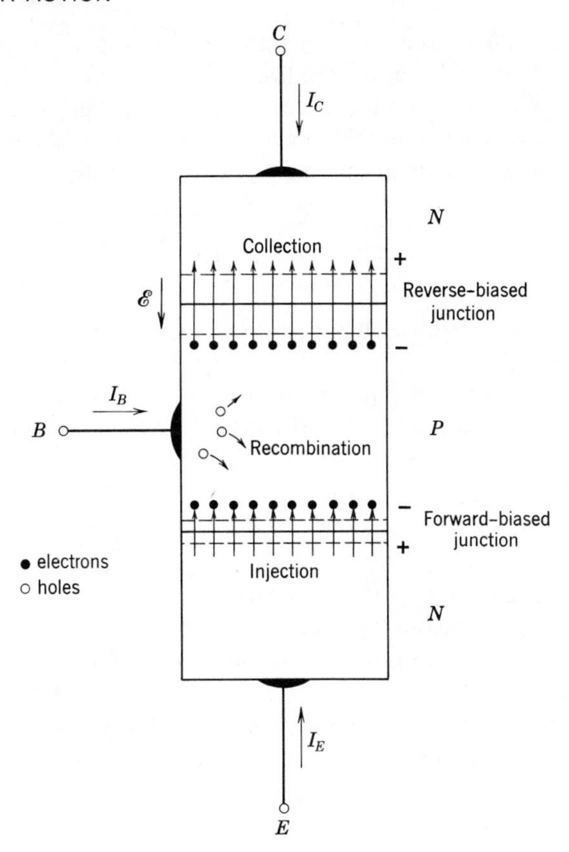

Figure 14.1 The *n-p-n* transistor.

Figure 14.1 shows an *n-p-n* transistor. In this device a forward-biased emitter-base junction serves to inject minority carriers (electrons) into the base region. Upon traveling through this region they arrive at the edge of the reverse-biased collector-base depletion layer. Both majority and minority carriers are present here; however, only minority carriers are swept through the depletion layer by the strong $\mathscr{E}$ field and are collected. The current flows in the external circuit via ohmic contacts to the emitter and collector regions.

An ohmic contact is also made to the base region and provides for the introduction of a control signal. This contact is capable of behaving as an efficient source or sink for majority carriers. Interaction between these majority carriers and the minority carriers injected from the emitter is a direct consequence of the *principle of space-charge neutrality*. Thus if majority carriers are introduced into the base region by means of the ohmic contact, the emitter current must adjust itself correspondingly to preserve space-charge neutral conditions in this base.

In operation there is a finite excess minority carrier concentration in the base. If n'_B is the steady-state value of this concentration, the magnitude of the excess charge in this region is given (for the one-dimensional case) by

$$Q_B = \int_0^{W_B} qn'_B \, dx, \tag{14.1}$$

where W_B is the effective base width between the edges of the two depletion layers.

These excess carriers will recombine every τ_{nB} seconds, where τ_{nB} is the electron lifetime in the base. Thus the rate of the flow of majority carriers into the base (i.e., the base current) required to preserve space-charge neutrality is given by

$$I_B = \frac{Q_B}{\tau_{nB}}, \tag{14.2}$$

where I_B is the magnitude of the base current.

Minority carriers in the base move by diffusion or by a combination of diffusion and drift. Let t_t be the transit time for these carriers from the emitter to the collector. On the average, then, a charge of Q_B is delivered to the external circuit in this period of time. Thus the magnitude of the output current is given by

$$I_C = \frac{Q_B}{t_t} \tag{14.3}$$

It follows therefore that

$$\frac{I_C}{I_B} = \frac{\tau_{nB}}{t_t} \tag{14.4}$$

Consequently, the low-frequency common-emitter current gain is given by

$$\beta_0 = \frac{\partial I_C}{\partial I_B} = \frac{\tau_{nB}}{t_t} \tag{14.5}$$

To obtain a reasonable value for this gain, it is necessary that $t_t \ll \tau_{nB}$. This can only be achieved if the base width W_B is made considerably shorter than L_{nB}, the diffusion length of minority carriers in the base. Thus for proper transistor action it is necessary that the emitter-base junction be an efficient injector of minority carriers into the base region and that the base region be narrow, that is, $W_B \ll L_{nB}$.

In normal operation the collector-base junction is reverse-biased. This prevents the injection of minority carriers into the base from the collector side. Thus the collector acts as an effective sink for minority carriers.

The low-frequency common-base current gain is defined by

$$\alpha_0 \equiv -\frac{\partial I_C}{\partial I_E} \tag{14.6a}$$

Since the sign convention for external currents has been taken as shown in Figure 14.1, $I_B + I_C + I_E = 0$. Thus

$$\alpha_0 = -\frac{\partial I_C}{\partial I_E} = \frac{\beta_0}{1+\beta_0} \tag{14.6b}$$

Figures 14.2*a* and 14.2*b* show the static characteristics of an *n-p-n* transistor in the common-base configuration. It is seen that the input characteristics closely resemble those of a forward-biased narrow-base diode. The output characteristics, on the other hand, are similar to those of a reverse-biased diode. The constant displacement between curves for equal increments of emitter current is due to the near-unity current gain for this configuration.

Figures 14.2*c* and 14.2*d* show the family of characteristics for the common-emitter configuration. These characteristics may be derived from those of Figures 14.2*a* and 14.2*b*; they are of a generally similar nature, the essential differences being caused by the large value of common-emitter current gain.

14.2. THE UNIFORM-BASE TRANSISTOR

This transistor is one in which the base region is uniformly doped. Thus there is no $\mathscr{E}$ field in this region for low-level injection conditions. For simplicity the emitter and collector regions are also considered uniformly doped, resulting in two abrupt *p-n* junctions placed back to back, with a narrow base region in common.

Figure 14.3*a* shows an *n-p-n* transistor of this type, biased into its active region (emitter-base diode forward-biased and collector-base diode reverse-biased). The various components of particle flow across the junctions are shown in this figure together with the associated current components. These components are as follows:

1. Electrons which are injected into the base region and give rise to a component of emitter current, I_{nE}.
2. Electrons which arrive at the edge of the collector-base depletion layer; are collected, and give rise to a component of collector current, I_{nC}.
3. Holes which are injected from the base into the emitter and result in a component of emitter current, I_{pE}.
4. Holes which are injected from the collector into the base. These result in the diffusion component of the reverse-biased collector-base

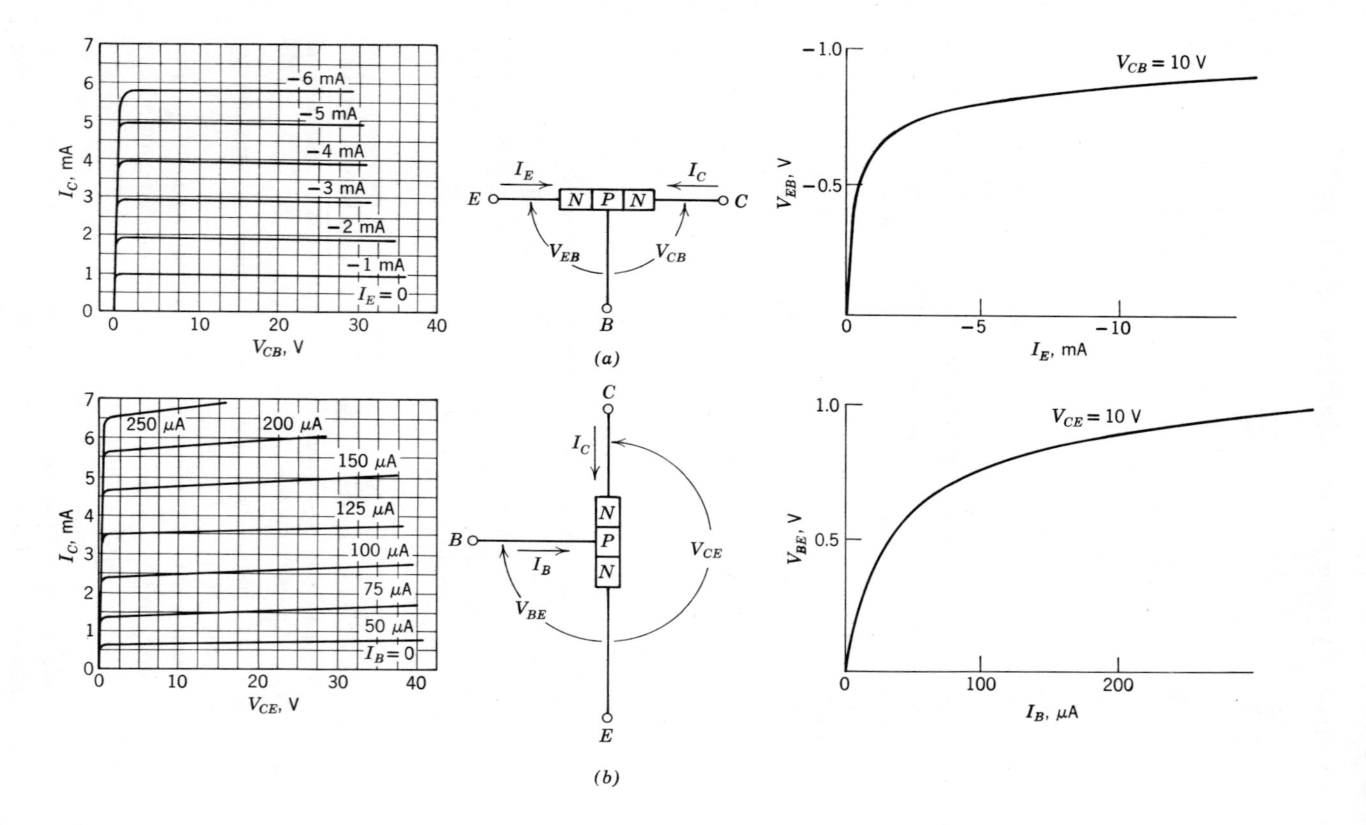

Figure 14.2 The *n-p-n* transistor characteristics.

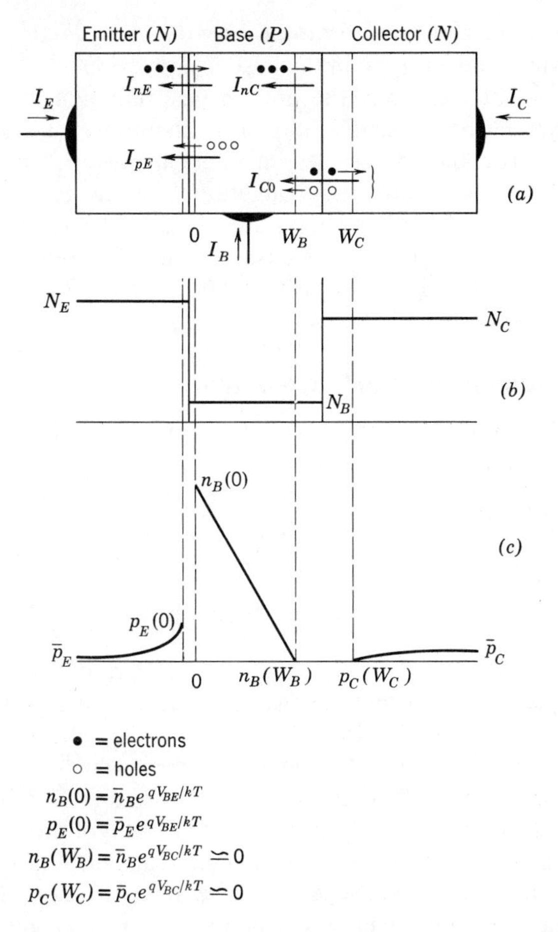

Figure 14.3 Currents and carrier concentrations in an *n-p-n* transistor.

junction current. As shown in Section 13.1.2.2, this component may be ignored in comparison with the charge-generation current.

5. A charge-generation current caused by the drift of electrons and holes generated in the collector-base depletion layer. This gives rise to a leakage-current term I_{C0}. Its magnitude is determined solely by the nearly constant collector-base voltage and by material properties [see (13.20)]. Thus it is not altered by changes in the diffusion-current components and does not affect the a-c characteristics of the transistor.

6. A charge-generation current is also established by the drift of carriers generated in the emitter-base depletion layer. This current is usually negligible for typical values of forward-current flow.

The minority carrier concentrations are shown in Figure 14.3*c*. Note that the carrier concentration in the base is approximately linear because this region is narrow. Thus the emitter-base junction is essentially a forward-biased narrow-base diode; many of its properties are identical to those of the narrow-base diodes described in Chapter 13.

Figure 14.3*a* also shows the convention for the terminal currents, which are defined as positive when flowing into the transistor. In normal operation the actual collector and base currents flow into an *n-p-n* transistor; thus, $I_B > 0$ and $I_C > 0$. On the other hand, the actual emitter current flows *out of* an *n-p-n* transistor; thus $I_E < 0$.

14.2.1. The Common-base Current Gain

The low-frequency current gain of the transistor in the common-base configuration is given by

$$\alpha_0 \equiv -\frac{\partial I_C}{\partial I_E} \tag{14.7}$$

$$= \frac{-\partial I_{nE}}{\partial I_E}\frac{\partial I_{nC}}{\partial I_{nE}}\frac{\partial I_C}{\partial I_{nC}}. \tag{14.8}$$

The first of these terms, $-\partial I_{nE}/\partial I_E$, is a measure of the ability of the emitter to inject electrons into the base and is termed the *emitter injection efficiency* γ. The second term, $\partial I_{nC}/\partial I_{nE}$, is called the *base transport factor* α_T and is a measure of the electron current which crosses the base without recombination. The third term, $\partial I_C/\partial I_{nC}$, is a measure of the *collection efficiency* of the collector; i.e. of the ratio of the total collector current to the electron component of collector current. This term is somewhat greater than unity, because of secondary ionization effects in the collector-base depletion layer. Its magnitude is denoted by M. The relative importance of these various terms is now considered.

14.2.1.1. The Emitter Injection Efficiency. The emitter injection efficiency is given by

$$\gamma = \frac{\partial I_{nE}}{\partial (I_{nE} + I_{pE})} \tag{14.9}$$

where I_{nE} and I_{pE} are the components of the emitter current due to electrons and holes respectively, as shown in Figure 14.3*a*. For a narrow-base diode biased in the forward direction.

$$I_{nE} = \frac{qAD_{nB}\bar{n}_B}{W_B}(e^{qV_{BE}/kT} - 1). \tag{14.10a}$$

The emitter region is long compared to the diffusion length of holes. Hence

$$I_{pE} = \frac{qAD_{pE}\bar{p}_E}{L_{pE}}(e^{qV_{BE}/kT} - 1), \tag{14.10b}$$

where A = cross-sectional area of the diode,
D_{nB} = diffusion constant of electrons in the base,
D_{pE} = diffusion constant of holes in the emitter,
$\bar{n}_B$ = equilibrium concentration of electrons in the base,
$\bar{p}_E$ = equilibrium concentration of holes in the emitter,
W_B = base width,
L_{pE} = diffusion length of holes in the emitter,
V_{BE} = voltage across the emitter-base junction (a positive quantity for operation in the active region).

Combining (14.10*a*) and (14.10*b*) gives

$$\frac{I_{nE}}{I_{nE}+I_{pE}} = \left(1 + \frac{D_{pE}}{D_{nB}}\frac{\bar{p}_E}{\bar{n}_B}\frac{W_B}{L_{pE}}\right)^{-1}, \tag{14.11}$$

so that

$$\gamma = \frac{\partial I_{nE}}{\partial(I_{nE}+I_{pE})}, \tag{14.12}$$

$$= \left(1 + \frac{D_{pE}}{D_{nB}}\frac{\bar{p}_E}{\bar{n}_B}\frac{W_B}{L_{pE}}\right)^{-1}. \tag{14.13}$$

If† the mobilities of the emitter and the base are assumed equal, i.e., if $\mu_{pE} \simeq \mu_{pB}$ and $\mu_{nE} \simeq \mu_{nB}$, (14.13) reduces to

$$\gamma \simeq \left(1 + \frac{\sigma_B}{\sigma_E}\frac{W_B}{L_{pE}}\right)^{-1}, \tag{14.14}$$

where σ_B and σ_E are the conductivities of the base and emitter respectively. Thus for high injection efficiency, it is necessary that $\sigma_E \gg \sigma_B$. In addition, the base width should be narrow compared to the diffusion length of holes in the emitter. The latter condition is not always readily met, since a high conductivity generally goes hand in hand with a short minority carrier diffusion length. In practice, however, typical values of γ range from 0.98 to 0.99.

†A reasonable assumption, since both regions are extrinsic.

14.2.1.2. The Base Transport Factor. Both I_{nC} and I_{nE} are proportional to the slope of the minority carrier concentration in the base at the edge of the respective depletion layers [see (13.54a)]. Hence

$$\frac{I_{nC}}{I_{nE}} = \frac{(\partial n_B/\partial x)_{x=W_B}}{(\partial n_B/\partial x)_{x=0}}. \tag{14.15}$$

The electron concentration in the base is given by [see (13.53)]

$$n_B - \bar{n}_B = \bar{n}_B(e^{qV_{BE}/kT} - 1)\,\frac{\sinh\{(W_B - x)/L_{nB}\}}{\sinh(W_B/L_{nB})} \tag{14.16}$$

Substituting into (14.15) results in

$$\frac{I_{nC}}{I_{nE}} = \left(\cosh\frac{W_B}{L_{nB}}\right)^{-1}, \tag{14.17}$$

Consequently the base transport factor α_T is given by

$$\alpha_T \equiv \frac{\partial I_{nC}}{\partial I_{nE}} = \left(\cosh\frac{W_B}{L_{nB}}\right)^{-1}, \tag{14.18}$$

$$\simeq \left(1 + \frac{W_B^2}{2L_{nB}^2} + \ldots\right)^{-1} \tag{14.19}$$

For $W_B \leqslant 0.1L_{nB}$, $\alpha_T \geqslant 0.995$. In double-diffused structures the base width is usually considerably shorter than L_{nB}; thus for all practical purposes $\alpha_T \simeq 1$.

14.2.1.3. The Collector Collection Efficiency. In an n-p-n transistor some of the injected electrons eventually arrive at the reverse-biased collector-base depletion layer where they come under the influence of a strong $\mathscr{E}$ field. Because of secondary ionization effects, the current arriving at the collector is M times that entering the depletion layer, where [see (13.31)],

$$M = \left[1 - \left(\frac{V_{CB}}{BV}\right)^m\right]^{-1}. \tag{14.20}$$

Here, BV is the breakdown voltage and V_{CB} is the voltage across the reverse-biased collector-base diode (a positive quantity for operation in the active region). The collection efficiency $\partial I_C/\partial I_{nC}$ is thus given by

$$\frac{\partial I_C}{\partial I_{nC}} = M = \left[1 - \left(\frac{V_{CB}}{BV}\right)^m\right]^{-1}. \tag{14.21}$$

In normal operation, at voltages well below the breakdown voltage, this term may be taken as unity. With increasing voltage, however, M increases very rapidly. Thus the current gain rises with increasing collector-base voltage until breakdown is ultimately reached.

14.2.2. The Common-emitter Current Gain

The low-frequency common-emitter current gain is obtained from (14.6*b*) as

$$\beta_0 = \frac{\alpha_0}{1-\alpha_0} \tag{14.22}$$

Since $M \simeq 1$, the quantity $1/\beta_0$ may be rewritten in the form

$$1/\beta_0 \simeq \frac{1-\gamma\alpha_T}{\gamma\alpha_T} \tag{14.23}$$

For values of γ and α_T close to unity, as is the case in practice,

$$\frac{1}{\beta_0} \simeq \frac{1-\gamma}{\gamma} + \frac{1-\alpha_T}{\alpha_T} \tag{14.24}$$

Direct substitution into (14.14) and (14.19) gives

$$\frac{1}{\beta_0} \simeq \frac{\sigma_B W_B}{\sigma_E L_{pE}} + \frac{W_B{}^2}{2L_{nB}^2} \tag{14.25}$$

In transistors where α_T approaches unity,

$$\beta_0 \rightarrow \frac{\sigma_E L_{pE}}{\sigma_B W_B} \tag{14.26}$$

The quantity $\sigma_B W_B$ is the sheet conductance† of the base region. In like manner, the quantity $\sigma_E L_{pE}$ is the sheet conductance of an emitter region for a thickness of one diffusion length (as measured from the edge of the depletion layer). Thus the common-emitter current gain cannot exceed the ratio of the sheet conductances of the effective emitter and base regions.

The reciprocal of the common-emitter current gain may also be written in terms of the individual current components. Since the leakage-current term can be neglected in comparison with other terms, and since both M and α_T are close to unity,

$$I_C \simeq I_{nC} \simeq I_{nE} \tag{14.27}$$

†The sheet resistance is the reciprocal of this term and has been defined in Section 4.9.1.

In addition, the base current is given by

$$I_B \simeq I_{pE} + I_{nE} - I_{nC}. \tag{14.28}$$

Thus

$$\frac{I_B}{I_C} \simeq \frac{I_B}{I_{nE}} \tag{14.29}$$

$$\simeq \frac{I_{pE}}{I_{nE}} + \frac{I_{nE} - I_{nC}}{I_{nE}}. \tag{14.30}$$

The first of the terms on the right-hand side is associated with the emitter injection efficiency, whereas the second is associated with the volume recombination current in the base. Writing this current as I_{rB},

$$\frac{I_B}{I_C} \simeq \frac{I_{pE}}{I_{nE}} + \frac{I_{rB}}{I_{nE}} \tag{14.31}$$

which gives the reciprocal of the common-emitter current gain upon differentiation.

The separation of the various contributing terms in this manner provides a useful means for visualizing the effect of each on the current gain. It addition, it allows the ready inclusion of other effects which may be important in specific situations.

14.2.2.1. *Fall-off at Low Current Levels.* The forward current of the emitter-base diode consists of a charge recombination term in addition to a diffusion term [see (13.66)]. Since the recombination current does not contribute to the minority carrier injection, we may write the common emitter current gain as

$$\frac{1}{\beta_0} = \frac{\partial I_B}{\partial I_{nE}}, \tag{14.32}$$

where

$$\frac{I_B}{I_{nE}} = \frac{I_{pE}}{I_{nE}} + \frac{I_{rE}}{I_{nE}} + \frac{I_{rB}}{I_{nE}}. \tag{14.33}$$

Here I_{rE} is the recombination current in the emitter depletion layer, and has been shown to be given by [see (13.66)]

$$I_{rE} = \frac{qAn_i x_0}{\tau_{p0} + \tau_{n0}} e^{qV_{BE}/2kT}, \tag{14.34}$$

where x_0 is the thickness of the depletion layer. The corresponding diffusion current is

$$I_{nE} = \frac{qAD_{nB}n_i^2}{W_B \bar{p}_B}(e^{qV_{BE}/kT} - 1). \tag{14.35}$$

Note that both of these current components are functions of the emitter-base voltage. At low forward currents, however, charge recombination in a silicon *p-n* junction dominates the forward-current characteristic [1]. Hence I_{re}/I_{nE} increases with falling current while the other terms remain constant. This results in a fall-off in the common-emitter current gain at low current levels.

Surface recombination effects are also present in transistors and contribute to this fall-off. The manner in which they affect the current gain will be described in Chapter 17.

14.2.2.2. *Fall-off at High Current Levels.* At high injection levels the minority carrier concentration in the base increases to the point where the concurrent change in majority carrier concentration becomes significant. This leads to an increase in the sheet conductance of the base region. A similar effect is present for the emitter region; however, this region is much more highly doped and the conductance change is insignificant. As a consequence, the injection efficiency of the transistor decreases with increasing forward current, leading to a fall-off in the current gain characteristic [2].

In addition to conductivity modulation effects an $\mathscr{E}$ field is established in the base region as a result of the variation of carrier concentration with distance. As explained in Section 13.1.2.4, this field aids the diffusion of minority carriers; in the limit it results in an effective doubling of the diffusion constant in the base region.

Finally, it is shown that the current distribution in the emitter becomes nonuniform at high injection levels. This crowding results in reinforcing the conductivity modulation effect in localized regions and produces a premature decrease in the current gain. We will consider these effects separately, recognizing that it is their combined behavior that results in the over-all variation of current gain at high injection levels.

Under low-level injection conditions,

$$\frac{I_B}{I_{nE}} \simeq \frac{I_{pE}}{I_{nE}} \tag{14.36}$$

$$\simeq \frac{\sigma_B W_B}{\sigma_E L_{pE}} \tag{14.37}$$

With increasing minority carrier injection into the base the majority carrier concentration rises to preserve charge neutrality. Thus for an injected excess electron concentration of n_B', the average majority carrier concentration is now given by

$$p_B \simeq N_B + \frac{n_B'}{2} \tag{14.38}$$

where N_B is the impurity concentration in the base. The average minority carrier concentration in the base is $n'_B/2$. Thus the effective base conductivity σ'_B is approximately given by

$$\sigma'_B \simeq \sigma_B\left(1+\frac{n'_B}{N_B}\right). \tag{14.39}$$

Ignoring for the present the concurrent change in the diffusion constant due to the increased injection level,

$$\frac{I_B}{I_{nE}} \simeq \frac{\sigma_B W_B}{\sigma_E L_{pE}}\left(1+\frac{n'_B}{N_B}\right) \tag{14.40}$$

A linear relation exists between I_{nE} and n'_B. Hence

$$\frac{\partial I_{nE}}{\partial n'_B} = \frac{I_{nE}}{n'_B} \tag{14.41}$$

Differentiating (14.40) and substituting (14.41),

$$\frac{1}{\beta_0} \simeq \frac{\sigma_B W_B}{\sigma_E L_{pE}}\left(1+\frac{2n'_B}{N_B}\right) \tag{14.42}$$

Thus the common-emitter current gain begins to fall off once the localized injected carrier concentration becomes comparable to one-half the impurity concentration in the base.

The effect of an altered diffusion constant may also be considered. On the one hand, the increase in the base carrier concentration results in a lower diffusion constant. On the other hand, however, the presence of the $\mathscr{E}$ field serves to increase its value. The combined effects are quite difficult to evaluate. It has been shown[2], however, that the error resulting from ignoring these effects is on the order of 20% or less in practical situations.

An important effect is that of emitter-current crowding[3]. In transistor operation the majority carrier base current flows laterally through this base, at right angles to the direction of minority carrier current flow. This is shown in Figure 14.4*a*. It is seen that the flow of this majority carrier current results in a potential drop in the lateral direction. Thus different regions of the emitter-base junction are biased to a different forward voltage. This, in turn, gives rise to uneven injection† over the area of the

†Note that a lateral ohmic drop of about 26 mV results in a reduction of the longitudinal injected current by a factor of 1/2.718.

emitter, with the minority carrier density falling off as we proceed from the edge of the emitter nearest the base contact and go deeper into the active region of the transistor. For a transistor geometry of the type shown in Figure 14.4*a*, the majority of the injection occurs from the edge of the emitter nearest to the base contact. In addition, the use of a second base contact effectively doubles the injecting area of the emitter.

The degree of emitter crowding may be determined by considering the highly idealized transistor of Figure 14.4*b*. Only one-half of the lateral dimensions of this transistor are shown. It is assumed that:

1. The emitter region represents an equipotential plane.
2. The entire base contact as well as the region between it and the edge of the emitter (shown cross-hatched in the figure) is considered to be an equipotential region.
3. The current gain is constant for the device over the whole region.

The emitter region is taken to have a length l_E and a half-width $w_E/2$. For high-level injection conditions the injected emitter current density at any point along the emitter width is given by

$$J(y) = J_0 e^{qV_{BE}(y)/2kT}, \tag{14.43}$$

where $V_{BE}(y)$ is the emitter-base voltage, and is a positive quantity. Differentiating,

$$dJ(y) = \frac{q}{2kT} J(y)\, dV_{BE}(y). \tag{14.44}$$

The effective conductivity of the base may be written by analogy with (14.39) as

$$\sigma_B' = \sigma_B \left[1 + \frac{J(y)}{J_1} \right] \tag{14.45}$$

where J_1 is the current density at which high-level injection begins. Since our interest is in the high-level injection case, (14.45) reduces to

$$\sigma_B' \simeq \frac{\sigma_B J(y)}{J_1}. \tag{14.46}$$

The base current at any point y is equal to the total base current minus the base current flowing up to this point. Thus

$$I_B(y) = I_B - \int_0^y (1-\alpha_0) J(y) l_E\, dy, \tag{14.47}$$

$$dI_B(y) = -(1-\alpha_0) J(y) l_E\, dy. \tag{14.48}$$

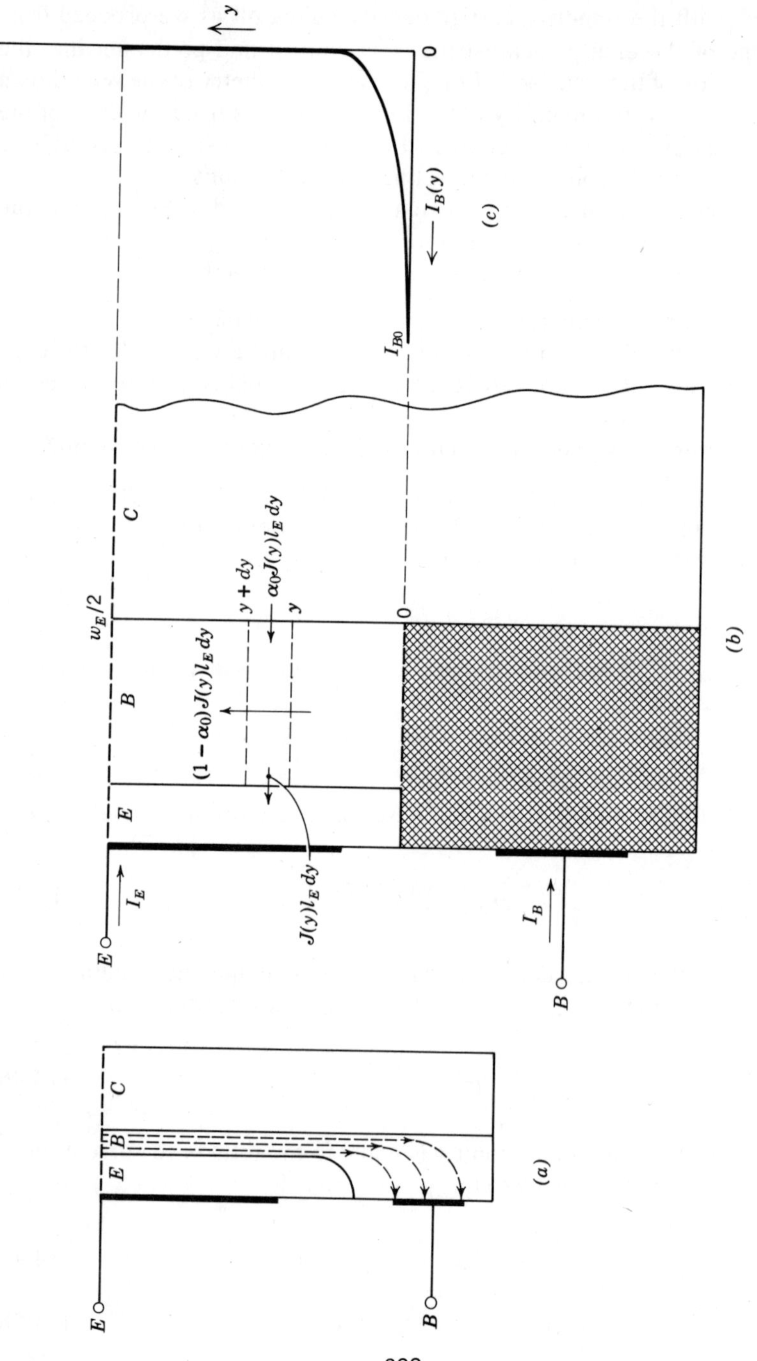

Figure 14.4 Emitter-crowding effects.

Finally, the change in the emitter-base voltage at any point is given by

$$dV_{BE}(y) = -\frac{I_B(y)}{\sigma_B' W_B l_E} dy. \tag{14.49}$$

This assumes a uniform lateral flow of current through the base region. In practice, the lateral current density is nonuniform, with the greater proportion of the current flow occurring near the emitter region. By writing the effective base width as W_B', (14.49) becomes

$$dV_{BE}(y) = -\frac{I_B(y)}{\sigma_B' W_B' l_E} dy. \tag{14.50}$$

Combining (14.44), (14.47), and (14.50),

$$\frac{dJ(y)}{dy} = -\frac{q}{2kT} \frac{I_B(y) J_1}{\sigma_B W_B' l_E} \tag{14.51}$$

Differentiating (14.48) and substituting into (14.51),

$$\frac{d^2 I_B(y)}{dy^2} = \frac{q}{2kT} \frac{J_1(1-\alpha_0)}{\sigma_B W_B'} I_B(y). \tag{14.52}$$

Solution of this equation results in a base current distribution of the exponentially decaying form shown in Figure 14.4*c*, with the majority of the current crowded into a region of thickness y_0, where

$$y_0 = \left[\frac{2kT}{q} \frac{W_B' \sigma_B}{J_1(1-\alpha_0)}\right]^{1/2} \tag{14.53}$$

Thus it follows that injection from the emitter will be predominantly an edge effect. It is interesting to note that, once high-level injection conditions are achieved, the effective width of the emitter injection region becomes independent of current density.

Most transistors are commonly operated under high-level injection conditions. For these conditions the effective width of the injection region is typically 1 to 5 μm for high-speed devices.

Although the above analysis is relatively crude, it serves to illustrate the following points:

1. Injection from the emitter is predominantly an edge effect and occurs from the edge that is closest to the edge of the base contact.
2. The concentration of current due to this edge injection results in high-level injection conditions being attained prematurely. Thus this effect

tends to cause an early decrease in current gain owing to a localized fall-off in injection efficiency.

3. In a well-designed transistor the emitter should be so shaped as to maximize its perimeter/area ratio. The simplest scheme consists of using a long, thin-strip structure, with one or more emitter strips alternating with base contact strips.† This is the well-known interdigitated structure in common use today.

A rough estimate may be made of the emitter current at which high-level injection starts. Consider a base impurity concentration of $10^{16}/\text{cm}^3$. Then the injected carrier concentration for high-level injection is about $5 \times 10^{15}/\text{cm}^3$. Assuming that current crowding is confined to a 5-μm wide edge region, that the base width is 0.5 μm, and that $D_{nB} \simeq 25$, high-level injection occurs when the forward current through the emitter-base diode exceeds 5 mA/mil of emitting periphery.

14.2.3. Parasitic Elements

Up to this point, device properties have been based on the idealized carrier concentrations shown in Figure 14.3*c*. We now consider a number of important elements which play a role in modifying these device properties. These elements are referred to as *parasitics*, since they are present in a real transistor and result in deteriorating its performance. They are a function of the physical dimensions and layout of the device, in addition to its doping profile. Thus the fabrication process used plays an important role in determining their values.

The inductive effects of the aluminum metallization are usually insignificant and are ignored. Inductive effects of the internal connections between the microcircuit and its package are either ignored or else combined with those of the external wiring. Thus the usual parasitics considered as part of the microcircuit are either resistive or capacitive in nature.

14.2.3.1. Capacitances. These are the transition capacitances of the emitter-base diode and the collector-base diode respectively. At high frequencies they serve as shunting paths across regions of the ideal transistor, leading to a degradation of device performance.

Consider, for example, the emitter-base transition capacitance C_{tE}. This capacitance provides a shunt path across the emitter-base junction hence reduces the injection efficiency at high frequencies. Consequently, its presence leads to a fall off in current gain at high frequencies. For a

†It is interesting to note that the circular emitter geometry is the worst from this point of view.

uniform-base *n-p-n* transistor, in which $\sigma_E \gg \sigma_B$, the emitter transition capacitance is given by [see (13.12*b*)]

$$C_{tE} = A\left(\frac{\epsilon\epsilon_0 q N_B}{2V_j}\right)^{1/2}, \tag{14.54}$$

where V_j is the emitter-base junction potential. Since this junction is forward-biased,

$$V_j = \Psi_E - V_{BE}, \tag{14.55}$$

where Ψ_E is the contact potential associated with the emitter-base junction, and ohmic drops are ignored.

The collector-base junction is operated under reverse-bias conditions. Again ignoring ohmic drops and assuming a contact potential Ψ_C for the collector-base junction, its transition capacitance is given by

$$C_{tC} = A\left(\frac{\epsilon\epsilon_0 q N_B}{2V_j}\right)^{1/2}, \tag{14.56}$$

where

$$V_j = V_{CB} + \Psi_C. \tag{14.57}$$

A heavily doped collector is assumed for this calculation, since this is the usual situation encountered in abrupt-junction, uniformly doped devices. The effect of the contact potential may be ignored in determining C_{tC} because V_{CB} is usually much larger than this term.

14.2.3.2. Resistances. A parasitic resistance is associated with each of the transistor regions. In general, this resistance combines with the parasitic capacitance in deteriorating the high-frequency performance.

The emitter is the most highly doped region of a transistor. Consequently, it is customary to ignore the parasitic resistance r_E associated with it, except in special low-level switching applications, where even the resistance of the bonding leads (0.1 to 0.3 Ω) must be considered.

The parasitic resistance of the collector, r_C, is the ohmic resistance associated with this region. For an epitaxial structure of the type shown in Figure 14.5*a*, it has been empirically found that a reasonable approximation to r_C results from assuming a uniform flow of current from the emitter to the collector, as shown. Thus

$$r_C \simeq \frac{\rho_C W_C}{A_E} \tag{14.58}$$

where ρ_C = the collector resistivity,
W_C = the distance from the collector-base junction to the edge of the heavily doped substrate,
A_E = the area of the emitter.

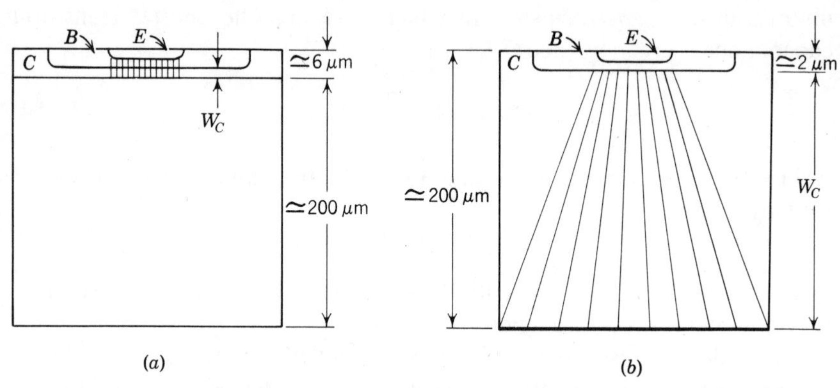

Figure 14.5 Parasitic collector resistance.

This expression ignores both emitter crowding effects and collector current spreading effects. Fortuitously, these effects tend to cancel in a structure of this type.

For the nonepitaxial structure shown in Figure 14.5*b* it is customary to assume a uniform spreading of current from the emitter to the collector, as shown in Figure 14.5*b*. For this situation

$$r_C \simeq \frac{\rho_C W_C}{(A_E + A_C)/2} \tag{14.59}$$

Here W_C is now the distance from the collector-base junction to the collector contact of area A_C.

The approximate nature of these expressions cannot be too highly emphasized. Although accurate spreading-resistance calculations have been made for specific geometries[4], they ignore the effects of emitter crowding and provide results that are no better than the crude approximation made here.

The parasitic resistance of the base r_B consists of two parts—the intrinsic base resistance r_{B1}, which comprises the region under the emitter in which both longitudinal and lateral currents flow, and the extrinsic part r_{B2}, in which only lateral current flows.

The computation of the extrinsic resistance is relatively straightforward. For the half-transistor shown in Figure 14.6, the extrinsic resistance to a single base contact is the resistance of the shaded region. This is given approximately by $d_{EB}/\sigma_B l_E W_2$, where σ_B is the conductivity of the base region, l_E is the length of the emitter, d_{EB} is the distance between the edge of the emitter and the base contact, and W_2 is the thickness of the base region between the base contact and the collector base junction.

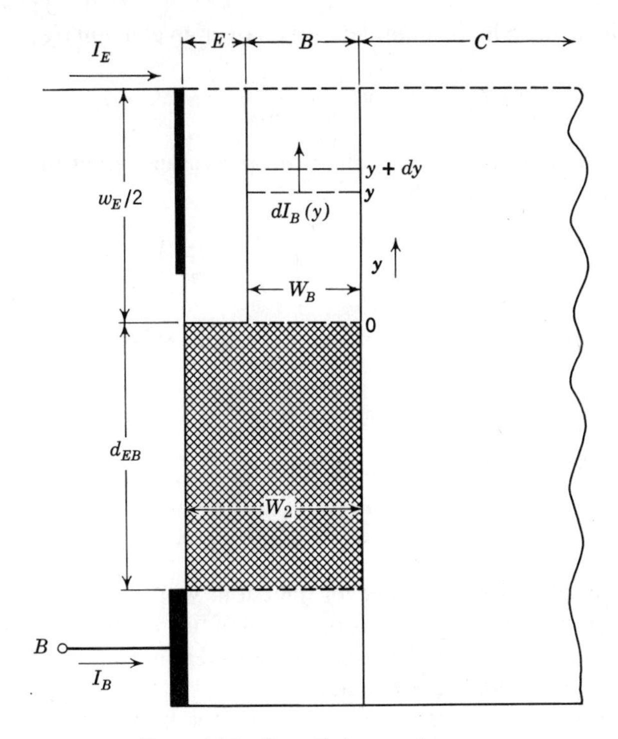

Figure 14.6 Parasitic base resistance.

If the complete transistor has two base contacts, this resistance is effectively halved, so that

$$r_{B2} \simeq \frac{d_{EB}}{2\sigma_B l_E W_2}. \tag{14.60}$$

For low-level injection emitter crowding does not occur and the entire emitter area behaves as an injection source for minority carriers. The intrinsic base resistance is now calculated for this condition of operation. Consider again[5] the half-transistor of Figure 14.6. The emitter and base currents flowing in this half are labeled I_E and I_B respectively, and injection from the emitter is assumed to be uniform.

The base current at any point y is given by

$$I_B(y) = I_B - \int_0^y - I_E(1-\alpha_0)\,dy. \tag{14.61}$$

Solving (14.61) subject to the condition that $I_B(y) = 0$ at $y = w_E/2$,

$$I_B(y) = I_B\left(1 - \frac{2y}{w_E}\right) \tag{14.62}$$

The voltage drop in the lateral direction over an element dy is

$$dV_B(y) \simeq \frac{I_B(y)}{\sigma_B W_B l_E} dy. \tag{14.63}$$

The mean value of this lateral voltage drop, averaged over the half-width of the emitter, is thus given by

$$\bar{V}_B = \frac{I_B}{\sigma_B W_B l_E (w_E/2)} \int_0^{w_E/2} \int_0^z \left(1 - \frac{2y}{w_E}\right) dy\, dz, \tag{14.64}$$

where z is introduced as a variable of integration. Solving,

$$\bar{V}_B = \frac{I_B w_E}{6\sigma_B W_B l_E}. \tag{14.65}$$

Thus the average intrinsic base resistance is

$$r_{B1} = \frac{\bar{V}_B}{I_B} = \frac{w_E}{6\sigma_B W_B l_E}. \tag{14.66}$$

This calculation assumes low-level injection, with no current crowding under the emitter. Note that r_{B1} is one-third of the intrinsic resistance that would be obtained if *all* the base current flowed uniformly from the *center* of the base region to the base contact.

The intrinsic base resistance of a transistor with a base contact on either side of the emitter region is one-half that given by (14.66). Thus the total base resistance of such a device is given, for low-level injection conditions, by

$$r_B \simeq \frac{w_E}{12\sigma_B W_B l_E} + \frac{d_{EB}}{2\sigma_B W_2 l_E}. \tag{14.67}$$

As the injection level rises, crowding effects under the emitter become increasingly important until eventually nearly all injection occurs from the emitter edges nearest the base contacts. Thus in the limit, the base resistance of a transistor with two base contacts is given by

$$r_B \simeq \frac{d_{EB}}{2\sigma_B W_2 l_E}. \tag{14.68}$$

In practice, the intrinsic base resistance is considerably larger than the extrinsic, typical values of r_{B1}/r_{B2} ranging from 4 to 8.

The foregoing computations may be readily extended to interdigitated structures with multiple emitter and base contacts. In all cases the analysis serves to emphasize the desireability of using a long-strip geometry, with the base contact as close as physically possible to the emitter region. In practice, this dimension may be kept as small as 3.0 μm ($\simeq$ 0.125 mil).

14.2.4. High-frequency Current Gain

The behavior of the transistor at high frequencies may be obtained by solving the time-varying form of the continuity and diffusion equations for minority carriers in the base. Direct solutions of this type are available in the literature; rather than repeat them here, we note [see (13.92)] that this equation is identical to its steady-state form, with the exception that the diffusion length L_{nB} is replaced by $L_{nB}/\sqrt{1+j\omega\tau_{nB}}$.

Both the injection efficiency and the collector multiplication factor are almost frequency independent at all frequencies of interest. Hence the frequency dependence of the current gain is primarily determined by the base transport factor. It follows therefore that the frequency-dependent form of the common-base current gain may be written as [see (14.18)]

$$\alpha \simeq \alpha_T = \left(\cosh\frac{W}{L_{nB}}\sqrt{1+j\omega\tau_{nB}}\right)^{-1}. \tag{14.69}$$

At $\omega\tau_{nB} \ll 1$, this reduces to its low-frequency value

$$\alpha_0 \simeq \left(\cosh\frac{W_B}{L_{nB}}\right)^{-1}. \tag{14.70}$$

The common-emitter current gain is given by

$$\beta = \frac{\alpha}{1-\alpha}, \tag{14.71}$$

$$\simeq \left(\cosh\frac{W_B}{L_{nB}}\sqrt{1+j\omega\tau_{nB}}-1\right)^{-1}. \tag{14.72}$$

This may be written in the form

$$\beta(s) = \frac{\beta_0}{(1+s/s_1)(1+s/s_2)\ldots} \tag{14.73}$$

where s is the differential operator, $s_1, s_2, s_3, \ldots$ are the poles of β, and β_0 is given as before by

$$\beta_0 = \frac{\alpha_0}{1-\alpha_0} \tag{14.74}$$

The poles $s_1, s_2, s_3, \ldots$ are roots of the equation

$$\cosh\frac{W_B}{L_{nB}}\sqrt{1+s_k\tau_{nB}}-1=0 \tag{14.75}$$

where k refers to the various roots. Setting

$$\frac{W_B}{L_{nB}}\sqrt{1+s_k\tau_{nB}}=jz \tag{14.76}$$

where z is a complex variable, (14.75) reduces to

$$\cosh jz=\cos z=1 \tag{14.77}$$

or

$$z=2\pi(k-1), \tag{14.78}$$

and k takes on the values 1, 2, 3, . . . corresponding to the various roots of (14.75). Substituting this equation into (14.76) gives

$$s_k=-\frac{1}{\tau_{nB}}\left[1+4\pi^2(k-1)^2\left(\frac{L_{nB}}{W_B}\right)^2\right]. \tag{14.79}$$

Consequently, the poles of β are given by

$$s_1=-\frac{1}{\tau_{nB}} \tag{14.80a}$$

$$s_2=-\frac{1}{\tau_{nB}}\left[1+4\pi^2\left(\frac{L_{nB}}{W_B}\right)^2\right] \tag{14.80b}$$

$$s_3=-\frac{1}{\tau_{nB}}\left[1+16\pi^2\left(\frac{L_{nB}}{W_B}\right)^2\right] \tag{14.80c}$$

and so on.

But $L_{nB}/W_B \gg 1$. Thus the poles s_2, s_3, . . . are removed from s_1 by orders of magnitude, and have little effect on the behavior of $\beta(s)$. As a consequence, the common-emitter current gain is accurately described by the single-pole expression

$$\beta=\frac{\beta_0}{1+s\tau_{nB}} \tag{14.81}$$

Thus the common-emitter current gain is constant at β_0 for low frequencies, is 3 dB down from its low-frequency value at $\omega = 1/\tau_{nB}$, and falls off at 6 dB/octave beyond this corner frequency. Note that the corner frequency is independent of base width†. This is a direct outcome of the principle of charge control, wherein it was shown that the common-emitter current gain was directly proportional to τ_{nB}.

†The formulation for the common-base current gain is somewhat more complex (see Problem 14.4).

It should be emphasized that (14.81) gives the current gain of the "ideal" transistor, in which the effect of the parasitic elements has been neglected. In general these elements will tend to lower the corner frequency to ω_β, where $\omega_\beta < 1/\tau_{nB}$.

14.2.5. The Gain-Bandwidth Product

The current gain-bandwidth product of a transistor provides a useful figure of merit for evaluating its high-frequency performance. At high frequencies, where $\omega \gg \omega_\beta$, (14.81) reduces to

$$\beta \simeq \frac{\beta_0}{j\omega/\omega_\beta} \tag{14.82}$$

and falls off inversely with frequency. The effect of the second and higher poles may be ignored since they are far removed from the first. Hence the current gain-bandwidth product ω_t is given by

$$\omega_t = \omega|\beta| \simeq \beta_0\omega_\beta \tag{14.83}$$

and is independent of the frequency at which it is measured. For this reason ω_t is sometimes defined as the frequency at which the magnitude of the common-emitter current gain becomes unity.

The significance of the gain-bandwidth product is seen by developing the equation which determines the response of a transistor amplifier to an arbitrary input current. Consider an amplifier stage in which for simplicity, it is assumed that the loading approximates a short circuit.† For this loading

$$\beta \simeq \frac{I_C(s)}{I_B(s)} = \frac{\beta_0}{1+s/\omega_\beta} \tag{14.84}$$

Rearranging terms, setting $\omega_t = \beta_0\omega_\beta$, and replacing s by the differential operator,

$$\frac{1}{\omega_t}\frac{dI_C(t)}{dt} + \frac{I_C(t)}{\beta_0} = I_B(t). \tag{14.85}$$

Solution of this equation yields the desired response.

The physical nature of the current gain-bandwidth product may be determined by use of the charge-control principle. The statement of this principle is established by setting up the continuity equation for minority carriers in the base and integrating over the base width.

†See Problem 14.6 for the more practical situation of a finite load resistance.

Figure 14.7 shows the time-varying junction and terminal currents in an *n-p-n* transistor as well as the currents through the depletion-layer capacitances. The leakage-current term has been ignored since it does not affect transistor behavior at high frequencies. Unit cross section is assumed for the device. For excess electrons in the base region,

$$\frac{\partial n'_B}{\partial t} = -\frac{n'_B}{\tau_{nB}} - \frac{1}{q}\frac{\partial I_{nB}(x,t)}{\partial x}, \qquad (14.86$$

where $I_{nB}(x,t)$ is the base current due to electron flow. Integrating over the base width gives

$$q\frac{\partial}{\partial t}\int_0^{W_B} n'_B dt = -\frac{q}{\tau_{nB}}\int_0^{W_B} n'_B dx - \int_0^{W_B} \frac{\partial I_{nB}(x,t)}{\partial x} dx. \qquad (14.87$$

The excess charge in the base is due to electrons. Since q denotes the magnitude of the charge on an electron, this excess charge is given by $Q_B(t)$, where

$$Q_B(t) = -q\int_0^{W_B} n'_B dt. \qquad (14.88$$

Substituting in (14.87),

$$-\frac{dQ_B(t)}{dt} = +\frac{Q_B(t)}{\tau_{nB}} - I_{nB}(W_B,t) + I_{nB}(0,t). \qquad (14.89$$

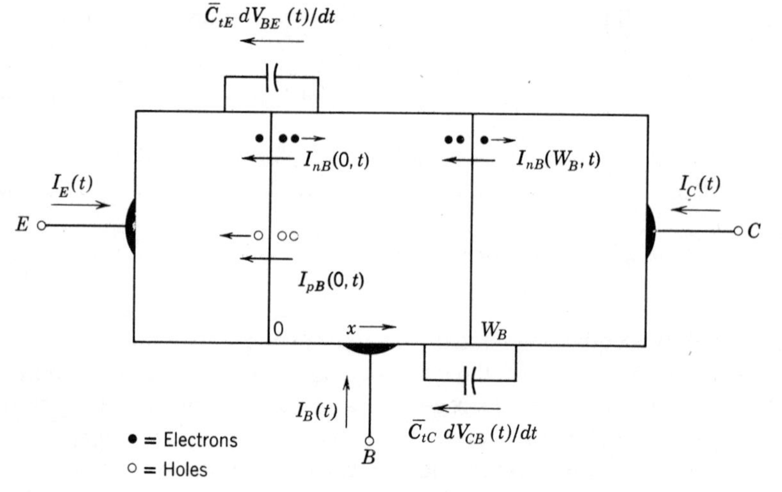

Figure 14.7 The *n-p-n* transistor.

From Figure 14.7 the emitter current is given by

$$-I_E(t) = I_{nB}(0, t) + I_{pB}(0, t) + \bar{C}_{tE}\frac{dV_{BE}(t)}{dt}, \tag{14.90}$$

where $\bar{C}_{tE}$ is the average transition capacitance of the emitter-base depletion layer. In like manner, the collector current is given by

$$I_C(t) = I_{nB}(W_B, t) + \bar{C}_{tC}\frac{dV_{CB}(t)}{dt}, \tag{14.91}$$

where $\bar{C}_{tC}$ is the average collector-base transition capacitance. Finally, $I_{pB}(0, t)$ is the hole component of the emitter current and may be ignored since $\gamma \simeq 1$.

Combining (14.89), (14.90), and (14.91), and noting that $I_B(t) + I_E(t) + I_C(t) = 0$,

$$I_B(t) = -\frac{dQ_B(t)}{dt} - \frac{Q_B(t)}{\tau_{nB}} + \bar{C}_{tE}\frac{dV_{BE}(t)}{dt} + \bar{C}_{tC}\frac{dV_{BC}(t)}{dt} \tag{14.92}$$

This equation restates the charge-control principle. Its various terms will now be evaluated.

14.2.5.1. *The Emitter Term.* The current through the emitter transition capacitance is $C_{tE}\, dV_{BE}(t)/dt$, and may be written as

$$\bar{C}_{tE}\frac{dV_{BE}(t)}{dt} = \bar{C}_{tE}\frac{dV_{BE}(t)}{dI_C(t)}\frac{dI_C(t)}{dt} \tag{14.93}$$

$$= \frac{\bar{C}_{tE}}{g_m}\frac{dI_C(t)}{dt}, \tag{14.94}$$

where g_m is the transconductance [6], representing the rate of change of the collector current with respect to the emitter-base voltage. The quantity $\bar{C}_{tE}/g_m$ has the dimensions of time. Thus we may write

$$\bar{C}_{tE}\frac{dV_{BE}(t)}{dt} = t_E\frac{dI_C(t)}{dt} \tag{14.95}$$

The transconductance may be evaluated by assuming a linear distribution of minority carriers in the base region, as shown in Figure 14.8. Then

$$I_C(t) \simeq I_{nB}(W_B, t) = -qD_{nB}\left(\frac{dn_B}{dx}\right)_{x=W_B} \tag{14.96}$$

$$= \frac{qD_{nB}\bar{n}_B}{W_B} e^{qV_{BE}(t)/kT}. \tag{14.97}$$

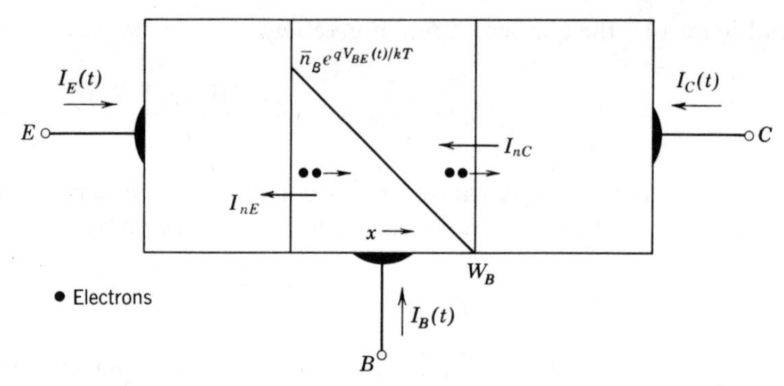

Figure 14.8 Carrier concentration in the base.

Differentiation gives

$$\frac{dI_C(t)}{dV_{BE}(t)} = g_m \tag{14.98}$$

$$= \frac{q}{kT} I_C(t). \tag{14.99}$$

14.2.5.2. *The Base Term.* With reference to Figure 14.8,

$$-I_E(t) \simeq I_{nB}(0, t) = -qD_{nB}\left(\frac{dn_B}{dx}\right)_{x=0} \tag{14.100}$$

$$= \frac{qD_{nB}\bar{n}_B}{W_B} e^{qV_{BE}(t)/kT}. \tag{14.101}$$

But

$$Q_B(t) = -\frac{q}{2} W_B \bar{n}_B e^{qV_{BE}(t)/kT}. \tag{14.102}$$

Therefore

$$Q_B(t) = \frac{W_B^2}{2D_{nB}} I_E(t) \tag{14.103}$$

$$\simeq -\frac{W_B^2}{2D_{nB}} I_C(t). \tag{14.104}$$

The quantity $W_B^2/2D_{nB}$ has been shown to be equal to t_B, the transit time of minority carriers through a narrow-base diode (see Section 13.1.4). Hence

$$\frac{dQ_B(t)}{dt} \simeq -t_B \frac{dI_C(t)}{dt} \tag{14.105}$$

14.2.5.3. *The Collector Term.* Consider the single-stage transistor amplifier of Figure 14.9 operating from a power supply of $+V_1$. Noting that the parasitic collector resistance is in series with the load resistor, the network equation may be written as

$$V_1 = I_C(t)(R_L + r_C) + V_{CB}(t) + V_{BE}(t), \tag{14.106}$$

so that

$$\partial V_{CB}(t) = -(R_L + r_C)\,\partial I_C(t) - \partial V_{BE}(t). \tag{14.107}$$

The emitter-base diode of a transistor is forward-biased. Consequently, small changes in $V_{BE}(t)$ give rise to large variations in $I_C(t)$. To an approximation, then,

$$dV_{CB}(t) \simeq -(R_L + r_C)\,dI_C(t). \tag{14.108}$$

Substituting t_C for $r_C\bar{C}_{tC}$,

$$\bar{C}_{tC}\frac{dV_{BC}(t)}{dt} = (R_L\bar{C}_{tC} + t_C)\frac{dI_C(t)}{dt} \tag{14.109}$$

14.2.5.4. The Depletion-layer Terms. The diffusion model of a transistor assumes depletion layers of zero width, and thus neglects the transport of charge carriers through these regions. Thus if the charge transported through the collector-base depletion layer at any given time is $Q_D(t)$, the base current required to support it is given by $dQ_D(t)/dt$. Furthermore, (14.92) must be modified to include this term.

If t_D is the transit time for charge carriers through this depletion layer, it follows that

$$Q_D(t) = t_D I_C(t), \tag{14.110}$$

so that

$$\frac{dQ_D(t)}{dt} = t_D\frac{dI_C(t)}{dt}. \tag{14.111}$$

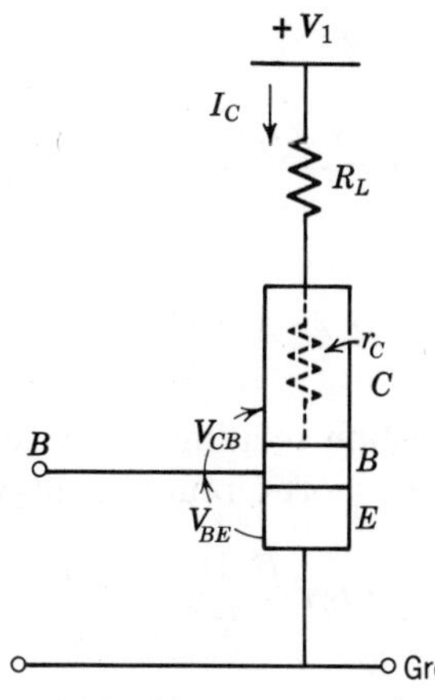

Figure 14.9 The common-emitter stage.

Note that this term is associated with the conduction current† through the depletion layer. Thus its presence is due to the fact that charge carriers (electrons) are actually transported through the collector-base depletion layer. At first glance, the arrival of these carriers at the depletion-layer edge would appear to be contrary to the assumption that the minority carrier concentration at this edge is zero. As shown in Section 13.1.4, minority carriers in a narrow, space-charge-neutral base region exhibit an increase in apparent velocity as they approach a sink. The upper limit for this velocity is the thermal velocity. Thus the excess carrier concentration must be finite (though small) at the edge of the depletion layer.

On arrival these carriers are transported through a depletion layer of thickness x_C at their limiting velocity ($v_{\text{lim}} \simeq 10^7$ cm/sec for electrons in silicon, as shown in Figure 11.19). This velocity is typically more than 100 times the average velocity of minority carriers in the base; thus the actual carrier concentration required to support the flow of collector current in the depletion layer is less than 1% of the average carrier concentration in the base. To an approximation, the transit time through the depletion layer is given by x_C/v_{lim}.

The contribution of the mobile charge in the emitter depletion layer may be treated in a similar manner. This region, however, is very narrow because it is forward-biased. As a result the transit time of carriers through it may be ignored.

14.2.5.5. The Transistor Equation. Combining the various terms of (14.95), (14.105), (14.109), and 14.111,

$$I_B(t) = \frac{-Q_B(t)}{\tau_{nB}} + (t_E + t_B + t_C + t_D + R_L\bar{C}_{tC})\frac{dI_C(t)}{dt} \qquad (14.112)$$

In steady state,

$$I_B(t) = \frac{-Q_B(t)}{\tau_{nB}} = +\frac{I_C(t)}{\beta_0}, \qquad (14.113)$$

so that

$$(t_E + t_B + t_C + t_D + R_L\bar{C}_{tC})\frac{dI_C(t)}{dt} + \frac{I_C(t)}{\beta_0} = I_B(t). \qquad (14.114)$$

This is the differential equation relating the base and collector currents for a resistive load R_L. Under short-circuit conditions it reduces to

$$(t_E + t_B + t_C + t_D)\frac{dI_C(t)}{dt} + \frac{I_C(t)}{\beta_0} = I_B(t). \qquad (14.115)$$

†The term involving $\bar{C}_{tC}$ is associated with the displacement current.

Comparison with (14.85) shows that

$$\frac{1}{\omega_t} = t_E + t_B + t_C + t_D \tag{14.116}$$

Thus the reciprocal gain-bandwidth product is given by the sum of the time constants associated with the emitter and collector transition capacitances, and the base and depletion-layer transit time. In terms of the device parameters.

$$\frac{1}{\omega_t} = \frac{\bar{C}_{tE}}{g_m} + \frac{W_B^2}{2D_{nB}} + \bar{C}_{tC}\, r_C + \frac{x_c}{v_{\text{lim}}}. \tag{14.117}$$

14.2.5.6. Fall-off at Low Current Levels. Inspection of (14.117) shows that the fall-off in ω_t at low current levels is due to the emitter charging-time constant, since g_m falls with decreasing collector current. Thus in the low current region ω_t varies inversely with current.

Recombination effects at the surface and in the emitter-base depletion layer may also be expected to deteriorate ω_t at low current levels. However, these effects are of a secondary nature and may be ignored.

14.2.5.7. Fall-off at High Current Levels [7]. In part, the fall-off in ω_t at high current levels is due to conductivity modulation of the base. As shown earlier, this leads to a reduction in the common-emitter current gain and hence, reduces the gain-bandwidth product. The primary reason for this fall-off, however, is the buildup of a minority carrier space charge in the collector-base depletion layer. As indicated in Section 14.2.5.4, minority carriers are indeed present in this region; their flow produces an effective increase in charge density in the part of the depletion layer that is within the base and a reduction in the part that is within the collector. To preserve over-all charge neutrality, the depletion layer shrinks in the former and expands in the latter. Thus the base width widens at high current levels.

Depending on the respective doping of the emitter and collector regions, the depletion-layer width may either increase or decrease. At any rate, since $t_D \ll t_B$, the net effect is a reduction in ω_t at high current levels.

Figure 14.10*a* shows the collector-base depletion layer of an *n-p-n* transistor in which current flow is assumed to be entirely by electrons. If n' is the concentration of electrons moving through the depletion layers, the magnitude of the collector current density is given by

$$J_C = qn'v_{\text{lim}} \tag{14.118}$$

where v_{lim} is the limiting velocity of electrons in silicon. Let

x_{B0}, x_{C0} = depletion-layer in the base and collector regions respectively, when no current is flowing,
x_B, x_C = depletion-layer width for a current density of J_C,
W_B = base width when no collector current is flowing,
V_j = voltage across the collector-base junction,
N_B, N_C = impurity concentration in the base and collector respectively.

Various dimensions are indicated in Figure 14.10*a*, and the charge density is shown in Figure 14.10*b*. On the base side there exists a space charge of $-qx_B(N_B+n')$ per unit area; the space charge on the collector side is $+qx_C(N_C-n')$ per unit area. For charge neutrality over the entire depletion layer,

$$qx_B(N_B+n') = qx_C(N_C-n').$$

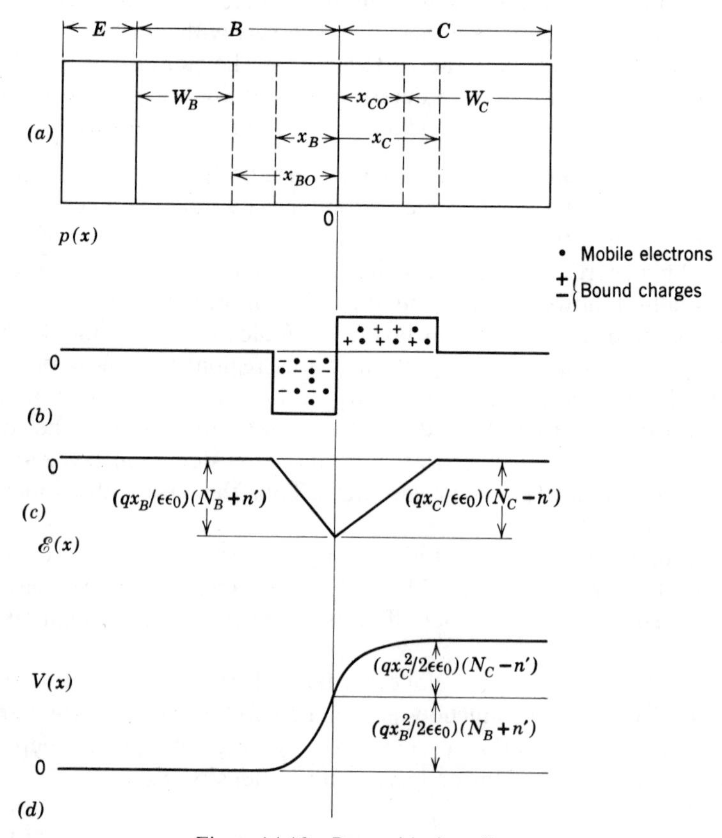

Figure 14.10 Base-widening effects.

Integration of Poisson's equation gives the $\mathscr{E}$ field shown in Figure 14.10*c*. A second integration gives the parabolic potential function of Figure 14.10*d*, such that

$$V_j = \frac{q}{\epsilon\epsilon_0}[(N_B + n')x_B^2 + (N_C - n')x_C^2]. \tag{14.120}$$

Combining (14.119) with (14.120),

$$x_C^2 = \frac{2\epsilon\epsilon_0 V_j N_B}{qN_C(N_B + N_C)} \frac{1 + n'/N_B}{1 - n'/N_C}. \tag{14.121}$$

For $n' = 0$, $x_C = x_{C0}$. Thus

$$x_C^2 = \frac{x_{C0}^2(1 + n'/N_B)}{1 - n'/N_C}. \tag{14.122a}$$

In like manner,

$$x_B^2 = \frac{x_{B0}^2(1 - n'/N_C)}{1 + n'/N_B}. \tag{14.122b}$$

These equations may be used to evaluate the behavior of $1/\omega_t$ for specific situations, as follows:

CASE 1. HIGHLY DOPED COLLECTOR

This situation is encountered in alloy transistors. Here, since $N_C \simeq \infty$, the depletion layer only moves within the base region. Thus $x_C = 0$ and

$$x_B^2 = \frac{x_{B0}^2}{1 + n'/N_B}. \tag{14.123}$$

With increasing current density the base width increases from W_B to $W_B + x_{B0} - x_B$. Simultaneously the depletion-layer width shrinks from x_{B0} to x_B. Making these substitutions and noting that r_C is negligible in this case,

$$\frac{1}{\omega_t} \simeq t_E + \frac{W_B^2}{2D_{nB}}\left[1 + \frac{x_{B0}}{W_B}\left(1 - \frac{1}{\sqrt{1 + n'/N_B}}\right)\right]^2 + \frac{x_{B0}}{v_{\lim}\sqrt{1 + n'/N_B}}. \tag{14.124}$$

As mentioned, the net result is a fall off in ω_t with increasing current level.

CASE 2. LIGHTLY DOPED COLLECTOR

A collector region of this type is encountered in diffused structures used in microcircuits. Here the body resistance of the collector cannot be ignored and results in a voltage drop. This reduces the actual voltage available across the collector-base depletion layer, so that

$$V_j \simeq V_{CB} + \Psi_C - J_C \rho_C W_C, \tag{14.125}$$

where

$$J_C = n' q v_{\text{lim}} \tag{14.126}$$

$$\rho_C \simeq \frac{1}{q \mu_C N_C} \tag{14.127}$$

and W_C is defined† as shown in Figure 14.10*a*.

For convenience we consider N_B to be much greater than N_C. Then

$$x_C^2 \simeq \frac{2\epsilon\epsilon_0 V_j}{q N_C (1 - n'/N_C)}. \tag{14.128}$$

Combining (14.125) and (14.128),

$$x_C^2 = \frac{x_{C0}^2 \{1 - [W_C v_{\text{lim}}/\mu_C (V_{CB} + \Psi_C)](n'/N_C)\}}{1 - n'/N_C}. \tag{14.129}$$

The gain-bandwidth product is subsequently given by

$$\frac{1}{\omega_t} = t_E + t_B + \frac{x_C}{v_{\text{lim}}} + \frac{\epsilon\epsilon_0 \rho_C W_C}{x_C}. \tag{14.130}$$

In practice, $W_C v_{\text{lim}} < \mu_C (V_{CB} + \Psi_C)$, so that the width of the depletion layer shrinks with increasing current density, and the collector time constant rises correspondingly. Ultimately, x_C approaches zero as

$$n'/N_C \to \frac{\mu_C (V_{CB} + \Psi_C)}{W_C v_{\text{lim}}}, \tag{14.131a}$$

resulting in a rapid increase in the collector charging time. The peak value of ω_t is thus attained at a current density of J_{max}, where

$$J_{\text{max}} \to \frac{q \mu_C N_C (V_{CB} + \Psi_C)}{W_C}. \tag{14.131b}$$

Kirk[8] has shown that, for a collector current in excess of that given by (14.131*b*), the position of the collector depletion-layer boundary adjacent to the base moves rapidly into the collector region, asymptoti-

†The collector width is considered to be infinitely long and is essentially unaltered by movement of the depletion layer.

cally approaching the collector contact with increasing current. Thus ω_t falls off very rapidly upon reaching its peak value.

A number of important approximations have been made in the development of the theory for the fall-off in ω_t at high current levels. For example, the shifting of the emitter-base depletion layer toward the emitter contact has been ignored. The inclusion of this effect results in reducing the current at which ω_t begins to fall off. The shifting of the base edge of the collector depletion layer (in case 2) also results in a fall-off at lower current levels than that given by the theory. Current-spreading effects from the emitter toward the collector tend to reduce the abrupt nature of the fall-off characteristic of ω_t at high current levels. Finally, the diffusion constant is effectively doubled because of high level injection effects (see Section 13.1.2.4). This extends the current level at which the fall-off occurs to a value beyond that given by (14.131*b*). In spite of these simplifications, experimental results have shown reasonable agreement with the theory, so that it provides a useful basis for device design.

14.2.6. Maximum Frequency of Oscillation

The gain-bandwidth product is an especially useful figure of merit in wide-band applications. For narrow-band situations, however, a far more meaningful parameter is the maximum frequency at which an amplifier can operate and still have a power gain in excess of unity. This is known as the maximum frequency of oscillation of the transistor.

By considering the circuit theory of tuned amplifier stages, it may be shown[9] that this frequency is given by

$$\omega_{\max} = \left(\frac{\omega_t}{4r_{B2}C_{tC}}\right)^{1/2} \tag{14.132}$$

It is interesting to note that the figures of merit for a transistor are directly related to the various parasitic terms. Thus the importance of minimizing these parasitics must be emphasized at this point.

14.2.7. Saturation Effects

Consider an *n-p-n* transistor the collector current of which is limited to I_{C1} by means of an appropriate suppy voltage and local resistance. For this condition it is necessary that a current I_{B1} be supplied to preserve charge neutrality in the base. The magnitude of this current is given by

$$I_{B1} = \frac{I_{C1}t_t}{\tau_{nB}} \tag{14.133a}$$

$$= \frac{I_{C1}}{\beta_0}. \tag{14.133b}$$

Let the base current be now raised to I_{B2}. For the preservation of charge neutrality it is essential that additional minority carriers be injected into the base region. This can only occur if the collector-base junction becomes *forward*-biased and behaves like an emitter! The transistor is now said to be operating in its *saturated* condition.

The forward-biased collector-base junction results in the injection of minority carriers into the base as well as into the collector regions. The resulting charge configuration for the saturated transistor is shown in Figure 14.11. The effect of the excess charge in the base may be taken into consideration by separating the base charge into $Q_B(t)$, the active base charge, and $Q_{sB}(t)$, the excess base charge. The excess collector charge during saturation is shown as $Q_{sC}(t)$. The emitter charge is ignored.

Associating recombination times of τ_{sB} and τ_{sC} with $Q_{sB}(t)$ and $Q_{sC}(t)$ respectively, the charge control equation for the base current may be modified to

$$I_B(t) = \frac{-dQ_B(t)}{dt} - \frac{Q_B(t)}{\tau_{nB}} + \bar{C}_{tE}\frac{dV_{BE}(t)}{dt} + \bar{C}_{tC}\frac{dV_{BC}(t)}{dt} + \frac{dQ_D(t)}{dt} - \frac{Q_{sB}(t)}{\tau_{sB}} + \frac{Q_{sC}(t)}{\tau_{sC}}. \tag{14.134}$$

The term τ_{sB} is a measure of the speed with which the excess charge in the base can be removed when the transistor is being pulled out of saturation. Since this region is narrow.

$$\tau_{sB} \simeq t_B \tag{14.135}$$

where t_B is the base transit time.

In like manner, τ_{sC} is a measure of the speed with which the excess charge can be removed from the collector region. In a non-epitaxial

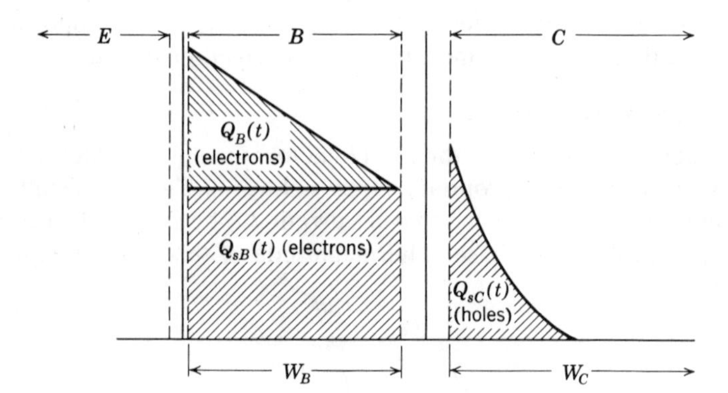

Figure 14.11 Stored charge in saturation.

transistor this region is wide, and the primary means of charge removal is by recombination. Thus

$$\tau_{sC} \simeq \tau_{pC} \tag{14.136}$$

where τ_{pC} is the minority carrier lifetime in the collector.

For an epitaxial transistor the width of the collector region may be comparable to its diffusion length. For this situation τ_{sC} may be determined by following closely the line of reasoning used to obtain (13.124), so that

$$\frac{1}{\tau_{sC}} = \frac{1}{\tau_{pC}} + \frac{1}{t_C} \tag{14.137a}$$

where t_C is the transit time of carriers through a narrow collector region. To an approximation,†

$$\tau_{sC} \simeq \frac{W_C^2}{2D_{pC}} \tag{14.137b}$$

Here W_C is the width of the collector region beyond the depletion-layer edge, and D_{pC} is the diffusion length of holes in the collector.

The amount of charge stored in the collector region is a function of the injection efficiency of the forward-biased collector-base junction. In alloy transistors where the collector is very heavily doped relative to the base, this injection efficiency is high ($\simeq$0.99) and the collector stored charge may be ignored.

14.2.8. The Breakdown Voltage

As the voltage across the reverse-biased collector-base junction is increased, the peak $\mathscr{E}$ field increases. For sufficiently high values of reverse bias the junction eventually breaks down by secondary ionization processes. The breakdown voltage for this junction, with the emitter open-circuited, is denoted by BV_{CB0}, and has been considered in detail in Section 13.1.1.6.

Avalanche breakdown may also occur if the transistor is operated in the common-emitter configuration. Here the breakdown voltage is due to the combined effects of secondary ionization and collector multiplication.

Consider a transistor in the common-emitter configuration, as shown in Figure 14.12. For this situation (with the base lead open) the breakdown voltage is denoted by BV_{CE0}. Referring to the figure,

$$I_C = -\alpha_0 I_E + I_{C0} \tag{14.138}$$

†Yanai[10] and co-workers have made a detailed analysis of this situation and concluded that $\tau_{sC} \simeq W_C^2/3D_{pC}$. However, the result of a (14.137b) is found to more accurate in practice.

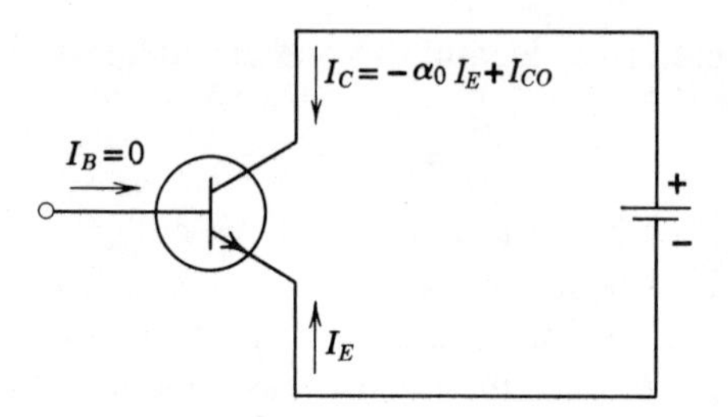

Figure 14.12 The common-emitter connection (open-base).

where I_{C0} is the reverse leakage current of the collector-base diode. Since the base is open $I_C = -I_E$. The current that flows in this circuit is thus given by

$$-I_E = \frac{I_{C0}}{1-\alpha_0} \tag{14.139}$$

The breakdown voltage BV_{CE0} is that voltage at which the denominator of this expression goes to zero, that is, at which $\alpha_0 = 1$. This occurs when

$$\alpha_0 = \gamma\alpha_T M = \frac{\gamma\alpha_T}{1-(BV_{CE0}/BV_{CB0})^m} = 1 \tag{14.140}$$

Solving,

$$\frac{BV_{CE0}}{BV_{CB0}} \simeq (1-\gamma\alpha_T)^{1/m}. \tag{14.141}$$

But $\gamma\alpha_T$ is equal to α_0 before collector multiplication effects become apparent. Hence

$$\frac{BV_{CE0}}{BV_{CB0}} = (1-\alpha_0)^{1/m} \tag{14.142}$$

$$\simeq \left(\frac{1}{\beta_0}\right)^{1/m} \tag{14.143}$$

The index m has been given as 4 for n-type silicon and 2 for p-type silicon (see Section 13.1.1.6). In double-diffused transistors the greater fraction of the depletion layer is in the n-type collector, so that

$$BV_{CB0} \simeq \sqrt[4]{\beta_0}\, BV_{CE0}. \tag{14.144}$$

A second possibility is that, with increasing collector supply voltage, the collector-base depletion layer widens until it eventually reaches through to the emitter. For a transistor in the common-emitter configuration of Figure 14.12, this causes the collector current to increase indefinitely. Thus BV_{CE0} is given by either (14.144) or by the "reach-through" voltage, whichever occurs earlier.

In a common base transistor reach-through does not cause the collector current to rise indefinitely because the emitter is left open. Consequently the breakdown voltage in this configuration is given by BV_{CB0}.

14.3. THE GRADED-BASE TRANSISTOR

As its name implies, the graded-base transistor is one in which the impurity concentration in the base is inhomogeneous. This grading is usually a byproduct of the diffusion process and is of the complementary error function (erfc) or gaussian type; in either case it is convenient to approximate it by a simple exponential.

The primary effect of an exponential grading in the base is that it gives rise to a constant electric field. If the emitter edge is more highly doped than the collector, this field aids the transport of minority carriers across the base. Thus the graded-base transistor should have a shorter base transit time than a comparable uniform-base structure and, hence, a higher gain-bandwidth product.

Figure 14.13*a* shows the impurity profile of a transistor with an exponentially graded base. Here the emitter is highly doped in order to obtain a high injection efficiency.

The doping profile of a double-diffused transistor is shown for comparison in Figure 14.13*b*. Here, too, the emitter region is highly doped

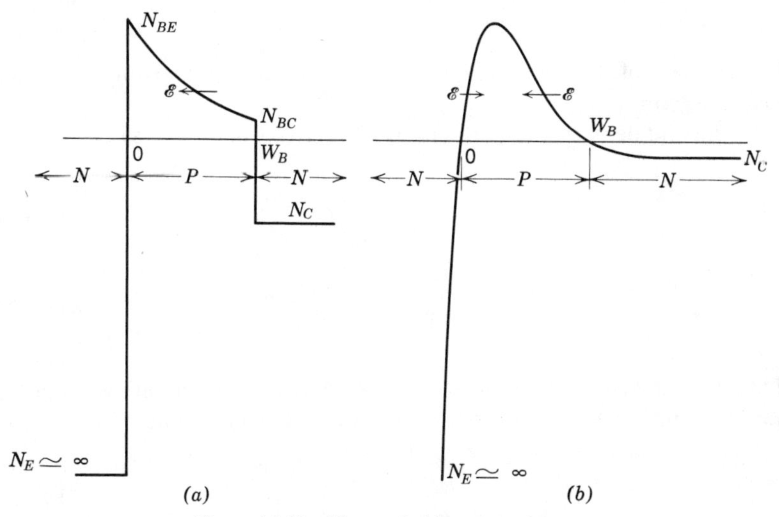

Figure 14.13 The graded-base transistor.

relative to the base region. As an outcome of the diffusion process, however, the collector region is relatively lightly doped. Finally, the impurity gradient in the base is exponential from some point beyond the edge of the emitter depletion layer. Thus injected minority carriers are first subjected to a retarding field and then to the aiding field. Qualitatively, the presence of the retarding field reduces the effect of the aiding field, although it has been indicated[11] that this effect can be ignored if the retarding field region is under 30 to 40% of the entire width.

Let the impurity concentration in the base be given by

$$N_B = N_{BE} e^{-\eta x/W_B}, \tag{14.145}$$

so that

$$N_B = N_{BE} \qquad \text{at } x = 0, \tag{14.146a}$$

$$N_B = N_{BC} \qquad \text{at } x = W_B, \tag{14.146b}$$

when

$$\eta = \ln \frac{N_{BE}}{N_{BC}}. \tag{14.147}$$

Thus η is a direct measure of the degree of grading in the base. In the absence of injection (or at low injection levels), the $\mathscr{E}$ field is given by

$$\mathscr{E}_0 = \frac{kT}{qN_B} \frac{dN_B}{dx} \tag{14.148}$$

$$= -\frac{kT}{q} \frac{\eta}{W_B}. \tag{14.149}$$

The direction of this field aids the transport of electrons from the emitter to the collector.

The current density due to the flow of electrons in the base is

$$J_{nE} = qn_B \mu_n \mathscr{E}_0 + qD_{nB} \frac{dn_B}{dx} \tag{14.150}$$

$$= qD_{nB} \left(\frac{dn_B}{dx} - \frac{n_B \eta}{W_B} \right), \tag{14.151}$$

where D_{nB} is assumed constant over the entire base region. Solving this equation, subject to the boundary condition that $n_B = 0$ at $x = W_B$, leads to

$$n_B = -\frac{J_{nE} W_B}{Qd_{nB}} \frac{1 - e^{-\eta(1 - x/W_B)}}{\eta}. \tag{14.152}$$

Figure 14.14 shows this minority carrier concentration for various values of η. The curve for $\eta = 0$ corresponds to uniform doping in the base and results in a linear change in carrier concentration with distance.

In typical graded structures, $\eta \simeq 6$ to 8. Thus the minority carrier concentration is approximately trapezoidal in character. Since the diffusion current is proportional to the slope of the minority carrier concentration, it is seen that current transport is initially almost entirely

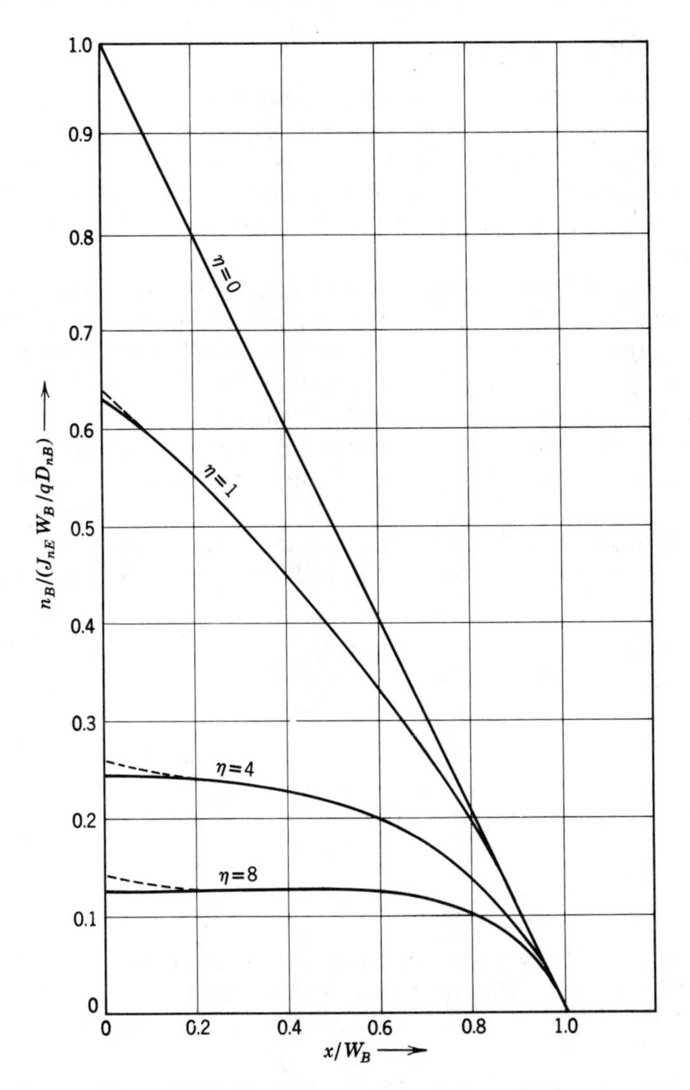

Figure 14.14 Carrier concentration in the graded-base transistor.

by drift (between $x = 0$ and $x = 0.5W_B$ for $\eta = 6$). As carriers approach the collector sink, however, current flow is almost entirely by diffusion.

In a practical situation this analysis must be modified to include the effects of the variation of the diffusion constant with impurity concentration. The net result is shown by the dashed line in Figure 14.14, which indicates that some diffusion effects do occur at the emitter edge of the base region.

14.3.1. Current-gain Theory

The current gain of a graded-base transistor is also related to the injection efficiency, the base transport factor, and the collection efficiency. Here the base transport factor is even higher than that for the uniform-base transistor because of the additional drift component. Thus its effect in determining the magnitude of the current gain can be safely ignored.

The collector-base depletion layer of a graded junction has been shown in Chapter 13 to be wider than that of a comparable abrupt junction for the same reverse voltage. Thus the corresponding $\mathscr{E}$ field in the depletion layer is smaller. As a result, multiplication effects in a graded transistor may also be ignored at typical operating voltages.

As a result of the above considerations, the magnitude of the current gain is entirely determined by the injection efficiency of the emitter-base diode.

The forward currents due to electron and hole flow in an n^+-p junction of arbitrary doping have been developed in Section 13.3.2, and are given by

$$-I_{nE} = \frac{qAD_{nB}n_i^2}{\int_0^{W_B} N_B dx} e^{qV_{BE}/kT} \tag{14.153a}$$

$$-I_{pE} = \frac{qAD_{pE}n_i^2}{\int_{L_{pE}}^{0} N_E dx} e^{qV_{BE}/kT} \tag{14.153b}$$

where N_B and N_E are the net impurity concentration in the base and emitter respectively. The common-emitter current gain is given by

$$\beta_0 \simeq \frac{\gamma}{1-\gamma} = \frac{I_{nE}}{I_{pE}} \tag{14.154}$$

so that

$$\beta_0 \simeq \frac{D_{nB} \int_{L_{pE}}^{0} N_E \, dx}{D_{pE} \int_0^{W_B} N_B \, dx}. \tag{14.155}$$

Thus the common-emitter current gain is directly proportional to the ratio of the number of ionized impurities within a diffusion length of the emitter to the number of ionized impurities within the base region.

14.3.1.1. Fall-off Effects. The fall-off of current gain at low injection levels is directly attributed to charge-recombination effects in the emitter-base depletion layer and to recombination effects at the surface. Thus the behavior of a graded-base transistor is very similar in this respect to that of its uniform-base counterpart.

With increasing current level the majority carrier concentration in the base is altered from N_B to $N_B + n_B'$ to maintain charge neutrality. Since (14.155) is applicable to the general situation, we may intuitively expect the common-emitter current gain to be given by β_0', where

$$\beta_0' \simeq \frac{D_{nB} \int_{L_{pE}}^{0} N_E \, dx}{D_{pE} \int_0^{W_B} (N_B + n_B') \, dx}. \tag{14.156}$$

In addition, the increase in majority carrier concentration in the base results in altering the electric field to an effective value $\mathscr{E}'$, so that

$$\mathscr{E}' = \frac{kT}{q(N_B + n_B')} \frac{d(N_B + n_B')}{dx}. \tag{14.157}$$

The current density due to electron flow is thus given under all levels of injection by

$$J_{nE} = q n_B \mu_n \mathscr{E}' + q D_{nB} \frac{dn_B}{dx} \tag{14.158}$$

$$= q n_B \mu_n \mathscr{E}_0 \frac{N_B}{N_B + n_B'} + q D_{nB} \frac{dn_B'}{dx} \frac{N_B + 2n_B'}{N_B + n_B'} \tag{14.159}$$

since $dn_B/dx = dn_B'/dx$ for charge neutrality. Thus the $\mathscr{E}$ field falls with increasing carrier concentration until it eventually drops off to zero.

As a result the effect of the grading is eliminated under high-level injection conditions, and the transistor behaves like a uniformly doped structure.

Crowding effects in the emitter will cause a premature fall off in current gain at high injection levels. The analysis is identical to that for the uniformly doped structure; however, the confinement of the lateral current flow to a width W'_B is more severe due to the varying impurity concentration in the base. Here, too, injection is seen to take place almost entirely at the emitter edges.

14.3.2. Parasitic Elements

Methods for determining the transition capacitances C_{tE} and C_{tC} have been described in Section 13.3.1 and will be briefly reviewed here. The collector-base junction may be treated as a linearly graded structure; alternately, its capacitance may be obtained with the aid of the computer derived curves of Figure 13.30. The emitter-base junction may also be treated as a linearly graded structure, provided the appropriate value of the built-in potential is used. For this diode this is given by the contact potential for the abrupt structure shown as *CDEF* in Figure 13.29.

Since the collector region is uniformly doped, its parasitic resistance r_C is given by the equations described in Section 14.2.3.2. The base resistance may again be subdivided into an intrinsic and an extrinsic part, and the evaluation carried out as before. The effect of the base impurity grading may be taken into consideration by noting that the average conductivity of the base region under the emitter is given by

$$\overline{\sigma}_B = q \int_{x_{EB}}^{x_{BC}} \mu_n N_B \, dx, \tag{14.160}$$

where x_{EB} and x_{BC} are the positions of the emitter-base and collector-base junctions respectively, and N_B is the net impurity concentration. Computer derived curves for this function are available[12] in the literature.

The conductivity of the base beyond the emitter region is given by $\sigma_{B\square} x_{BC}$, where $\sigma_{B\square}$ is the sheet conductance of the base region, and is given in Figure 4.31 for p-type gaussian diffusion into n-type silicon. Typical values of $1/\sigma_{B\square}$ range from 100 to 200 Ω/square.

14.3.3. High-frequency Current Gain

At high frequencies the short-circuit forward-current gain of a graded-base transistor is primarily due to the frequency characteristics of the base transport factor α_T. This factor may be determined by solving the time-varying form of the continuity and diffusion equations, as shown in Appendix B. This solution results in

$$\alpha \simeq \alpha_T = \frac{Ze^{\eta/2}}{Z\cosh Z + (\eta/2)\sinh Z} \tag{14.161}$$

where

$$Z = \left[\frac{\eta^2}{4} + \left(\frac{W_B}{L_{nB}}\right)^2 (1 + j\omega\tau_{nB})\right]^{1/2}. \tag{14.162}$$

For a uniform-base transistor, $\eta = 0$, and

$$Z \simeq \frac{W_B}{L_{nB}}\sqrt{1 + j\omega\tau_{nB}} \tag{14.163}$$

The common-emitter current gain may be written by setting $\beta = \alpha/(1-\alpha)$. Kelly and Ghausi[13] have shown that the poles of β are given by this procedure as

$$s_k = -\frac{1}{\tau_{nB}}\left\{1 + \left(\frac{L_{nB}}{W_B}\right)^2 [4\pi^2(k-1)^2 \pm j4\pi(k-1)\cosh^{-1} e^{\eta/2}]\right\}, \tag{14.164}$$

where $k = 1, 2, 3, \ldots$. The first of these poles is thus

$$s_1 = -\frac{1}{\tau_{nB}} \tag{14.165}$$

Since $(L_{nB}/W_B)^2 \gg 1$, all other poles are many decades beyond the first pole and may be ignored. Thus the common-emitter current gain of a graded-base transistor may be accurately represented by a single pole, with the magnitude of β down to 0.707 of its low-frequency value at $\omega = 1/\tau_{nB}$. Note that this is the corner frequency of the ideal transistor, free of parasitics. For the real transistor this corner frequency is written as ω_β, where $\omega_\beta < 1/\tau_{nB}$. Consequently, the gain-bandwidth product for the graded-base transistor is given by

$$\omega_t = \beta_0\omega_\beta \tag{14.166}$$

14.3.4. The Gain-Bandwidth Product

Since the charge-control principle is applicable to a transistor regardless of the nature of the doping profile, the gain-bandwidth product may be written directly as

$$\frac{1}{\omega_t} = \frac{\bar{C}_{tE}}{g_m} + t_B + \bar{C}_{tC}r_C + \frac{x_C}{v_{\text{lim}}} \tag{14.167}$$

The base transit time may be obtained[15] by setting v_B as the average velocity of electrons at a position x in the base. Then

$$t_B = \int_0^{W_B} \frac{1}{v_B} dx. \tag{14.168}$$

Assuming that current flow is entirely by electrons,

$$J_{nE} = -qn_B v_B \tag{14.169}$$

Combining (14.169) with (14.152) for the current density due to electrons,

$$\frac{1}{v_B} = \frac{W_B}{\eta D_{nB}} (1 - e^{-\eta/(1-x/W_B)}). \tag{14.170}$$

Substituting into (14.168) and integrating over the base width,

$$t_B = \frac{W_B^2}{D_{nB}} \left(\frac{\eta + e^{-\eta} - 1}{\eta^2} \right) \tag{14.171}$$

$$= \frac{W_B^2}{kD_{nB}} \tag{14.172}$$

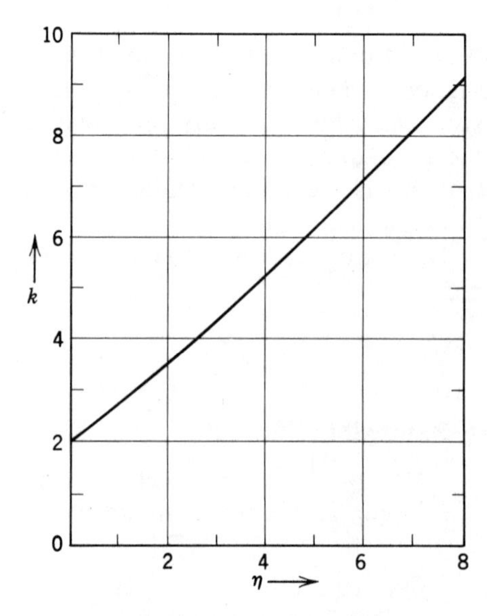

Figure 14.15 Correction factor.

where k is shown in Figure 14.15 for different values of η. At $\eta = 0$ this expression reduces to the transit time for a uniform-base device.

It is important to note that this computation applies only to low-level injection conditions. Examination of (14.159) for the electron current shows that:

1. The $\mathscr{E}$ field falls with increasing injection level until its effect is no longer apparent.
2. The effective diffusion constant increases with injection level until it is eventually twice its value at low levels. Thus, in the limit, the base transit time for all devices (graded or otherwise) is given by

$$t_B \rightarrow \frac{W_B{}^2}{4D_{nB}} \tag{14.173}$$

For design purposes, the value of k in (14.172) is conveniently chosen as 4 to 5 for a graded-base transistor, and assumed to be constant over the full range of current levels.

For graded devices the fall in ω_t at low current levels is dominated by the increase in the emitter charging-time constant. At high current levels the fall-off characteristics approach closely those described in case 2 of Section 14.2.5.7, since the effects of a varying diffusion constant can be ignored.

14.3.5. Saturation Effects

Saturation effects in uniformly doped devices have been discussed in Section 14.2.7, and were shown to be more closely related to the nature of the collector region than to the impurity grading in the base. Thus we will confine our attention to the graded-base transistor made by the DDE process. Here the features of significance are as follows:

1. The collector region is lightly doped,
2. The collector-base junction is linearly graded,
3. The physical volume of the base region is small, whereas that of the collector region is large.

As a consequence, the injection efficiency of the collector is very low ($\simeq$0.2). This, in turn, leads to the copious injection of minority carriers (holes) into the collector when the transistor is saturated. To an approximation, the charge stored in the base region is significantly less than that stored in the collector, and may be ignored. Thus the storage-time constant for a non-epitaxial structure is given by $\tau_{sC} \simeq \tau_{pC}$.

Modern high-speed microcircuits are almost exclusively epitaxial in nature, so that the width of the collector region is often comparable to, or less than, the hole diffusion length. For this case, then, the collector storage-time constant is approximately given by $\tau_{sC} \simeq W_C^2/2D_{pC}$ and is equal to the transit time of minority carriers through this region.

14.3.6. The Breakdown Voltage

The avalanche breakdown voltage of a diffused junction has been treated in Section 13.3.1, and the results are shown in Figure 13.32. In general, the diffused junction has a higher breakdown voltage than its linearly graded counterpart.

Under reverse bias the depletion layer extends into both the collector and the base region. The movement of this depletion layer, however, into the base is considerably restricted since it penetrates into a region of continually increasing impurity concentration. Computer-derived solutions of this problem have been made by Lawrence and Warner[15], and show that the reach-through voltage for diffused structures is well in excess of the avalanche breakdown voltage. Thus breakdown of the collector-base diode of double-diffused structures is primarily caused by secondary ionization processes.

As before, the breakdown voltage in the common-emitter configuration (with the base left open) is due to both secondary ionization and collector multiplication. The breakdown voltage is thus given by

$$BV_{CB0} \simeq \sqrt[4]{\beta_0} BV_{CE0} \tag{14.174}$$

14.4. REFERENCES

[1] C. T. Sah et al., "Carrier-Generation and Recombination in *p-n* Junctions and *p-n* Junction Characteristics," *Proc. IRE*, **45**(9), 1228–1243(1957).

[2] W. M. Webster, "On the Variation of Junction Transistor Current Amplification Factor with Emitter Current," *Proc. IRE* **42**(6), 914–920 (1954).

[3] N. H. Fletcher, "Some Aspects of the Design of Power Transistors," *Proc. IRE*, **43**(5), 551–559 (1955).

[4] D. P. Kennedy, "Spreading Resistance in Cylindrical Semiconductors," *J. Appl. Phys.*, **31**(8), 1490–1497 (1960).

[5] A. B. Phillips, *Transistor Engineering*, McGraw–Hill, New York, 1962.

[6] P. E. Gray et al., *Physical Electronics and Circuit Models of Transistors – SEEC*, Vol. 2, Wiley, New York, 1964.

[7] C. T. Kirk, Jr., "A theory of Transistor Cutoff Frequency Falloff at high Current Densities," *IEEE Trans. on Electron Devices*, **ED-9**(2), 164–174 (1962).

[8] R. H. Lloyd, "The Movement of the Depletion Layer Boundaries at High Current Densities," *IEEE Trans. on Electron Devices*, **ED-13**(12), 991–992 (1966).

[9] C. L. Searle et al., *Elementary Circuit Properties of Transistors – SEEC*, Vol. 3, Chap. 8, Wiley, New York, 1964.

[10] H. Yanai et al., "The Minority Carrier Storage Effect in the Collector Region and the Storage Time of Transistors," *Proc. IEEE*, **52**(3), 312–314 (1964).

[11] M. Tannenbaum and D. E. Thomas, "Diffused Emitter and Base Silicon Transistors," *Bell System Tech. J.*, **35**(1), 1–22 (1956).
[12] J. C. Irvin, "Resistivity of Bulk Silicon and Diffused Layers in Silicon," *Bell System Tech. J.*, **41**, 387–408 (1962).
[13] J. J. Kelly, and M. S. Ghausi, "Poles of the Alpha of Drift Transistors," *IEEE Trans. on Circuit Theory*, **CT-12**(4), 593–595 (1965).
[14] J. L. Moll, and I. M. Ross, "Dependence of Transistor Parameters on the Distribution of Base Layer Resistivity," *Proc. IRE*, **44**(1), 72–78 (1956).
[15] H. Lawrence, and R. M. Warner, "Diffused Junction Depletion Layer Calculations," *Bell System Tech. J.*, **39**(2), 389–403 (1960).

14.5. PROBLEMS

1. Show that the common-emitter current gain increases with increasing voltage. An *n-p-n* transistor has a collector-base breakdown of 60 V and a common-emitter current gain of 40 at $V_{CB} = 5$ V. Determine the collector voltage at which $\beta_0 = 80$.

2. A silicon transistor has an emitter concentration of 10^{21} atoms/cm^3, a base concentration of 10^{19} atoms/cm^3, and a collector concentration of 10^{16} atoms/cm^3. The emitter-base junction is at 2 μm, an the collector-base junction is at 3 μm. Determine the common emitter current gain at -55, 25, and 150°C respectively. Assume that the minority carrier lifetimes are relatively independent with temperature and are both equal to 0.5 μs.

3. A uniformly doped transistor is made with circular geometry. The radius of the emitter region is r_1 and the inner radius of the base contact is r_2. Show that the extrinsic base resistance is equal to $(1/2\pi\sigma_B W_2)\ \ln\ (r_2/r_1)$ and the intrinsic base resistance is equal to $1/8\pi\sigma_B W_B$. Note that W_2 is the base-region thickness under the base contact.

4. Show that, for a uniform-base transistor,

$$\alpha = \frac{1}{\cosh\ (W_B/L_{nB})}\ \frac{1}{(1+s/\omega_0)\,(1+s/9\omega_0)\,(1+s/25\omega_0)\cdots}$$

where $\omega_0 = (\pi^2/8)/t_B$. Determine the phase shift of this function at $\omega = \omega_0$, and show that the common base current gain may be approximated by

$$\alpha = \frac{\alpha_0 e^{-jm\omega/\omega_0}}{1+j\omega/\omega_0}$$

where $m \simeq 0.22$.

5. From Problem 4 it is seen that the common-base cutoff frequency is only a function of the transit time of carriers through the base while the common - emitter cutoff frequency is only a function of minority carrier lifetime. Explain these facts on a purely physical basis. (*Hint:* Use the charge-control principle.)

6. A transistor amplifier has a load resistance of R_L. Show that the differential equation relating collector and base currents is

$$\left(R_L\bar{C}_{tC}+\frac{1}{\omega_t}\right)\frac{dI_C(t)}{dt}+\frac{I_C(t)}{\beta_0}=I_B(t),$$

where $\bar{C}_{tC}$ is the collector transition capacitance and ω_t is the gain-bandwidth product.

7. Show that ω_t falls monotonically with decreasing collector voltage for a constant collector current. What physical processes cause this to occur?

8. A transistor is fabricated with a linearly graded base. Determine the base transit time for minority carriers.

9. A graded-base transistor has a highly doped collector. Show that the collector depletion layer extends x_0 into the base region, where

$$x_0^2=\frac{\epsilon\epsilon_0 V_j}{qN_{BE}}\frac{\eta^2}{\eta-1+e^{-\eta}},$$

and V_j is the voltage across the collector-base junction.

Chapter 15

Monolithic Transistors and Diodes

Individual transistors and diodes are fabricated in a microcircuit chip by means of diffusion processes. Thus we may expect the general characteristics of these devices to be essentially similar to those of the diffused (or graded) structures described in the preceding chapters.

In this chapter we shall consider transistors and diodes fabricated by the double-diffused epitaxial (DDE) process which is in general use today. In particular, emphasis will be placed on those characteristics in which the monolithic devices differ from their discrete counterparts.

15.1. THE MONOLITHIC TRANSISTOR

Figure 15.1 shows a typical transistor made by the DDE process. From this figure it is seen that this transistor differs from is discrete counterpart in a number of ways, as follows:

1. Both the emitter-base junction and the collector-base junction have curved sidewall areas in addition to their plane floor areas. To an approximation, the sidewall may be assumed to be cylindrical in shape, even though the lateral penetration of the diffusion is only 75 to 85% of the penetration normal to the surface (see Section 4.6.1).
2. The collector region also has a floor and a sidewall; here, too, a cylindrical shape factor may be assumed in computing the area of the sidewall.
3. The collector contact is made on the same side as the emitter and base contacts. Thus the current flow in the collector of a monolithic transistor is significantly different from that in its discrete counterpart.

The effect of these various features is now considered.

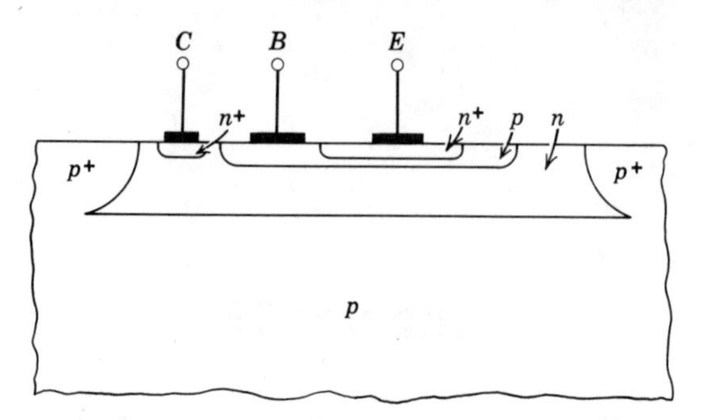

Figure 15.1 The monolithic transistor.

15.1.1. Emitter Capacitance

Figure 15.2 shows the doping profile of a monolithic transistor in a direction normal to the surface. As described in Section 13.3, the emitter floor may be treated as a linearly graded structure, provided that the effect of the built-in potential due to the $\mathscr{E}$ field is taken into consideration in determining the voltage across the junction.

The sidewall capacitance is considerably more difficult to calculate. Although the problem may be solved by a piecewise linear approximation technique, an alternate method is to obtain upper and lower bounds for this capacitance and take the average. The upper bound for this capacitance is obtained by treating the sidewall as an abrupt junction with a surface concentration of N_{OE} on one side and N_{OB} on the other. The lower

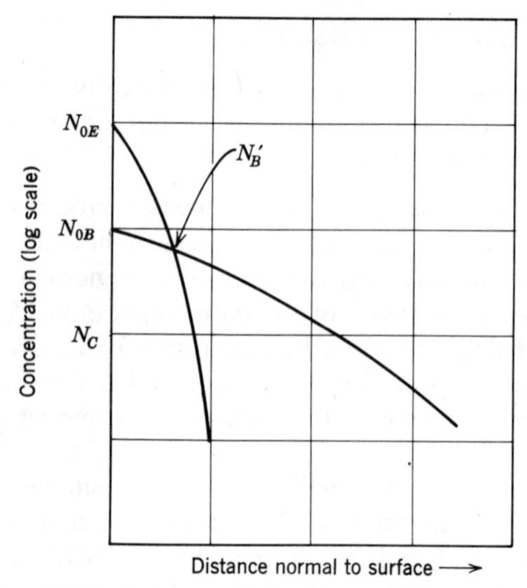

Figure 15.2 Doping profile for the monolithic transistor.

bound is given by treating the sidewall in the same manner as the floor. Finally, the total capacitance C_{tE} is given by the sum of the capacitances of the floor and the sidewall.

15.1.2. Collector Capacitance

Both floor and sidewall of the collector-base junction consist of a p-type gaussian diffusion into an n-type uniform background. Thus these regions may be considered together and the value of C_{tC} determined in the manner outlined in Section 13.3.

15.1.3. Substrate Capacitance

Here the floor and sidewall must be considered separately. The floor junction may be approximated by an abrupt doping profile, since the collector region is epitaxially grown.

The resistivity of the substrate is typically 10 Ω-cm. Consequently, this junction may be treated as an n^+-p structure, with the depletion layer confined to the p-side of the junction. Alternately, the resistivity of both collector and substrate may be considered in computing this component of the transition capacitance (see Problem 15.1).

The sidewall component of this capacitance is more closely approximated by a p-type gaussian diffusion into an n-type collector of uniform background concentration, with the majority of the depletion layer extending into the collector side. The curves of Figure 13.30 may be used to determine the value of this capacitance. Finally, the total capacitance C_{tS} is given by the sum of the floor and sidewall components.

15.1.4. Base Resistance

The base resistance of a monolitic transistor is very similar to that of its discrete counterpart and has been discussed in Section 14.3.2.

15.1.5. The Collector Resistance

Figure 15.3 shows the layout and cross section of a microcircuit transistor made by the DDE process. Here an estimate of the colletor parasitic resistance may be made by assuming uniform current flow in the collector region, from the edge of the emitter to the edge of the n^+ diffusion to which the collector contact is made. This trapezoidal current path is shown shaded in Figure 15.3 and has a resistance given by

$$r_C \simeq \frac{2\rho_C d_{EC}}{(l_E + l_C)(x_{CS} - x_{BC})} \tag{15.1}$$

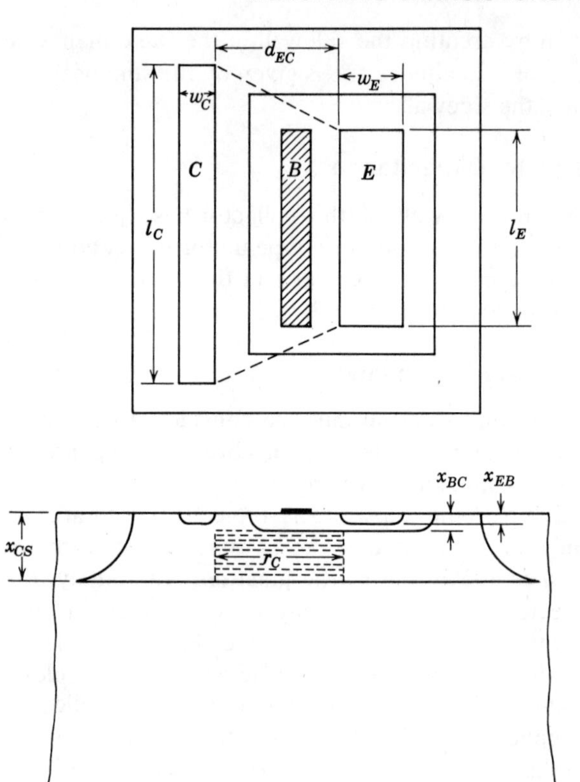

Figure 15.3 Parasitic collector resistance.

where ρ_C is the collector resistivity and the various dimensions are shown in the figure. Here the trapezoid is approximated by an equivalent rectangle. The error involved in not using a more correct conformal[1] mapping procedure is well within the limits of the other approximations made. Thus the added complexity is not warranted in this situation.

Equation 15.1 assumes a single current path between emitter and collector. In structures with more than one path (for example, if two collector contacts are used), the resistances of these paths must be separately determined, the total resistance being given by their parallel combination.

15.1.6. The Collector Equivalent Circuit

Figure 15.4*a* shows a possible equivalent circuit for the parasitic elements on the collector side of a monolithic transistor of the nonepitaxial type. Here $\overline{C}_{ts}$ and r_C are shown as distributed elements of a

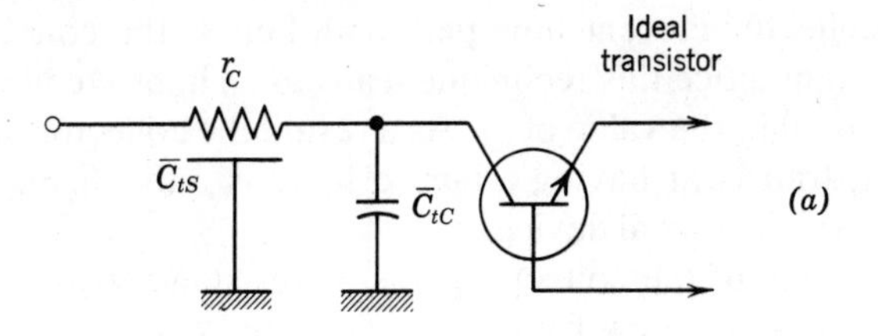

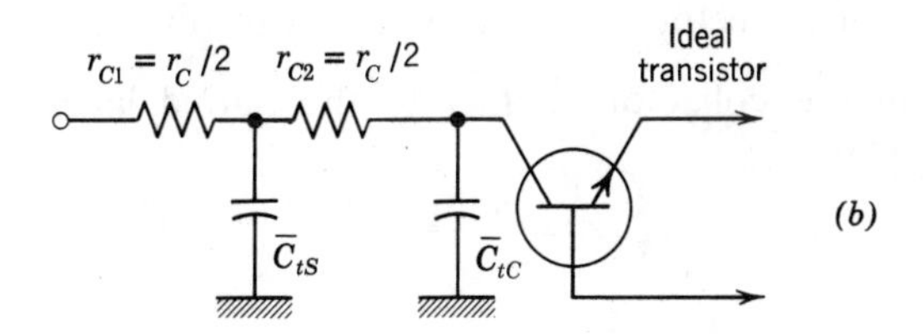

Figure 15.4 Collector equivlaent circuit.

transmission line. This line may be approximated by the T network shown in Figure 15.4*b* such that

$$r_{C1} = r_{C2} \simeq 0.5 r_C \tag{15.2}$$

15.1.7. Gain-Bandwidth Product

Having established an equivalent circuit that takes into consideration the substrate capacitance, it is possible to write the gain-bandwidth product directly in the form

$$\frac{1}{\omega_t} \simeq \frac{\bar{C}_{tE}}{g_m} + \frac{W_B{}^2}{4D_{nB}} + \bar{C}_{tC} r_{C2} + \frac{x_C}{v_{\lim}} + \bar{C}_{tS} r_{C1} \tag{15.3}$$

where $\bar{C}_{tS}$ is the average value of the substrate capacitance. A comparison of (15.3) with (14.167) for a discrete double-diffused transistor shows that the monolithic structure will have a lower value of ω_t. In part this is caused by the additional substrate charging term. The contribution of this term is important, since the collector parasitic resistance of a microcircuit transistor is considerably larger than that of a discrete device.

15.1.8. Effect of the Buried Layer

In saturated switching applications it is especially important to reduce the value of r_C. Here a buried layer of low sheet resistance is used, as described in Section 10.4. The presence of this buried layer drastically alters the collector current flow path and, hence, the collector parasitic resistance. In practice this technique leads to an improvement of as much as a factor of 10 in the value of r_C. As a result, the collector resistance of a microcircuit transistor having a buried layer is only slightly larger than that of a discrete epitaxial device.

The magnitude of the collector parasitic resistance may be determined by assuming uniform flow from the emitter and collector contacts down to the buried layer and a trapezoidal current flow path across this buried layer. The total collector resistance consists of three parts, as follows:

1. r'_C—from the collector contact to the buried layer, as shown in Figure 15.5. Here

$$r'_C \simeq \frac{\rho_C(x_{CS}-x_{EB})}{l_C w_C} \tag{15.4}$$

where l_C and w_C are the length and width of the collector contacts respectively.

2. r''_C—from the emitter to the buried layer. Here the assumption of uniform current flow is extremely inaccurate owing to crowding effects. However, this is somewhat offset by the fact that current spreading from the emitter to the low-resistivity buried layer has also been ignored. To an approximation, then,

$$r''_C \simeq \frac{\rho_C(x_{CS}-x_{BC})}{l_E w_E} \tag{15.5}$$

where l_E and w_E are the length and width of the emitter region respectively.

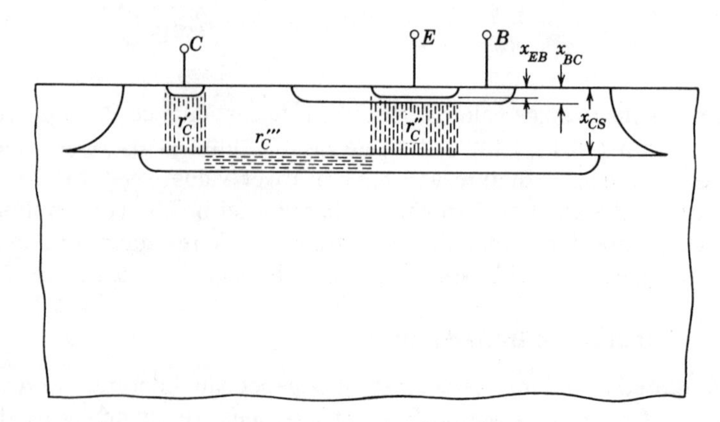

Figure 15.5 The buried-layer transistor.

3. r_C''' – the resistance to the flow of current in the buried layer between the edges of the emitter and the collector. This flow takes the form of a trapezoidal path, with conducting parallel edges of l_C and l_E respectively. Consequently,

$$r_C''' \simeq \frac{2\rho_\square d_{EC}}{l_E + l_C} \tag{15.6}$$

where $\rho_\square$ is the sheet resistance of the buried layer. In typical microcircuits this resistance is approximately 10 to 15 Ω/square.

The equivalent circuit at the collector side may be drawn as shown in Figure 15.6*a*. To an approximation, this circuit can be redrawn in the form of Figure 15.6*b*, such that

$$r_{C1} = \frac{r_C' + r_C'''}{2} \tag{15.7}$$

and

$$r_{C2} = \frac{r_C' + r_C'''}{2} + r_C'' \tag{15.8}$$

Using these values of r_{C1} and r_{C2}, the gain-bandwidth product may be obtained by substitution in (15.3).

It should be noted here that the use of the buried layer greatly increases the gain-bandwidth product of a monolithic transistor because of the

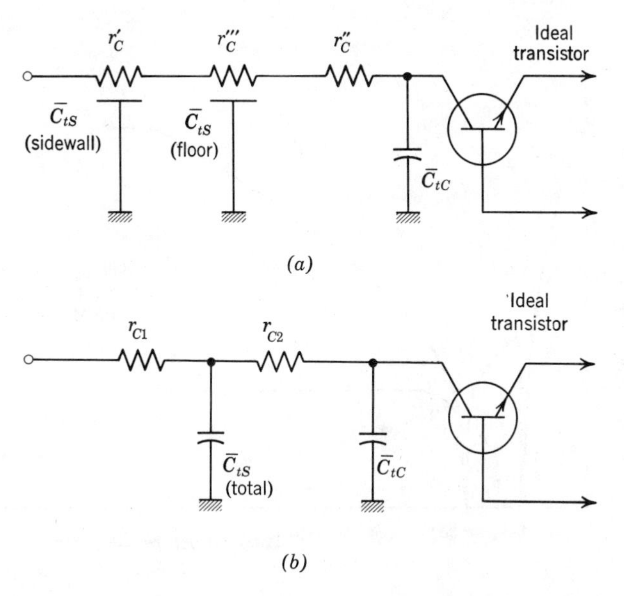

Figure 15.6 Collector equivalent circuit for buried-layer devices.

reduction in r_C. However, ω_t is still slightly lower than that obtained in a discrete epitaxial device because of the additional substrate parasitic capacitance.

15.2. SPECIAL TRANSISTOR STRUCTURES

A number of different transistor configurations may be fabricated to meet the needs of special situations. A few of these will be considered in the following.

15.2.1. Symmetrical Transistors

In low-level chopper applications it is important that the reverse-current gain† of the transistor be as high as possible. This requirement may be met by using a thin epitaxial layer and a buried layer impurity which is subject to large out-diffusion or etch-back effects. For example, phosphorus may be used as the buried layer dopant since its diffusion constant is ten times that of antimony. Alternately, the use of arsenic combined with a low epitaxial growth rate will lead to enhanced etch-back effects. In either case these techniques alter the diffusion profile of the transistor to that shown in Figure 15.7 and result in an increased

†This requires that the emitter-base junction be reverse-biased and the collector-base junction forward-biased.

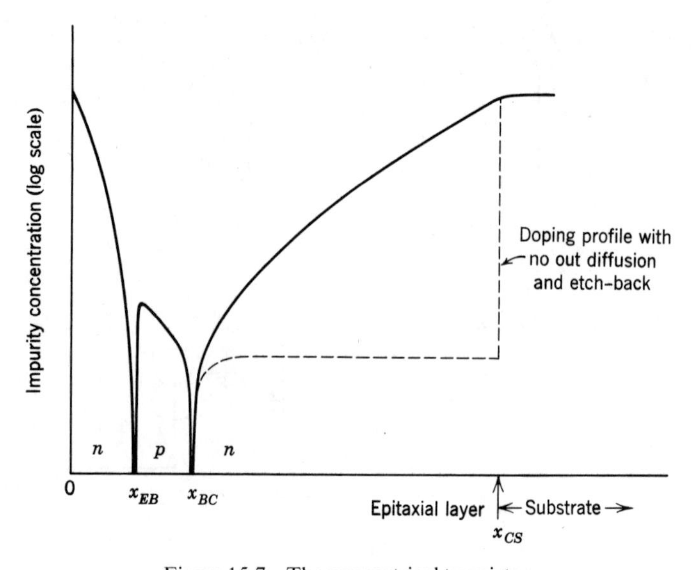

Figure 15.7 The symmetrical transistor.

injection efficiency in the reverse direction. The doping profile, in the absence of out-diffusion or etch-back effects, is shown by a dashed line in this figure. As a result of this technique the collector-base breakdown voltage is reduced. This is generally not important in low-level chopping applications.

A significant disadvantage of this technique lies in the fact that the entire microcircuit is now subject to this type of doping profile. This can be prevented by using buried layers of both antimony and phosphorus. Unfortunately this increases the number of steps required in the fabrication process. In general, a topological solution is always to be preferred, since it involves no additions or changes to the monolithic process.

15.2.2. Unsymmetrical Transistors

In yet other situations it is necessary to make the reverse common-base current gain as small as possible (≤ 0.01). This may be done by making the collector area considerably larger than the emitter area. Since injection is primarily an edge effect, this serves to make the effective base width in the reverse direction considerably larger than that in the forward direction, as shown for the idealized structure of Figure 15.8. Next the device is gold doped to reduce the diffusion length of minority carriers in the base. Since $\alpha_T/(1-\alpha_T) \simeq (L_{nB}/W_B)^2$, the combined effect is to reduce drastically the reverse-current gain while only slightly affecting the gain in the forward direction.

An alternate method[2], suitable for even lower values of reverse-current gain (≤ 0.001), is shown in Figure 15.9. Here the base region is

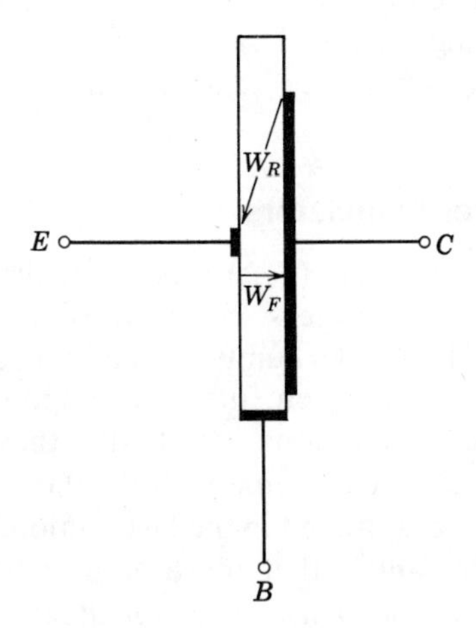

Figure 15.8 The unsymmetrical transistor.

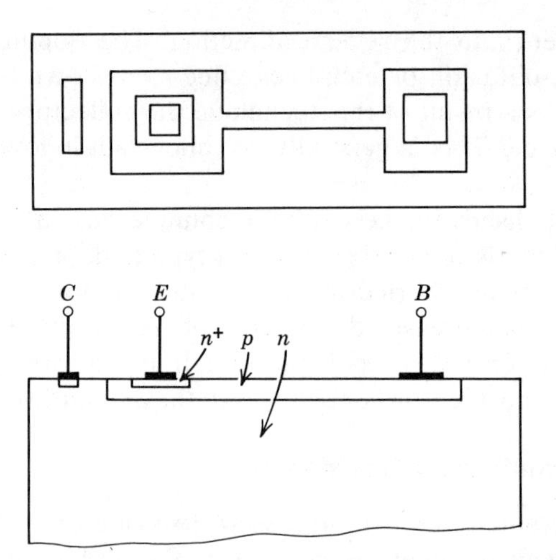

Figure 15.9 Transistor with extremely low inverse gain.

shaped so as to have a relatively high lateral resistance. In the forward direction the device behaves like a normal transistor with a relatively high extrinsic base resistance. In the reverse direction, however, with the collector acting as an emitter, the voltage drop due to the high lateral resistance of the base region causes almost all injection to occur near the base contact. Thus there is negligible transistor action in the reverse direction.

The scheme of Figure 15.9 is superior to that of Figure 15.8 in its ability to obtain a lower value of inverse-current gain. It has the disadvantage, however, of occupying a considerably larger area of the microcircuit. In addition, the base resistance is inherently larger than that for the structure of Figure 15.8.

15.2.3. Multi-emitter Transistors

In transistor-transistor logic (T^2L) it is necessary to use a single transistor with a number of emitters. Such transistors consist of a single collector and base pocket, and a number of emitter regions.

There are two special requirements placed upon a device of this type as a result of circuit considerations. The first of these is that its common-base reverse-current gain be extremely low (≤ 0.01). This may be accomplished by the techniques outlined in the last section.

The second requirement is that there be no interaction between any two inputs at all times. Thus parasitic *n-p-n* effects between two emitters and the base region must be negligible. This requirement is met by physically placing the emitters as far apart as possible (so as to make the base width of the parasitic transistor large) and by subsequent gold doping. Two commonly used configurations for a structure of this type are shown in Figure 15.10.

15.2.4. Complementary Transistors

In some microcircuits *p-n-p* transistors are required in addition to *n-p-n* devices. As mentioned in Section 10.6, the lateral *p-n-p* is perhaps the simplest device of this type[2, 3], requiring no additional fabrication steps beyond those of the basic monolithic process.

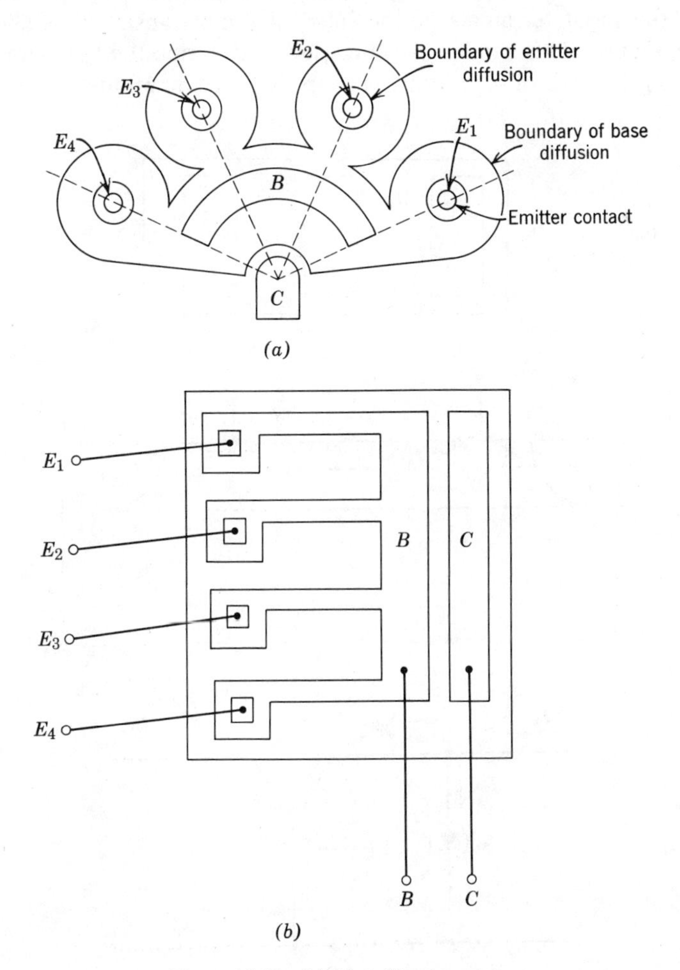

Figure 15.10 Multi-emitter transistors.

Figures 15.11*a* and *b* show a lateral transistor having strip geometry. Since this device is fabricated at the same time as the rest of the microcircuit, its emitter and collector are the usual *p*-type gaussian "base" diffusions, whereas the base contact is made to an *n*-type erfc-diffused region of the type normally used for emitters. In operation minority carriers (holes) are injected at the emitter and transported through the base region to the collector. In addition, holes diffuse toward the epi-substrate interface and do not contribute to the lateral transistor action.

The distance between the lower edge of the diffusion pocket and the epi-substrate is large compared with a hole diffusion length L_{pB}. Consequently this path behaves as a parasitic wide-base diode, placed in shunt across the input terminals of the internal *p-n-p* transistor, as shown in Figure 15.11*c*. This model may be used to describe the behavior of the device on the basis of the following simplifying assumptions:

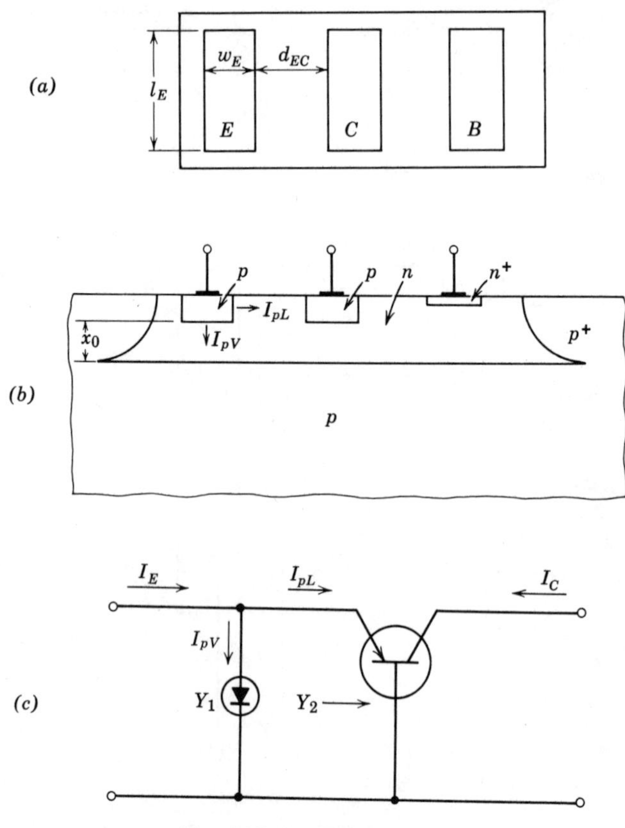

Figure 15.11 The lateral transistor.

1. The emitter diffusion pocket is rectangular in cross section, of width w_E, length l_E, and depth x_0.
2. The diffused region is very heavily doped so that there is no electron component of injected current.
3. Carriers injected from the emitter sidewall are assumed to cooperate in *p-n-p* action, whereas those injected from the emitter floor represent the parasitic *p-n* diode.

For these assumptions, the lateral hole current is given by

$$I_{pL} = -\frac{q l_E x_0 D_{pB} p'_B(0)}{d_{EC}} \tag{15.9}$$

where $p'_B(0)$ is the excess hole concentration at the edge of the emitter-base depletion layer, and the base region between emitter and collector is assumed to be narrow.

In the vertical direction the base region is wide compared to L_{pB}.

Thus

$$I_{pV} = -\frac{q l_E w_E D_{pB} p'_B(0)}{L_{pB}} \tag{15.10}$$

Consequently, the injection efficiency is

$$\gamma = \frac{\partial I_{pV}}{\partial (I_{pV} + I_{pL})} \tag{15.11}$$

so that

$$\frac{1-\gamma}{\gamma} = \frac{\partial I_{pV}}{\partial I_{pL}} \tag{15.12}$$

$$= \frac{w_E d_{EC}}{x_0 L_{pB}} \tag{15.13}$$

Thus

$$\frac{1}{\beta_0} = \frac{w_E d_{EC}}{x_0 L_{pB}} + \frac{d_{EC}^2}{2L_{pB}^2} \tag{15.14}$$

where the second term is the contribution of the base transport factor.

The degradation effect of the parasitic diode is significant in linear microcircuit applications. Here typical values for the various parameters are $L_{pB} = 30$ to $60\ \mu\text{m}$, $w_E = d_{EC} = 3\ \mu\text{m}$, and $x_0 = 1.5\ \mu\text{m}$, resulting in values† of β_0 between 5 and 10.

†A second collector region increases this value by a factor of 2.

In digital saturated-logic microcircuits, where gold doping is used, values of L_{pB} are typically under 3 μm, with a corresponding reduction in β_0 to about 0.5. Thus lateral transistor applications in digital circuits are generally limited to situations where the device is not required to have gain.

High-frequency performance may be determined with the aid of the equivalent circuit of Figure 15.11*c*. Here the d-c emitter current for the entire device is labeled I_E and that for the internal *p-n-p* transistor is labeled I_{pL}. The d-c current in the parasitic diode is labeled I_{pV}. Finally, the d-c collector current is labeled I_C. The various a-c components are i_E, i_{pL}, i_{pV}, and i_C respectively.

For the over-all device,

$$\alpha = \frac{i_C}{i_E} = \frac{i_C}{i_{pL}}\frac{i_{pL}}{i_E} \tag{15.15}$$

$$= \frac{i_C}{i_{pL}}\frac{Y_2}{Y_1+Y_2} \tag{15.16}$$

where Y_1 is the admittance of the parasitic diode and Y_2 is the input admittance of the ideal *p-n-p* transistor.

The parasitic diode is a wide-base structure. Consequently [see (13.111)],

$$Y_1 = \frac{qI_{pV}}{kT}(1+j\omega\tau_{pB})^{1/2}. \tag{15.17}$$

This admittance may also be approximated by [see (13.112)]

$$Y_1 = \frac{qI_{pV}}{kT}\left(1+\frac{j\omega}{2D_{pB}/L_{pB}^2}\right). \tag{15.18}$$

The emitter-base diode of the internal *p-n-p* transistor is a short-base structure. Hence [see (13.107)]

$$Y_2 = \frac{qI_{pL}}{kT}\left(1+\frac{j\omega}{3D_{pB}/d_{EC}^2}\right). \tag{15.19}$$

Since $I_{pL} > I_{pV}$ and $3D_{pB}/d_{EC}^2 \gg 2D_{pB}/L_{pB}^2$,

$$\frac{Y_2}{Y_1+Y_2} \simeq \frac{Y_2}{Y_1} \tag{15.20}$$

The corner frequency of a transistor in the common-emitter connection is given by the reciprocal of the minority carrier lifetime in the base [see (14.81)]. Writing β_0' as the common-emitter current gain of the internal *p-n-p*, ($\beta_0' \simeq 2L_{pB}^2/d_{EC}^2$), the corner frequency in the common-base connection is seen to be given by

$$\omega_\alpha \simeq \frac{\beta_0'}{\tau_{pB}} \simeq \frac{2D_{pB}}{d_{EC}^2} \tag{15.21}$$

Thus

$$\frac{i_C}{i_E} \simeq \frac{\alpha_0'}{1+j\omega/(2D_{pB}/d_{EC}^2)} \tag{15.22}$$

where α_0' is the low-frequency current gain of the internal *p-n-p* transistor. Finally

$$\frac{i_C}{i_E} = \alpha \simeq \frac{\alpha' Y_2}{Y_1} \tag{15.23}$$

A study of the equations for these various terms shows that the corner frequency for Y_1 is considerably below that for Y_2 or α'. Consequently, the behavior of α is dominated by this term; to an approximation,

$$\alpha \simeq \frac{\alpha_0}{Y_1} = \frac{\alpha_0}{(1+j\omega\tau_{pB})^{1/2}} \tag{15.24}$$

At low frequencies (15.24) reduces to

$$\alpha \simeq \frac{\alpha_0}{1+\frac{1}{2}j\omega\tau_{pB}+\ldots} \tag{15.25}$$

Thus the current gain is down to 0.707 of its d-c value at $\omega = 2/\tau_{pB}$. At high frequencies the magnitude of the current gain varies inversely with the square root of frequency, that is, at 3dB/octave. In contrast, the magnitude of the current gain of a double-diffused transistor falls off at 6 dB/octave.

15.3. THE MONOLITHIC DIODE

A monolithic diode may be fabricated by making a single *p*-type base diffusion into the *n*-type epitaxial layer. Alternately, the diode may be obtained by connecting together the various regions of a transistor. Figure 15.12*a* shows the different combinations possible. Their characteristics may be evaluated by setting up the equations for the d-c terminal voltages and currents of a transistor under any bias condition.

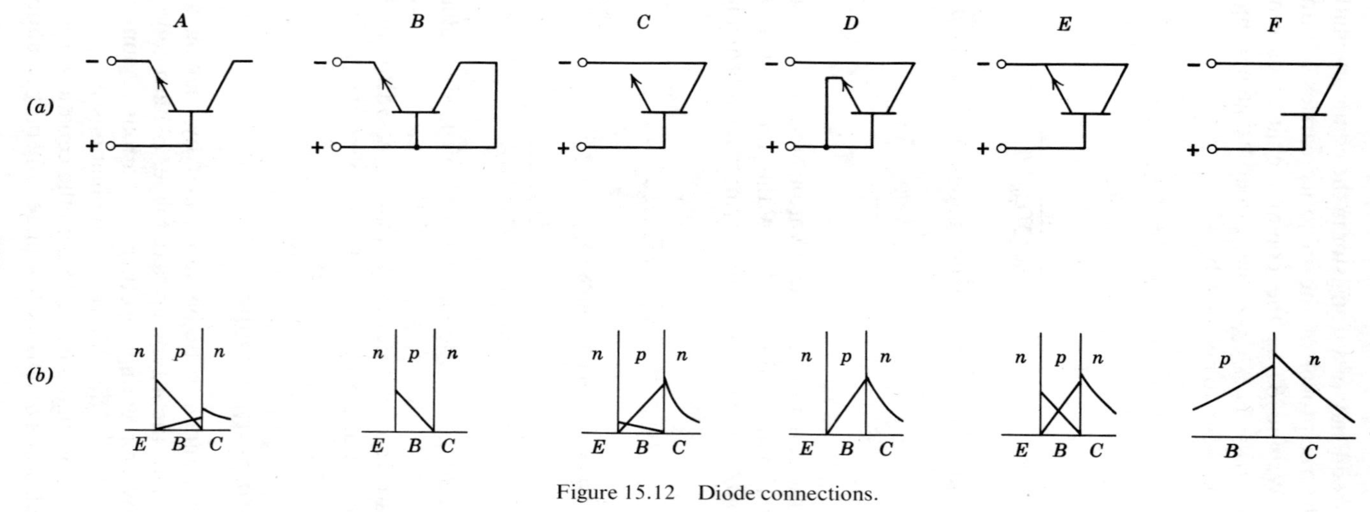

Figure 15.12 Diode connections.

For the usual sign convention† the steady-state solution of the continuity and diffusion equations for an *n-p-n* transistor may be written[5] in the form

$$\begin{bmatrix} -I_E \\ -I_C \end{bmatrix} = \begin{bmatrix} a_{11}, & a_{12} \\ a_{21}, & a_{22} \end{bmatrix} \begin{bmatrix} e^{qV_{BE}/kT} - 1 \\ e^{qV_{BC}/kT} - 1 \end{bmatrix} \tag{15.26}$$

A physical interpretation may be given to the various matrix coefficients, as follows:

$$-I_E = a_{11}(e^{qV_{BE}/kT} - 1) \qquad \text{when } V_{BC} = 0. \tag{15.27}$$

Thus a_{11} is the coefficient in the voltage-current relation of the emitter-base diode when the collector is short-circuited to the base, i.e., for case *B* in Fig. 15.12.

$$-I_C = a_{22}(e^{qV_{BE}/kT} - 1) \qquad \text{when } V_{BE} = 0. \tag{15.28}$$

Thus a_{22} is the coefficient associated with the equation for the collector-base diode when the emitter is short-circuited to the base; i.e., for case *D*.

$$-I_C = a_{21}(e^{qV_{BE}/kT} - 1) \qquad \text{when } V_{BC} = 0. \tag{15.29}$$

Here I_C is the collector current that flows when the transistor is connected as in the case *B*. Writing α_F as the d-c common-base current gain in the forward direction, (15.27) and (15.29) may be combined to obtain

$$a_{21} = -\alpha_F a_{11}, \tag{15.30}$$

$$-I_E = a_{12}(e^{qV_{BC}/kT} - 1) \qquad \text{when } V_{BE} = 0. \tag{15.31}$$

Writing α_R as the d-c common-base current gain in the reverse direction, that is, if the collector is operated as an emitter and vice versa,

$$a_{12} = -\alpha_R a_{22}. \tag{15.32}$$

Thus (15.26) may be rewritten as

$$\begin{bmatrix} -I_E \\ -I_C \end{bmatrix} = \begin{bmatrix} a_{11}, & -\alpha_R a_{22} \\ -\alpha_F a_{11}, & a_{22} \end{bmatrix} \begin{bmatrix} e^{qV_{BE}/kT} - 1 \\ e^{qV_{BC}/kT} - 1 \end{bmatrix} \tag{15.33}$$

where a_{11}, a_{22}, α_F and α_R are all positive quantities. For double-diffused transistors, α_F is typically 0.98 to 0.99, whereas α_R varies from 0.2 to 0.3. In heavily gold-doped devices, $\alpha_R \simeq 0.01$ while the value of α_F is essentially unchanged.

†Currents positive when going into the transistor.

15.3.1. Recovery Characteristics

Equation 15.33 may be used to determine the junction potentials for the various diode connections of Figure 15.12*a*. Thus it may be shown that, under conditions of forward-current flow, the collector-base junction is reverse-biased for case B, slightly forward-biased for case A, and heavily forward-biased for case C, D and E. With this in mind, the charge stored in the various transistor regions is sketched as in Figure 15.12*b*. Note that a negligible amount of charge can be stored in the emitter at any time, since this region is highly doped.

Inspection of Figure 15.12*b* shows that the amount of stored charge in the transistor configurations is the least for case B, and increases in the following order: B, A, D, C, E. The charge stored in the simple diode configuration of case F is the most of all, since the decay of this charge is exponential in both the n as well as the p region.

We may expect the recovery times of the various diodes to be placed in this same order. Typical values, given by Baker and Herr[6] for devices which are not gold doped, are as follows: Case B, 15 nsecs; Case A, 20 nsecs; Case D, 100 nsecs; Case C, 125 nsecs.

In gold-doped devices the same general trend occurs; however, the difference between these various connections is not as large since the collector stored charge is greatly reduced by the introduction of the gold.

15.3.2. The Forward Conductance

The forward-conductance characteristics of the various transistor configurations may also be compared by solving (15.33). At any forward current I_F the forward-voltage drop at low current levels is given by V_F', as follows:

CASE A.

$$V_F' = \frac{kT}{q} \ln\left[\frac{I_F}{a_{11}(1-\alpha_F\alpha_R)} + 1\right]. \tag{15.34a}$$

CASE B.

$$V_F' = \frac{kT}{q} \ln\left(\frac{I_F}{a_{11}} + 1\right). \tag{15.34b}$$

CASE C.

$$V_F' = \frac{kT}{q} \ln\left[\frac{I_F}{a_{22}(1-\alpha_F\alpha_R)} + 1\right]. \tag{15.34c}$$

CASE D.

$$V_F' = \frac{kT}{q} \ln\left(\frac{I_F}{a_{22}} + 1\right) \tag{15.34d}$$

CASE E.

$$V_F' = \frac{kT}{q} \ln\left[\frac{I_F}{a_{11}(1-\alpha_R) + a_{22}(1-\alpha_F)} + 1\right]. \tag{15.34e}$$

In these equations we note that

$$1-\alpha_F\alpha_R < 1. \tag{15.35}$$

In addition, $a_{22} > a_{11}$, since the collector area is larger than the emitter area. As a result, the forward conductance of the diode of case D is the highest, and the order of falling conductance is D, C, E, B, A. This is borne out in experimental data taken by Lin[7] for a forward current in 1 mA, as follows: Case D, 605 mV; Case C, 605 mV; Case E, 609 mV; Case B 677 mV; Case A, 694 mV.

The forward conductance of the diode of case F may be expected to be somewhat lower than that for case C (collector-base diode with emitter open). This is due to the fact that the slope of the minority carrier concentration in the p region is steeper in case C because the presence of the emitter region. In effect, the configuration of case C is very similar to that of a short-base diode, whereas case F represents a wide-base structure.

At high current levels, the ohmic voltage drop across the parasitic resistance dominates the forward-conductance characteristic. Noting that $r_E \simeq 0$ (since the emitter is highly doped), this ohmic drop is given as follows by:

CASE A. $$V_F'' = I_F r_B. \tag{15.36a}$$

CASE B. $$V_F'' = I_F r_B(1-\alpha_F). \tag{15.36b}$$

CASE C. $$V_F'' = I_F(r_B + r_C). \tag{15.36c}$$

CASE D. $$V_F'' = I_F[r_B(1-\alpha_R) + r_C]. \tag{15.36d}$$

CASE E. $$V_F'' \simeq I_F r_B(1-\alpha_F). \tag{15.36e}$$

CASE F. $$V_F'' \simeq I_F(r_C + r_B). \tag{15.36f}$$

Thus the highest conductance is exhibited by case B for high current levels, the order of falling conductance being given by B, E, D, A, C, F. Warner and Fordemwalt[8] have quoted the following data for the voltage drop across a typical diode operated at a forward current of 10 mA (see Problem 2): Case B, $V_F = 850$ mV; Case E, $V_F = 920$ mV; Case D, $V_F = 940$ mV; Case A, $V_F = 960$ mV; Case C, $V_F = 950$ mV.

15.3.3. Breakdown Voltage

The breakdown voltage of the diodes of case A and B is that associated with the emitter-base diode. Typically this is on the order of 6 V. On the other hand, the breakdown voltage in cases C, D, and F is that associated with the collector-base junction. This is usually considerably in excess of 6 V, its precise value being a function of the collector resistivity and the impurity grading constant $\mathscr{A}$.

In case E, where both diodes are connected in parallel, the breakdown voltage is given by that of the emitter-base diode. In all of the foregoing situations, breakdown is by means of secondary ionization processes and not by reach-through, as described in Section 14.3.6.

15.3.4. Substrate Capacitance Effects

Figure 15.13 shows the various diode configurations together with the structures used to implement them. The diode equivalent circuits may be drawn by inspection of these structures, as shown. In each case the diode is seen to have a parasitic capacitance across its terminals. In addition, one side of the diode is bypassed to ground via a shunting capacitance. This shunting capacitance appears at the p side of the diode in A and B, and at the n side in C, D, E, and F. Its magnitude is equal to C_{ts} in all cases except case A, where it is given by $C_{ts}\,C_{tc}/(C_{ts}+C_{tc})$.

15.3.5. Active Parasitics

In microcircuit structures the p-type substrate is tied to the most negative point in the circuit to insure that the collector-substrate junction is reverse-biased. Thus under certain conditions of circuit operation it is possible to obtain transistor action in the p-n-p device comprising the base, collector, and substrate regions shown in Figure 15.14a.

The doping profile of this parasitic p-n-p transistor is shown in Figure 15.14b. Here the emitter-base junction approximates a linearly graded structure, and has an injection efficiency of about 0.5 to 0.7. Thus, even if a base transport factor of unity is assumed, the device has a maximum common-emitter current gain of about 2! Furthermore, its base width is many times larger than that of the n-p-n transistor and hence its high-frequency performance is considerably poorer.

In circuit operation the presence of the p-n-p parasitic transistor may give rise to three different types of effects, as follows:

1. As an active device it may introduce a variable parasitic impedance in the microcircuit of which it is a part.
2. It may provide a "sneak" path by which unwanted coupling occurs between two otherwise isolated components.
3. It may provide a shunt path which diverts part of the current in the circuit.

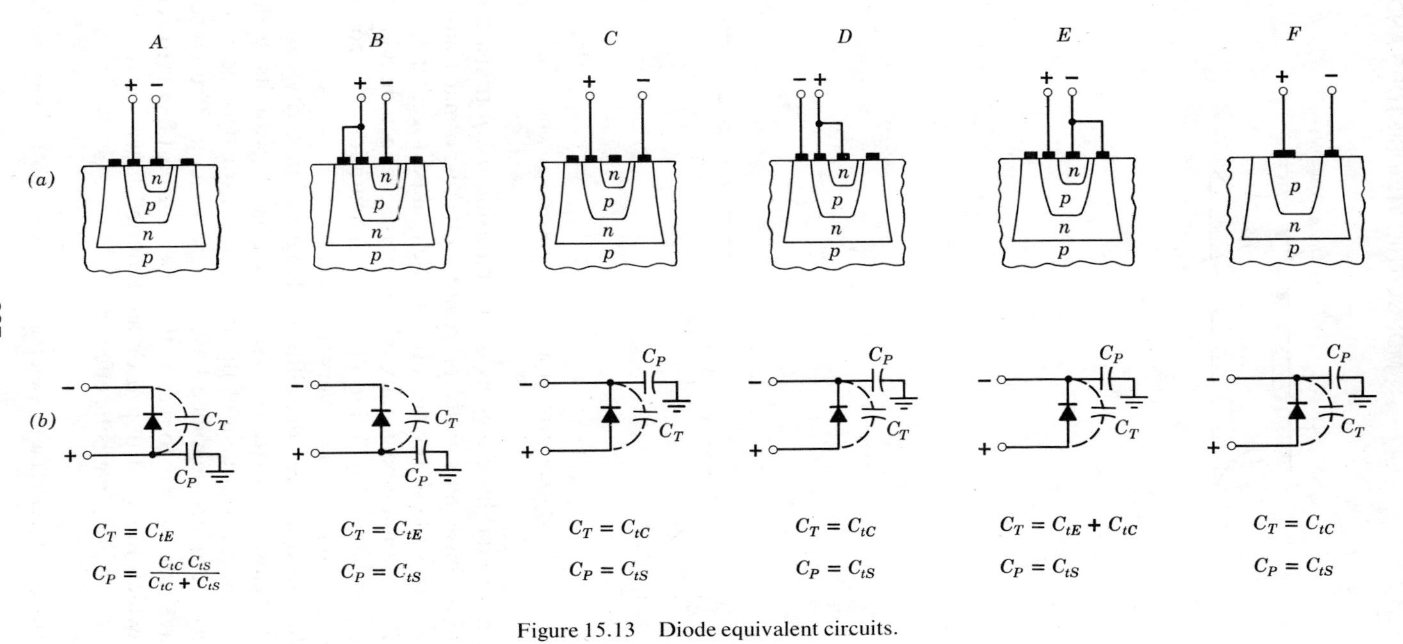

Figure 15.13 Diode equivalent circuits.

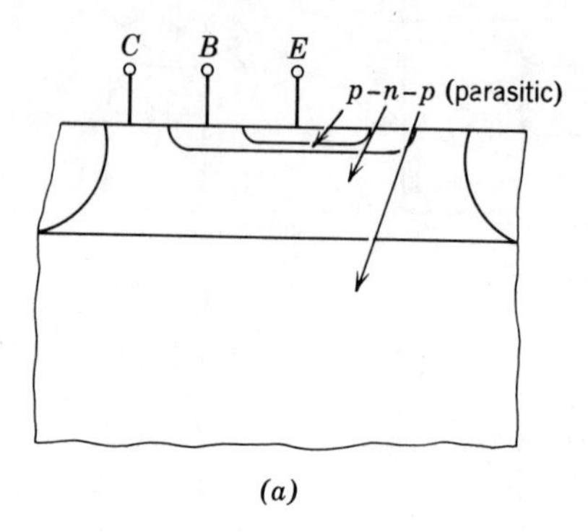

(a)

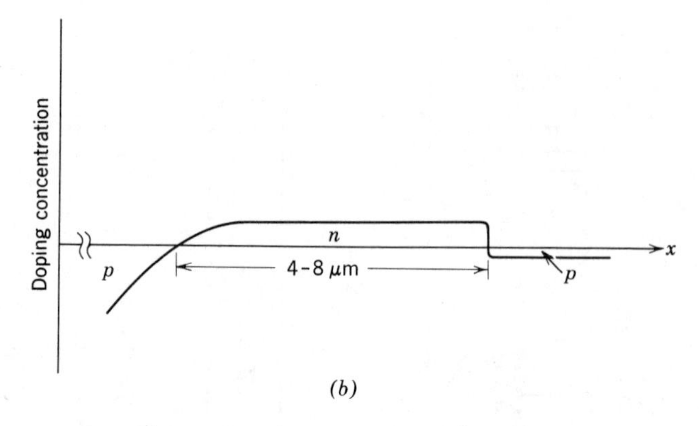

(b)

Figure 15.14 The parasitic *p-n-p* transistor.

The emitter and base regions of the parasitic *p-n-p* transistor are the same as the base and collector regions of the normal transistor respectively. Since these regions are short-circuited together in the diode of case *B*, there is no possibility of active *p-n-p* effects in this configuration (see Figure 15.15). With all other configurations the possibility of obtaining an active *p-n-p* is a very real one.

In gold-doped devices the diffusion length of minority carriers in the collector region† is extremely short compared with its width. Consequently, the current gain of the *p-n-p* transistor becomes vanishingly small because of its degraded base transport factor. The presence of a buried layer also serves to reduce the base transport factor to zero. Thus devices which are gold doped or have buried layers (or both) are not prone to active *p-n-p* parasitic effects.

†That is, in the base region of the parasitic transistor *p-n-p* transistor.

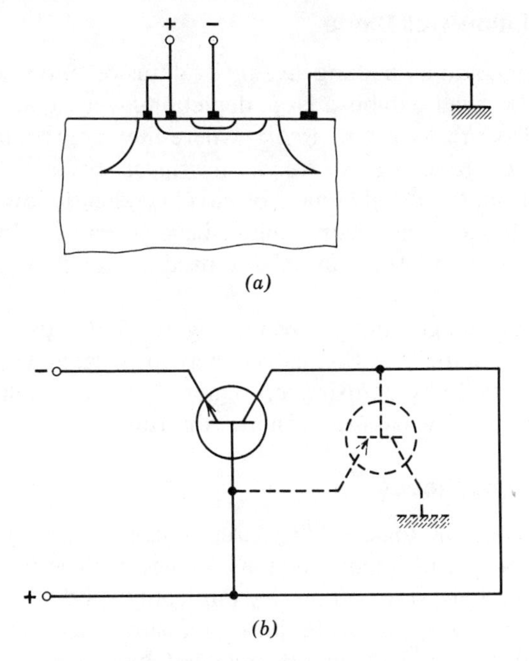

Figure 15.15 A diode configuration.

15.3.6. Choice of Diode Configuration

In view of the foregoing considerations, it is concluded that the diode configuration of case *B* is the best one for general circuit applications because of the following:

1. It has the best recovery characteristics.
2. It has the highest forward conductance at reasonable levels of operating current.
3. It is free from parasitic *p-n-p* effects.

Its single disadvantage is the fact that it has a breakdown voltage of about 6 V. In the event that a higher breakdown voltage is required, it is necessary to use the configurations of either case *C* or *D*. The configurations of cases *E* and *F* have no special advantage and are not used in practice.

15.4. SPECIAL DIODE STRUCTURES

As with transistors, a number of special diode structures are also possible. Some of these are described in the following.

15.4.1. Isolation-wall Diode

Diodes fabricated by making an emitter diffusion into the highly doped p-type isolation wall exhibit a large depletion-layer capacitance per unit area. Thus they are useful in circuits where they are required to provide temporary storage capability (e.g., in trigger circuits for flip-flops). They suffer from the disadvantage of having a slightly lower breakdown voltage than diodes made from emitter-base structures. In addition, the structure of Figure 15.16*a* can only be used in circuits where the p side is grounded.

An alternate configuration, shown in Figure 15.16*b*, uses a buried layer to stop the p^+ diffusion. In this manner an isolated high-capacitance diode may be made by diffusing a p region during the isolation diffusion step and subsequently making an emitter diffusion within it.

15.4.2. Multiple Diodes

In diode-transistor logic (DTL) circuits it is necessary to provide a number of diodes with their p regions connected together, as shown in Figure 15.17*a*. This may be done by fabricating all diodes in a common isolation pocket with separate base diffusions, as shown in Figure 15.17*b*; a further area saving is obtained by using a common base pocket, as shown in Figure 15.17*c*. In this last configuration care must be taken to avoid active n-p-n transistor action between adjacent emitters.

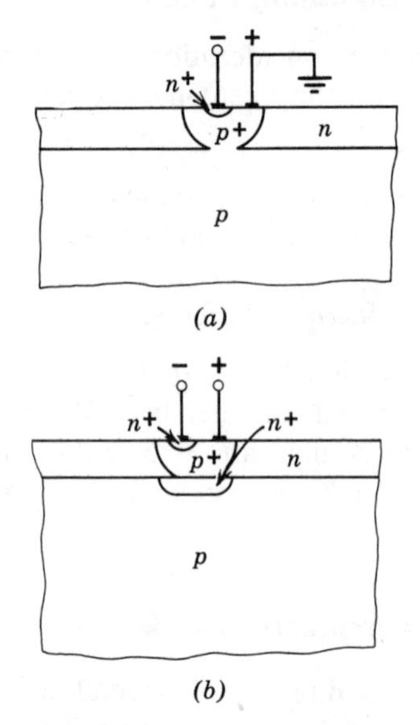

Figure 15.16 Isolation-wall diodes.

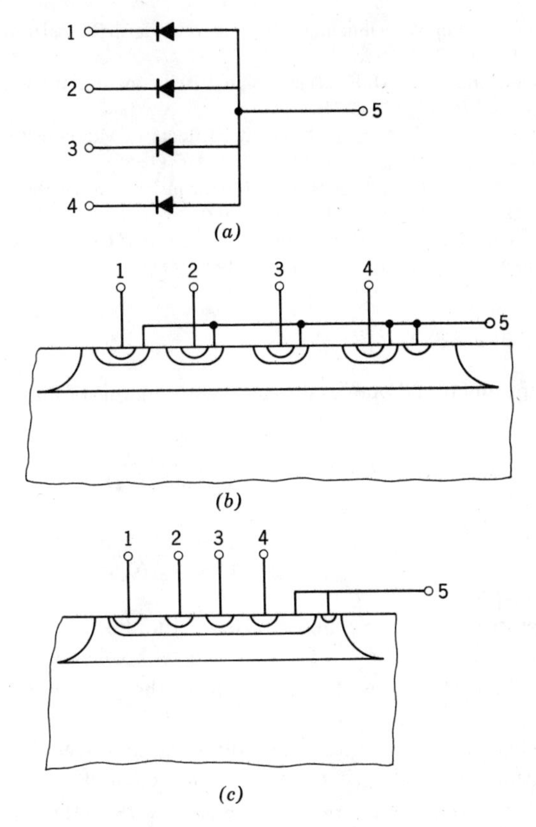

Figure 15.17 Multiple diodes.

This is done by fabricating the structure with a large lateral spacing between the various emitter regions and using gold doping to degrade the minority carrier lifetime. One such structure takes the form of Figure 15.10*b*, with the collector and base regions connected to form the common positive terminal.

15.5. REFERENCES

[1] J. B. Compton, and W. W. Happ, "Capacitive Design Considerations for Film-Type and Integrated Circuits," *IEEE Trans. on Electron Devices*, **ED-11** (10), 447–455 (1964).

[2] D. F. Allison, "Transistor Structure with Small Inverse Gain," *Patent No. 3,315,138*, assigned to Signetics Corp., California, (April 18, 1967).

[3] H. C. Lin et al., "Lateral Complementary Transistor Structure for the Simultaneous Fabrication of Functional Blocks," *Proc. IEEE*, **52** (12), 1491–1495 (1964).

[4] J. Lindmayer, and W. Schneider, "Theory of Lateral Transistors," *Solid State Electron.*, **10** (3), 225–234 (1967).
[5] J. J. Ebers, and J. L. Moll, "Large-Signal Behavior of Junction Transistors," *Proc. IRE*, **42** (12), 1761–1762 (1954).
[6] D. W. Baker, and E. A. Herr, "Parasitic Effects in Microelectronic Circuits," *IEEE Trans. on Electron Devices*, **ED-12** (4), 161–167 (1965).
[7] H. C. Lin, "Diode Operation of a Transistor in Functional Blocks," *IEEE Trans. on Electron Devices*, **ED-10** (3), 189–194 (1963).
[8] R. M. Warner, Jr., and J. N. Fordemwalt, *Integrated Circuits—Design Principles and Fabrication*, McGraw-Hill, New York, 1965.

15.6. PROBLEMS

1. An abrupt-junction diode is fabricated with regions of finite impurity. Show that

$$x_p^2 = \frac{2\epsilon\epsilon_0 V_j}{qN_a(1+N_a/N_d)}$$

and

$$x_n^2 = \frac{2\epsilon\epsilon_0 V_j}{qN_d(1+N_d/N_a)}.$$

2. The measured data of Section 15.3.2 for the diode conductance at high level do not concur with the predictions of the theory. Explain why this is the case.

3. I_{CO} has been defined as the coefficient in the voltage-current relation of the collector-base diode when the emitter is left open. In like manner, I_{EO} is defined as the coefficient in the emitter-base diode equation when the collector is left open. Determine I_{CO} and I_{EO} in terms of a_{11}, a_{22}, α_F, and α_R.

4. The collector region of a transistor made by the DDE process is 6 μm long and has a resistivity of 0.5 Ω-cm. Determine the base transport factor of the parasitic *p-n-p* transistor for (*a*) a gold-doped structure, with $\tau_p \simeq 10^{-8}$ sec. (*b*) a non-gold-doped structure, with $\tau_p \simeq 10^{-6}$ sec. Hence determine the common-emitter current gain of the parasitic *p-n-p* device. Assume $\gamma = 0.7$.

5. A monolithic transistor has the following specifications:

Collector resistivity, 0.25 Ω-cm,
Emitter depth, 1.5 μm,
Base depth, 2 μm,
Substrate depth, 15 μm,
Emitter area, 0.25 $\times$ 2 mils,
Collector contact area, 0.25 $\times$ 3 mils,
Number of collectors, 2
Distance between emitter and collector edge, 0.5 mil.

Determine the collector parasitic resistance, assuming a buried layer having a sheet resistance of (*a*) 15 Ω/square; (*b*) 5 Ω/square.

6. Repeat Problem 5, assuming an epitaxial layer thickness of 5 μm. What conclusions can you draw from the results of Problems 5 and 6?

Chapter 16

Monolithic Resistors and Capacitors

In addition to transistors and diodes, microcircuits make extensive use of resistors and, to a lesser extent, of capacitors. Monolithic capacitors are quite inferior in the characteristics to commonly available discrete units and take up considerably more space than active elements. Consequently, it is not surprising that the microcircuit designer seeks to avoid their use whenever possible.

In this chapter the properties of monolithic resistors and capacitors are described, and their limitations considered in detail. The treatment of thin-film components is a subject in itself and is adequately treated in the references [1–3] at the end of this chapter.

16.1. MONOLITHIC RESISTORS

Over the years a number of different techniques have been proposed for the fabrication of microcircuit resistors. Of these, the only method in use today is one in which the resistor is formed by diffusion into a background region of opposite conductivity type. These resistors are far from ideal circuit elements; however, their almost universal adoption is based on the fact that no additional processes are needed for their fabrication. Furthermore, they are adequate in the overwhelming majority of both digital and linear microcircuit applications.

Figure 16.1 shows a resistor comprising a *p*-type diffusion into an *n*-type background. Resistors of this type are made simultaneously with the base diffusion; consequently, many of their characteristics are determined by the requirements of the associated active devices which form the basis of the microcircuit design.

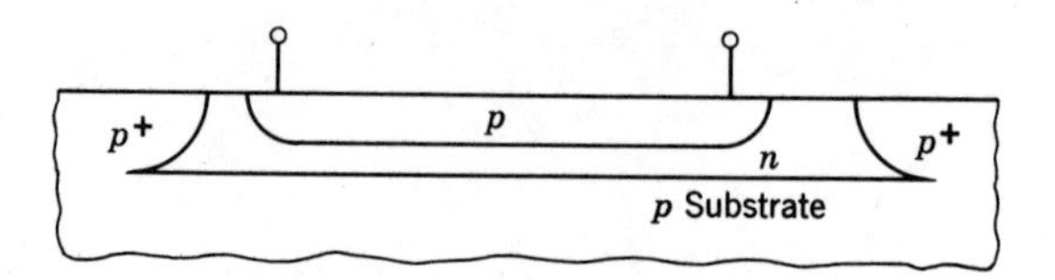

Figure 16.1 The diffused resistor.

16.1.1. Resistor Values and Tolerances

The sheet resistance of a base diffusion is typically set at a value between 100 to 200 Ω/square. Thus the base diffusion results in reasonably proportioned resistors from about 50 Ω to 10 kΩ and is useful for the majority of circuit applications.

The tolerance on the initial value of the resistor is a function of a number of factors, as follows:

The Diffusion Process. In practice, the sheet resistance is monitored during this process (as described in Section 4.9.4) and can be held to a tolerance of ±5% of its room temperature value.

Mask Dimensions. Drawings are normally cut on a drafting machine with a ±0.1-mil error. Upon reduction by a factor of 100 or more, the resulting error in the actual mask becomes vanishingly small.

Oxide-Cut Dimensions. An important source of error lies in the difficulty of carrying over the mask dimensions to the dimensions of the cut in the oxide window. The edges of such a window tend to be uneven, because of limitations in the photographic process as well as in the etching process. This error may be minimized by using as wide a line width as possible.

End Effects. These are difficult to calculate and lead to an uncertainty in the value of the resistor. Their effect may be minimized by making the resistor reasonably long compared to its width.

In practice, using line-widths of 0.25 mil or more, a tolerance of ±10% of the initial value is generally possible for resistors of this type. At both extremes of resistor range, however, this tolerance is difficult to achieve.

The temperature characteristics of a diffused resistance are somewhat difficult to compute accurately because of the variation of its doping concentration with diffusion depth. To an approximation, however, its characteristics are dominated by the region near the surface where the doping concentration is the greatest.

Curves of normalized resistance, based on the variation of mobility with temperature, are shown in Figure 16.2 for p-type material with

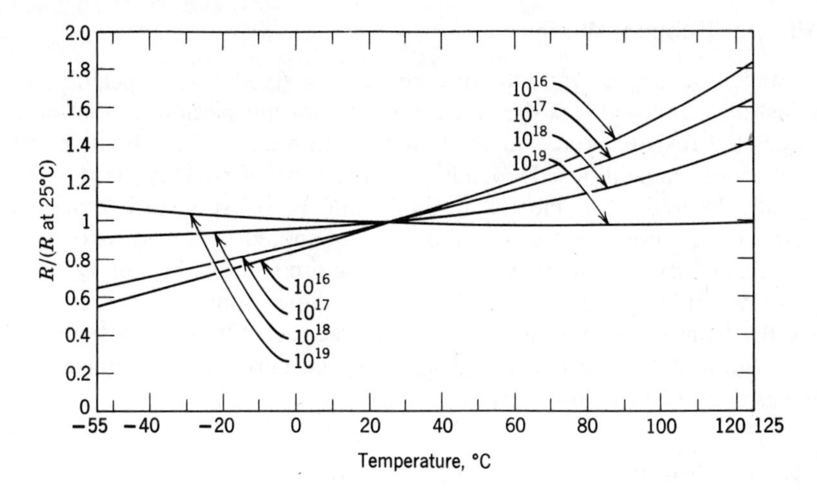

Figure 16.2 Temperature coefficients.

different carrier concentrations. These curves are approximate, in that they do not take into consideration the variation in the ionized impurity concentration. This variation is particularly significant at high doping concentrations where resistance changes due to mobility are very small. In addition, it is a sensitive function of the activation energy of the dopant that is used.

In practice, most base diffusions are performed with surface concentrations between 10^{18} and 10^{19} atoms/cm^3. For a typical boron-diffused resistor with a surface concentration of 10^{19} atoms/cm^3, this results in temperature variations (from the nominal value at 25°C) of approximately +4% at −55°C, and +8% at +125°C.

16.1.2. Isolation

A diffused resistor is isolated from its background by the contact potential of the associated *p-n* junction or by a higher reverse voltage if the background region is appropriately biased. As a consequence, a number of resistors may be placed in the same isolation pocket, resulting in a physical saving of space on the silicon slice. In practice, it is customary to lump as many resistors as conveniently possible within an individual pocket.

The breakdown voltage between resistors is equal to the collector-base breakdown voltage of transistors made on the same slice. In addition, the reverse-leakage-current density is also the same as that of the associated collector-base junction.

16.1.3. Resistor Width

Since the aspect ratio of any resistor is fixed for a specific sheet resistance, the width is the single most important parameter that determines the resistor area. The lower limit to this area is fixed by its power-dissipation capabilities. Depending on the type of package used, this is usually limited to at most 2 to 5 mW/mil^2 of resistor surface area. In many cases, however, where the power handled by the resistor is insignificant, its minimum width is usually limited by the ability to cut windows in the oxide with satisfactory dimensional tolerances. Line widths from 0.25 to 1 mil are used in practice. Wherever possible, the line width should be made as large as space permits to minimize tolerances due to dimensional errors.

16.1.4. Contacts

Ohmic contacts are made to the resistor by aluminum evaporation, followed by subsequent alloying as described in Section 8.2.1. Sometimes a p^+ region is diffused into the resistor and contact is made to this region. This results in a more linear voltage-current characteristic; however, it involves an additional masking and diffusion step and is only used in special situations.

The presence of the contact results in distortion of the current flow lines in its vicinity and alters its effective resistance. Many configurations have been attempted to minimize these effects. However, the practical approach is to use a contact scheme that is reasonably uniform in its current flow characteristics and, at the same time, convenient to implement. Two such schemes are shown in Figure 16.3. Empirically it has been found that the contact of Figure 16.3*a* results in an effective extension of length of the uniform section of the resistor by 0.65 square. The scheme of Figure 16.3*b* occupies less space and extends the resistor length by 0.35 square.

In applications in which the resistor value is critical, taps may be placed at $\pm 5\%$ of the design value, and the final metallization pattern may be selected after an experimental choice of the correct tap has been made. Sometimes the ohmic contacts are made to all of the taps, even though the final metallization path connects to only one of them.

16.1.5. Resistor Shape

The microcircuit pattern is usually cut on a drafting machine. As such, all cuts are generally restricted to straight lines, circular arcs, and right-angle bends. Both the circular arc and the right-angle bend are used

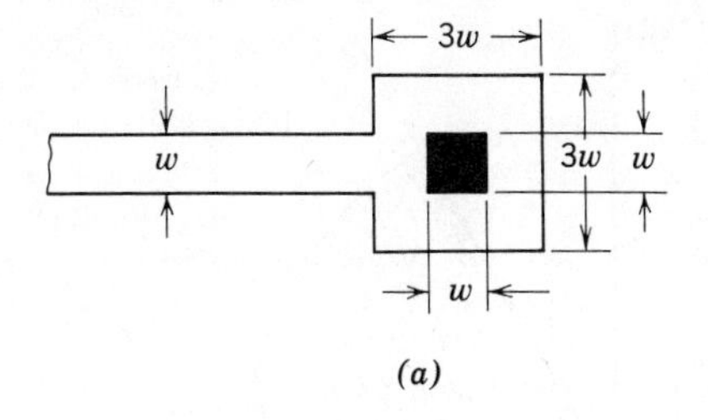

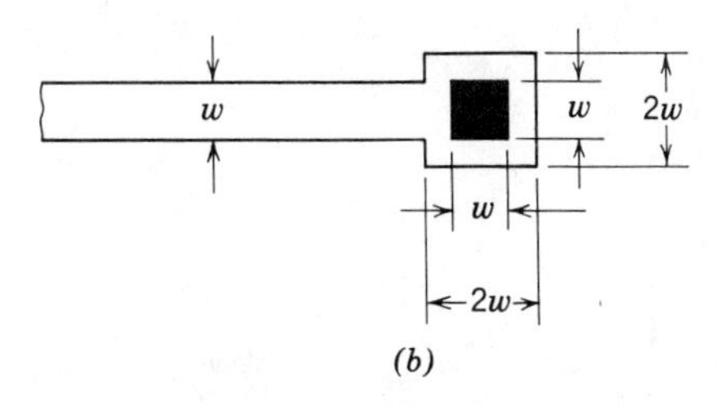

Figure 16.3 End effects.

for accommodating high-value resistances into a reasonable space factor.

A conformal mapping technique may be used to determine the resistance of a circular arc, of inner and outer radii of r_i and r_0 respectively, as shown in Figure 16.4*a*. The angle subtended by this element of the resistor is θ rads. The resistor is shown in the complex z plane, where

$$z = x + jy \tag{16.1}$$

The transformation is made into the w plane by means of the mapping function

$$w = \ln z = u + jv \tag{16.2}$$

Writing z in polar coordinates,

$$w = \ln re^{j\theta}, \tag{16.3}$$

$$= \ln r + j\theta, \tag{16.4}$$

hence

$$u = \ln r, \tag{16.5}$$

$$v = \theta. \tag{16.6}$$

Thus the resistor of Figure 16.4*a* maps conformally into a rectangle in the w plane, with AB mapping into $A'B'$, and CD mapping into $C'D'$.

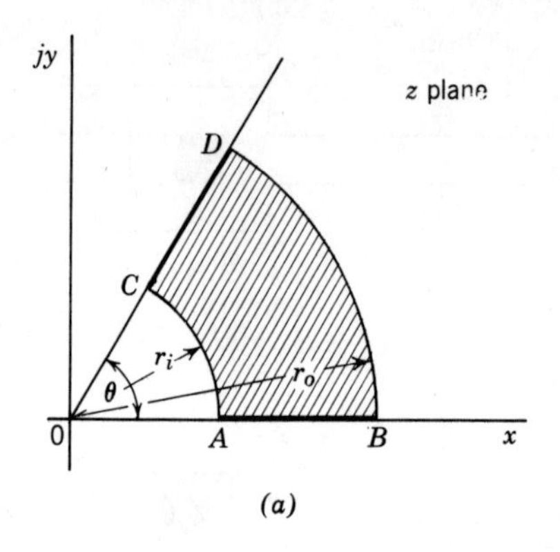

(a)

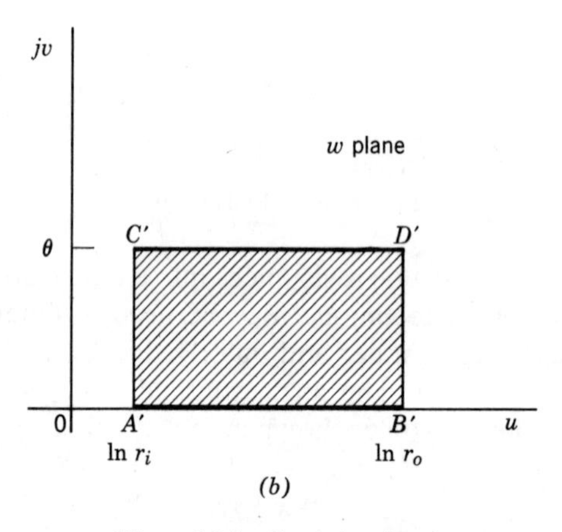

(b)

Figure 16.4 Curved resistors.

The dimensions of this rectangle are

$$A'B' = C'D' = \ln r_0 - \ln r_1 \tag{16.7}$$

$$= \ln r_0/r_1 \tag{16.8}$$

and

$$A'C' = B'D' = \theta. \tag{16.9}$$

The value of the resistance is given by $\rho_\square(A'C')/A'B'$, where $\rho_\square$ is its sheet resistance. In the z plane, the resistance is

$$R = \frac{\rho_\square \theta}{\ln (r_0/r_i)} \tag{16.10}$$

Although conformal mapping techniques have been used to determine the resistance of a single right-angle bend [4], practical situations usually call for a double right-angle bend, which is best handled empirically. It has been found [5] that a bend with the proportions shown in Figure 16.5 has an effective resistance of one square between *AB* and *CD*, provided it is followed on either side by a minimum of three squares of resistor length.

It may also be shown that a sharp right-angle bend leads to an infinite concentration of current at its inside corner. In spite of this fact, however, it is extremely popular in low-level situations because of the ease with which it is drawn. On occasion a small fillet is added to relieve this current concentration.

16.1.6. Parasitic Effects

Associated with the diffused resistance are a number of parasitic terms which must be considered in high-frequency applications. If the resistor is made positive relative to the *n* region, the over-all structure will exhibit *p-n-p* transistor action of the type described in Section 15.3.5. The equivalent circuit of this element is shown in three-lumped form in Figure 16.6*b*. A simplified version of this circuit is drawn in Figure 16.6*c*, from which it is seen that the resistor behaves as an active *RC* transmission line. Each capacitive element of this line is equal to the collector-base capacitance of the parasitic transistor, multiplied by its common-emitter current gain (which may be as high as 1 to 2). This is highly undesirable, since the characteristics of the resistor are now dependent on the gain characteristics of the transistor as well as on its collector-base capacitance.

The parasitic transistor does not present a problem in gold-doped circuits in which its current gain is essentially zero. In linear circuits, however (or current-mode logic circuits), the gold-doping step is omitted. Here the *n*-type epitaxial region must be tied to the most positive point in the system, so that this transistor is cut off at all times. This technique

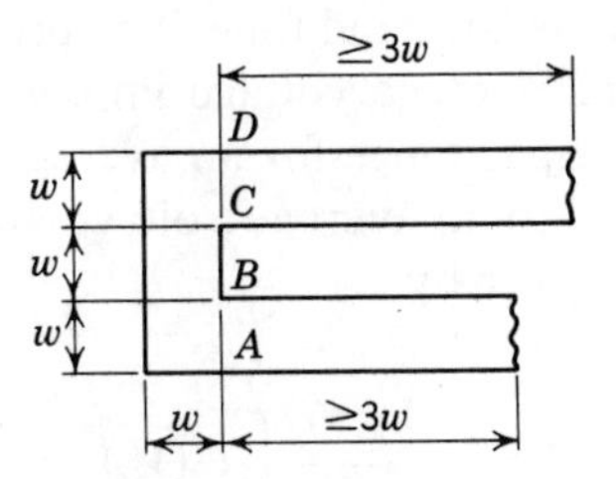

Figure 16.5 Folded resistors.

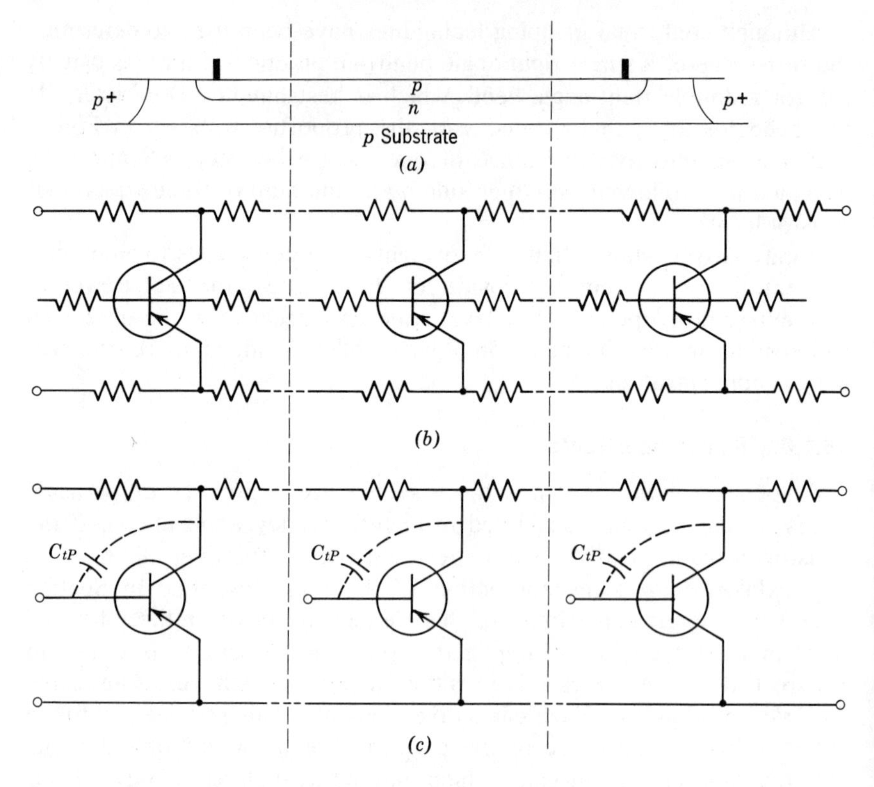

Figure 16.6 Parasitic effects.

is so simple and effective that it is commonly used even with gold-doped circuits, thus providing an additional margin of safety.

With the parasitic transistor cut off, the equivalent circuit of the resistor reduces to that of a passive RC transmission line. The shunt capacitance of this line is given by the capacitance of a linearly graded junction. In a practical situation, where a time-varying signal is applied to the resistor, the capacitance of any elemental section is a function of both distance along the resistor and time. The problem may be simplified by considering only the average voltage impressed across the resistor. Thus let $C(V_j)$ be its capacitance for an average voltage of V_j. Then, if the resistor is subjected to an average voltage change from V_a to V_b, its effective capacitance is given by

$$\bar{C} = \frac{1}{V_b - V_a} \int_{V_a}^{V_b} C(V_j)\, dV_j. \tag{16.11}$$

The equivalent circuit of the resistor may now be drawn in lumped form, as shown in Figure 16.7*a*. Alternately, it may also be drawn in distributed network form as a *RC* transmission line with total capacitance† of $(\pi^2/8)\bar{C}$, as shown in Figure 16.7*b*. The validity of this approximation may be estimated by comparing the response of these networks to a step of input current. For the T network, the short-circuit current gain transform is given by

$$-h_{21}(s) = -\frac{I_2(s)}{I_1(s)} = \frac{1}{1+s\bar{C}R/2}, \tag{16.12}$$

where s is the differential operator. Setting $\omega_0 = 2/\bar{C}R$,

$$-h_{21}(s) = \frac{1}{1+s/\omega_0} \tag{16.13}$$

and the response of this network is thus

$$\left|\frac{I_2(t)}{I_1(t)}\right| = 1 - e^{-\omega_0 t} \tag{16.14}$$

This response is shown in Figure 16.8.

†Although the factor $\pi^2/8$ is carried through the mathematical treatment, it is worth noting that the degree of approximation in determining $\bar{C}$ admits of the further approximation that $(\pi^2/8)\bar{C} \simeq \bar{C}$.

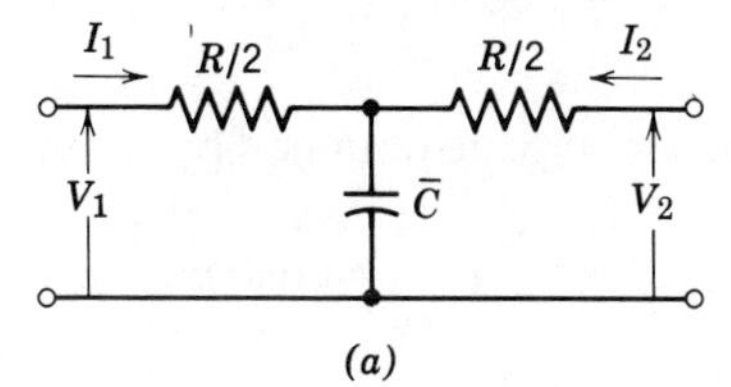

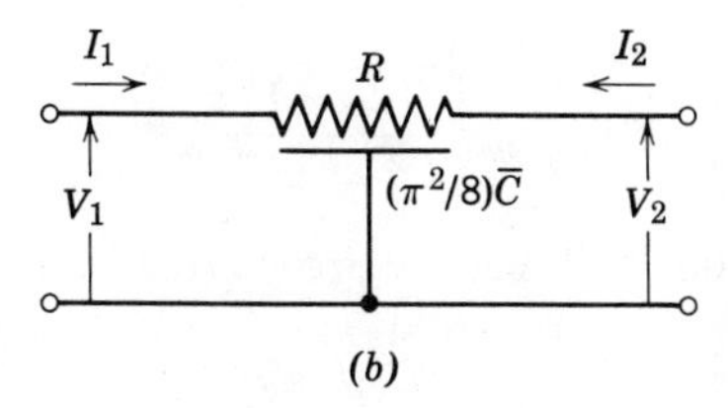

Figure 16.7 Equivalent circuit of a diffused resistor.

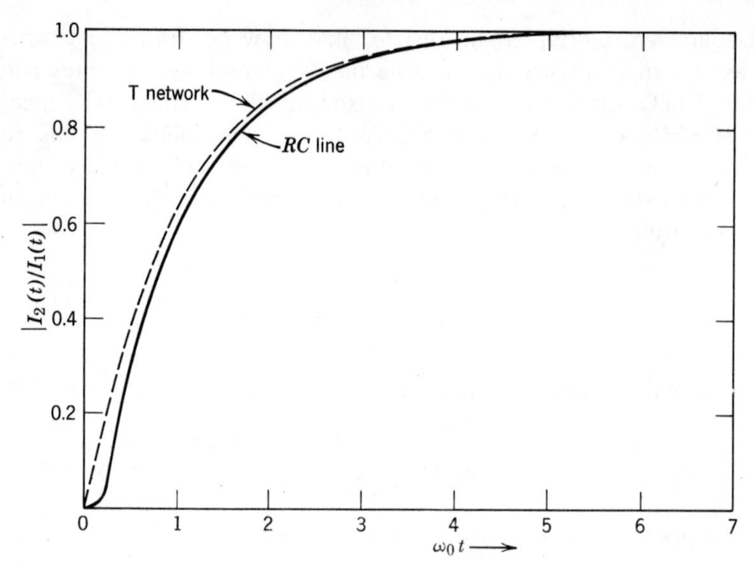

Figure 16.8 Transient response.

For the transmission-line equivalent circuit[6], the current gain transform is given by

$$-h_{21}(s) = -\frac{I_2(s)}{I_1(s)} = \left(\cosh\sqrt{\frac{\pi^2}{8}s\bar{C}R}\right)^{-1}, \tag{16.15}$$

$$= \left[\cosh\frac{\pi}{2}\sqrt{\frac{s}{\omega_0}}\right]^{-1}. \tag{16.16}$$

The poles of this function may be obtained by setting

$$\cosh\frac{\pi}{2}\sqrt{\frac{s}{\omega_0}} = \cosh jz = 0, \tag{16.17}$$

where z is the complex variable. Then

$$\cosh jz = \cos z = 0 \tag{16.18}$$

and

$$z_k = \frac{(2k-1)\pi}{2}, \tag{16.19}$$

where k takes on values $1, 2, 3, \ldots$. From (16.17),

$$jz_k = \frac{\pi}{2}\sqrt{\frac{s_k}{\omega_0}}. \tag{16.20}$$

Therefore

$$s_k = -(2k-1)^2\omega_0 . \tag{16.21}$$

The current gain transform is thus given by

$$-h_{21}(s) = \frac{1}{(1+s/\omega_0)(1+s/9\omega_0)(1+s/25\omega_0)\ldots} \tag{16.22}$$

so that

$$-\frac{I_2(s)}{I_1(s)} = \frac{1}{s(1+s/\omega_0)(1+s/9\omega_0)(1+s/25\omega_0)\ldots} \tag{16.23}$$

$$\simeq \frac{1}{s} - \frac{9}{8(s+\omega_0)} + \frac{1}{8(s+9\omega_0)} \cdots \tag{16.24}$$

The response to a step input current is given by the inverse Laplace transform as

$$\left|\frac{I_2(t)}{I_1(t)}\right| \simeq 1 - \tfrac{9}{8}e^{-\omega_0 t} + \tfrac{1}{8}e^{-9\omega_0 t} - \ldots . \tag{16.25}$$

This function is also shown in Figure 16.8. Comparison with the transient response for the T network shows a nearly identical characteristic except that the response of the transmission line is delayed by $\omega_0 t \simeq 0.22$ radians.

The transmission-line equivalent circuit may also be used to determine the small-signal characteristics of this resistor at any fixed value of d-c bias. Thus for the transmission line shown in Figure 16.7*b*,

$$\begin{bmatrix} I_1 \\ I_2 \end{bmatrix} = \begin{bmatrix} y_{11} & y_{12} \\ y_{12} & y_{22} \end{bmatrix} \begin{bmatrix} V_1 \\ V_2 \end{bmatrix} \tag{16.26}$$

where

$$y_{11} = y_{22} = \frac{1}{R}\frac{\pi}{2}\sqrt{\frac{s}{\omega_0}} \coth \frac{\pi}{2}\sqrt{\frac{s}{\omega_0}} \tag{16.27}$$

and

$$y_{12} = -\frac{1}{R}\frac{\pi}{2}\sqrt{\frac{s}{\omega_0}} \operatorname{cosech} \frac{\pi}{2}\sqrt{\frac{s}{\omega_0}} \tag{16.28}$$

The input impedance of the resistor, with output shorted, is thus given by (see Problem 16.8)

$$h_{11} = \frac{1}{y_{11}} = R\frac{(1+s/4\omega_0)(1+s/16\omega_0)(1+s/36\omega_0)\ldots}{(1+s/\omega_0)(1+s/9\omega_0)(1+s/25\omega_0)\ldots}. \tag{16.29}$$

Figure 16.9 shows the magnitude and phase of h_{11}/R. It is seen that the magnitude of this impedance falls off with frequency at 3 dB/octave while the phase asymptotes to 45°.

The short-circuit current gain of the transmission line is given by

$$-h_{21} = -\frac{y_{21}}{y_{11}} = \left[\cosh \frac{\pi}{2}\sqrt{\frac{s}{\omega_0}}\right]^{-1}. \tag{16.30}$$

Figure 16.9 also shows the attenuation and phase characteristics of h_{21}. Here the magnitude of h_{21} falls off continaully with increasing frequency and the phase lag increases without limit.

At frequencies up to $\omega = \omega_0$ the current gain of the transmission line may be approximated by that of a simple RC network with an excess phase-delay term. The magnitude of this phase delay is about 11° at $\omega = \omega_0$. Assuming a linear phase-delay characteristic,†

$$-h_{21} = \frac{e^{-jm\omega/\omega_0}}{1 + j\omega/\omega_0} \tag{16.31}$$

where m $\simeq$ 0.22 radians.

†Note the similarity between this result and that obtained for the current gain of a uniform-base transistor.

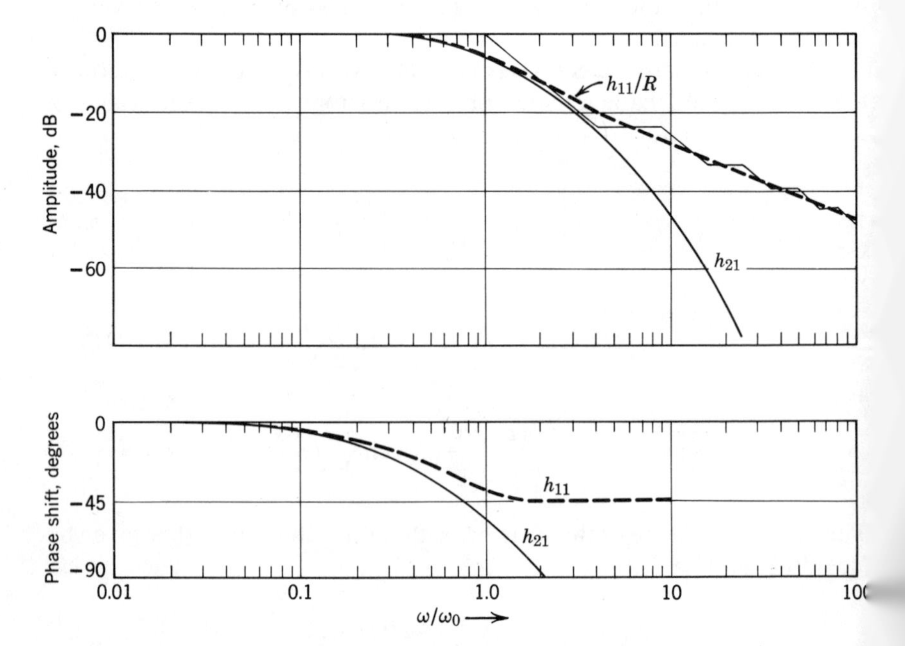

Figure 16.9 Impedance characteristics.

For any given fabrication process the cutoff frequency is a function of the desired resistor value and the resistor width. Thus

$$R = \frac{\rho_{\square} l}{w} \tag{16.32}$$

and

$$\bar{C} = c_0 l w, \tag{16.33}$$

where c_0 is the capacitance per unit area. Combining these equations,

$$\omega_0 = \frac{2}{\bar{C}R} = \frac{2\rho_{\square}}{c_0 R^2 w^2} \tag{16.34}$$

Consequently, for any specified resistance value, the cutoff frequency varies inversely with the square of the resistor width. This is an important criterion in selecting the dimensions of a resistor; in essence, it states that, for high-frequency operation, resistor widths should be kept as small as possible, commensurate with achieving the desired tolerance.

16.1.7. Low-value Resistors

Low values of resistors may be obtained by using the properties of the diffused emitter region. Here a sheet resistance of 2 to 10 Ω/square is typical and results in a reasonable aspect ratio for resistor values as low as 1 Ω.

Figure 16.10 shows the cross section of a resistor of this type. The resistor is fabricated by making an n-type emitter diffusion into a p-type base region obtained by a prior base-diffusion step. It is especially important that this base region be tied to the most negative point in the circuit to ensure that the emitter-base diode is cut off. The reason for this

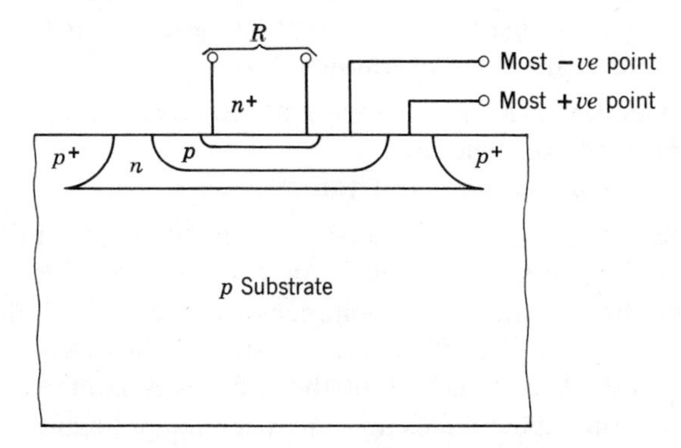

Figure 16.10 The emitter-diffused resistor.

lies in the fact that the "parasitic" *n-p-n* transistor in this situation has a gain of 40 or more, even in gold-doped circuits. In addition, a parasitic *p-n-p-n* structure is present here; under certain conditions [7] the structure can latch up, leading to circuit failure.

The low-frequency characteristics of emitter-diffused resistors are essentially identical to those of resistors made by the base-diffusion step. However, because of the low resistance values involved, end effects are generally more important in determining the resistor value. In addition, the breakdown voltage of an emitter-diffused resistor is the same as that of an emitter-base junction, and is limited to about 6 V.

High-frequency effects may usually be ignored because the cutoff frequency is generally a decade or more beyond that of other resistors in a typical microcircuit.

16.2. MONOLITHIC CAPACITORS

The requirements on a capacitor are that it have a high cutoff frequency, and that its capacitance value be independent of the amplitude of the signals that are impressed across it. The only capacitors satisfactorily meeting these requirements are those made by the vacuum evaporation of thin conducting films on either side of an insulating material. These have been treated extensively in the literature [1–3] and are not considered here.

Two types of capacitors can be fabricated by monolithic processes: *p-n* junction capacitors and metal-oxide-semiconductor (MOS) capacitors. These are now described together with their characteristics.

16.2.1. The *p-n* Junction Capacitor

The properties of this type of capacitor are similar to those of the various *p-n* junctions used in a microcircuit. Figure 16.11 shows three different fabrication schemes, as follows:

Type 1. Collector-base structure with buried layer.
Type 2. Emitter-base structure.
Type 3. Emitter-p^+ isolation structure.

All of these structures have parasitic resistances which degrade their cutoff frequency (since ω_0 varies inversely with the *CR* product). In addition, some have parasitic capacitances associated with them. All are voltage dependent and alter the characteristics of the circuit in which the capacitor is used. A comparison of these types is made in Table 16.1.

The *p-n* junction capacitances are only used in applications where their nonlinear behavior can be tolerated. Thus they are suitable for use as

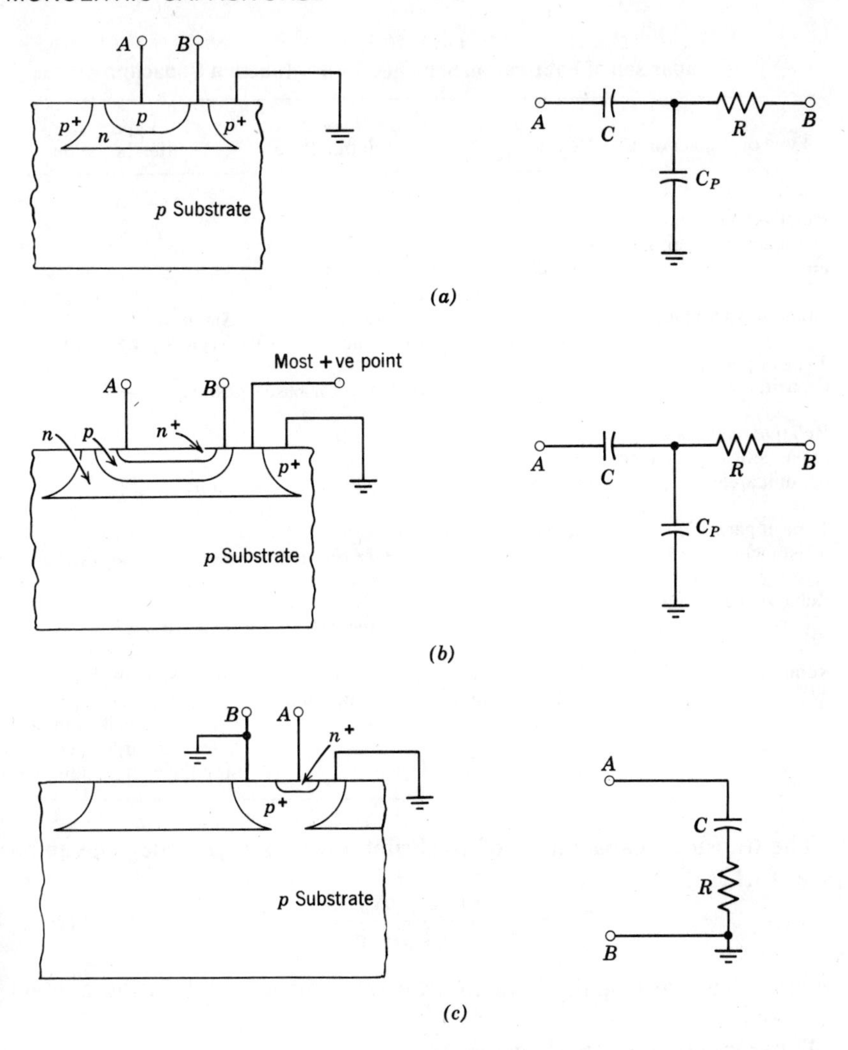

Figure 16.11 Monolithic capacitors.

speedup capacitances in digital circuits, temporary storage devices in triggering schemes, and as bypass capacitances.

The temperature variation of a *p-n* junction capacitance is caused by two effects. First, the dielectric constant of silicon is temperature dependent and increases at the rate of about 200 ppm/°C (parts per million per degree Centigrade). Second, the Fermi level moves towards the intrinsic level with increasing temperature. As a result the contact potential falls with increasing temperature.

Table 16.1
Comparison of Fabrication Schemes for *p-n* Junction Capacitor

Type of Capacitor, C	Type 1 Collector-Base	Type 2 Emitter-Base	Type 3 Emitter-Isolation
Relative magnitude of capacitance per unit area	Smallest	Medium	Largest
Breakdown voltage	Largest (typically 20–50 V)	Medium (typically 5–6 V)	Smallest (typically 4.5–5 V)
Type of parasitic capacitance, C_p	Collector-substrate	Collector-base	None
Relative magnitude of parasitic capacitance per unit area	Smallest	Medium	None
Type of parasitic resistance	Collector series resistance	Base resistance	Isolation-wall resistance
Relative magnitude of parasitic resistance	Medium	Largest	Smallest
Remarks	Used in series or in shunt	Used in series or in shunt	Only useful in shunt applications; can be used in series if a buried layer is incorporated to stop the p^+ diffusion

The transition capacitance of an abrupt junction n^+-p diode is given by [see (13.12*b*)]

$$C_t = A\left[\frac{\epsilon\epsilon_0 q N_a}{2(V_R+\Psi)}\right]^{1/2}, \tag{16.35}$$

where V_R is the magnitude of the reverse voltage and Ψ is the contact potential.

Differentiation of (16.35) gives

$$\frac{dC_t}{C_t} = \frac{1}{2}\left[\frac{d\epsilon}{\epsilon} + \frac{d(V_R+\Psi)}{V_R+\Psi}\right]. \tag{16.36}$$

Thus a fractional change in the capacitance is equal to one-half the sum of a fractional change in the relative permittivity and a fractional change in the junction voltage. Writing T as the temperature, (16.36) may be rewritten as

$$\frac{dC_t/dT}{C_t} = \frac{1}{2}\left[\frac{d\epsilon/dT}{\epsilon} + \frac{d(V_R+\Psi)/dT}{V_R+\Psi}\right]. \tag{16.37}$$

In this equation

$$\frac{d\epsilon/dT}{\epsilon} \simeq 200 \text{ ppm/°C}. \quad (16.38)$$

Note also that $d(V_R + \Psi) = d\Psi$.

The temperature variation of the contact potential may be approximately determined by assuming that the entire variation is due to the changing value of the intrinsic carrier concentration. In practical situations this potential is found to fall at the rate of approximately 2 mV/°C. Thus

$$\frac{d(V_R + \Psi)/dT}{V_R + \Psi} \simeq \frac{2000}{V_R + \Psi} \text{ ppm/°C}. \quad (16.39)$$

Combining (16.37), (16.38), and (16.39), the temperature variation of a depletion-layer capacitance is given by

$$\frac{dC_t/dT}{C_t} \simeq \left(\frac{1000}{V_R + \Psi} + 100\right) \text{ ppm/°C}. \quad (16.40)$$

This expression may be used to estimate the temperature variation of the emitter-base junction.

16.2.2. The MOS Capacitor

The MOS capacitance is formed by using an emitter region as one plate of the capacitor, the aluminum metallization as the second plate, and the intervening oxide layer as the dielectric. The construction of this capacitor is shown in Figure 16.12. No useful purpose is served by making a base diffusion in this case; in fact, such a diffusion would necessitate operating the emitter-base diode so formed at cutoff and results in an increased parasitic shunt capacitance. As a consequence, the emitter diffusion is made directly into the collector.

Although its capacitance per unit area is well below that of a *p-n* junction, the MOS device is superior in all other respects, as follows:

1. Its capacitance is given by the well-known formula for a parallel-plate structure ($C = \epsilon\epsilon_0 A/x$, where $\epsilon \simeq 3.9$ for silicon dioxide, and x is the thickness of the oxide layer). Since the lower plate of this capacitor is made of highly doped material, charge-depletion effects[8] in this region are insignificant. Hence the capacitance is essentially independent of voltage, and the device does not distort signals applied to it.

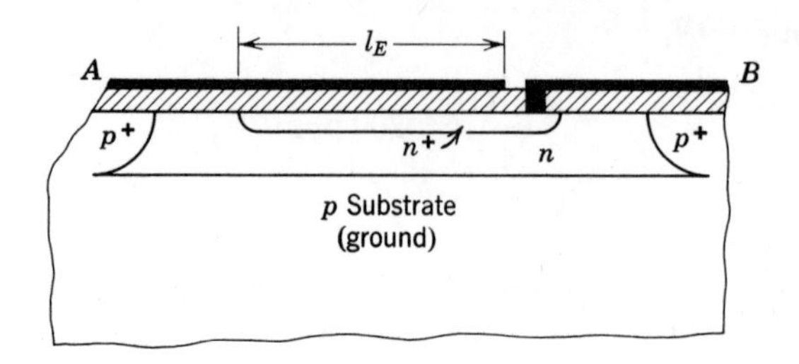

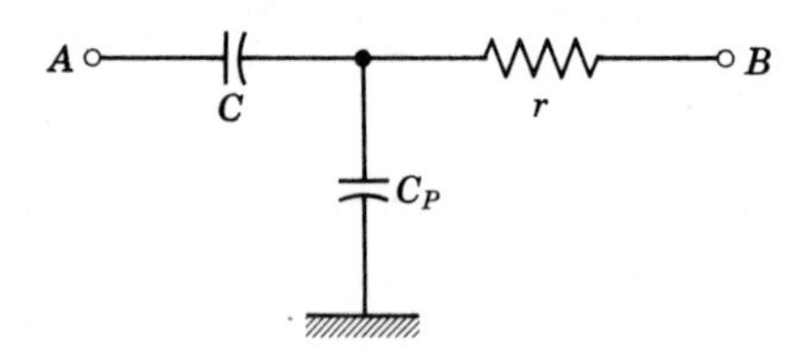

Figure 16.12 The MOS capacitor.

2. Since it is larger in area than a *p-n* junction of equal capacitance, its dimensions can be held to tighter tolerances.

3. The series parasitic resistance associated with it (see Figure 16.12) is that of a diffused emitter layer, and is considerably lower than that obtained with the *p-n* junctions of Figure 16.11. If the length and width of the upper plate are l_E and w_E respectively, this series resistance is approximately equal† to $\rho_{\square E} l_E / 3w_E$, where $\rho_{\square E}$ is the emitter sheet resistance (about 2 to 10 Ω/square).

4. The shunt parasitic capacitance is that of the collector-substrate junction and is lower in magnitude than that of other junctions in the microcircuit.

5. The temperature variation of a MOS capacitance is caused by the change in the dielectric constant of the oxide layer. Typically this is about 20 ppm/°C.

The MOS capacitor is extensively used in those situations where a tolerance requirement is placed on the value of the capacitance, and where it is important that distortion products are not introduced into the circuit. Thus its main uses are in linear circuits.

†The derivation of this expression follows closely the calculation of base resistance in Section 14.2.3.2.

16.3. REFERENCES

[1] L. Holland (Ed.), *Thin-Film Microelectronics*, *Wiley*, New York, 1965.
[2] L. Holland, *The Vacuum Deposition of Thin Films*, Chapman and Hall, London, 1963.
[3] T. V. Sikina, "High Density Tantalum Film Microcircuits," *Proc. 1962 Electronic Components Conference*, Washington, D. C, 1962.
[4] R. P. Donovan et al., "Integrated Silicon Device Technology, Vol. 1 – Resistance," *Technical Documentary Report No. ASD-TDR-63-316*; Research Triangle Institute, Durham, N. Carolina 1963.
[5] R. M. Warner, Jr., and J. N. Fordemwalt, *Integrated Circuits*, McGraw-Hill, New York, 1965.
[6] S. S. Hakim, *Junction Transistor Circuit Analysis*, Wiley, New York, 1962.
[7] J. L. Moll et al., "*P-n-p-n* Transistor Switches," *Proc. IRE*, **44** (9), 1174–82 1956.
[8] L. M. Terman, "An Investigation of Surface States at a Silicon/Silicon Oxide Interface Employing M-O-S Diodes," *Solid State Electron.* **5**, 285–299 1962.

16.4. PROBLEMS

1. A microcircuit resistor is made on a 150-Ω/square layer, with an average capacitance of 0.7 pF/mil². Assuming a 0.25-mil cut in the oxide, determine the cutoff frequency of resistors of 100, 1000, and 10,000 Ω respectively. Neglect end effects.

2. Determine the sheet resistance of a diffused emitter layer over a temperature range of −55 to +150°C. Assume the doping configuration of Figure 10.5*b*.

3. Compare the resistance of the foldover shown in Figure 16.5 with that of a semicircular bend used to provide the same function. Assume identical separations between resistor strips.

4. A 3000-Ω resistor is fabricated with the doping profile of Figure 10.5*b*. Assuming an oxide cut of 0.25 mil, design this resistor and determine its cutoff frequency for a reverse bias of 5 V. Include the effect of the contacts and all folds, and maintain a relatively square shape for the over-all resistor.

5. A linearly graded *p-n* junction has a capacitance of 10 pF at a reverse bias of 1 volt. Determine its average capacitance when it is swept over a voltage range of zero to V. Plot this capacitance as a function V. Assume $\Psi = 0.8$ V.

6. A coupling capacitor, comprising a base-collector junction, is used with its base region tied to a signal source and the collector tied to a low-impedance load. A step of voltage is applied to the circuit. Assuming that the capacitors are voltage independent, sketch the waveform of the load current and compute the charge delivered if: (*a*) the parasitic capacitance and the coupling capacitance are equal in magnitude; (*b*) the parasitic capacitance is ignored.

7. Determine the input impedance of an *RC* transmission line whose output is open-circuited. What is its cutoff frequency?

8. Show that

$$\frac{\sinh (\pi/2)\sqrt{s/\omega_0}}{(\pi/2)\sqrt{s/\omega_0}} = \left(1+\frac{s}{4\omega_0}\right)\left(1+\frac{s}{16\omega_0}\right)\left(1+\frac{s}{36\omega_0}\right)\cdots$$

9. An 3pF capacitor is made with an oxide thickness of 3000 Å and an emitter layer with a sheet resistance of 5 Ω/square. Assuming a 1×10 shape factor, with contact made to a short side, determine a transmission line type of equivalent circuit for this component. Hence determine its cutoff frequency. What is the cutoff frequency if the contact is made to a long side?

Chapter 17

Surface Phenomena

The surface of a semiconductor presents a major discontinuity in an otherwise reasonably periodic crystal lattice. Its electronic properites are thus largely determined by the defect nature of this discontinuity. Even if atomically clean surfaces were possible, their behavior would be governed by the large number of dangling bonds at the silicon-gas interface.

In this chapter the nature of the real surface of the semiconductor is described, together with the manner in which it controls the properties of the material below it. In addition, the effect of surface phenomena on device behavior is discussed in some detail.

17.1. SURFACE STATES

From purely quantum-mechanical considerations [1], it can be shown that the discontinuity in the periodic potential at a clean semiconductor surface gives rise to a number of allowed states within the forbidden gap. These states, the so-called *Tamm* or *Shockley states,* are associated with unsaturated covalent bonds at the surface discontinuity and are acceptor-like in their behavior. In freshly cleaned silicon their density is roughly equal to the density of dangling bonds on the silicon surface ($\simeq 10^{15}/cm^2$). These unsaturated bonds exhibit a strong attraction for foreign impurities, which become rapidly attached to the surface on exposure by either absorption or chemisorption processes. In microcircuits, in which an oxide layer is grown out of this silicon surface, a number of surface atoms are bound to the oxygen in the form of silica polyhedra. In either case the net result is a partial coherence of the silicon lattice and, thus, a reduction in the density of Tamm-Shockley states to about 10^{11} to $10^{12}/cm^2$.

The time constant associated with these states is on the order of 1 μsec or less. Consequently they are referred to as *fast states*. A large fraction of these fast states is deep lying, and is responsible for generation and recombination effects at the surface. Thus these states play an important role in determining the behavior of nonequilibrium processes at the surface.

Slow states are also found to exist on a semiconductor surface. With chemically etched silicon, these states are present on the surface of the adsorbed layer and may have a positive- or negative-charge state. In thermally oxidized surfaces[2], the first 100 to 200 Å of surface layer represents transition between the silicon and the oxide. Consequently, there is a net immobile positive charge associated with the excess silicon ions in this layer.

The density of slow surface states is on the order of $10^{11}/cm^2$ in thermally grown oxides, and about $10^{12}/cm^2$ for chemically treated silicon surfaces. These states are primarily traps in their behavior, with lifetimes on the order of a few seconds to months. Consequently they are responsible for establishing the surface potential by pinning the Fermi level at the surface trap level.

The remaining bulk of the oxide layer is occupied by various ionic species and by oxygen-ion vacancies. Their presence degrades the long-term stability of the microcircuit (see Section 17.4.5).

17.2. THE SPACE-CHARGE REGION

Figure 17.1 shows a schematic view of the thermally oxidized silicon surface. Note that the surface states of the semiconductor give rise to the formation of a space-charge layer in the bulk material beneath this surface. This is a direct consequence of the charge neutrality principle. If Q_{ss} is the total charge in the surface states, and Q_{sc} is the charge in the space charge layer, then

$$Q_{ss} = -Q_{sc} \tag{17.1}$$

for no applied voltage.

The depth of penetration of the space-charge layer into a semiconductor is considerably higher than that caused by surface charges on a metal, because the free-carrier concentration in a semiconductor is comparably lower. In a typical silicon structure this layer may be 0.1 to 1.0 μm deep, and can represent a considerable fraction of the depth in which the entire microcircuit is fabricated. Consequently the effects of this space-charge region cannot be ignored in the evaluation of microcircuit device characteristics.

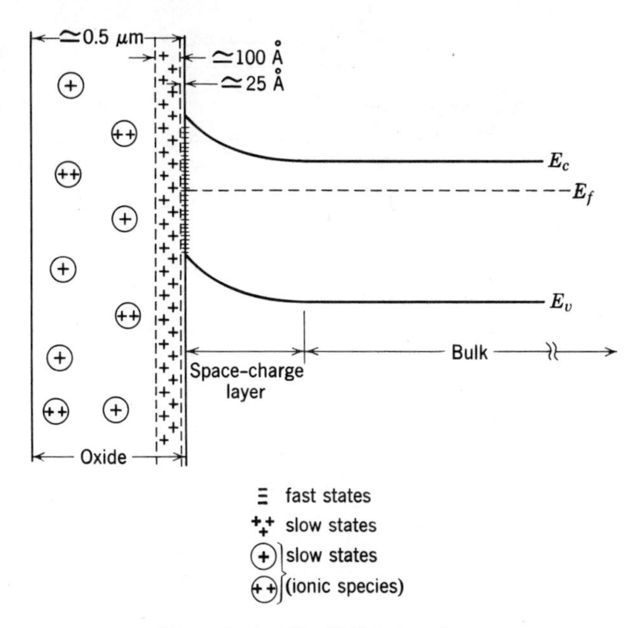

Figure 17.1 The Si-SiO_2 interface.

Surface states may be either acceptor-like or donor-like in character. The nature of the underlying space-charge layer depends on their type and density, and on the properties of the semiconductor material. Various possibilities exist, as follows:

CASE 1. DONOR SURFACE STATES†

The presence of donor states on the surface results in a net positive surface charge. As a consequence of charge neutrality, the space charge in the material below the surface is negative.

1. *Donor states on n-type silicon.* This situation is illustrated in Figure 17.2*a*. Here the semiconductor material near the surface is seen to become more *n* type as a result of charge distribution in the bulk. Hence an *accumulation* or *enhancement* layer is said to exist in the semiconductor.

2. *Donor states on p-type silicon.* Once again, the space-charge distribution is such as to make the material below the surface become more *n* type, that is, less *p* type. This situation is shown in Figure 17.2*b*, where a *depletion* or *exhaustion* layer is said to exist. Figure 17.2*c* shows the situation where the surface layers are actually inverted to *n*

†This represents the situation in thermally oxidized silicon surfaces.

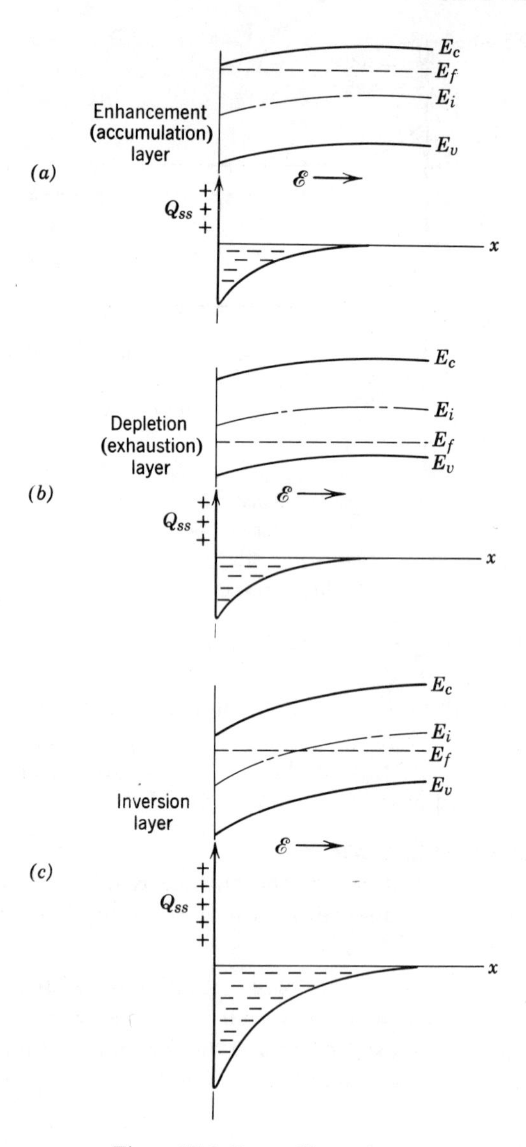

Figure 17.2 Donor-like surface states.

type. For this case an *inversion* layer is said to be present in the semiconductor. Note that this inversion layer is subsequently followed by a depletion layer before the material reestablishes its bulk properties. In all of these cases the direction of the electric field vector is into the bulk of the silicon.

CASE 2. ACCEPTOR SURFACE STATES

The surface tends to become negative in the presence of these states. The various situations that may arise are shown in Figure 17.3.

The presence of acceptor states of p-type silicon gives rise to an accumulation layer, as shown in Figure 17.3a. Acceptor states on n-type silicon give rise to a depletion layer, as shown in Figure 17.3b, or to an

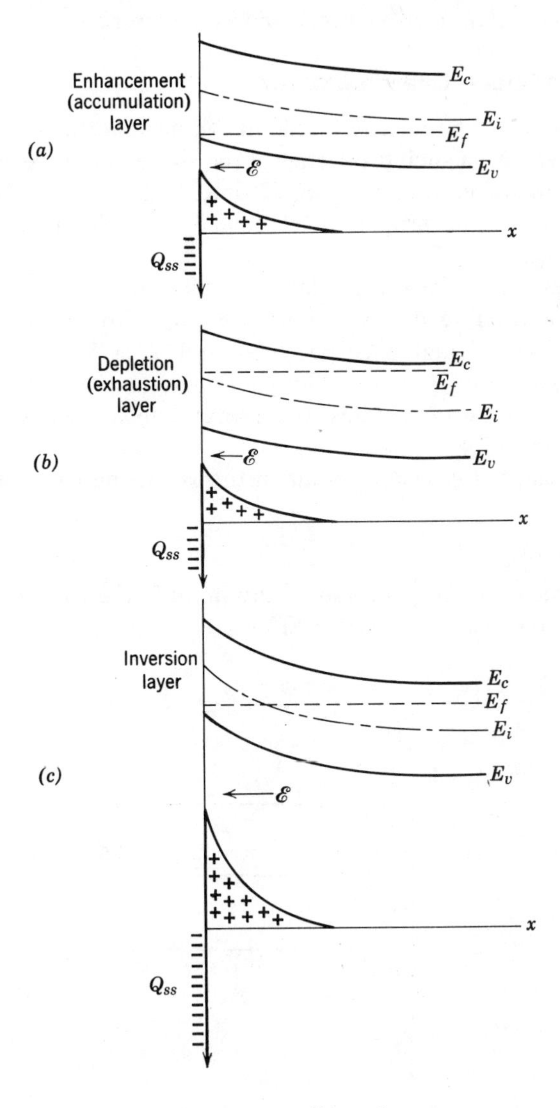

Figure 17.3 Acceptor-like surface states.

inversion layer, as shown in Figure 17.3c. In all cases the electric field vector points towards the surface of the silicon.

CASE 3. NO SURFACE STATES

For this situation the energy-level diagram is unchanged in going from the bulk to the surface. Thus the properties of the material just below the surface are the same as for the semiconductor bulk. This is known as the *flat-band condition* and is often used as a reference

17.2.1. The Space-Charge Density

The behavior of the energy bands in the neighborhood of the surface has a direct effect on such properties of the space-charge region as mobile carrier concentration, electric field, lifetime, and mobility. These, in turn, strongly influence the properties of devices in which this space-charge region is contained.

The properties of the surface layer[3] may now be determined. Consider a nondegenerate n type semiconductor, with donor and acceptor concentrations of N_d and N_a respectively, and $N_d \gg N_a$. It is assumed that all impurities are fully ionized. Although no assumptions are made about the details of the surface states, the energy bands are assumed bent as shown in Figure 17.4.

Let the potential ϕ at any point in the semiconductor be defined by

$$q\phi = E_f - E_i \tag{17.2}$$

In addition let V be the potential at any point in the space-charge region with respect to its value in the bulk. Thus

$$V = \phi - \phi_b \tag{17.3}$$

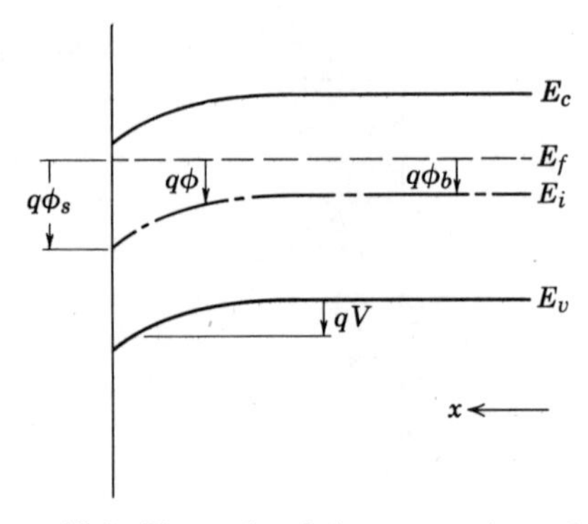

Figure 17.4 Energy-band picture near the surface.

For the situation of an n-type semiconductor as shown both ϕ and V are positive.

Let $\bar{n}$ and $\bar{p}$ be the mobile carrier concentrations in thermal equilibrium. Finally, write

$$u = \frac{q\phi}{kT} \tag{17.4}$$

and

$$v = \frac{qV}{kT} \tag{17.5}$$

The subscript b will be used to refer to material within the bulk where surface effects are absent. In like manner, the subscript s will refer to the surface.

In the bulk the equilibrium electron and hole concentrations are given by

$$\bar{n}_b = n_i e^{q\phi_b/kT} = n_i e^{u_b} \tag{17.6}$$

and

$$\bar{p}_b = n_i e^{-q\phi_b/kT} = n_i e^{-u_b}. \tag{17.7}$$

Charge neutrality conditions prevail in the bulk region. Hence

$$N_d - N_a = \bar{n}_b - \bar{p}_b \tag{17.8}$$

$$= 2n_i \sinh u_b \tag{17.9}$$

At any point in the semiconductor the free electron and hole concentrations are

$$\bar{n} = n_i\, e^{q\phi/kT} = n_i e^{u} \tag{17.10}$$

and

$$\bar{p} = n_i\, e^{-q\phi/kT} = n_i\, e^{-u} \tag{17.11}$$

respectively.

Thus the net space-charge density at any point is given by

$$\rho(x) = q(N_d - N_a + \bar{p} - \bar{n}), \tag{17.12}$$

$$= 2qn_i(\sinh u_b - \sinh u). \tag{17.13}$$

Substituting (17.4) and (17.3) into Poisson's equation $[\partial^2\phi/\partial x^2 = -\rho(x)/\epsilon\epsilon_0]$ gives

$$\frac{\partial^2 u}{\partial x^2} = \frac{1}{L_i^2}(\sinh u - \sinh u_b), \tag{17.14}$$

where

$$L_i = \left(\frac{\epsilon\epsilon_0 kT}{2q^2 n_i}\right)^{1/2}. \tag{17.15}$$

The quantity L_i is often referred to as the *intrinsic Debye length*.

17.2.2. The $\mathscr{E}$ Field

Poisson's equation may be solved by integrating from the bulk toward the surface. Thus

$$\int_0^{\partial u/\partial x} \frac{\partial u}{\partial x} \, d\frac{\partial u}{\partial x} = \frac{1}{L_i^2} \int_{u_b}^{u} (\sinh u - \sinh u_b) \, du, \tag{17.16}$$

Solving,

$$\frac{\partial u}{\partial x} = \pm \frac{\sqrt{2}}{L_i} [(u_b - u) \sinh u_b - (\cosh u_b - \cosh u)]^{1/2} \tag{17.17}$$

Noting that $u = q\phi/kT$, the electric field at the surface is given by

$$\mathscr{E}_s = \left(\frac{-d\phi}{dx}\right)_{\phi=\phi_s}, \tag{17.18}$$

$$= \pm \frac{\sqrt{2}kT}{qL_i} [(u_b - u_s) \sinh u_b - (\cosh u_b - \cosh u_s)], \tag{17.19}$$

$$= \frac{kT}{qL_i} F(u_s, u_b), \tag{17.20}$$

$$= 12F(u_s, u_b) \text{ volts/cm} \tag{17.21}$$

for silicon at 300°K.

The function $F(u_s, u_b)$ is shown in Figure 17.5 for various values [4] of u_b and u_s. Also shown is the sign of the space charge in the semiconductor. Thus the situation of Figure 17.4 (where u_b and u_s are positive,

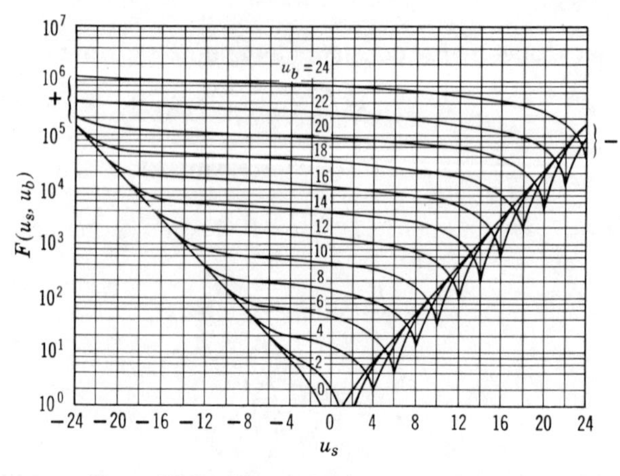

Figure 17.5 Curves of $F(u_s, u_b)$ (Young [4]).

with $u_s > u_b$) results in a negative value of $F(u_s, u_b)$ i.e., the electric field vector pointing into the bulk. This can be verified by physical reasoning, as shown in Figure 17.2*a*.

From symmetry considerations,

$$F(u_s, u_b) = -F(-u_s, -u_b). \tag{17.22}$$

17.2.3. Effective Width of the Space-charge Layer

The magnitude of the space-charge density that is required to support the $\mathscr{E}$ field is given by Q_{sc}, where

$$Q_{sc} = \epsilon\epsilon_0 \mathscr{E}_s \tag{17.23}$$

$$= \frac{kT\epsilon\epsilon_0}{qL_i} F(u_s, u_b), \tag{17.24}$$

$$= 1.3 \times 10^{-11} F(u_s, u_b) \text{ coulombs/cm}^2 \tag{17.25}$$

for silicon at 300°K.

The barrier height is V_s. Thus the surface space-charge capacitance C_{sc} is given by

$$C_{sc} = \frac{Q_{sc}}{V_s} \tag{17.26}$$

$$= \frac{\epsilon\epsilon_0 F(u_s, u_b)}{L_i v_s} \tag{17.27}$$

where $v = qV/kT$.

The effective width of the space-charge layer may be defined as that width which would result in a parallel-plate capacitor of capacitance C_{sc}. Writing this width as L_{sc} gives

$$C_{sc} = \frac{\epsilon\epsilon_0}{L_{sc}}, \tag{17.28}$$

so that

$$L_{sc} = \frac{L_i v_s}{F(u_s, u_b)} \tag{17.29}$$

This parameter may be written in terms of the *extrinsic Debye length* L_e, defined by

$$L_e \equiv \left[\frac{\epsilon\epsilon_0 kT}{q^2(\bar{p}_b + \bar{n}_b)}\right]^{1/2}. \tag{17.30}$$

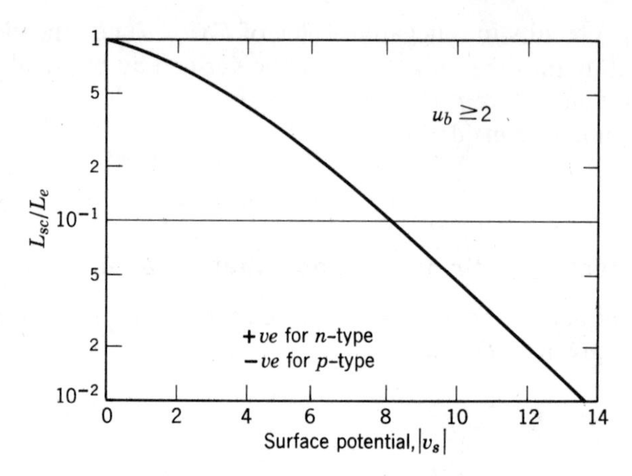

Figure 17.6 Effective length of the space-charge layer (adapted from Many *et al.* [1]).

Direct substitution of (17.6) and (17.7) into this equation results in

$$L_e = \frac{L_i}{\sqrt{\cosh u_b}} \tag{17.31}$$

Therefore

$$\frac{L_{sc}}{L_e} = \frac{v_s\sqrt{\cosh u_b}}{F(u_s, u_b)} \tag{17.32}$$

Figure 17.6 shows this relation to accumulation layers where $u_b \geqslant 2$. It is seen from this curve that the effective width of the space-charge layer falls with increasing barrier height.

17.2.4. Excess Carrier Concentrations

The excess concentration of mobile carriers in the space-charge layer may now be determined. Let ΔN and ΔP refer to these carriers, expressed as an excess over the value for the flat-band condition, per unit surface area. Then

$$\Delta P = \int_0^\infty (\bar{p} - \bar{p}_b)\, dx, \tag{17.33}$$

$$= n_i \int_0^\infty (e^{-u} - e^{-u_b})\, dx. \tag{17.34}$$

Combining with (17.17) and setting

$$F(u, u_b) = \pm\sqrt{2}[(u_b - u)\sinh u_b - (\cosh u_b - \cosh u)]^{1/2} \tag{17.35}$$

gives the excess hole concentration as

$$\Delta P = n_i L_i \int_{u_s}^{u_b} \frac{e^{-u} - e^{-u_b}}{F(u, u_b)} \, du \tag{17.36}$$

$$= n_i L_i G(u_s, u_b) \tag{17.37}$$

$$= 4.2 \times 10^7 G(u_s, u_b) \text{ holes/cm}^2 \tag{17.38}$$

for silicon at 300°K. In like manner,

$$\Delta N = \int_0^\infty (\bar{n} - \bar{n}_b) \, dx \tag{17.39}$$

$$= n_i L_i \int_{u_s}^{u_b} \frac{e^{u} - e^{-u_b}}{F(u, u_b)} \, du \tag{17.40}$$

$$= n_i L_i G(-u_s, -u_b) \tag{17.41}$$

$$= 4.2 \times 10^7 G(-u_s, -u_b) \text{ electrons/cm}^2, \tag{17.42}$$

for silicon at 300°K.

The functions $G(u_s, u_b)$ is shown in Figure 17.7 for various values of u_s and u_b. Here, too, the sign of G is indicated, and positive values of G result in an increase in the carrier density concerned. For the situation of Figure 17.4 the sign of ΔN is positive while that of ΔP is negative, i.e. the surface layers become more n type, as seen directly from the energy-level diagram of Figure 17.4.

17.2.5. Surface Conductance

The change in carrier concentration associated with bending of the energy bands results in a change in the surface conductivity. The extent of this change may be determined, taking the flat-band condition as a reference.

Figure 17.8 shows a semiconductor, having an upper surface, up to a thickness t, which constitutes the region in which the energy bands are bent.

The change in conductance (as measured across the faces of width w and depth d) is given by

$$\Delta G = \Delta\sigma t \frac{w}{l} = \int_0^t q[\mu_n(x) n'(x) + \mu_p(x) p'(x)] \, dx, \tag{17.43}$$

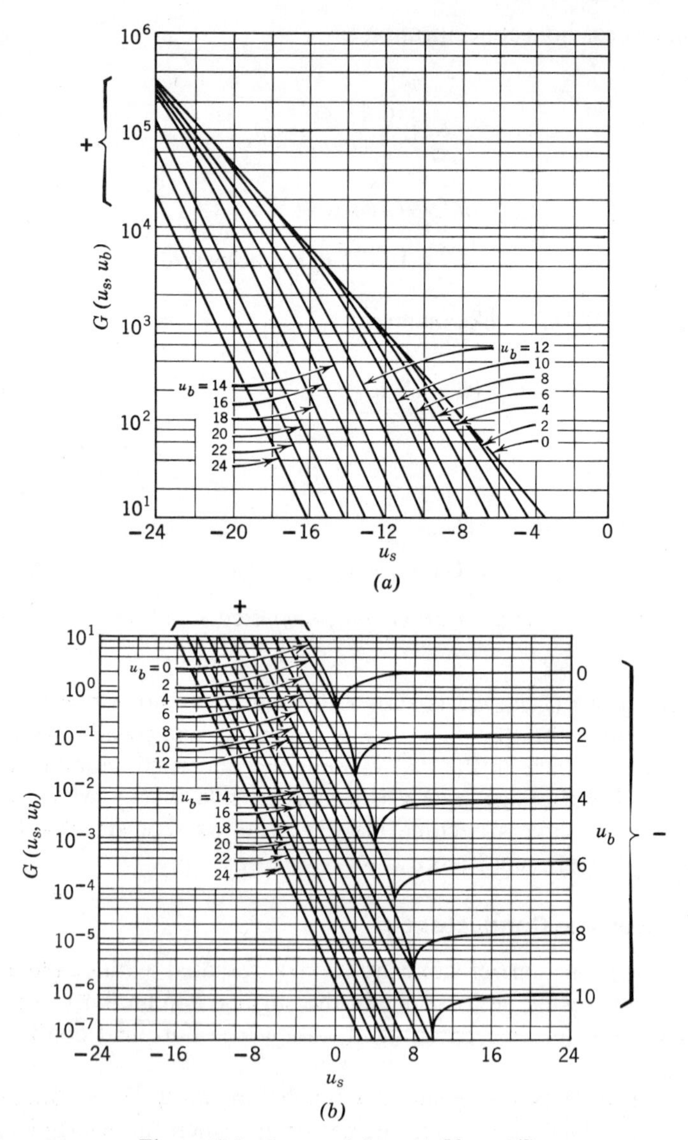

Figure 17.7. Curves of $G(u_s, u_b)$ (Young[4]).

where $n'(x)$ and $p'(x)$ are electron and hole carrier concentrations in excess of the flat-band condition. The change in sheet conductance of the layer is then given by

$$\Delta\sigma_{\square} = q\mu_p \Delta P + q\mu_n \Delta N, \tag{17.44}$$

provided that the mobilities are assumed constant throughout the material.

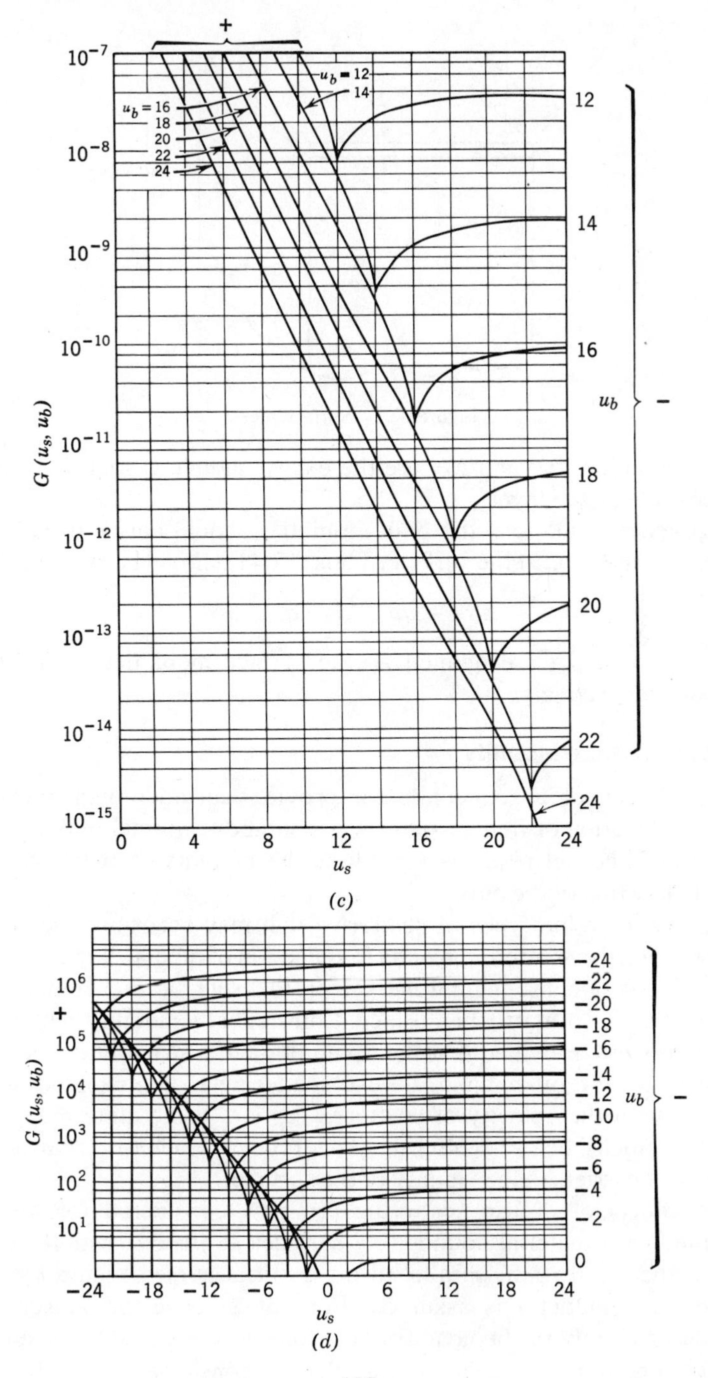

+
$u_b = 12$
14
$u_b = 16$
18
20
22
24
10^{-7}
10^{-8}
10^{-9}
10^{-10}
10^{-11}
10^{-12}
10^{-13}
10^{-14}
10^{-15}
$G(u_s, u_b)$
0
4
8
12
16
20
24
u_s
12
14
16
18
20
22
24
u_b
−

(c)

10^6
+
10^5
10^4
10^3
10^2
10^1
$G(u_s, u_b)$
−24
−18
−12
−6
0
6
12
18
24
u_s
−24
−22
−20
−18
−16
−14
−12
−10
−8
−6
−4
−2
0
u_b
−

(d)

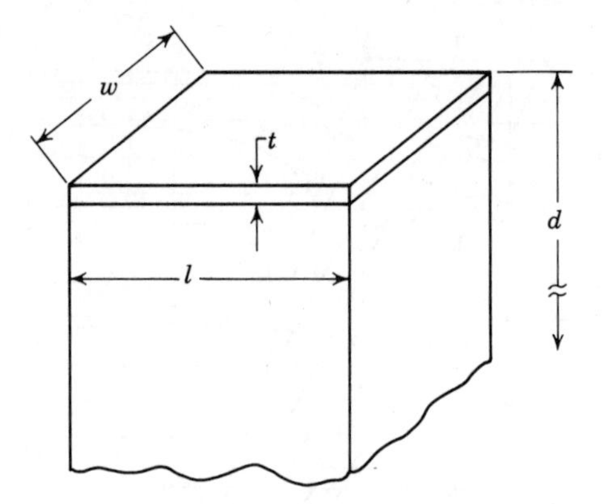

Figure 17.8 Surface layers.

In this equation, ΔP and ΔN are the excess mobile carrier densities per unit area of surface layer.

In practical situations the bulk mobilities should not be used because of carrier scattering at the surface. Thus (17.41) should be replaced by

$$\Delta\sigma_{\square} = q\mu_{ps}\,\Delta P + q\mu_{ns}\,\Delta N, \tag{17.45}$$

where μ_{ps} and μ_{ns} are defined as the surface mobilities for holes and electrons respectively.

17.2.6. Surface Mobility

The surface of a semiconductor provides an additional mechanism for the scattering of free carriers over and above normal bulk-scattering processes. The net result is to reduce the mobility at the surface to a value below that of the bulk.

The two extreme types of scattering that may occur are specular and diffuse. Specular scattering results in reflection of mobile carriers from the surface, and a consequent reversal of the sign of the component of velocity that is normal to this surface. Consequently mobility is unchanged by this process, provided that the assumption of a scalar effective mass is justified. On the other hand, diffuse scattering is a completely random process, with the velocity of scattered carriers being quite independent of their velocity before scattering. This process leads to a change in the mobility of carriers in the vicinity of the surface.

The effects of diffuse scattering are now considered for a slab of semiconductor material bounded by surfaces at $z = \pm d$, with the current flow in the x direction, parallel to these surfaces (see Figure 17.9). An n-type semiconductor is assumed. For convenience the subscript n is dropped, and only the behavior of electrons is considered in the analysis. In addition, only one side of the slab is considered, i.e., the region from $z = 0$ to $z = +d$.

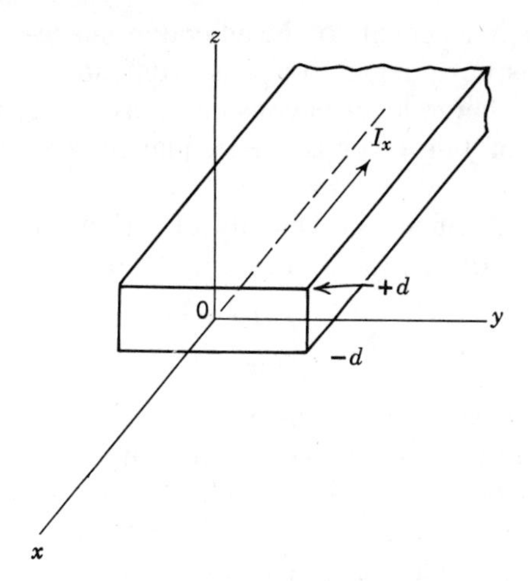

Figure 17.9 Current flow in a semiconductor.

Let τ_s and τ_b be the average collision times associated with carrier scattering in the surface and in the bulk respectively. If l is the mean free path for carriers in the bulk and d is the average carrier distance from the surface of the thin slab, then

$$\frac{\tau_s}{\tau_b} = \frac{d}{l} \tag{17.46}$$

If surface- and bulk-scattering processes are assumed to act independently, the effective collision time τ is given by

$$\frac{1}{\tau} \simeq \frac{1}{\tau_s} + \frac{1}{\tau_b} \tag{17.47}$$

But

$$\mu = \frac{q\tau}{m^*} \tag{17.48}$$

The mobility is thus given by

$$\mu/\mu_b = \frac{1}{1 + \tau_b/\tau_s} \tag{17.49}$$

or by

$$\mu = \frac{\mu_b}{1 + l/d} \tag{17.50}$$

This is the effective mobility for free carriers in a thin slab. It is important to note, however, that this is *also* the surface mobility of a thick semiconductor in which carriers are confined to move in a thin region near the surface. Such a situation exists for accumulation layers on a thick semiconductor slab. Here the effect of surface scattering may be determined by considering the mobile carriers to be restrained within a potential well, the width of which is equal to the effective space-charge layer width L_{sc}. The surface is considered to act as a diffuse scatterer, whereas the edge of the space-charge layer is considered as a specular reflector. The average mobility of this layer is then equal to the surface mobility of the thick slab.

In the absence of a field the velocity of arrival of electrons from the interior of a semiconductor to its surface at $z=+d$ is given† by

$$c_{zb}=\left(\frac{kT}{2\pi m^*}\right)^{1/2}, \tag{17.51}$$

where c_{zb} is defined as the unilateral mean velocity in the z direction.

In an accumulation layer, however, majority carriers are accelerated toward the surface because of the $\mathscr{E}$ field. Their increase in kinetic energy is associated with some average value of the barrier height, a convenient estimate being $qV_s/2$. Thus the velocity of carriers in the surface layer may be written as c_{zs}, where

$$c_{zs}=(\pi m^*)^{-1/2}\left(\frac{kT}{2}+\frac{qV_s}{2}\right)^{1/2}, \tag{17.52}$$

$$=c_{zb}\sqrt{1+v_s}. \tag{17.53}$$

Now,

$$\tau_s=\frac{L_{sc}}{c_{zs}} \tag{17.54a}$$

and

$$\tau_b=\frac{l}{c_{zb}} \tag{17.54b}$$

Combining these equations with (17.53) gives

$$\frac{\tau_s}{\tau_b}=\frac{L_{sc}}{l\sqrt{1+v_s}} \tag{17.55}$$

The surface mobility of the thick slab (which is equal to the average mobility of the space-charge layer) is thus given by

$$\frac{\mu_s}{\mu_b}=\left(1+\frac{l}{L_{sc}}\sqrt{1+v_s}\right)^{-1}. \tag{17.56}$$

Combining (17.15), (17.29), (17.51), (17.54*b*), and (17.48) gives

$$\frac{\mu_s}{\mu_b}=\left[1+\mu_b\left(\frac{m^*n_i}{\pi\epsilon\epsilon_0}\right)^{1/2}\frac{F(u_s,u_b)}{v_s}\sqrt{1+v_s}\right]^{-1} \tag{17.57}$$

†See Appendix C.

As seen from Figure 17.6, L_{sc} falls with increasing barrier height, resulting in a fall of the surface mobility [see (17.56)]. This is to be expected, since carriers are constrained to move within a narrower potential well as the accumulation layer becomes stronger.

The foregoing analysis excludes consideration of that part of the excess mobile carrier concentration which moves outside of the potential well. Consequently its range of validity is mainly confined to strong accumulation layers. The analysis for weak accumulation layers is more complex and will not be considered here.

In evaluating the surface mobility of depletion and inversion layers, it is only necessary to consider those carriers which have sufficient energy to surmount the retarding field in the semiconductor and reach the surface. The rest of the carriers are reflected in a specular manner and do not contribute to the scattering process.

The density of carriers that will surmount the surface barrier is $\bar{n}_b e^{v_s}$. Thus $\bar{n}_b(1-e^{v_s})$ carriers are not subject to surface scattering.

The current I_x is made up of carriers moving with their bulk mobility as well as carriers moving with their average mobility. This current is thus given by

$$I_x = q\mu_b \mathscr{E}_x[\Delta N + \bar{n}_b(1-e^{v_s})d] + q\mu \mathscr{E}_x \bar{n}_b e^{v_s} d. \tag{17.58}$$

Here, too, only one side of the semiconductor is considered, between $z=0$ and $z=+d$. For the flat-band condition,

$$I_{x0} = q\mu \mathscr{E}_x \bar{n}_b d. \tag{17.59}$$

Thus the current increment due to a departure from the flat-band condition is

$$\Delta I_x = I_x - I_{x0}, \tag{17.60}$$

$$= q\mu_b \mathscr{E}_x[\Delta N + \bar{n}_b(1-e^{v_s})\, d] - q\mu \mathscr{E}_x[\bar{n}_b(1-e^{v_s})\, d]. \tag{17.61}$$

But

$$\mu = \frac{\mu_b}{1+l/d} \tag{17.62}$$

$$\simeq \mu_b\left(1-\frac{l}{d}\right) \tag{17.63}$$

for a thick slab. Consequently,

$$I_x = qu_b \mathscr{E}_x[\Delta N + \bar{n}_b l(1-e^{v_s})]. \tag{17.64}$$

The surface mobility may be written as

$$\mu_s = \frac{\Delta I_x}{q \mathscr{E}_x \Delta N} \tag{17.65}$$

Therefore,

$$\frac{\mu_s}{\mu_b} = 1 + \frac{\bar{n}_b l}{\Delta N} (1 - e^{v_s}) \tag{17.66}$$

Combining (17.6), (17.15), (17.41), (17.48), (17.51), and (17.54*b*) gives

$$\frac{\mu_s}{\mu_b} = 1 + \frac{\mu_b e^{\mu_b}(1 - e^{v_s})}{G(-u_s, -u_b)} \left(\frac{m^* n_i}{\pi \epsilon \epsilon_0}\right)^{1/2}. \tag{17.67}$$

Note that $G(-u_s, -u_b)$ is negative for depletion layers. Thus the surface mobility is seen to approach the bulk mobility for large (negative) values of v_s.

In general there appears to be good agreement between the simple theory outlined here[1] and the experimental results. In those situations in which this is not the case it is usually possible to account for the discrepancy by postulating that the scattering is, in part, specular.

If k is the probability of an electron being scattered by the diffuse mechanism, then the surface mobility for accumulation layers is directly given by

$$\frac{\mu_s}{\mu_b} = \left[1 + k\mu_b \left(\frac{m^* n_i}{\pi \epsilon \epsilon_0}\right)^{1/2} \frac{F(u_s, u_b)}{v_s} \sqrt{1 + v_s}\right]^{-1}. \tag{17.68}$$

The surface mobility for depletion and inversion layers is given by

$$\frac{\mu_s}{\mu_b} = 1 + \frac{k \mu_b e^{u_b}(1 - e^{v_s})}{G(-u_s, -u_b)} \left(\frac{m^* n_i}{\pi \epsilon \epsilon_0}\right)^{1/2}. \tag{17.69}$$

17.3. SURFACE RECOMBINATION VELOCITY

The presence of fast deep-lying surface states leads to the enhancement of recombination processes at the surface. To characterize this effect it is useful to think of the surface as a sink for minority carriers that approach it at a velocity S, the surface recombination velocity. Then S is defined as the ratio of the number of minority carriers recombining per second per unit surface area to the excess carrier concentration in the bulk just beneath the surface.

The magnitude of S is a function of the detailed nature of the fast surface states and of the surface potential. Analysis is carried out along the lines outlined in Section 12.6. Because the detailed nature of the fast states (type, density, or distribution) is not known, a very simple model is considered, with N_{rs} recombination centers/cm^2, all located at a discrete energy level E_r. The capture cross section to electrons and holes is taken as A_n and A_p respectively.

The number of holes recombining per second per unit surface area may be written directly as [by analogy with (12.35)]:

$$U_s = \frac{p_s n_s - n_i^2}{(n_1 + n_s)/S_{p0} + (p_1 + p_s)/S_{n0}} \tag{17.70}$$

where p_s and n_s are the mobile carrier densities at the surface, n_1 and p_1 are the equilibrium values of these carrier densities if the Fermi level is at the recombination level, and

$$S_{p0} = A_p v_{tp} N_{rs}, \tag{17.71a}$$

$$S_{n0} = A_n v_{tn} N_{rs}. \tag{17.71b}$$

For silicon typical values of A_p and A_n are both 10^{-15} cm^2, while the number of recombination centers in thermally oxidized silicon is about 10^{10} to 10^{11}/cm^2 of surface.

For small disturbances from equilibrium,

$$n_s = \bar{n}_b + \delta n_b, \tag{17.72a}$$

$$p_s = \bar{p}_b + \delta p_b. \tag{17.72b}$$

In addition, for charge neutrality,

$$\delta n_b = \delta p_b. \tag{17.73}$$

Substituting these equations into (17.79) gives

$$U_s = \frac{(\bar{n}_b + p_b)\delta p_b\, S_{p0}\, S_{n0}}{n_1 S_{n0} + p_1 S_{p0} + n_s\, S_{n0} + p_s\, S_{p0}}. \tag{17.74}$$

Defining a u_0 such that $u_0 = 1/2\ ln\,(S_{n0}/S_{p0})$, gives

$$S_{n0} = \sqrt{S_{p0} S_{n0}}\, e^{+u_0}, \tag{17.75a}$$

$$S_{p0} = \sqrt{S_{p0} S_{n0}}\, e^{-u_0}. \tag{17.75b}$$

In addition,

$$n_1 = n_i e^{(E_r - E_i)/kT} \equiv n_i e^{u_r}, \tag{17.76a}$$

$$p_1 = n_i e^{-(E_r - E_i)/kT} \equiv n_i e^{-u_r}. \tag{17.76b}$$

Substituting (17.75) and (17.76) into (17.74) gives

$$S = \frac{U_s}{\delta p_b} = \frac{(\bar{n}_b + \bar{p}_b)\sqrt{S_{p0}S_{n0}}}{2n_i[\cosh(u_r - u_0) + \cosh(u_s - u_0)]} \tag{17.77}$$

The surface recombination velocity is maximized at $u_s = u_0$. Thus

$$\frac{S}{S_{\max}} = \frac{1 + \cosh(u_r - u_0)}{\cosh(u_r - u_0) + \cosh(u_s - u_0)} \tag{17.78}$$

This function is shown in Figure 17.10 for various values of $u_r - u_0$.

Note that u_0 is a measure of the inequality of the capture cross sections of electrons and holes. The general character of this function may be understood if it is recognized that the recombination process depends on the capture of both electrons and holes by the recombination center. Thus the slower of these processes is the rate-limiting one. Recombination centers are most effective if the electron and hole capture rates are identical, resulting in the peak value of $S/S_{\max}$ at $u_0 = 0$. This is indeed the case for silicon.

Typical values of S range from as high as 10^5 cm/sec on sandblasted surfaces to 10 to 100 cm/sec on properly etched or otherwise treated silicon surfaces. The value of S for thermally oxidized surfaces is typically

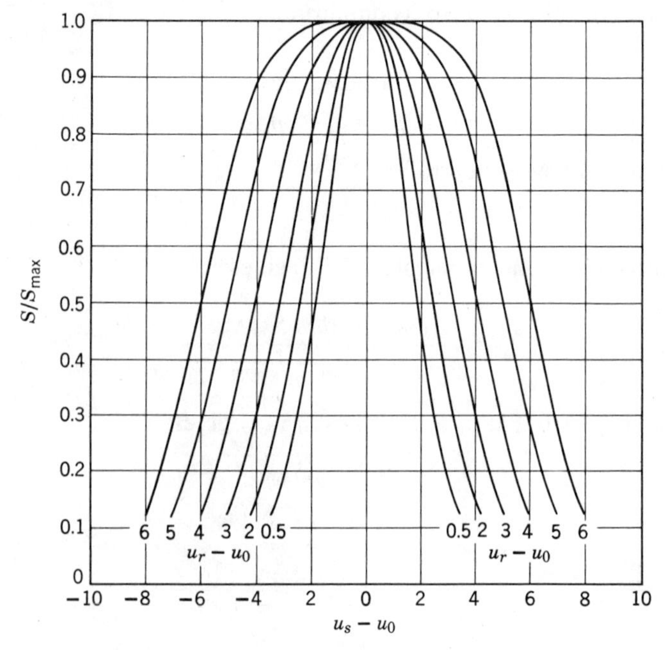

Figure 17.10 Surface recombination velocity curves (Many *et al.* [1]).

10 cm/sec. Note that since S is the velocity with which minority carriers approach the surface (and are recombined) its upper value is given by the unilateral mean velocity $\sqrt{kT/2\pi m^*}$.

A number of experiments have established the validity of this model. Some of these consist of measurements made on surfaces with various etch treatments and gas ambients. Yet others consist of measurements of thermally oxidized surfaces, on which the surface potential is altered by placing a charge on the oxide-gas interface by means of an electrode. In all cases excellent correlation has been found with theory. Thus it is unnecessary to use a more complex model for this process.

17.3.1. The Effective Lifetime

An effective lifetime may now be determined. Consider an n-type semiconductor, shown in Figure 17.11 in the form of a rectangular parallelopiped, bounded in the y and z directions at $y = \pm A$ and $z = \pm B$ respectively. Consider that there is a uniform minority carrier density in the bar. At time $t = 0$ carrier generation ceases, so that the minority carriers decay to their equilibrium value. This decay is characterized by an effective time constant τ_{eff}.

The minority carrier concentration is given by the three-dimensional form of the continuity and diffusion equations, as follows:

$$\frac{\partial p'}{\partial t} = -\frac{p'}{\tau_p} + D_p\left(\frac{\partial^2 p'}{\partial x^2} + \frac{\partial^2 p'}{\partial y^2} + \frac{\partial^2 p'}{\partial z^2}\right), \tag{17.79}$$

where p' is the excess hole concentration.

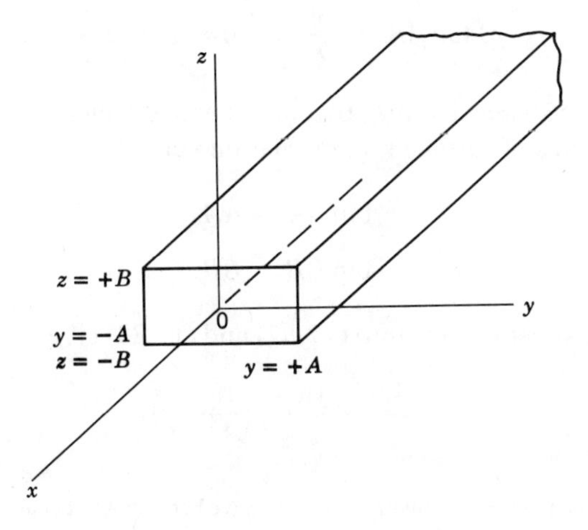

Figure 17.11 A semiconductor slab.

Excess holes diffuse to the surface as fast as they can recombine at this sink. Thus this equation is subject to the boundary condition that the flux density of excess holes in the y direction be given by

$$\pm Sp' = D_p \frac{\partial p'}{\partial y} \quad \text{at} \quad y = \pm A. \tag{17.80a}$$

In like manner, the flux density of excess holes in the z direction is given by

$$\pm Sp' = D_p \frac{\partial p'}{\partial z} \quad \text{at} \quad z = \pm B. \tag{17.80b}$$

An infinite-series solution† of (17.70) may be assumed, the first term taking the form

$$p' = e^{-\nu t} \cos \alpha y \cos \beta z. \tag{17.81}$$

Only this term need be considered, since it results in the slowest decay.

Substituting (17.81) in (17.79) gives

$$\frac{1}{\tau_{\text{eff}}} = \nu = \frac{1}{\tau_p} + D_p(\alpha^2 + \beta^2), \tag{17.82}$$

where τ_{eff} is the effective minority carrier lifetime for the bar.

Furthermore, substituting (17.81) into (17.80) gives

$$\alpha \tan \alpha A = \frac{S}{D_p}, \qquad 0 \leqslant \alpha A \leqslant \frac{\pi}{2}, \tag{17.83a}$$

$$\beta \tan \beta B = \frac{S}{D_p}, \qquad 0 \leqslant \beta B \leqslant \frac{\pi}{2}. \tag{17.83b}$$

Again, only the smallest values of α and β are considered.

For low values of surface recombination velocity

$$\tan \alpha A \simeq \alpha A, \tag{17.84a}$$

$$\tan \beta B \simeq \beta B. \tag{17.84b}$$

Making these substitutions into (17.82) and (17.83) gives

$$\frac{1}{\tau_{\text{eff}}} \to \frac{1}{\tau_p} + S\left(\frac{1}{A} + \frac{1}{B}\right) \tag{17.85}$$

†Note that terms exhibiting symmetry in y and z have been chosen because of the symmetric nature of the problem.

For high values of surface recombination velocity both αA and βB approach $\pi/2$. Thus

$$\frac{1}{\tau_{\text{eff}}} \to \frac{1}{\tau_p} + \frac{\pi^2 D_p}{4}\left(\frac{1}{A^2} + \frac{1}{B^2}\right). \tag{17.86}$$

Note that for high values of surface recombination velocity the effective lifetime is independent of the nature of the surface. This is because the rate limiting mechanism is the velocity with which the carriers can diffuse towards the surface where they may recombine. As mentioned earlier, the upper limit of this velocity is given by $\sqrt{kT/2\pi m^*}$.

17.4. INFLUENCE ON DEVICE PARAMETERS

Having described the influence of the surface states on the physical properties of the underlying semiconductor material, it is now necessary to determine the manner in which they affect the device properties. This is done by considering separately the various parameters of interest. In each case the role of the surface states is determined on the assumption that all other effects are absent.

17.4.1. Reverse Current

Figure 17.12*a* shows an oxide-masked diffused *p-n* junction in the absence of slow surface states (i.e., the flat-band condition). Here the depletion layer extends up to the surface, where the density of fast, deep-lying states is considerably higher than in the bulk.

Assuming recombination centers in the middle of the energy gap, the reverse current in the volume of this depletion layer is given by [see (13.20)]

$$I'_R = \frac{qn_i}{\tau_{p0} + \tau_{n0}} x_0 A, \tag{17.87}$$

where A is the cross-section area of the junction and x_0 is the depletion-layer width. At the surface the steady-state recombination rate per unit area is given by

$$U = n_i \sqrt{S_{p0} S_{n0}} \tag{17.88}$$

If A_s is the area of the depletion layer that is exposed to the surface, the leakage current associated with it is given by

$$I''_R = qn_i \sqrt{S_{p0} S_{n0}} A_s. \tag{17.89}$$

Thus the total reverse current for the junction is given by $I'_R + I''_R$.

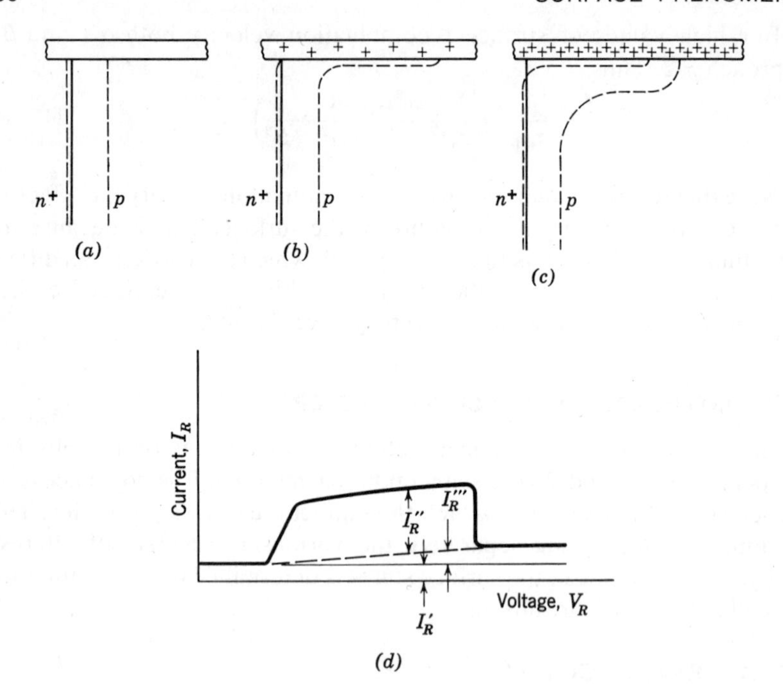

Figure 17.12 Leakage current in *p-n* junctions.

Figure 17.12*b* shows the junction in the presence of positively charged slow-surface states. The effect of these states is to produce an *n*-type shift in the surface layers of the semiconductor. Consequently, the junction behaves as if it were more highly doped on the *n* side and more lightly doped on the *p* side, thus altering the shape of the depletion layer to form a channel region.

The presence of the channel increases the effective area of the depletion layer at the surface by orders of magnitude. This, in turn, results in a corresponding increase of leakage current I_R''.

The formation of the surface channel also results in the creation of a field-induced depletion-layer region. The reverse current associated with this region is given by

$$I_R''' = \frac{qn_i}{\tau_{p0}' + \tau_{n0}'} x_0' A', \tag{17.90}$$

where the $'$ refer to parameters of the field-induced region.

Figure 17.12*c* shows the effect of increasing the surface concentration of positively charged surface states to the point where inversion occurs in the *p*-type material. This causes the movement of the depletion layer over to the p side of the metallurgical junction, resulting in a reduction of its exposed area. Thus the reverse-current term I''_R falls until it is approximately equal to its value for the flat-band condition. In addition, the volume of the field-induced depletion layer is now greatly increased, resulting in an increase of I''_R. The over-all behavior of these three current terms is shown in Figure 17.12*d*.

It is worth noting that the base region of an *n-p-n* transistor made by the DDE process is highly doped so that these effects are not sufficiently serious to impair device performance. In *p-n-p* transistors, however, the collector region is the most lightly doped. Consequently, the fabrication of these devices has always been plagued by channel formation, with high leakage and premature breakdown characteristics. These problems have been solved by guard-ring and surface-doping methods; however, they require additional steps in the microcircuit fabrication process, and are not commonly used.

17.4.2. The Forward-diode Characteristics

At low values of applied voltages, the forward current in a diode is dominated by recombination processes within the depletion layer. In the bulk the voltage-current relation of the diode is given by [see (13.66)]

$$I_F = \frac{qn_i}{\tau_{p0} + \tau_{n0}} x_0 A e^{qV_F/2kT}. \tag{17.91}$$

The surface component of the forward current may be written by analogy with the aid of (17.89). Thus

$$I_{Fs} = qA_s n_i \sqrt{S_{p0} S_{n0}} e^{qV_F/2kT} \tag{17.92}$$

where A_s is the surface area of the depletion layer.

With increasing applied voltage, diffusion processes dominate the forward-current characteristic. Hence surface effects may be neglected at these levels.

17.4.3. Breakdown Voltage

Figure 17.13 shows a cross section of an n^+-p diode in the vicinity of the surface, for four different surface conditions. The situation for negatively charged surface states is shown in Figure 17.13*a*, the flat-band condition is shown in Figure 17.13*b*, and conditions for positively charged surface states are indicated in Figures 17.13*c* and 17.13*d*. The following features may be deduced from this figure:

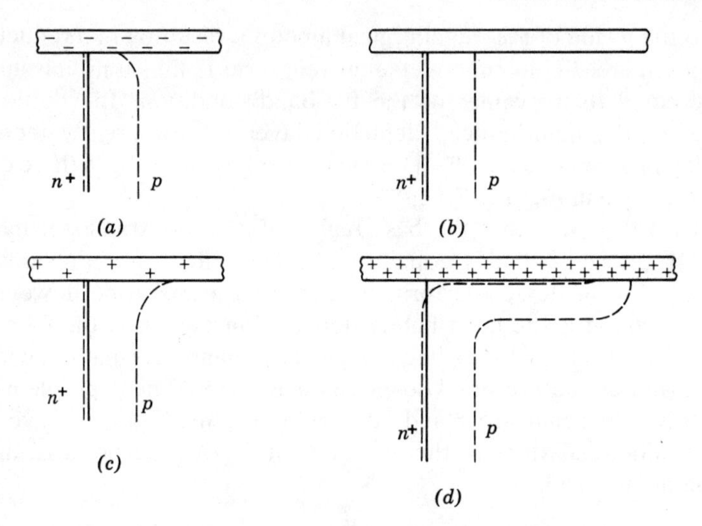

Figure 17.13 Depletion-layer variations in p-n junctions.

1. The flat-band condition is the one where the field intensity in the depletion layer is the same at the surface as at the bulk. Thus for the junction of Figure 17.13*b*, the surface breakdown voltage is equal to the bulk breakdown voltage.

2. The presence of an accumulation layer on the surface causes the depletion layer to be curved, as shown in Figure 17.13*a*. The situation is very similar to that of a cylindrical junction described in Section 13.3.1.1. For this case it was shown that the peak field is higher than that of a comparably doped parallel-plane structure. Hence breakdown is initiated at the surface where the curvature is a maximum and at a voltage below the breakdown voltage in the bulk.

3. Figure 17.13*c* illustrates the presence of a depletion layer on the surface of a p region. This case corresponds to a cylindrical junction, the depletion layer of which curves *away* from the n^+ region. By a reasoning similar to that of Section 13.3.1.1, it can be shown that the surface breakdown voltage is higher than that in the bulk. Thus breakdown will occur in the bulk region at a voltage given by the bulk parameters.

4. Figure 17.13*d* shows the presence of a strong inversion layer, resulting in a large field-induced region. Here the edge of the depletion layer curves toward the n^+ region, resulting in a premature breakdown at the surface. This results in a greatly increased reverse current, which flows along the field-induced channel. The maximum current density that can be supported by this channel is given by

$$J_{\text{lim}} = nqv_{\text{lim}}, \tag{17.93}$$

where $v_{\lim}$ is the terminal velocity of carriers. Consequently, the reverse current limits to some finite value $I_{R,\lim}$, as shown in Figure 17.14. Eventually bulk breakdown occurs, and the current increases without limit.

17.4.4. Transistor Current Gain

The effect of surface states on the current gain of an *n-p-n* transistor may be estimated[2] by modifying (14.33) for the current ratio I_B/I_C to include a surface recombination current term I_{sB}. Thus

$$\frac{I_B}{I_C} \simeq \frac{I_{pE}}{I_{nE}} + \frac{I_{rB}}{I_{nE}} + \frac{I_{sB}}{I_{nE}} \tag{17.94}$$

where

$$I_{sB} = qA_sSn_B. \tag{17.95}$$

Here A_s is the surface area and S is the surface recombination velocity.

The electron concentration varies over the base width, from a value $n_B(0)$ at the edge of the emitter depletion layer to zero at the collector-base junction. However most of the recombination has been shown[6] to take place where the oxide-masked junction intersects the surface, because this is a region of high-impurity density with much lattice damage. To an approximation, then,

$$I_{sB} \simeq qA_sSn_B(0). \tag{17.96}$$

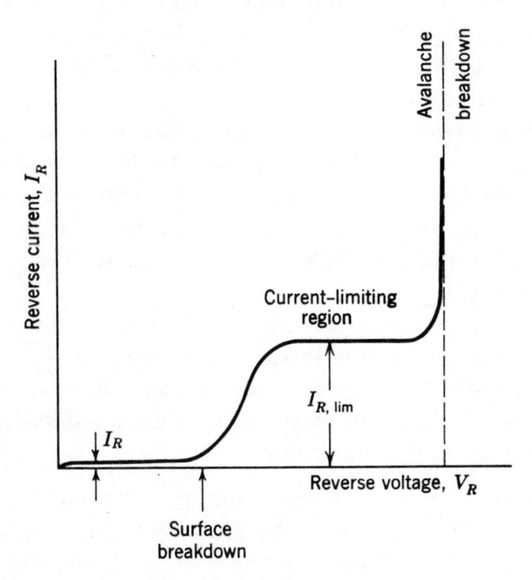

Figure 17.14 Channeling characteristics.

Assuming a uniform-base concentration, the electron current may be written as

$$I_{nE} = \frac{qAD_{nB}n_B(0)}{W_B}, \tag{17.97}$$

where A is the cross-sectional area of the transistor. Combining these terms, making the substituting $L = \sqrt{D\tau}$, and differentiating,

$$\frac{\partial I_B}{\partial I_C} = \frac{1}{\beta_0} = \frac{\sigma_B W_B}{\sigma_E L_{pE}} + \frac{W_B^2}{2D_{nB}}\left(\frac{1}{\tau_{nB}} + \frac{2A_s S}{AW_B}\right). \tag{17.98}$$

Thus the surface recombination term reduces the lifetime of minority carriers to an effective value $\tau_{nB,\text{eff}}$ such that

$$\frac{1}{\tau_{nB,\text{eff}}} \simeq \frac{1}{\tau_{nB}} + \frac{2A_s S}{AW_B}. \tag{17.99}$$

This, in turn, leads to a reduction of the common emitter current gain to the value given by (17.98).

17.4.5. Long-term Stability

In addition to the various electronic phenomena associated with fast surface states, oxide-coated silicon is found to develop semipermanent changes in its surface potential over a period of time. These changes affect the long-term stability of associated devices, and are particularly severe when the oxide is subjected for periods of time to external fields at elevated temperatures.

In recent years many studies have been made to understand the causes of these changes and, hopefully, to eliminate them. Much of the recent impetus has come from the development of the surface-controlled MOS transistor. Although the present theories are not conclusive, they go a long way toward explaining these phenomena as well as suggesting techniques for reducing their effect.

Figure 17.15 shows the surface region of a piece of silicon. A thermally grown oxide is present on this surface and is covered by a metallic film. Semipermanent shifts that appear in the surface potential of this structure may be explained in terms of an oxygen vacancy model[8] in which the following events are thought to occur:

1. The superoxide ion O_2^- (see Section 6.3.3), which is considered to be the oxidizing species in thermally grown SiO_2, is free to move in the silica layer under the influence of the $\mathscr{E}$ field. Layers of this type are known to have significant oxygen-ion vacancy concentrations, and the

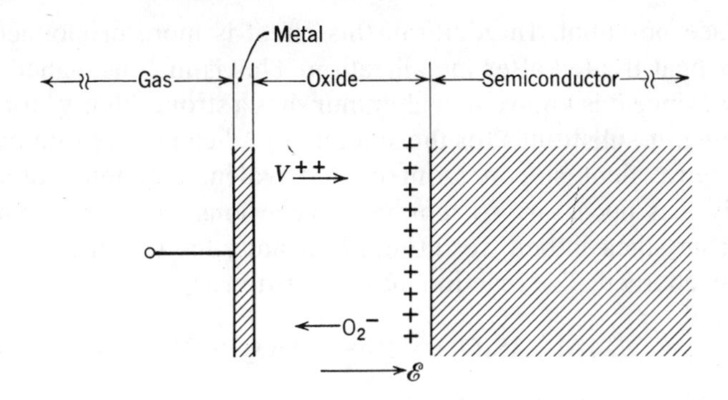

Figure 17.15 Ionic motion in oxides.

movement of the O_2^- is accomplished by the process of jumping from one such oxygen vacancy to the next. Thus for an electric field directed into the semiconductor surface, the oxygen ions move toward the oxide-metal interface, resulting in the formation of new vacancies closer to the silicon. This is equivalent to the migration of oxygen-ion vacancies to the oxide-silicon interface. Since these vacancies represent the absence of otherwise filled oxygen sites, they carry with them a net positive charge, and are designated V^{++}.

2. Oxygen ions may be removed at the outside surface of the silicon dioxide by reaction with the surface metallization or by the application of an $\mathscr{E}$ field. This results in leaving behind oxygen vacancies in the SiO_2. On the other hand, experimental evidence shows that an excess of oxygen, beyond the requirements of stoichiometry, cannot occur.

The net result of applying a field across the oxide in the direction shown is an accumulation of a positive space-charge layer close to the silicon-oxide interface and a corresponding *n*-type shift in the surface potential. Measurements on structures of this type have shown this layer to be about 25 Å thick.

The amount of positive space-charge accumulation is a function of the magnitude and time duration of the field impressed across the oxide and of the ambient temperature. It has been experimentally observed that significant charge migration can occur at temperatures as low as 125°C. On removal of the stress field and reduction of the temperature, the space charge becomes immobilized, resulting in a semipermanent change in the properties of the underlying material. Shifts in the surface potential are not irreversible; that is, the space-charge layer, hence the surface potential, can be altered by varying the stress and temperature conditions.

It has been shown that the use of aluminum for the metallization tends to increase the amount of semipermanent shift that can occur in the surface potential. In addition, this effect is more pronounced if the circuit is heat treated after metallization. This is in consonance with the theory[9], since it is known that aluminum has a strong affinity† for oxygen. In so doing, it substitutes for the silicon in a silica tetrahedron by replacing the tetravalent silicon atom by a trivalent aluminum atom. Consequently it is necessary to remove oxygen ions from some part of the neighboring silica lattice, resulting in vacant sites which are effectively positively charged. The reaction may be written as

$$4Al + 3SiO_2 \rightarrow 3Si + 4AlO_2^- + 2V^{++}. \quad (17.100)$$

Since the migration of oxygen ions in SiO_2 is vacancy induced, passivation of the silicon surface may be achieved by reducing the vacancy concentration in the oxide layer. This may be accomplished[7] by depositing a layer of P_2O_5 on the oxide. This layer provides an excess of oxygen to the silica structure and greatly reduces the number of oxygen-ion vacancies in its vicinity. Yamin[10] has shown that it is not necessary for the entire SiO_2 layer to approach stoichiometry in order to inhibit this migration. In his experiments a thin layer, relatively free from vacancies, was found quite effective. In line with the above theory it is reasonable to expect that a surface layer of B_2O_3 cannot be useful in this application because it does not reduce the oxygen decifiency in the oxide. This has indeed been found to be the case in practice.

An alternate theory[11, 12] for long-term shifts in the surface potential is based on the presence of the highly mobile sodium ion Na^+ as a contaminant in the oxide layer. The presence of this ion is readily explained since sodium is a constitutent of the quartz tubes used for diffusion. Its presence has been detected in the oxide layers, and experiments have confirmed its ability to reduce drastically the long-term stability of oxide-coated devices.

The use of a P_2O_5 glass layer has been found to be effective in passivating sodium-contaminated silicon surfaces. It has been proposed that this is caused, in part, by the gettering action of sodium by P_2O_5.

Techniques for surface passivation are far from perfect. The use of P_2O_5 is quite popular, since a layer of this glass is formed during the final diffusion step (the *n*-type emitter diffusion) in present-day microcircuit fabrication. On the other hand, the fabrication of a *p-n-p* transistor results in a final surface layer of B_2O_3. This layer must be removed before surface coating with P_2O_5 if passivation is desired.

Other surface coatings have also been used with varying degrees of

†Indeed, the choice of aluminum for interconnections is primarily based on its excellent bonding properties of the oxide as a result of this affinity.

success. Notable among these are PbO and MnO_2. In addition, attempts have been made to use oxides formed by deposition techniques in order to reduce the vacancy concentration.

Gaseous annealing cycles have also been used. As can be expected, reducing gases such as hydrogen greatly degrade the long-term stability of the surface potential. Most success to date has been obtained with nitrogen; however, the reasons for this success are not fully established.

Finally, attempts have been made to cover the entire surface oxide with other materials. One such, silicon nitride[13], has been found to be most promising for this application. At the present, nitride technology is considerably more complex than that of thermal oxidation, and its use is restricted to special situations. There appears to be no doubt, however, that it will be more extensively employed in the coating of microcircuits; present indications are that its use will eliminate the need for a package, thus leading to a considerable reduction in cost.

17.5. CONCLUDING REMARKS

By way of conclusion it is well to summarize the detailed nature of the thermally oxidized silicon surface[2] insofar as it fits the various observed experimental data.

1. The silicon surface presents a boundary in an otherwise regular crystal latice. This break in periodicity gives rise to a number of states distributed within the forbidden gap. The theories of Tamm and Shockley predict the nature of these states for an atomically clean, damage-free, abrupt silicon-vacuum transition.
2. The surface of thermally grown silicon represents a relatively gradual termination of the silicon single crystal. This leads to a reduction of the density of Tamm-Shockley states. The exact nature of these states is not known, but evidence indicates that they have a density of 10^{10} to $10^{11}/cm^2$ and capture cross sections of 10^{-15} cm^2 to both electrons and holes. They are fast states and determine the surface lifetime.
3. The first 100 to 200 Å of thermally grown material represents the silicon-rich transition between silicon and oxygen. Thus it contributes a fixed, net positive charge and results in bending the energy bands beneath the surface to cause an n-shift in the silicon. The net positive charge Q_{ss} is about $10^{11}q$ coulombs/cm^2. The capture cross section of these states is estimated as being many orders of magnitude below that of the fast states. In addition their time constants range from seconds to months.

The charge associated with these fixed states corresponds to unfilled silicon bonds in the SiO_2 tetrahedra. Thus it is stable at all operating or storage temperatures.

4. The bulk of the SiO_2 layer contains superoxide-ion molecules O_2^-, oxygen-ion vacancies V^{++}, and metal contamination ions such as Na^+. All of these are free to migrate through the oxide under the influence of a field at elevated temperatures. They all provide slow surface states, and their net effect is to degrade the long-term stability of the surface potential.

17.6. REFERENCES

[1] A. Many et al., *Semiconductor Surfaces*, North Holland, Amsterdam 1965.

[2] A. S. Grove, *Physics and Technology of Semiconductor Devices*, Wiley, New York, 1966.

[3] R. H. Kingston, and S. F. Neustadter, "Calculation of the Space Charge, Electric Field, and Free Carrier Concentration at the Surface of a Semiconductor," *J. Appl. Phys.*, **26**(6), 718–720, 1955.

[4] C. E. Young, "Extended Curves of Space Charge, etc.," *J. Appl. Phys.*, **32**(3) 329–332 1961.

[5] A. B. Philips, *Transistor Engineering*, McGraw-Hill, New York, 1962.

[6] P. J. Coppen, and W. T. Matzen, "Distribution of Recombination Current in Emitter-Base Junctions of Silicon Transistors," *IEEE Trans. on Electron Devices*, **ED-9**(1), 75–81 (1962).

[7] C. T. Sah, "Effect of Surface Recombination and Channel on *P-N* Junction and Transistor Characteristics," *IEEE Trans. on Electron Devices*, **ED-9**(1) 94–108 (1962).

[8] J. E. Thomas, Jr., and D. R. Young, "Space Charge Model for Surface Potential Shifts in Silicon Passivated with Thin Insulating Layers," *IBM J. Res. Dev.*, **8**(4) 368–375 (1964).

[9] D. P. Seraphim et al., "Electrochemical Phenomena in Thin Films of Silicon Dioxide on Silicon," IBM *J. Res. Dev.*, **8**(4), 400–409 (1964).

[10] M. Yamin, "Observations on Phosphorus Stabilized SiO_2 Films," *IEEE Trans. on Electron Devices*, **ED-13**(2) 256–259 (1966).

[11] E. Yon et al., "Sodium Distribution in Thermal Oxide on Silicon by Radiochemical and MOS Analysis," *IEEE Trans. on Electron Devices*, **ED-13**(2), 276–280(1966).

[12] E. Kooi and M. V. Whalen, "On the Roles of Na and H_2 in the Si–SiO_2 System," *Appl. Phys. Letters*, **9**, 314 (1966).

[13] V. Y. Doo et al., "Preparation and Properties of Pyrolytic Si_3N_4," *J. Electrochem Soc.*, **113**, 1279 (1966).

17.7. PROBLEMS

1. A piece of 0.5 Ω-cm *n*-type silicon has 10^{11} positively charged surface states per cm^2. Compute the magnitude of the $\mathscr{E}$ field and the surface potential.

2. Repeat Problem 1 for a piece of 5-Ω-cm *p*-type silicon.

3. A piece of 1-Ω-cm *p*-type silicon is thermally oxidized. What is the maximum concentration of surface charge that can be present before inversion occurs? For this situation determine the magnitude and direction of the $\mathscr{E}$ field.

4. A n^+-p junction is built with a 1-Ω-cm p-type region. The junction is reverse-biased at 10 V. What is the approximate width of the depletion layer for: (*a*) the flat band condition; (*b*) 10^{12} donor-like surface charges/cm^2; (c) 10^{12} acceptor-like surface charges/cm^2?

5. A n-type semiconductor bar has a square cross section. Assuming a resistivity of 1 Ω-cm and a bulk lifetime of 10 μ secs., determine the effective lifetime as the recombination velocity is varied from zero to its maximum value.

6. What is the maximum attainable value of surface recombination velocity for an n-type piece of silicon?

7. The surface potential of a 1-Ω-cm n-type silicon slice is varied over the range of values of u_s from -20 to $+20$. What is the sheet conductance for this range?

8. The experiment described in Problem 7 is conducted by bringing a parallel plate close to the semiconductor surface to form a capacitor, and altering the voltage across this capacitor. What is the surface-charge-density range that is required to conduct this experiment?

Appendix A

The Mathematics of Diffusion

A. 1. GENERAL SOLUTION OF THE DIFFUSION EQUATION FOR AN INFINITE BODY

This solution follows closely the method of Boltaks [1]. The statement of Fick's second law is

$$\frac{\partial N}{\partial t} = D\frac{\partial^2 N}{\partial x^2} \tag{A.1}$$

The initial distribution of the solute is given by

$$N(x, 0) = f(x). \tag{A.2}$$

One solution of this equation may be written as the product of two functions, one of time and one of space. Thus let

$$N(x, t) = X(x)T(t). \tag{A.3}$$

Substituting in (A.1), and rearranging terms,

$$\frac{1}{DT}\frac{dT}{dt} = \frac{1}{X}\frac{d^2X}{dx^2} \tag{A.4}$$

Since the left-hand side of this equation is only a function of t and the right-hand side only of x, each side must be equal to some constant which is independent of both t or x. Writing this constant as $-\lambda^2$, (A.3) can be broken into

$$\frac{dT}{T} = -\lambda^2 D\, dt \tag{A.5}$$

and

$$\frac{d^2X}{dx^2} = -\lambda^2 X. \tag{A.6}$$

Solving,

$$T(t) = \gamma e^{-\lambda^2 Dt}, \tag{A.7}$$

$$X(x) = \alpha \cos \lambda x + \beta \sin \lambda x, \tag{A.8}$$

and

$$N(x,t) = \lambda e^{-\lambda^2 Dt} (A \cos \lambda x + B \sin \lambda x), \tag{A.9}$$

where γ, α, β are constants of integration, $A = \alpha\gamma$, and $B = \beta\gamma$.

The general solution of (A.1) can be written as a sum of partial solutions of this type because it is a linear equation. In addition, since the body has infinite dimensions, the choice of λ is quite arbitrary and the summation over discrete values of λ can be replaced by an integral. Hence this solution may be written as

$$N(x,t) = \int_{-\infty}^{+\infty} e^{-\lambda^2 Dt} (A \cos \lambda x + B \sin \lambda x)\, d\lambda. \tag{A.10}$$

It is now necessary to solve for the various constants of integration. Fourier's integral theorem states that

$$f(x) = \frac{1}{2\pi} \int_{-\infty}^{+\infty} \int_{-\infty}^{+\infty} f(\xi) \cos [\lambda(\xi - x)]\, d\xi\, d\lambda, \tag{A.11}$$

$$= \frac{1}{2\pi} \int_{-\infty}^{+\infty} \left\{ \left[\int_{-\infty}^{+\infty} f(\xi) \cos \lambda\xi\, d\xi \right] \cos \lambda x\, dx \right.$$

$$\left. + \left[\int_{-\infty}^{+\infty} f(\xi) \sin \lambda\xi\, d\xi \right] \sin \lambda x\, dx \right\} d\lambda. \tag{A.12}$$

At time $t = 0$ the initial concentration is given by $f(x)$.

Substituting in (A.10),

$$f(x) = N(x,0) = \int_{-\infty}^{\infty} (A \cos \lambda x + B \sin \lambda x)\, d\lambda. \tag{A.13}$$

Comparing (A.12) and (A.13),

$$A = \frac{1}{2\pi} \int_{-\infty}^{\infty} f(\xi) \cos \lambda\xi \, d\xi, \tag{A.14}$$

$$B = \frac{1}{2\pi} \int_{-\infty}^{\infty} f(\xi) \sin \lambda\xi \, d\xi. \tag{A.15}$$

Substituting these values of A and B in (A.10) gives

$$N(x, t) = \frac{1}{2\pi} \int_{-\infty}^{\infty} f(\xi) \left[\int_{-\infty}^{\infty} e^{-\lambda^2 Dt} \cos \lambda (\xi - x) \, d\lambda \right] d\xi. \tag{A.16}$$

But

$$\int_{-\infty}^{\infty} e^{-\lambda^2 Dt} \cos \lambda (\xi - x) \, dx = \sqrt{\pi/Dt} \, e^{-(\xi - x)^2/4Dt}, \tag{A.17}$$

hence

$$N(x, t) = (2\sqrt{\pi Dt})^{-1} \int_{-\infty}^{\infty} f(\xi) \, e^{-(\xi - x)^2/4Dt} d\xi. \tag{A.18}$$

This is the general solution of the diffusion equation for an infinite body. Here

$$f(\xi) = N(\xi, 0). \tag{A.19}$$

A.2. DIFFUSION FROM AN INFINITELY THICK LAYER INTO AN INFINITE BODY

Consider an infinitely thick body, thc initial impurity concentration of which is given by

$$N(x, 0) = N_0 \qquad \text{for} \quad -\infty < x \leqslant 0, \tag{A.20a}$$

$$N(x, 0) = 0 \qquad \text{for} \quad 0 \leqslant x < \infty. \tag{A.20b}$$

Equation A.19 can be rewritten as

$$N(x, t) = (2\sqrt{\pi Dt})^{-1} \left[\int_{-\infty}^{0} f(\xi) e^{-(\xi - x)^2/4Dt} \, d\xi + \int_{0}^{\infty} f(\xi) \, e^{-(\xi - x)^2/4Dt} \, d\xi \right] \tag{A.21}$$

where

$$f(\xi) = 0 \qquad \text{for } \xi \geqslant 0, \tag{A.22a}$$

$$f(\xi) = N_0 \qquad \text{for } \xi \leqslant 0 \tag{A.22b}$$

Hence

$$N(x,t) = \frac{N_0}{2\sqrt{\pi\, Dt}} \int_{-\infty}^{0} e^{-(\xi - x)^2/4Dt}\, d\xi. \tag{A.23}$$

Setting $z = (\xi - x)/2\sqrt{Dt}$, (A.23) reduces to

$$N(x,t) = \frac{N_0}{\sqrt{\pi}} \int_{-\infty}^{-x/2\sqrt{Dt}} e^{-z^2} dz \tag{A.24}$$

$$= \frac{N_0}{\sqrt{\pi}} \left(\int_{-\infty}^{0} e^{-z^2} dz - \int_{-x/2\sqrt{Dt}}^{0} e^{-z^2}\, dz \right). \tag{A.25}$$

Since e^{-z^2} is symmetrical in z, (A.25) reduces to

$$N(x,t) = \frac{N_0}{\sqrt{\pi}} \left(\int_{0}^{\infty} e^{-z^2} dz - \int_{0}^{x/2\sqrt{Dt}} e^{-z^2} dz \right). \tag{A.26}$$

But

$$\int_0^\infty e^{-z^2}\, dz = \frac{\sqrt{\pi}}{2}, \tag{A.27}$$

and

$$\int_0^y e^{-z^2}\, dz = \frac{\sqrt{\pi}}{2} \operatorname{erf} y. \tag{A.28}$$

Hence

$$N(x,t) = \frac{N_0}{2} \left(1 - \operatorname{erf} \frac{x}{2\sqrt{Dt}} \right), \tag{A.29}$$

$$= \frac{N_0}{2} \operatorname{erfc} \frac{x}{2\sqrt{Dt}} \tag{A.30}$$

Figure A.1 shows a sketch of this impurity distribution at various points in time. Table A.1 gives values of $\operatorname{erf} z$ for different values of z.

Table A1
Error Function erf(z)†

z	erf(z)	z	erf(z)	z	erf(z)	z	erf(z)
0.00	0.000 000	0.50	0.520 500	1.00	0.842 701	1.50	0.966 105
0.01	0.011 283	0.51	0.529 244	1.01	0.846 810	1.51	0.967 277
0.02	0.022 565	0.52	0.537 899	1.02	0.850 838	1.52	0.968 413
0.03	0.033 841	0.53	0.546 464	1.03	0.854 784	1.53	0.969 516
0.04	0.045 111	0.54	0.554 939	1.04	0.858 650	1.54	0.970 586
0.05	0.056 372	0.55	0.563 323	1.05	0.862 436	1.55	0.971 623
0.06	0.067 622	0.56	0.571 616	1.06	0.866 144	1.56	0.972 628
0.07	0.078 858	0.57	0.579 816	1.07	0.869 773	1.57	0.973 603
0.08	0.090 078	0.58	0.587 923	1.08	0.873 326	1.58	0.974 547
0.09	0.101 281	0.59	0.595 936	1.09	0.876 803	1.59	0.975 462
0.10	0.112 463	0.60	0.603 856	1.10	0.880 205	1.60	0.976 348
0.11	0.123 623	0.61	0.611 681	1.11	0.883 533	1.61	0.977 207
0.12	0.134 758	0.62	0.619 411	1.12	0.886 788	1.62	0.978 038
0.13	0.145 867	0.63	0.627 046	1.13	0.889 971	1.63	0.978 843
0.14	0.156 947	0.64	0.634 586	1.14	0.893 082	1.64	0.979 622
0.15	0.167 996	0.65	0.642 029	1.15	0.896 124	1.65	0.980 376
0.16	0.179 012	0.66	0.649 377	1.16	0.899 096	1.66	0.981 105
0.17	0.189 992	0.67	0.656 628	1.17	0.902 000	1.67	0.981 810
0.18	0.200 936	0.68	0.663 782	1.18	0.904 837	1.68	0.982 493
0.19	0.211 840	0.69	0.670 840	1.19	0.907 608	1.69	0.983 153
0.20	0.222 703	0.70	0.677 801	1.20	0.910 314	1.70	0.983 790
0.21	0.233 522	0.71	0.684 666	1.21	0.912 956	1.71	0.984 407
0.22	0.244 296	0.72	0.691 433	1.22	0.915 534	1.72	0.985 003
0.23	0.255 023	0.73	0.698 104	1.23	0.918 050	1.73	0.985 578
0.24	0.265 700	0.74	0.704 678	1.24	0.920 505	1.74	0.986 135
0.25	0.276 326	0.75	0.711 156	1.25	0.922 900	1.75	0.986 672
0.26	0.286 900	0.76	0.717 537	1.26	0.925 236	1.76	0.987 190
0.27	0.297 418	0.77	0.723 822	1.27	0.927 514	1.77	0.987 691
0.28	0.307 880	0.78	0.730 010	1.28	0.929 734	1.78	0.988 174
0.29	0.318 283	0.79	0.736 103	1.29	0.931 899	1.79	0.988 641
0.30	0.328 627	0.80	0.742 101	1.30	0.934 008	1.80	0.989 091
0.31	0.338 908	0.81	0.748 003	1.31	0.936 063	1.81	0.989 525
0.32	0.349 126	0.82	0.753 811	1.32	0.938 065	1.82	0.989 943
0.33	0.359 279	0.83	0.759 524	1.33	0.940 015	1.83	0.990 347
0.34	0.369 365	0.84	0.765 143	1.34	0.941 914	1.84	0.990 736
0.35	0.379 382	0.85	0.770 668	1.35	0.943 762	1.85	0.991 111
0.36	0.389 330	0.86	0.776 100	1.36	0.945 561	1.86	0.991 472
0.37	0.399 206	0.87	0.781 440	1.37	0.947 312	1.87	0.991 821
0.38	0.409 009	0.88	0.786 687	1.38	0.949 016	1.88	0.992 156
0.39	0.418 739	0.89	0.791 843	1.39	0.950 673	1.89	0.992 479
0.40	0.428 392	0.90	0.796 908	1.40	0.952 285	1.90	0.992 790
0.41	0.437 969	0.91	0.801 883	1.41	0.953 852	1.91	0.993 090
0.42	0.447 468	0.92	0.806 768	1.42	0.955 376	1.92	0.993 378
0.43	0.456 887	0.93	0.811 564	1.43	0.956 857	1.93	0.993 656
0.44	0.466 225	0.94	0.816 271	1.44	0.958 297	1.94	0.993 923
0.45	0.475 482	0.95	0.820 891	1.45	0.959 695	1.95	0.994 179
0.46	0.484 655	0.96	0.825 424	1.46	0.961 054	1.96	0.994 426
0.47	0.493 745	0.97	0.829 870	1.47	0.962 373	1.97	0.994 664
0.48	0.502 750	0.98	0.834 232	1.48	0.963 654	1.98	0.994 892
0.49	0.511 668	0.99	0.838 508	1.49	0.964 898	1.99	0.995 111

Table A1 *(continued)*

2.00	0.995 322	2.50	0.999 593	3.00	0.999 977 91	3.50	0.999 999 257
2.01	0.995 525	2.51	0.999 614	3.01	0.999 979 26	3.51	0.999 999 309
2.02	0.995 719	2.52	9.999 634	3.02	0.999 980 53	3.52	0.999 999 358
2.03	0.995 906	2.53	0.999 654	3.03	0.999 981 73	3.53	0.999 999 403
2.04	0.996 086	2.54	0.999 672	3.04	0.999 982 86	3.54	0.999 999 445
2.05	0.996 258	2.55	0.999 689	3.05	0.999 983 92	3.55	0.999 999 485
2.06	0.996 423	2.56	0.999 706	3.06	0.999 984 92	3.56	0.999 999 521
2.07	0.996 582	2.57	0.999 722	3.07	0.999 985 86	3.57	0.999 999 555
2.08	0.996 734	2.58	0.999 736	3.08	0.999 986 74	3.58	0.999 999 587
2.09	0.996 880	2.59	0.999 751	3.09	0.999 987 57	3.59	0.999 999 617
2.10	0.997 021	2.60	0.999 764	3.10	0.999 988 35	3.60	0.999 999 644
2.11	0.997 155	2.61	0.999 777	3.11	0.999 989 08	3.61	0.999 999 670
2.12	0.997 284	2.62	0.999 789	3.12	0.999 989 77	3.62	0.999 999 694
2.13	0.997 407	2.63	0.999 800	3.13	0.999 990 42	3.63	0.999 999 716
2.14	0.997 525	2.64	0.999 811	3.14	0.999 991 03	3.64	0.999 999 736
2.15	0.997 639	2.65	0.999 822	3.15	0.999 991 60	3.65	0.999 999 756
2.16	0.997 747	2.66	0.999 831	3.16	0.999 992 14	3.66	0.999 999 773
2.17	0.997 851	2.67	0.999 841	3.17	0.999 992 64	3.67	0.999 999 790
2.18	0.997 951	2.68	0.999 849	3.18	0.999 993 11	3.68	0.999 999 805
2.19	0.998 046	2.69	0.999 858	3.19	0.999 993 56	3.69	0.999 999 820
2.20	0.998 137	2.70	0.999 866	3.20	0.999 993 97	3.70	0.999 999 833
2.21	0.998 224	2.71	0.999 873	3.21	0.999 994 36	3.71	0.999 999 845
2.22	0.998 308	2.72	0.999 880	3.22	0.999 994 73	3.72	0.999 999 857
2.23	0.998 388	2.73	0.999 887	3.23	0.999 995 07	3.73	0.999 999 867
2.24	0.998 464	2.74	0.999 893	3.24	0.999 995 40	3.74	0.999 999 877
2.25	0.998 537	2.75	0.999 899	3.25	0.999 995 70	3.75	0.999 999 886
2.26	0.998 607	2.76	0.999 905	3.26	0.999 995 98	3.76	0.999 999 895
2.27	0.998 674	2.77	0.999 910	3.27	0.999 996 24	3.77	0.999 999 903
2.28	0.998 738	2.78	0.999 916	3.28	0.999 996 49	3.78	0.999 999 910
2.29	0.998 799	2.79	0.999 920	3.29	0.999 996 72	3.79	0.999 999 917
2.30	0.998 857	2.80	0.999 925	3.30	0.999 996 94	3.80	0.999 999 923
2.31	0.998 912	2.81	0.999 929	3.31	0.999 997 15	3.81	0.999 999 929
2.32	0.998 966	2.82	0.999 933	3.32	0.999 997 34	3.82	0.999 999 934
2.33	0.999 016	2.83	0.999 937	3.33	0.999 997 51	3.83	0.999 999 939
2.34	0.999 065	2.84	0.999 941	3.34	0.999 997 68	3.84	0.999 999 944
2.35	0.999 111	2.85	0.999 944	3.35	0.999 997 838	3.85	0.999 999 948
2.36	0.999 155	2.86	0.999 948	3.36	0.999 997 983	3.86	0.999 999 952
2.37	0.999 197	2.87	0.999 951	3.37	0.999 998 120	3.87	0.999 999 956
2.38	0.999 237	2.88	0.999 954	3.38	0.999 998 247	3.88	0.999 999 959
2.39	0.999 275	2.89	0.999 956	3.39	0.999 998 367	3.89	0.999 999 962
2.40	0.999 311	2.90	0.999 959	3.40	0.999 998 478	3.90	0.999 999 965
2.41	0.999 346	2.91	0.999 961	3.41	0.999 998 582	3.91	0.999 999 968
2.42	0.999 379	2.92	0.999 964	3.42	0.999 998 679	3.92	0.999 999 970
2.43	0.999 411	2.93	0.999 966	3.43	0.999 998 770	3.93	0.999 999 973
2.44	0.999 441	2.94	0.999 968	3.44	0.999 998 855	3.94	0.999 999 975
2.45	0.999 469	2.95	0.999 970	3.45	0.999 998 934	3.95	0.999 999 977
2.46	0.999 497	2.96	0.999 972	3.46	0.999 999 008	3.96	0.999 999 979
2.47	0.999 523	2.97	0.999 973	3.47	0.999 999 077	3.97	0.999 999 980
2.48	0.999 547	2.98	0.999 975	3.48	0.999 999 141	3.98	0.999 999 982
2.49	0.999 571	2.99	0.999 976	3.49	0.999 999 201	3.99	0.999 999 983

†For a more complete table, see L. J. Comrie, *Chambers Six Figure Mathematical Tables*, Vol. 2, W. & R. Chambers, Edinburgh, 1949.

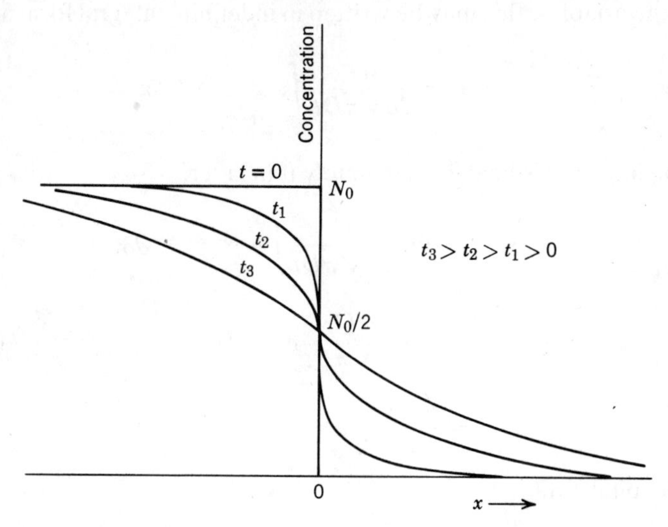

Figure A.1 Complementary error function diffusion.

A.3. DIFFUSION FROM AN INFINITELY THIN LAYER INTO AN INFINITE BODY

Consider an infinite body, of unit cross section, with an impurity located in an infinitely thin layer within it. Let Q be the quantity of matter in this layer. It is required to find the impurity distribution as a function of time.

This problem may be solved by setting the internal-layer thickness as $\pm\delta$, with an impurity concentration N_0. Then

$$N(x,0) = 0, \qquad -\infty < x \leqslant -\delta, \tag{A.31a}$$

$$= N_0, \qquad -\delta \leqslant x \leqslant \delta, \tag{A.31b}$$

$$= 0, \qquad \delta \leqslant x < \infty, \tag{A.31c}$$

and

$$Q = 2N_0. \tag{A.32}$$

Substituting into (A.18) gives

$$N(x,t) = \frac{N_0}{2\sqrt{\pi D t}} \int_{-\delta}^{+\delta} e^{-(\xi - x)^2/4Dt}\, d\xi, \tag{A.33}$$

$$= \frac{Q}{2\delta\sqrt{\pi D t}} \int_{0}^{\delta} e^{-(\xi - x)^2/4Dt}\, d\xi. \tag{A.34}$$

Changing variables, this may be written in indefinite integral form as

$$N(x,t) = \frac{Q}{2\delta\sqrt{\pi Dt}} \int e^{-(\delta-x)^2/4Dt}\, d\delta. \tag{A.35}$$

Since the impurity is held in an infinitely thin layer,

$$N(x,t) = \lim_{\delta\to 0} \frac{Q}{2\delta\sqrt{\pi Dt}} \int e^{-(\delta-x)^2/4Dt}\, d\delta, \tag{A.36a}$$

$$= \lim_{\delta\to 0} \frac{Q}{2\sqrt{\pi Dt}} e^{-(\delta-x)^2/4Dt}, \tag{A.36b}$$

$$= \frac{Q}{2\sqrt{\pi Dt}} e^{-x^2/4Dt}, \tag{A.37}$$

by L' Hospital's rule.

Figure A.2 shows a sketch of this distribution at various points in time.

A.4. DIFFUSION INTO A SEMI-INFINITE BODY

A semi-infinite body is one in which one boundary is at $x = 0$ and the other is at $x = \infty$. The solution of this case is obtained from the infinite-body situation by considerations of symmetry. Various examples follow.

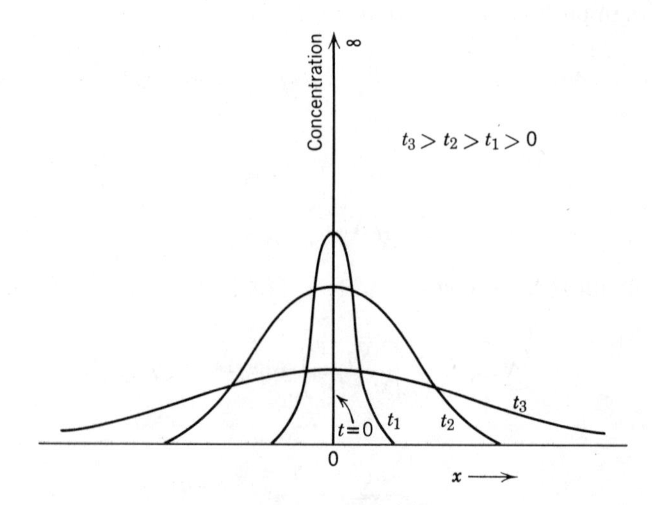

Figure A.2 The gaussian diffusion.

A.4.1. Diffusion from a Constant Source into a Semi-infinite Body

For this situation the concentration at $x = 0$ is N_0. Direct comparison with the case discussed in Section A.2 results in a distribution given by [compare with (A.29)]

$$N(x, t) = N_0\left(1 - \operatorname{erf}\frac{x}{2\sqrt{Dt}}\right), \tag{A.38}$$

$$= N_0 \operatorname{erfc}\frac{x}{2\sqrt{Dt}}. \tag{A.39}$$

A.4.1.1. Quantity of Impurity Transported Through the Plane x = 0.

The quantity of impurity that has been transported into the region where $x \geqslant 0$ may be obtained as follows:

The flux density is given by Fick's first law as

$$j(x, t) = -D\frac{\partial N}{\partial x}. \tag{A.40}$$

The flux density of the material passing through the plane $x = 0$ is thus

$$j(0, t) = -D\left(\frac{\partial N}{\partial x}\right)_{x=0} \tag{A.41}$$

But

$$N(x, t) = N_0 \operatorname{erfc}\frac{x}{2\sqrt{Dt}}. \tag{A.42}$$

Substituting in (A.41),

$$j(0, t) = N_0\sqrt{D/\pi t}. \tag{A.43}$$

Then the quantity of material transported through the plane at $x = 0$ at any time t is given by

$$Q = \int_0^t j(0, t)\, dt \tag{A.44}$$

$$= 2N_0\sqrt{Dt/\pi} \tag{A.45}$$

A.4.2. Diffusion from an Infinitely Thin Layer into a Semi-infinite Body

This situation is similar to that described in Section A.3. For the case of a quantity of matter Q diffusing in both directions into an infinite body the impurity distribution was given by [see (A.43)]

$$N(x, t) = \frac{Q}{2\sqrt{\pi Dt}}\, e^{-x^2/4Dt}. \tag{A.46}$$

This is the same result as would be obtained if half this matter diffused into a semi-infinite body. Hence, if the entire amount of matter Q diffused into the semi-infinite body, the impurity distribution would be twice as large, as given by

$$N(x, t) = \frac{Q}{\sqrt{\pi D t}} e^{-x^2/4Dt}. \tag{A.47}$$

A.5. DIFFUSION FROM A FINITE BODY WITH CAPTURING BOUNDARIES

For this situation we consider a body with an initial concentration $N(x, 0)$. The thickness of the body is a and the boundaries act as perfect sinks for the impurity so that (see Figure A.3)

$$N(0, t) = N(a, t) = 0 \qquad \text{for } t > 0. \tag{A.48}$$

Since the boundaries are finite, the choice of λ in (A.9) cannot be arbitrarily made. Hence the general solution of the diffusion equation must be written in summation form as

$$N(x, t) = \sum_{m=0} e^{-\lambda_m{}^2 Dt} (A_m \cos \lambda_m x + B_m \sin \lambda_m x), \tag{A.49}$$

where A_m, B_m, and λ_m are appropriate parameters the values of which must be determined. Now

$$N(0, t) = \sum_0^\infty A_m e^{-\lambda_m{}^2 Dt} = 0. \tag{A.50}$$

Since this holds for all t,

$$A_m = 0. \tag{A.51}$$

Application of the second boundary condition gives

$$N(a, t) = \sum_0^\infty B_m e^{-\lambda_m{}^2 Dt} \sin \lambda_m a = 0. \tag{A.52}$$

B_m cannot be zero, for otherwise the solution would be degenerate; hence

$$\lambda_m a = m\pi, \tag{A.53}$$

where m is an integer.

Inserting (A.51) and (A.53) into (A.49) gives

$$N(x, t) = \sum_0^\infty B_m e^{-(m\pi/a)^2 Dt} \sin \frac{m\pi x}{a} \tag{A.54}$$

The final boundary value is used to evaluate B_m. Thus

$$N(x,0) = \sum_{0}^{\infty} B_m \sin \frac{m\pi x}{a} \tag{A.55}$$

Multiplying both sides by $\sin(n\pi x/a)$, where n is an integer, and integrating from 0 to a,

$$\int_0^a N(x,0) \sin \frac{n\pi x}{a}\, dx = \sum_0^{\infty} B_m \int_0^a \sin \frac{m\pi x}{a} \sin \frac{n\pi x}{a}\, dx. \tag{A.56}$$

But

$$\int_0^a \sin \frac{m\pi x}{a} \sin \frac{n\pi x}{a}\, dx = 0 \tag{A.57}$$

for $m \neq n$, and $a/2$ for $m = n$. Thus (A.56) reduces to

$$\int_0^a N(x,0) \sin \frac{m\pi x}{a}\, dx = B_m \frac{a}{2} \tag{A.58}$$

since $n = m$.

Substituting into (A.54) gives the final solution as

$$N(x,t) = \frac{2}{a} \sum_0^{\infty} \left\{ e^{-(m\pi/a)^2 Dt} \sin m\pi \frac{x}{a} \left[\int_0^a N(x,0) \sin \frac{m\pi x}{a}\, dx \right] \right\}. \tag{A.59}$$

A.5.1. The Constant Initial Concentration Case

Consider the case of out-diffusion from a uniformly doped finite body. The boundaries are again considered to be infinite sinks. Then

$$N(x,0) = N_0 \qquad \text{for } 0 \leqslant x \leqslant a. \tag{A.60}$$

Direct substitution into (A.58) gives

$$\frac{N_0 a}{m\pi} [1 - (-1)^m] = B_m \frac{a}{2} \tag{A.61}$$

Thus

$$B_m = 2N_0 \frac{1-(-1)^m}{m\pi}. \tag{A.62}$$

Substituting into (A.59) gives

$$N(x,t) = \frac{2N_0}{\pi} \sum_0^{\infty} \left[\frac{1-(-1)^m}{m} \right] e^{-(m\pi/a)^2 Dt} \sin \frac{m\pi x}{a} \tag{A.63}$$

This function is zero for even values of m, and converges very rapidly for odd values. Boltaks[1] has shown that the error in ignoring all terms other than $m = 1$ is only 1% for values of $t > 0.045a^2/D$. Hence,

$$N(x, t) \simeq \frac{4N_0}{\pi} e^{-\pi^2 Dt/a^2} \sin \frac{\pi x}{a} \tag{A.64}$$

Figure A.3 shows this distribution for various points in time.

A.5.2. Bilateral Diffusion from a Constant Concentration into a Finite Body

Consider the bilateral diffusion of an impurity into a finite body of width a. For this case

$$N(0, t) = N(a, t) = N_0 \qquad \text{for all } t. \tag{A.65}$$

This situation can be compared directly with that described in Section A.5.1 for out-diffusion from a finite body. The result may be written by inspection as

$$N(x, t) = N_0\left(1 - \frac{4}{\pi} e^{-\pi^2 Dt/a^2} \sin \frac{\pi x}{a}\right). \tag{A.66}$$

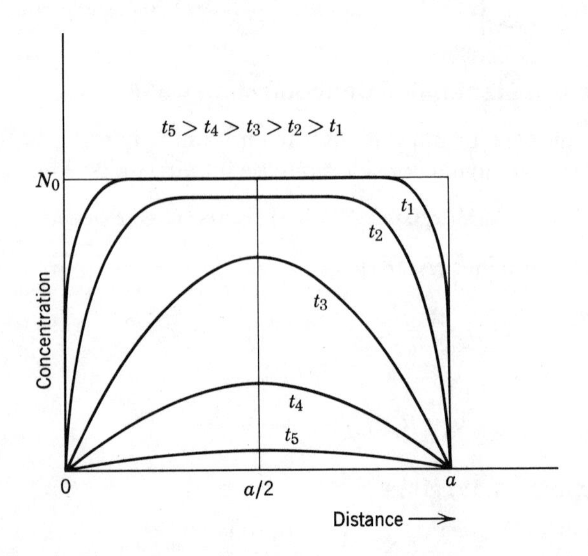

Figure A.3 Out-diffusion at a surface.

A.6. SOME USEFUL ERROR-FUNCTION RELATIONS

1. $$\int_0^\infty e^{-z^2}\,dz = \frac{\sqrt{\pi}}{2}.$$

2. $$\int_0^y e^{-z^2}\,dz = \frac{\sqrt{\pi}}{2}\operatorname{erf} y.$$

3. $$\int_0^y e^{-z^2}\,dz = y - \frac{y^3}{3\times 1!} + \frac{y^5}{5\times 2!}\cdots.$$

4. $$\frac{d}{dy}\operatorname{erf} y = \frac{2}{\sqrt{\pi}}e^{-y^2}.$$

5. $$\frac{d^2}{dy^2}\operatorname{erf} y = -\frac{4}{\sqrt{\pi}}ye^{-y^2}.$$

6. $$\operatorname{erf} -y = -\operatorname{erf} y.$$

7. $$\operatorname{erf} 0 = 0.$$

8. $$\operatorname{erf} \infty = 1.$$

In addition, a short list of error functions is given in Table A.1.

A.7. REFERENCES

[1] B. I. Boltaks, *Diffusion in Semiconductors*, Academic, New York, 1963.
[2] L. A. Pipes, *Applied Mathematics for Engineers and Physicists*, McGraw-Hill, New York, 1958.
[3] W. Jost, *Diffusion in Solids, Liquids, and Gases*, Academic, New York, 1962.

Appendix B

Current Gain of a Graded-base Transistor

Consider a graded-base *n-p-n* transistor [1, 2] of unit cross section. For this transistor (see Figure B.1) let

$$N_B = N_{BE} e^{-\eta x / W_B}, \tag{B.1}$$

where

$$\eta = \ln \frac{N_{BE}}{N_{BC}} \tag{B.2}$$

and quantities have been defined as in Section 14.3.

The $\mathscr{E}$ field in the base is given by

$$\mathscr{E}_0 = -\frac{kT}{q} \frac{\eta}{W_B} \tag{B.3}$$

and aids the flow of carriers from the emitter toward the collector.

The continuity and diffusion equation for excess minority carriers (electrons) may be written as

$$\frac{\partial n_B'}{\partial t} = D_{nB} \frac{\partial^2 n_B'}{\partial x^2} + \mu_n \mathscr{E}_0 \frac{\partial n_B'}{\partial x} - \frac{n_B'}{\tau_{nB}} \tag{B.4}$$

Assume a solution in the form

$$n_B' = n_{\text{dc}}'(x) + n_{\text{ac}}'(x)\, e^{j\omega t}. \tag{B.5}$$

Substituting (B.5) into (B.4), the time-varying form of this equation is obtained as

$$\frac{d^2 n_{\text{ac}}'}{dx^2} + \frac{\mu_n \mathscr{E}_0}{D_{nB}} \frac{d n_{\text{ac}}'}{dx} - \frac{n_{\text{ac}}'}{(L_{nB}')^2} = 0, \tag{B.6}$$

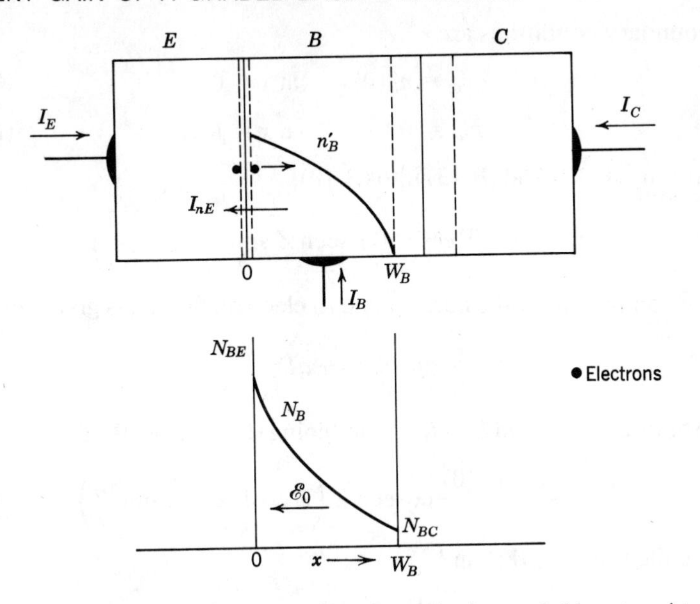

Figure B.1 Doping profile and carrier concentration in a graded-base transistor.

where

$$L'_{nB} = \frac{L_{nB}}{\sqrt{1 + j\omega\tau_{nB}}}. \tag{B.7}$$

As shown in Section 13.1.3, this is of the same form as the d-c equation, except for the replacement of L_{nB} by L'_{nB}.

From (B.3),

$$\eta = -\frac{\mathscr{E}_0 \mu_n W_B}{D_{nB}} \tag{B.8}$$

Hence, (B.6) reduces to

$$\frac{d^2 n'_{\text{ac}}}{dx^2} - \frac{\eta}{W_B}\frac{dn'_{\text{ac}}}{dx} - \frac{n'_{\text{ac}}}{(L'_{nB})^2} = 0. \tag{B.9}$$

The solution of this equation is

$$n'_{\text{ac}} = A_1 e^{a_1 x} + A_2 e^{a_2 x}, \tag{B.10}$$

where a_1 and a_2 are given by

$$a_1, a_2 = \frac{1}{W_B}\left[\frac{\eta}{2} \pm \left\{\frac{\eta^2}{4} + \left(\frac{W_B}{L_{nB}}\right)^2 (1 + j\omega\tau_{nB})\right\}^{1/2}\right], \tag{B.11}$$

$$= \frac{1}{W_B}\left(\frac{\eta}{2} \pm Z\right) \tag{B.12}$$

The boundary conditions are

$$n'_{\rm ac} = n'_B(0) \qquad \text{at } x = 0, \tag{B.13a}$$

$$n'_{\rm ac} = 0 \qquad \text{at } x = W_B. \tag{B.13b}$$

Substituting (B.12) and (B.13) into (B.10),

$$n'_{\rm ac} = n'_B(0)e^{\eta x/2W_B} \operatorname{cosech} Z \sinh Z\left(1-\frac{x}{W_B}\right) \tag{B.14}$$

The a-c component of the current due to electron flow i_n, is given by

$$i_n = q\mu_n n'_{\rm ac}\mathscr{E}_0 + qD_{nB}\frac{\partial n'_{\rm ac}}{\partial x}. \tag{B.15}$$

At the emitter $x = 0$ and $i_n = i_{nE}$. Combining (B.14) and (B.15),

$$i_{nE} = \frac{qD_{nB}n'_B(0)}{W_B} \operatorname{cosech} Z\left(Z\cosh Z + \frac{\eta}{2}\sinh Z\right). \tag{B.16}$$

At the collector, $x = W_B$, and

$$i_n = i_{nC} = \frac{qD_{nB}n'_B(0)}{W_B} e^{\eta/2} Z \operatorname{cosech} Z, \tag{B.17}$$

hence

$$\alpha_T = \frac{i_{nC}}{i_{nE}} \tag{B.18}$$

$$= \frac{e^{\eta/2}}{Z\cosh Z + (\eta/2)\sinh Z} \tag{B.19}$$

where

$$Z = \left\{\frac{\eta^2}{4} + \left(\frac{W_B}{L_{nB}}\right)^2 (1 + j\omega\tau_{nB})\right\}^{1/2}, \tag{B.20}$$

and

$$\eta = \ln\frac{N_{BE}}{N_{BC}}. \tag{B.21}$$

Combining the effect of the emitter injection efficiency, the common-base current gain of a graded-base transistor may be written as

$$\alpha = \frac{\alpha_0 e^{\eta/2}}{Z\cosh Z + \eta/2 \sinh Z} \tag{B.22}$$

Figure B.2 shows attenuation and phase characteristics of α. The attenuation function is seen to approximate a simple pole behavior closely (as shown by the dashed line) over the useful range of frequencies,

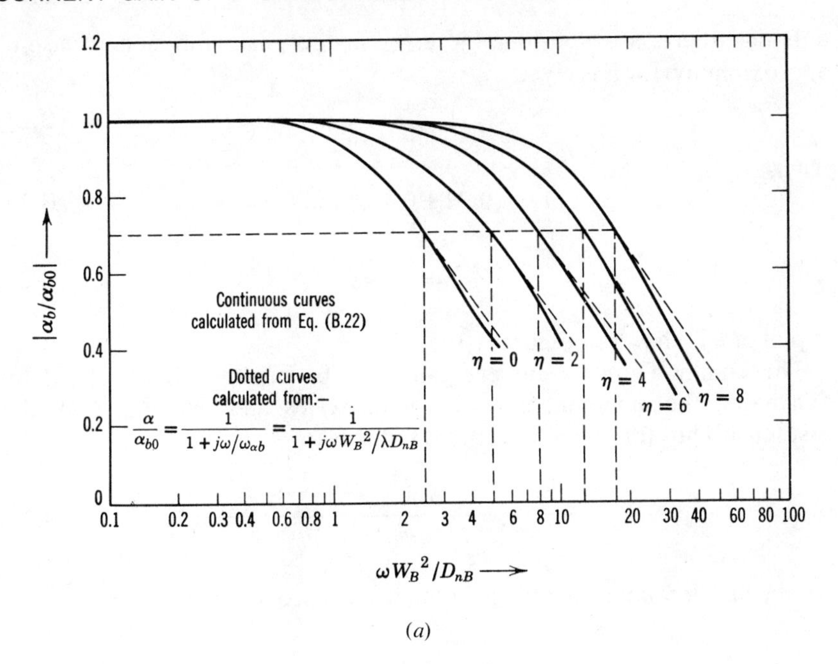

(a)

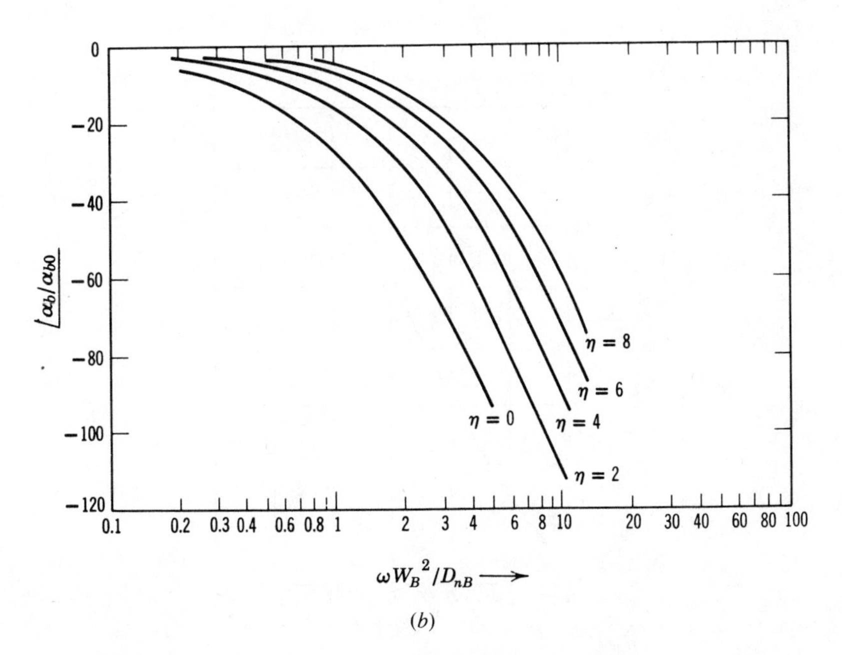

(b)

Figure B.2 Frequency and phase characteristics (Hakim[1]).

although there is a considerable error in the phase function. A useful approximation for α is given by

$$\alpha \simeq \frac{\alpha_0 e^{-jm\omega/\omega_\alpha}}{1+j\,\omega/\omega_\alpha} \tag{B.23}$$

where

$$m \simeq (0.22+0.1\,\eta)\text{ radians} \tag{B.24}$$

and

$$\omega_\alpha = \frac{\lambda D_{nB}}{W_B{}^2}. \tag{B.25}$$

A plot of λ is shown in Figure B.3.

The common-emitter current gain may be obtained from (B.23) by expanding the exponential in a Taylor series and retaining only the first term. Thus for $m\omega < \omega_\alpha$, (B.23) reduces to

$$\beta \simeq \frac{\alpha_0 e^{-jm\omega/\omega_\alpha}}{(1-\alpha_0)+j\,(\omega/\omega_\alpha)\,(1+m\alpha_0+\ldots)} \tag{B.26}$$

from which it follows that the gain-bandwidth product is given by

$$\omega_t \simeq \frac{\alpha_0\omega_\alpha}{1+m} \tag{B.27}$$

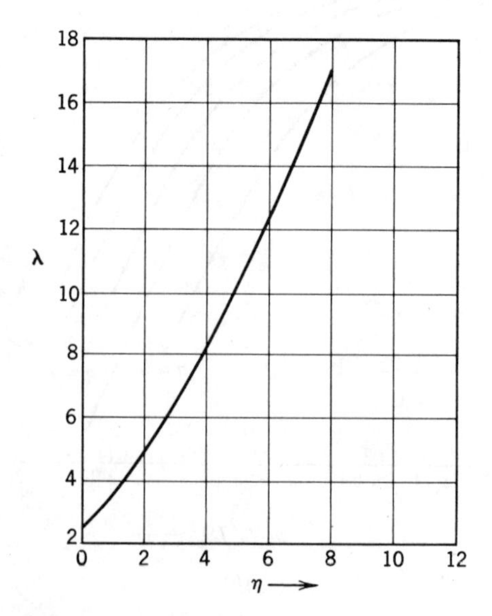

Figure B.3 Correction factor for the graded-base transistor (Hakim [1]).

B.1. REFERENCES

[1] S. S. Hakim, *Junction Transistor Circuit Analysis*, Chap. 2. Wiley, New York, 1962.
[2] M. S. Ghausi, *Principles and Design of Linear Active Circuits*, Chap. 9, McGraw-Hill, New York, 1965.

Appendix C

The Unilateral Mean Velocity

A thick semiconductor is bounded by a surface at $z = +d$. It is required to determine the unilateral mean velocity with which mobile electrons approach this surface from within the semiconductor. This velocity may be determined by noting that carrier motion gives rise to a unilateral thermal current density j_z.

Consider electrons in the conduction band. If $g(E)$ is the density-of-states function for carriers in the energy range from E to $E+dE$, and $f(E)$ is the probability of occupation of these states, then

$$dn(E) = f(E)\, g(E)\, dE. \tag{C.1}$$

Combining (11.34) and (11.38), and noting that

$$1 + e^{(E-E_f)/kT} \simeq e^{(E-E_f)/kT}, \tag{C.2}$$

one obtains

$$d\bar{n} = \frac{2\bar{n}}{\sqrt{\pi}}\left(\frac{E-E_c}{kT}\right)^{1/2} e^{-(E-E_c)/kT}\, d\left(\frac{E-E_c}{kT}\right), \tag{C.3}$$

where $\bar{n}$ is the equilibrium concentration of electrons. For electrons at an energy level E within the conduction band,

$$E - E_c = \frac{P_r^2}{2m^*} \tag{C.4}$$

where P_r is the momentum, and m^* is the effective mass of the electron. Thus

$$d\bar{n} = \frac{\bar{n}}{(2\pi m^* kT)^{3/2}}\, e^{-P_r^2/2m^*kT}\, 4\pi P_r^2\, dP_r \tag{C.5}$$

In momentum space, as shown in (11.25),

$$4\pi P_r^2 \, \Delta P_r = \Delta V_p \tag{C.6}$$

from which it follows that

$$4\pi P_r^2 \, dP_r = d(m^* v_z)\, d(m^* v_y)\, d(m^* v_x), \tag{C.7}$$

$$= m^{*3} dv_z dv_y dv_x \tag{C.8}$$

In addition,

$$P_r^2 = P_x^2 + P_y^2 + P_z^2, \tag{C.9}$$

$$= m^{*2}(v_z^2 + v_y^2 + v_x^2). \tag{C.10}$$

Substituting these equations into (C.5) gives

$$d\bar{n} = \bar{n}\left(\frac{m^*}{2\pi kT}\right)^{3/2} e^{-m^*(v_x^2+v_y^2+v_z^2)/2kT} dv_z \, dv_y \, dv_x$$

The unilateral thermal current density j_z is given by

$$j_z = q \int_{\text{all } n} v_z d\bar{n}. \tag{C.11}$$

Substituting (C.5), (C.8), and (C.10) into (C.11) gives

$$j_z = \bar{n}q\left(\frac{m^*}{2\pi kT}\right)^{3/2} \int_{v_z=0}^{\infty} \int_{v_y=-\infty}^{\infty} \int_{v_x=-\infty}^{\infty} v_z e^{-m^*(v_x^2+v_y^2+v_z^2)/2kT} dv_z dv_y dv_x. \tag{C.12}$$

Making a change of variables,

$$p = v_x\left(\frac{m^*}{2kT}\right)^{1/2}, \tag{C.13}$$

$$r = v_y\left(\frac{m^*}{2kT}\right)^{1/2}, \tag{C.14}$$

$$s = v_z\left(\frac{m^*}{2kT}\right)^{1/2}, \tag{C.15}$$

(C.12) reduces to

$$j_z = \frac{\bar{n}q}{\pi}\left(\frac{2kT}{\pi m^*}\right)^{1/2} \int_{v_s=0}^{\infty} \int_{v_r=-\infty}^{\infty} \int_{v_p=-\infty}^{\infty} s e^{-s^2} e^{-r^2} e^{-p^2} \, dp \, dr \, ds. \tag{C.16}$$

But

$$\int_{-\infty}^{\infty} e^{-p^2}\,dp = \int_{-\infty}^{\infty} e^{-r^2}\,dr = \sqrt{\pi} \tag{C.17}$$

and

$$\int_{0}^{\infty} se^{-s^2}\,ds = \tfrac{1}{2}\,. \tag{C.18}$$

Hence

$$j_z = \bar{n}q\left(\frac{kT}{2\pi m^*}\right)^{1/2}. \tag{C.19}$$

The unilateral mean velocity in the z direction is thus given by $j_z/\bar{n}q$, i.e., by $(kT/2\pi m^*)^{1/2}$.

List of Symbols

Symbol	Description	Section
A	Cross-sectional area	2.1
A_E	Area of the emitter	14.2.3.2
A_C	Area of the collector	14.2.3.2
A_n	Electron capture cross section	12.5
A_p	Hole capture cross section	12.5
A_s	Surface area	17.4.2
a	Cube edge	1.2
$a_{11}, a_{12}, a_{21}, a_{22}$	Matrix elements in the transistor equation	15.3
$\mathscr{A}$	Linear grade constant	13.2
BV	Breakdown voltage	13.1.1.6
BV_{CBO}	Collector-base junction breakdown voltage, with the emitter open	14.2.8
BV_{CEO}	Collector-base junction breakdown voltage, emitter shorted to base	14.2.8
b	Amount of shear	1.4.2.1
C_d	Diffusion capacitance	13.1.3
C_I	Concentration of solute in the initial charge, by weight	2.3.2
C_L	Concentration of solute in a melt, by weight	2.2
C_L'	Interface concentration of solute in a melt, by weight	2.2.2
C_M	Initial concentration of solute in a melt, by weight	2.2.1
C_S	Concentration of solute in a crystal, by weight	2.2
C_{S1}	Concentration of solute in a crystal at some specific point	2.3.2

Symbol	Description	Section
C_t	Transition capacitance	13.1.1.2
C_{tC}	Transition capacitance of the collector-base junction	14.2.3.1
$\bar{C}_{tC}$	Average value of C_{tC}	14.2.5
C_{tE}	Transition capacitance of the emitter-base junction	14.2.3.1
$\bar{C}_{tE}$	Average value of C_{tE}	14.2.5
C_{tS}	Transition capacitance of the collector-substrate junction	15.1.4
$\bar{C}_{tS}$	Average value of C_{tS}	15.1.7
c_0	Capacitance per unit area	16.1.6
c_{zb}	Unilateral mean velocity in the bulk	17.2.6
c_{zs}	Unilateral mean velocity at the surface	17.2.6
D	Diffusion constant	2.2.2
D', D''	Effective diffusion constants	4.3.2.1
D_E	Diffusion constant of an epitaxial layer	5.3.2
D_i	Interstitial diffusion constant	4.3.2.1
D_n	Diffusion constant for electrons	11.7
D_{nB}	Diffusion constant for electrons in the base	14.2.1.1
D_p	Diffusion constant for holes	11.7
D_{pB}	Diffusion constant for holes in the base	15.2.4
D_{pE}	Diffusion constant for holes in the emitter	14.2.7
D_{pC}	Diffusion constant for holes in the collector	14.2.1.1
D_s	Substitutional diffusion constant	4.3.2.1
D_S	Diffusion constant for the substrate	5.3.2
D_0	Diffusion coefficient	4.3
d	Thickness or separation	1.3
d_{EB}	Distance between the edges of the emitter region and the base contact	14.2.3.2
d_{EC}	Distance between the edge of the emitter region and the collector contact	15.1.5
E	Internal energy	1.4.1.1
E_c	Conduction band energy level	11.1.4
E_d	Donor energy level	11.3.2
E_f	Fermi level	11.1.3
E_i	Intrinsic level	11.2
E_j, E_k	Energy levels	11.1.3
E_m	Interaction energy	4.1.1
E_n	Height of a potential barrier (in electron volts)	4.1.2
E_p	Activation energy of permeation	6.1

Symbol	Description	Section
E_r	Recombination level	12.4
E_s	Energy of formation of a Schottky defect	1.4.1.1
E_v	Valence band energy level	11.1.4
E_1, E_2	Energy levels	11.1.1
E_μ	Elastic strain energy	1.4.2.1
$\mathscr{E}$	Electric field intensity	4.2.1
$\mathscr{E}'$	Effective value of the electric field intensity	14.3.1.1
$\mathscr{E}_{\text{peak}}$	Peak electric field intensity	13.1.1.1
$\mathscr{E}_x$	Electric field intensity in the x-direction	17.2.6
$\mathscr{E}_0$	Intensity of the built in electric field	14.3
F	Free energy	1.4.1.1
F	Force on an electron	11.5
$F_{1/2}(\eta_f)$	The Fermi integral	11.3.6.2
f_j	Probability of electron occupation at the jth level	11.1.3
G, G_0	Generation rates	12.1
g	Degeneracy factor	11.3.1
g_a	Degeneracy factor at the acceptor level	11.3.3
g_d	Degeneracy factor at the donor level	11.3.2
g_m	Transconductance	14.2.5.1
$g(E)$	Density of states function	11.1.3
H_i	Heat input	2.1
H_0	Heat loss	2.1
h	Planck's constant	11.1.4
h	Gas phase mass transfer coefficient	5.1.2
h_{11}	Input impedance, output shorted	16.1.6
h_{21}	Forward current gain, output shorted	16.1.6
I	Current	4.9.1
I_B	Base current	14.1
I_C	Collector current	14.1
I_{C0}	Reverse current of the collector-base junction	14.2
I_E	Emitter current	14.1
I_F	Forward current	13.1.3
$I_{F,s}$	Surface component of the forward current	17.4.2
I_n	Current due to electron flow	13.1.2.2
I_{nB}	Current due to electron flow in the base	14.2.5
I_{nC}	Current due to electron flow in the collector	14.2
I_{nE}	Current due to electron flow in the emitter	14.2
I_p	Current due to hole flow	13.1.2.2
I_{pB}	Current due to hole flow in the base	14.2.5
I_{pE}	Current due to hole flow in the emitter	14.2

Symbol	Description	Section
I_{pL}	Lateral component of current due to hole flow	15.2.4
I_{pV}	Vertical component of current due to hole flow	15.2.4
I_R	Reverse current	13.1.3
I_{rB}	Generation-recombination current in the base	14.2.2
I_{rE}	Generation-recombination current in the emitter	14.2.2.1
I_{rec}	Recombination current	13.1.1.4
I_{sB}	Surface recombination current in the base	17.4.4
I_0	Reverse current of a *p-n* junction	13.1.2.2
i	Time-varying component of current	13.1.5
i_C	Time-varying component of collector current	15.2.4
i_E	Time-varying component of emitter current	15.2.4
i_{nC}	Time-varying component of current due to electron flow in the collector	B
i_{nE}	Time-varying component of current due to electron flow in the emitter	B
i_{pL}	Time-varying part of the lateral component of hole current	15.2.4
i_{pV}	Time-varying part of the vertical component of hole current	15.2.4
J	Current density	11.5
J_C	Collector current density	14.2.5.7
J_{diff}	Diffusion component of the current density	13.1.2
J_{max}	Maximum current density	14.2.5.7
J_n	Current density due to electrons	11.5
J_{nB}	Current density due to the flow of electrons in the base	14.3
J_p	Current density due to holes	11.5
j	Flux density	4.2
K	Permeability	6.1
K_0	Permeability constant	6.1
k	Boltzmann's constant	1.4.1.1
k	Surface reaction rate constant	5.1.2
k	Distribution coefficient	2.2
k_e	Effective distribution coefficient	2.2.2
L	Length of a molten zone	2.3.1
L	Liquid phase in a system	3.2
L_e	Extrinsic Debye length	17.2.3
L_i	Intrinsic Debye length	17.2.1
L_n	Diffusion length for electrons	13.1.2.2

Symbol	Description	Section
L_n'	Effective diffusion length for electrons	13.1.3
L_{nB}	Diffusion length for electrons in the base	14.1
L_p	Diffusion length for holes	13.1.2.2
L_{pB}	Diffusion length for holes in the base	14.1
L_{pE}	Diffusion length for holes in the emitter	14.2.1.1
L_{sc}	Effective length of the space charge layer	17.2.3
l_C	Length of the collector region	15.1.5
l_E	Length of the emitter region	14.2.2.2
M	Multiplication factor	13.1.1.6
M	Number of electrons in a system	11.1.3
M^{s+}	Impurity atom with a net charge of $s+$	11.3.1
M^{x+}	Oxidation state of a semiconductor	7.1
m^*	Effective mass	4.2.1
m_n^*	Effective electron mass	11.1.4
m_p^*	Effective hole mass	11.1.4
m_0	Mass of an electron	11.1.4
N	Number of atoms per unit volume	1.4.1.1
N^0	Concentration of unionized impurities	11.3.1
N^{s+}	Concentration of impurities in the $(s+)^{\text{th}}$ charge state	11.3.1
N'	Concentration of available interstitial sites	1.4.1.1
N_a	Acceptor concentration	11.3.2
N_B	Impurity concentration in the base	14.2.2.2
N_C	Impurity concentration in the collector	4.6
N_c	Density of states at the conduction-band edge	11.1.5
N_d	Donor concentration	11.3.2
N_E	Impurity concentration in the emitter	14.3.1
N_g	Concentration of reacting species in the gas phase	5.1.2
N_r	Concentration of recombination centers	12.6
N_{rs}	Surface concentration of recombination centers	17.3
N_S	Substrate concentration	5.3.2
N_{S1}, N_{S2}	Number of available states	11.1.1
N_{sa}	Concentration of shallow acceptors	11.4.2
N_{sd}	Concentration of shallow donors	11.4.2
N_t	Concentration of trapping centers	12.8
N_v	Density of states at the valence-band edge	11.1.5
N_x	Impurity concentration at a point	5.3.1
N_∞	Impurity concentration at infinity	5.3.1
N_0	Surface concentration	4.5.1

Symbol	Description	Section
N_0	Equilibrium concentration of a reacting species	5.1.2
N_{0B}	Surface concentration of the base diffusion	4.6
N_{0E}	Surface concentration of the emitter diffusion	4.6
n	Number of atoms in a unit volume	5.1.2
$\bar{n}$	Equilibrium electron concentration	11.1.5
n'	Excess electron concentration	12.1
n'_{ac}	Time varying component of the excess electron concentration	13.1.3
n_B	Concentration of electrons in the base	14.2.1.2
$\bar{n}_B$	Equilibrium concentration of electrons in the base	14.2.1.1
n'_B	Excess concentration of electrons in the base	14.1
n'_{dc}	Steady state component of the excess electron concentration	13.1.3
n_f	Volume concentration of Frenkel defects	1.4.1.1
n_i	Intrinsic carrier concentration	11.2
n_N	Electron concentration in a n-type region	13.1.2
n_P	Electron concentration in a p-type region	13.1.2
n_s	Volume concentration of Schottky defects	1.4.1.1
n_s	Electron concentration at the surface	17.3
P_r	Momentum vector (radial)	11.1.4
P_x, P_y, P_z	Momentum vectors (rectangular coordinates)	11.1.4
p'	Excess hole concentration	12.1
$\bar{p}$	Equilibrium hole concentration	11.1.5
p_B	Hole concentration in the base region	14.2.2.2
$\bar{p}_E$	Equilibrium hole concentration in the emitter region	14.2.1.1
p_N	Hole concentration in the n-region	13.1.2.1
$\bar{p}_N$	Equilibrium hole concentration in a n-region	13.1.2.1
p_P	Hole concentration in a p-region	13.1.2.1
$\bar{p}_P$	Equilibrium hole concentration in the p-region	13.1.2.1
p_s	Hole concentration at the surface	17.3
Q	Quantity of matter	4.5.2
Q_B	Excess stored base charge	13.1.4
Q_D	Charge transported through the collector-base depletion layer	14.2.5.4
Q_{sB}	Excess stored charge in the base of a saturated transistor	14.2.7
Q_{sC}	Excess stored charge in the collector of a transistor	14.2.7

Symbol	Description	Section
Q_{sc}	Charge in the space charge layer in a semiconductor	17.2
Q_{ss}	Charge in the surface states	17.2
q	Magnitude of charge on the electron	4.2.1
R	Crystal growth rate	2.2.2
R	Recombination rate	12.1
R	Resistance	4.9.1
R_{cn}	Rate of electron capture	12.4
R_{cp}	Rate of hole capture	12.4
R_{en}	Rate of electron emission	12.4
R_{ep}	Rate of hole emission	12.4
R_i	Inner radius	1.4.2.1
R_L	Load resistance	14.2.5.3
R_o	Outer radius	1.4.2.1
r	Radius	1.4.2.1
r_B	Parasitic base resistance	14.2.3.2
r_{B1}	Intrinsic base resistance	14.2.3.2
r_{B2}	Extrinsic base resistance	14.2.3.2
r_C	Parasitic collector resistance	14.2.3.2
r_d	Dynamic resistance	13.1.3
r_0	Tetrahedral radius	1.2
r_1	Radius at the edge of the depletion layer	13.3.1.1
r_2	Radius of n^+ region	13.3.1.1
S	Weight of solute in the melt at any given point during crystal growth	2.2.1
S	Entropy	1.4.1.1
S	Surface recombination velocity	17.3
S_{n0}	Surface recombination constant for electrons	17.3
S_{max}	Maximum value of the surface recombination velocity	17.3
S_{p0}	Surface recombination constant for holes	17.3
s	Amount of solute in a molten zone	2.3.1
s	Differential operator	14.2.4
s_1, s_2	Simple poles	14.2.4
T	Absolute temperature	1.4.1.1
T_A	Melting point of A	3.3
T_B	Melting point of B	3.3
T_E	Eutectic temperature	3.3
t	Time	2.1
t	Thickness	4.9.1

Symbol	Description	Section
t_B	Base transit time	13.1.4
t_C	Collector time constant	14.2.5.3
t_D	Transit time through the depletion layer	14.2.5.4
t_E	Emitter time constant	14.2.5.1
t_f	Fall time	13.1.5
t_s	Storage time	13.1.5
t_t	Transit time	14.1
U	Dimensionless parameter	4.5.3
U_n	Generation rate for electrons	13.1.2.1
U_p	Generation rate for holes	13.1.2.1
U_s	Charge generation rate at the surface	17.3
u	Dimensionless parameter	17.2.1
u_b	Magnitude of u in the bulk	17.2.1
u_s	Magnitude of u at the surface	17.2.2
V	Voltage	4.9.1
V	Volume	1.4.2.1
V_A	Applied voltage	13.1.2
V_B	Lateral voltage drop in the base	14.2.3.2
$\bar{V}_B$	Mean value of lateral voltage drop in the base	14.2.3.2
V_{BE}	Base-emitter voltage	14.2.1.1
V_{CB}	Collector-base voltage	14.2.1.3
V_D	Voltage drop across a diode	13.1.5
V_F	Forward Voltage	13.1.3
V'_F	Forward voltage	15.3.1
V''_F	Forward voltage	15.3.1
V_R	Reverse voltage	13.1.1.1
V_j	Junction voltage	13.1.1.1
V_p	Volume in momentum space	11.1.4
V_0	Transition voltage	13.1.2.4
V_1	Power supply voltage	14.2.5.3
$V_{\mathscr{E}}$	Voltage supported by the $\mathscr{E}$ field	13.1.2.4
v	Dimensionless parameter	17.2.1
v	Carrier velocity	11.5
$v(t)$	Time varying component of voltage	13.1.3
v_d	Drift velocity of carriers	4.2.1
$v_{\lim}$	Limiting velocity of carriers	14.2.5.4
v_s	Magnitude of v at the surface	17.2.3
v_t	Thermal velocity	11.5
v_{tn}	Thermal velocity for electrons	12.5
v_{tp}	Thermal velocity for holes	12.5

Symbol	Description	Section
W	Weight of a crystal during growth	2.2.1
W	Width	13.1.2.2
W_B	Base width	14.1
W'_B	Effective base width	14.2.2.2
W_C	Width of the collector region	14.2.3.2
W_L	Weight of liquid	3.2
W_M	Initial weight of a melt	2.2.1
W_N	Width of the *n*-region	13.3.2
W_P	Width of the *p*-region	13.3.2
W_S	Weight of a solid	3.2
W_1	Weight of a crystal at a specific point in its growth	2.2.2
W_2	Thickness of the base region measured from the surface	14.2.3.2
w	Width	4.9.1
w	Depletion layer width at breakdown	13.1.1.6
w_B	Width of the base-diffused area	14.2.2.2
w_E	Width of the emitter-diffused area	14.2.2.2
w_C	Width of the collector pocket	14.2.2.2
x	Distance	2.1
x_{BC}	Position of the base-collector junction	13.3.1
x_C	Collector depletion layer thickness	14.2.5.4
x_{CS}	Position of the collector-substrate junction	13.3.1
x_{EB}	Position of the emitter-base junction	13.3.1
x_j	Junction depth	4.6
x_{j1}	Junction depth after predeposition	4.9.4
x_N	Depletion layer width in the *n*-region	13.1.1.1
x_P	Depletion layer width in the *p*-region	13.1.1.1
x_0	Depletion layer thickness	13.2
x_0	Diffusion depth in a lateral transistor	15.2.4
Y_1	An admittance	15.2.4
y_d	Dynamic admittance	13.1.3
Z	Number of electrons associated with an atom	4.2.1
z	The complex variable	14.2.4
α	A component in a binary system	3.2
α	Common base current gain	14.2.4
α_F	Common base forward current gain	15.3
α_i	Ionization constant	13.1.1.6
α_n	Recombination rate constant for electrons	12.5
α_{nt}	Trapping rate constant for electrons	12.8

Symbol	Description	Section
α_R	Common base reverse current gain	15.3
α_T	Base transport factor	14.2.1
α_0	Low frequency common base current gain	14.1
α_0'	Low frequency common base current gain of the internal *p-n-p* transistor	15.2.4
β	A component in a binary system	3.2
β	Common emitter current gain	15.3
β_n	Recombination rate constant for electrons	12.7.1
β_p	Recombination rate constant for holes	12.7.1
β_0	Low frequency common emitter current gain	14.1
β_0'	Effective low frequency common emitter current gain	14.3.1.1
β_0'	Low frequency common emitter current gain of the internal *p-n-p* transistor	15.2.4
Γ_y	The gamma function of y	13.2.3
γ	A component in a binary system	3.2
γ	Injection efficiency	14.2.1
δ	Stagnant layer thickness	2.2.2
δ	Half-width of the depletion layer of a linearly graded junction	13.2.1
ϵ	Misfit factor	1.2
ϵ	Relative permittivity	11.3
ϵ_0	Permittivity of free space	11.3
η	Grading constant for impurity concentrations in the base	14.3
θ	Angle between Miller planes	1.3
λ	Wavelength	11.5.1
λ	An integration variable	A.1
λ_k	Interference wavelength	6.7
μ	Shear modulus	1.4.2.1
μ	Mobility	4.2.1
μ_b	Bulk mobility	17.2.6
μ_C	Mobility of majority carriers in the collector region	14.2.5.7
μ_I	Impurity mobility	11.5.2
μ_L	Lattice mobility	11.5.1
μ_n	Electron mobility	11.5
μ_{nB}	Electron mobility in the base	14.2.1.1
μ_{nE}	Electron mobility in the emitter	14.2.1.1
μ_p	Hole mobility	11.5

Symbol	Description	Section
μ_{pB}	Hole mobility in the base	14.2.1.1
μ_{pE}	Hole mobility in the emitter	14.2.1.1
μ_s	Surface mobility	17.2.6
ν	Jump frequency	4.1.1
ν_{phonon}	Phonon frequency	11.5.1
ν_{photon}	Photon frequency	11.5.1
ν_0	Frequency of lattice vibrations	4.1.1
ρ	Space charge density	13.1.1
ρ	Resistivity	4.9.1
$\overline{\rho}$	Average resistivity	5.5.3
ρ_C	Collector resistivity	14.2.3.2
ρ_s	Specific gravity of silicon	2.1
$\rho_{\square}$	Sheet resistance	4.9.1
$\overline{\sigma}$	Average conductivity	4.9.3
σ_B	Conductivity of the base	14.2.1.1
$\overline{\sigma}_B$	Average conductivity of the base	14.3.2
σ'_B	Effective conductivity of the base	14.2.2.2
σ_E	Conductivity of the emitter	14.2.1.1
$\sigma_{\square B}$	Sheet conductance of the base	14.3.2
$\sigma_{\square E}$	Sheet conductance of the emitter	16.2.2
τ	Effective minority carrier lifetime	13.1.5
τ_b	Bulk lifetime	17.2.1
τ_C	Collision time	11.5
$\langle \tau_C \rangle$	Average collision time	11.5
τ_{eff}	Effective lifetime	17.3.1
τ_F	Effective lifetime under forward bias	13.1.5
τ_g	Trap lifetime	12.8
τ_n	Electron lifetime	12.2
τ_{nB}	Electron lifetime in the base	14.1
τ_{n0}	Electron lifetime	12.6
τ_p	Hole lifetime	12,2
τ_{pB}	Hole lifetime in the base	15.2.4
τ_{pC}	Hole lifetime in the collector	14.2.7
τ_{p0}	Hole lifetime	12.6
τ_R	Effective lifetime under reverse bias	13.1.5
τ_s	Surface lifetime	17.2.6
τ_{sB}	Recombination time constant for excess base stored charge during saturation	14.2.7
τ_{sC}	Recombination time constant for excess collector stored charge during saturation	14.2.7

Symbol	Description	Section
ϕ	Growth factor	5.3.1
ϕ	Potential	17.2.1
ϕ_b	Potential in the bulk	17.2.2
ϕ_s	Potential at the surface	17.2.2
Ψ	Contact potential	13.1
Ψ_C	Contact potential across the collector-base depletion layer	14.2.5.7
Ψ_n	Contact potential of an *n*-type depletion region	13.1
Ψ_p	Contact potential of a *p*-type depletion region	13.1
ω	Angular frequency	13.1.3
$\omega_{\max}$	Maximum frequency of oscillation	14.2.6
ω_t	The gain-bandwidth product	14.2.5
ω_0	A specific angular frequency	16.1.6
ω_α	Cut-off frequency for α	B
ω_β	Cut-off frequency for β	14.2.4

INDEX